A COURSE IN

mathematics

FOR STUDENTS OF
PHYSICS: 2

PAUL BAMBERG
SHLOMO STERNBERG

The right of the University of Cambridge to print and sell all manner of books was granted by Henry VIII in 1534. The University has printed and published continuously since 1584.

CAMBRIDGE UNIVERSITY PRESS
Cambridge
New York Port Chester
Melbourne Sydney

Published by the Press Syndicate of the University of Cambridge
The Pitt Building, Trumpington Street, Cambridge CB2 1RP
40 West 20th Street, New York, NY 10011, USA
10 Stamford Road, Oakleigh, Melbourne 3166, Australia

First Published 1990

Printed in Great Britain at the University Press, Cambridge

British library cataloguing in publication data

Bamberg, Paul
A course in mathematics for students of
physics: 2
1. Mathematical physics.
I. Title. II. Sternberg, Shlomo
510′.2453 QC20

Library of Congress cataloguing in publication data

Bamberg, Paul G.
A course in mathematics for students of physics: 2
Bibliography
Includes index.
1. Mathematics–1961–. I. Sternberg, Shlomo.
II. Title.
QA37.2.B36 1990 510 86–2230

ISBN 0 521 33245 1

TM

CONTENTS OF VOLUME 2

CONTENTS OF VOLUME 1

PREFACE

This book, with apologies for the pretentious title, represents the text of a course we have been teaching at Harvard for the past eight years. The course is aimed at students with an interest in physics who have a good grounding in one-variable calculus. Some prior acquaintance with linear algebra is helpful but not necessary. Most of the students simultaneously take an intensive course in physics and so are able to integrate the material learned here with their physics education. This also is helpful but not necessary. The main topics of the course are the theory and physical application of linear algebra, and of the calculus of several variables, particularly the exterior calculus. Our pedagogical approach follows the 'spiral method' wherein we cover the same topic several times at increasing levels of sophistication and range of application, rather than the 'rectilinear approach' of strict logical order. There are, we hope, no vicious circles of logical error, but we will frequently develop a special case of a subject, and then return to it for a more general definition and setting only after a broader perspective can be achieved through the introduction of related topics. This makes some demands of patience and faith on the part of the student. But we hope that, at the end, the student is rewarded by a deeper intuitive understanding of the subject as a whole.

Here is an outline of the contents of the book in some detail. The goal of the first four chapters is to develop a familiarity with the algebra and analysis of square matrices. Thus, by the end of these chapters, the student should be thinking of a matrix as an object in its own right, and not as a square array of numbers. We deal in these chapters almost exclusively with 2×2 matrices, where the most complicated of the computations can be reduced to solving quadratic equations. But we always formulate the results with the higher-dimensional case in mind. We begin Chapter 1 by explaining the relation between the multiplication law of 2×2 matrices and the geometry of straight lines in the plane. We develop the algebra of 2×2 matrices and discuss the determinant and its relation to area and orientation. We define the notion of an abstract vector space, in general, and explain the concepts of basis and change of basis for one- and two-dimensional vector spaces.

In Chapter 2 we discuss conformal linear geometry in the plane, that is, the geometry of lines and angles, and its relation to certain kinds of 2×2 matrices. We also discuss the notion of eigenvalues and eigenvectors, so important in quantum mechanics. We use these notions to give an algorithm for computing the powers of a matrix. As an application we study the basic properties of Markov chains.

The principal goal of Chapter 3 is to explain that a system of homogeneous linear differential equations with constant coefficients can be written as $d\mathbf{u}/dt = A\mathbf{u}$ where A is a matrix and $\mathbf{u}$ is a vector, and that the solution can be written as $e^{At}\mathbf{u}_0$ where $\mathbf{u}_0$ gives the initial conditions. This of course requires us to explain what is meant by the exponential of a matrix. We also describe the qualitative behavior of solutions and the inhomogeneous case, including a discussion of resonance.

Chapter 4 is devoted to the study of scalar products and quadratic forms. It is rich in physical applications, including a discussion of normal modes and a detailed treatment of special relativity.

Chapters 5 and 6 present the basic facts of the differential calculus. In Chapter 5 we define the differential of a map from one vector space to another, and discuss its basic properties, in particular the chain rule. We give some physical applications such as Kepler motion and the Born approximation. We define the concepts of directional and partial derivatives, and linear differential forms.

In Chapter 6 we continue the study of the differential calculus. We present the vector versions of the mean-value theorem, of Taylor's formula and of the inverse function theorem. We discuss critical point behavior and Lagrange multipliers.

Chapters 7 and 8 are meant as a first introduction to the integral calculus. Chapter 7 is devoted to the study of linear differential forms and their line integrals. Particular attention is paid to the behavior under change of variables. Other one-dimensional integrals such as arc length are also discussed.

Chapter 8 is devoted to the study of exterior two-forms and their corresponding two-dimensional integrals. The exterior derivative is introduced and invariance under pullback is stressed. The two-dimensional version of Stokes' theorem, i.e. Green's theorem, is proved. Surface integrals in three-space are studied.

Chapter 9 presents an example of how the results of the first eight chapters can be applied to a physical theory – optics. It is all in the nature of applications, and can be omitted without any effect on the understanding of what follows.

In Chapter 10 we go back and prove the basic facts about finite-dimensional vector spaces and their linear transformations. The treatment here is a straightforward generalization, in the main, of the results obtained in the first four chapters in the two-dimensional case. The one new algorithm is that of row reduction. Two important new concepts (somewhat hard to get used to at first) are introduced: those of the dual space and the quotient space. These concepts will prove crucial in what follows.

Chapter 11 is devoted to proving the central facts about determinants of $n \times n$

matrices. The subject is developed axiomatically, and the basic computational algorithms are presented.

Chapters 12–14 are meant as a gentle introduction to the mathematics of shape, that is, algebraic topology. In Chapter 12 we begin the study of electrical networks. This involves two aspects. One is the study of the 'wiring' of the network, that is, how the various branches are interconnected. In mathematical language this is known as the topology of one-dimensional complexes. The other is the study of how the network as a whole responds when we know the behavior of the individual branches, in particular, power and energy response. We give some applications to physically interesting networks.

In Chapter 13 we continue the study of electrical networks. We examine the boundary-value problems associated with capacitive networks and use these methods to solve some classical problems in electrostatics involving conductors.

In Chapter 14 we give a sketch of how the one-dimensional results of Chapters 12 and 13 generalize to higher dimensions.

Chapters 15–18 develop the exterior differential calculus as a continuous version of the discrete theory of complexes. In Chapter 15 the basic facts of the exterior calculus are presented: exterior algebra, k-forms, pullback, exterior derivative and Stokes' theorem.

Chapter 16 is devoted to electrostatics. We suggest that the dielectric properties of the vacuum give the continuous analog of the capacitance of a network, and that these dielectric properties are what determine Euclidean geometry in three-dimensional space. The basic facts of potential theory are presented.

Chapter 17 continues the study of the exterior differential calculus. The main topics are vector fields and flows, interior products and Lie derivatives. These are applied to magnetostatics.

Chapter 18 concludes the study of the exterior calculus with an in-depth discussion of the star operator in a general context.

Chapter 19 can be thought of as the culmination of the course. It applies the results of the preceding chapters to the study of Maxwell's equations and the associated wave equations.

Chapters 20 and 21 are essentially independent of Chapters 9–19 and can be read independently of them. They are not usually included in our one-year course. But Chapters 1–9, 20 and 21 would form a self-contained unit for a shorter course.

The material in Chapter 20 is a relatively standard treatment of the theory of functions of a complex variable, suitable for students at the level of this book.

Chapter 21 discusses some of the more elementary aspects of asymptotics.

Chapter 22 shows how the exterior calculus can be used in classical thermodynamics, following the ideas of Born and Carathéodory.

The book is divided into two volumes, with Chapters 1–11 in volume 1.

Most of the mathematics and all of the physics presented in this book were developed by the first decade of the twentieth century. The material is thus at least seventy-five years old. Yet much of the material is not yet standard in the elementary courses (although most of it with the possible exception of network theory must be learned for a grasp of modern physics, and is studied at some stage of the physicist's career). The reasons are largely historical. It was apparent to Hamilton that the real and complex numbers were insufficient for the deeper study of geometrical analysis, that one wants to treat the number pairs or triplets of the Cartesian geometry in two and three dimensions as objects in their own right with their own algebraic properties. To this end he developed the algebra of quaternions, a theory which had a good deal of popularity in England in the middle of the nineteenth century. Quaternions had several drawbacks: they more naturally pertained to four, rather than to three dimensions – the geometry of three dimensions appeared as a piece of a larger theory rather than having a natural existence of its own; also, they have *too much* algebraic structure, the relation between quaternion multiplication, for example, and geometric constructions in three dimensions being somewhat complicated. (The first of these objections would, of course be regarded far less seriously today. But it would be replaced by an objection to a theory that is *limited* to four dimensions.) Eventually, the three-dimensional *vector algebra* with its scalar and vector products was distilled from the theory of quaternions. It was conjoined with the necessary differential operations, and give rise to the *vector analysis* as finally developed by Gibbs and promulgated by him in a famous and very influential text.

So vector analysis, with its grad, div, curl etc. became the standard language in which the geometric laws of physics were taught. Now while vector analysis is well suited to the geometry of three-dimensional Euclidean space, it has a number of serious drawbacks. First, and least serious, is that the essential unity of the subject is obscured. Thus the fundamental theorem of the calculus, Green's theorem, Gauss' theorem and Stokes' theorem are all aspects of the same theorem (now called Stokes' theorem). But this is not at all clear in the vector analysis treatment. More serious is that the fundamental operators involve the Euclidean structure (for example grad and div) or the three-dimensional structure and orientation as well (for example curl). Thus the theory is wedded to a three-dimensional orientated Euclidean space. A related problem is that the operators do not behave nicely under general changes of coordinates – their expression in non-rectangular coordinates being unwieldy. Already Poincaré, in his fundamental scientific and philosophical writings which led to the theory of relativity, stressed the need to distinguish between those laws of geometry and physics which are 'topological', i.e. depend only on the differential structure of space and so are invariant under smooth deformations, and those which depend on more geometrical structure such as the notion of distance. One of the major impacts of the theory of relativity on mathematics was to encourage the study of higher-dimensional spaces, a study which had existed in the previous mathematical literature, but was not regarded as central to the study of geometry. Another was to emphasize general coordinate changes. The vector analysis was not up to these two tasks and so was supplemented in the more advanced literature by *tensor analysis*. But tensor analysis with its jumble of indices has a number of serious drawbacks, the most serious of which being that it is extraordinarily difficult to tell which operations have any geometric significance and which are artifacts of the coordinate system. Thus, while it is reasonably well-suited for computation, it is hard to assess exactly what it is that one is computing. The whole purpose of the development initiated by Hamilton – to have a calculus whose objects have a perceived geometrical significance – was vitiated. In order to make the theory work one had to introduce a relatively sophisticated geometrical construct, such as an affine connection. Even with such constructs the geometric meanings of the operations are obscure. In fact tensor analysis never displaced the intuitively clear vector analysis from the elementary curriculum.

It is generally accepted in the mathematics community, and gradually being accepted in the physics community, that the most suitable framework for geometrical analysis is the exterior differential calculus of Grassmann and Cartan. This calculus has the advantage that its computational rules are simple and concise, that its objects have a transparent geometrical significance, that it works in all

Maxwell's equations in the course of history
The constants c, μ_0, and ε_0 are set to 1.

The homogeneous equation	**The inhomogeneous equation**
Earliest form	
$\frac{\partial B_x}{\partial x}+\frac{\partial B_y}{\partial y}+\frac{\partial B_z}{\partial z}=0$	$\frac{\partial E_x}{\partial x}+\frac{\partial E_y}{\partial y}+\frac{\partial E_z}{\partial z}=\rho$
$\frac{\partial E_z}{\partial y}-\frac{\partial E_y}{\partial z}=-\dot{B}_x$	$\frac{\partial B_z}{\partial y}-\frac{\partial B_y}{\partial z}=j_x+\dot{E}_x$
$\frac{\partial E_x}{\partial z}-\frac{\partial E_z}{\partial x}=-\dot{B}_y$	$\frac{\partial B_x}{\partial z}-\frac{\partial B_z}{\partial x}=j_y+\dot{E}_y$
$\frac{\partial E_y}{\partial x}-\frac{\partial E_x}{\partial y}=-\dot{B}_z$	$\frac{\partial B_y}{\partial x}-\frac{\partial B_x}{\partial y}=j_z+\dot{E}_z$
At the end of the last century	
$\nabla\cdot\mathbf{B}=0$ $\nabla\times\mathbf{E}=-\dot{\mathbf{B}}$	$\nabla\cdot\mathbf{E}=\rho$ $\nabla\times\mathbf{B}=\mathbf{j}+\dot{\mathbf{E}}$
At the beginning of this century	
${}^*F^{\beta\alpha}{}_{,\alpha}=0$	$F^{\beta\alpha}{}_{,\alpha}=j^\beta$
Mid-twentieth-century	
$\mathrm{d}F=0$	$\delta F=J$

dimensions, that it behaves well under maps and changes of coordinates, that it has an essential unity to its principal theorems and that it clearly distinguishes between the 'topological' and 'metrical' properties. The geometrical laws of physics take on a simple and elegant form in terms of the exterior calculus. To emphasize this point, it might be useful to reproduce the above table, taken from Thirring's *Course on Mathematical Physics.*

Hermann Grassmann (1809–77) published his *Ausdehnungslehre* in 1844. It was not appreciated by the mathematical community and was dismissed by the leading German mathematicians of his time. In fact, Grassmann was never able to get a university position in mathematics. He remained a high-school teacher throughout his career. (Nevertheless, he seemed to have a happy and productive life. He raised a large family and was recognized as an expert on Sanskrit literature.) Towards the end of his life he tried again, with another edition of his *Ausdehnungslehre*, but this fared no better than the first. Only one or two mathematicians of his time, such as Möbius, appreciated his work. Nevertheless, the *Ausdehnungslehre* (or calculus of extension) contains for the first time many of the notions central to modern mathematics and most of the algebraic structures used in this book. Thus vector spaces, exterior algebra, exterior and interior products and a form of the generalized Stokes' theorem all make their appearance.

Elie Cartan (1869–1951) is now universally recognized as the leading geometer of our century. His early work, of such overwhelming importance for modern mathematics, on Lie groups and on systems of partial differential equations was done in relative obscurity. But, by the 1920s, his work became known to the broad mathematical community, due, in part, to the writings of Hermann Weyl who presented novel expositions of his work at a time when the theory of Lie groups began to play a central role in mathematics and in physics. Cartan's work on the theory of principal bundles and connections is now basic to the theory of elementary particles (where it goes under the generic name of 'gauge theories'). In 1922 Cartan published his book *Leçons sur les invariants intégraux* in which he showed how the exterior differential calculus, which he had invented, was a flexible tool, not only for geometry but also for the variational calculus and a wide variety of physical applications. It has taken a while, but, as we have mentioned above, it is now recognized by mathematicians and physicists that this calculus is the appropriate vehicle for the formulation of the geometrical laws of physics. Accordingly, we feel that it should displace the 'vector calculus' in the elementary curriculum and have proceeded accordingly.

Some explanation is in order for the time and effort devoted to the theory of electrical networks, a subject not usually considered as part of the elementary curriculum. First of all there is a purely pedagogical justification. The subject always goes over well with the students. It provides a down-to-earth illustration of such concepts as dual space and quotient space, concepts which frequently seem overly abstract and not readily accepted by the student. Also, in the discrete, algebraic setting of network theory, Stokes' theorem appears as essentially a definition, and a natural one at that. This serves to motivate the d operator and Stokes' theorem in the setting of the exterior calculus. There are deeper, more philosophical reasons for our decision to emphasize network theory. It has been recognized for about a century that the forces that hold macroscopic bodies together are essentially electrical in character. Thus (in the approximation where the notion of rigid body and Euclidean geometry makes sense, that is, in the non-relativistic realm) the concept of a rigid body, and hence of Euclidean geometry, derives from electrostatics. The frontiers of physics, both in the very small (the study of elementary particles) and the very large (the study of cosmology) have already begun to reopen fundamental questions as to the geometry of space and time. We thought it wise to bring some of the issues relating geometry to physics before the student even at this early stage of the curriculum. The advent of the computer, and also some of the recent theories of physics will, no doubt, call into question the discrete versus the continuous character of space and time (an issue raised by Riemann in his dissertation on the foundations of geometry). It is to be hoped that our discussion may be of some use to those who will have to deal with this problem in the future.

Of course, we have had to omit several important topics due to the limitation of a one-year course. We do not discuss infinite-dimensional vector spaces, in particular Hilbert spaces, nor do we define or study abstract differentiable manifolds and their properties. It has been our experience that these topics make too heavy a demand on the sophistication of the student, and the effort involved in explaining them is best expended elsewhere. Of course, at various places in the text we have to pay the price for not having these concepts at our disposal. More serious is the omission of a serious discussion of Fourier analysis, classical mechanics and probability theory. These topics are touched upon but not presented as a coherent subject of study. Our only excuse is that a thorough study of each would probably require a semester's course, and substantive treatments from the modern viewpoint are available elsewhere. A suggested guide to further reading is given at the end of the book.

We would like to thank Prof. Daniel Goroff for a careful reading of the manuscript and for making many corrections and fruitful suggestions for improvement. We would also like to thank Jeane Morris for her excellent typing and her devoted handling of the production of the manuscript from the inception of the project to its final form, over a period of eight years.

12

The theory of electrical networks

Chapters 12–14 are meant as a gentle introduction to the mathematics of shape, that is, algebraic topology. In Chapter 12 we begin the study of electrical networks. This involves two aspects. One is the study of the 'wiring' of the network, how the various branches are interconnected. In mathematical language this is known as the topology of one-dimensional complexes. The other is the study of how the network as a whole responds, when we know the behavior of the individual branches, in particular as regards power and energy. We give some applications to physically interesting networks.

Introduction

Electrical circuit theory is an approximation to electromagnetic theory in which it is assumed that the interesting phenomena can all be described in terms of what is happening along the wires and other parts of an electrical circuit. It further assumes that the circuit can be decomposed into various components, each with a specified mode of behavior, and the problem is to predict how the system as a whole will behave when the components are interconnected in various ways.

The basic variables of circuit theory are familiar from household appliances; they are current, voltage and power. A fundamental unit in electromagnetic theory is the charge of the electron. As of this writing (prior to the discovery of quarks) no known particle has a charge that is a fractional part of the charge of the electron. The practical unit of charge is the coulomb, which represents the negative of the charge of 6.24×10^{18} electrons. In 1819 Oersted observed that a flow of electric charge produced a force on a magnetic needle, and that the force was proportional to the rate of flow of charge. The measurement of this effect is much easier than the measurements of electrostatic forces that are needed to measure charge. For this reason, *current*, rather than charge, is a basic variable of circuit theory. The unit of current is the ampere, where 1 ampere = 1 coulomb/second. We shall use I to denote current. Thus

$$I = \frac{dQ}{dt}$$

where I is current, measured in amperes, Q is charge, measured in coulombs, and t is time, measured in seconds. In general, we would expect that the current flowing through a circuit would depend on position. In circuit theory it is assumed that the current takes on a definite value at each component (but may be time-dependent). If α denotes a branch of the circuit, then we let $I_\alpha(t)$ denote the current flowing through α at time t.

An element of charge may gain or lose energy as it passes through a portion of a circuit. The energy change (measured in joules) per unit charge (measured in coulombs) is known as the *voltage*. Thus 1 volt = 1 joule/coulomb. We will denote voltage by V. The voltage difference at time t between the two end points of the branch α will be denoted by $V^\alpha(t)$.

The product of current and voltage has the units of energy/time which is known as *power*. The unit of power corresponding to the units that we have introduced above is called the watt. Thus

$$1 \text{ watt} = 1 \text{ volt} \times \text{ampere} = 1 \text{ joule/second}.$$

In circuit theory, it is assumed that there are three kinds of branches: *inductors*, *capacitors* and *resistors*. Each of these types has a different kind of relationship between voltage and current. In an inductor, the voltage is proportional to the rate of change of the current: if α is an inductor the relation is

$$L_\alpha \frac{dI_\alpha}{dt} = V^\alpha$$

where the constant L_α is known as the *inductance* of the inductor. The unit of inductance corresponding to the units introduced above is the henry, defined by

$$1 \text{ henry} = 1 \text{ volt}/(1 \text{ ampere/second}) = 1 \text{ volt-second/ampere}.$$

In a capacitor, the current is proportional to the rate of change of voltage: if α is a capacitor the relation is

$$C^\alpha \frac{dV^\alpha}{dt} = I_\alpha$$

where the constant C^α, known as the *capacitance* of the capacitor, is measured in farads. In terms of units already introduced,

$$1 \text{ farad} = 1 \text{ coulomb/volt}.$$

In a resistor, there is a functional relation involving the current and the voltage (and not their derivatives directly). In the classical theory of *linear passive* circuits, the relation is given in the form $V = RI$ where R is a constant, measured in ohms,

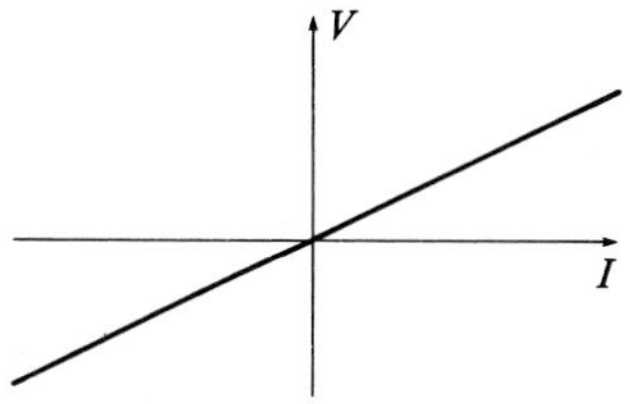

resistor only

Figure 12.1

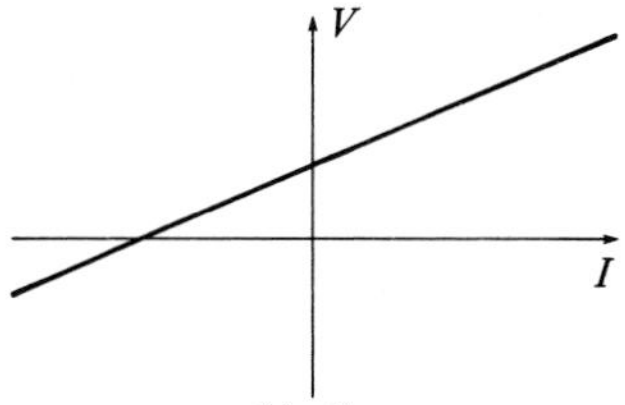

real battery

Figure 12.2

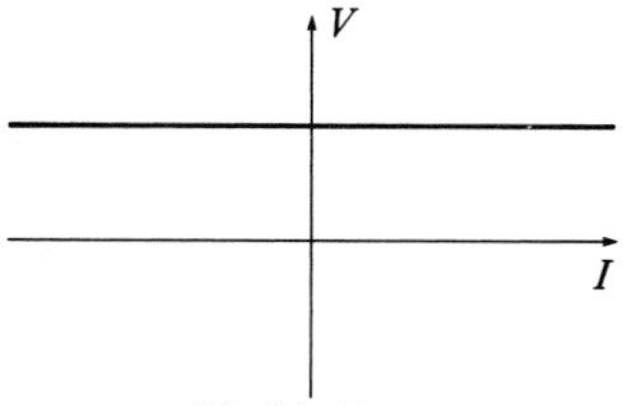

ideal battery

Figure 12.3

called the *resistance* of the branch. Its graph in the IV-plane is a straight line passing through the origin. More generally, we shall regard as a *linear resistor* any device which can be described by an (inhomogeneous) linear equation in I and V. For example, a real battery, with *internal resistance* R^α, can be described by $V^\alpha = W^\alpha + R^\alpha I_\alpha$, where the constant W^α is called the *emf* of the battery. An ideal battery, which provides voltage W^α no matter how much current is drawn, is described by $V^\alpha = W^\alpha$, whose graph is a horizontal straight line, while an ideal current source, which provides current K_α no matter what the voltage across its terminals may be, is described by $I_\alpha = K_\alpha$, whose graph is a vertical straight line.

In analyzing circuits in which the voltages and currents change with time, we must consider sources of voltage and current which vary with time. In this case, the most general linear resistive branch is described by

$$V^\alpha(t) - W^\alpha(t) = R_\alpha(I_\alpha(t) - K_\alpha(t))$$

where $W^\alpha(t)$ and $K_\alpha(t)$ are specified functions of time.

More generally, we might consider nonlinear resistors, inductors, or capacitors. A device such as a diode or thermistor may be regarded as a *nonlinear resistor*

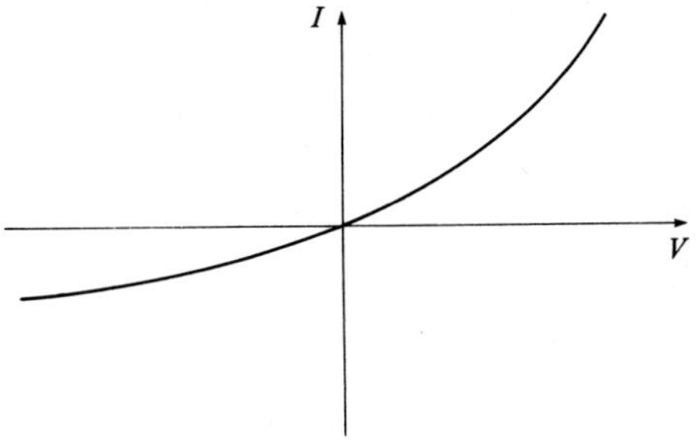

Figure 12.4

described by a nonlinear function $I_\alpha = f(V^\alpha)$ as suggested in figure 12.4. The most general resistive branch is characterized by a functional relationship $R(I,V,t) = 0$, where the only restriction is that no derivatives of I or V appear.

Generalizing the notion of inductor or capacitor, we might imagine an inductor whose inductance is a function of I, so that

$$V = L(I)\frac{\mathrm{d}I}{\mathrm{d}t}.$$

This would be the case, for example, if an inductor with an iron core were used for large currents. Similarly, we might consider a nonlinear capacitor, described by

$$I = C(V)\frac{\mathrm{d}V}{\mathrm{d}t}$$

which could be the result of using a dielectric with a nonlinear response.

Little more will be said about nonlinear devices, but it is important to recognize that most of the theory which follows, which is concerned primarily with setting up rather than solving the equations for electrical networks, applies with equal validity to linear and nonlinear elements.

We now assume that our electrical circuit is built out of b branches of the three types just mentioned. The branches are connected at their end points to one another in some way. We wish to determine the currents and voltages in all the branches, $2b$ unknowns in all. Each branch gives one equation (either a differential equation or a functional relation involving the current and voltage through that branch). We need b more equations. These are given by what are now known as *Kirchhoff's laws.*

Kirchhoff, as a student in Neumann's seminar, made the first comprehensive study of the network problem. He published his results in 1845 and 1847. He proved the existence of a solution to the network problem for a passive linear purely resistive network; i.e., for one in which there are only passive linear resistors. In solving this problem, he was one of the first to study the algebraic properties of shape. This abstract study of shape was created, as a mathematical discipline, some fifty years later by Poincaré, who called the subject *analysis situs.* Today the subject is called algebraic topology. The natures of the methods of algebraic topology make themselves apparent in the case of passive linear resistive networks, and so we shall occupy a considerable amount of space studying these networks before returning to the general case. In treating these networks, we shall follow a 1923 paper by Weyl, in which a proof of Kirchhoff's results is presented in a fashion that explains more directly the relations with algebraic topology.

Kirchhoff's laws, as restated by Maxwell, have a very simple formulation. *Kirchhoff's current law* asserts that since charge cannot be created or destroyed, and since no charge can be stored at an ideal point, the algebraic sum of all the currents entering or leaving a junction of branches must be zero. *Kirchhoff's voltage law* asserts that there exists a function, Φ, called the electrostatic potential, such that the voltage across each branch is given by the difference of the values of Φ at the end points, i.e., the two junctions of the branch. Maxwell devised two methods of solving the resistive network problem which are known as Maxwell's *mesh-current* and *node-potential* methods. We shall begin by working some examples to illustrate Maxwell's methods, and, in the process, set up some of the language of algebraic topology.

12.1. Linear resistive circuits

Resistors connected by metal wires of negligible resistance, as shown in figure 12.5, are said to be connected *in parallel*. Suppose that a battery supplying a constant voltage, V, is connected across the group of resistors. The ith resistor has resistance R_i and we set $G_i = R_i^{-1}$. (G_i is called the *conductance* of the ith resistor.) We are interested in calculating the total current delivered by the battery and the current in each branch after the current has become steady. The connection between all the upper terminals of the resistors ensures that in the steady state all these terminals are at the same potential, and the same applies to the lower terminal. Hence the voltage across each resistor is the same, and is equal to the voltage, V, of the battery. From the point of view of circuit theory, we can imagine all of the upper wire shrunk to a point, and similarly all of the wire connecting the lower terminals. We may thus replace figure 12.5 by figure 12.6 in which there are two vertices A and B, the top and the bottom, and $n + 1$ branches, of which one is the

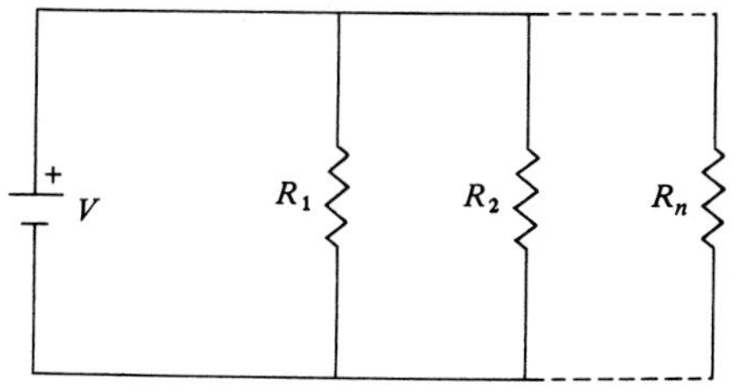

Figure 12.5

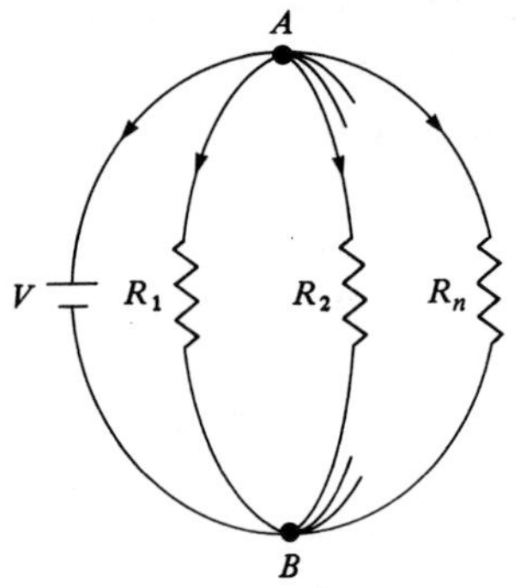

Figure 12.6

battery and the rest are linear passive resistors. Since the same voltage, V, is applied across all the resistors, it follows from the equation $V = RI$ or $I = GV$ that the currents through the resistors are $G_1V, G_2V, \ldots, G_nV$. By Kirchhoff's current law, the current flowing through the battery branch must be equal and opposite to the sum of the currents flowing through the resistors, and is therefore

$$-(G_1 + G_2 + \cdots + G_n)V.$$

(In deriving this result, we are making the sign convention that all branches in figure 12.6 are given similar orientations, so that, for example, all currents flowing from top to bottom are counted as positive, and from bottom to top are considered as negative.) We have completely solved this trivial network problem in that we now know the voltage across and the current through each branch. It follows from the above result that the total current supplied by the battery is the same as would be supplied if the battery were connected across a single resistor of conductance $G = G_1 + G_2 + \cdots + G_n$, i.e., of resistance R where

$$\frac{1}{R} = \frac{1}{R_1} + \frac{1}{R_2} + \cdots + \frac{1}{R_n}.$$

This is, of course. the well-known result, taught in all elementary courses, which states that, if a number of resistors are connected in parallel, they are

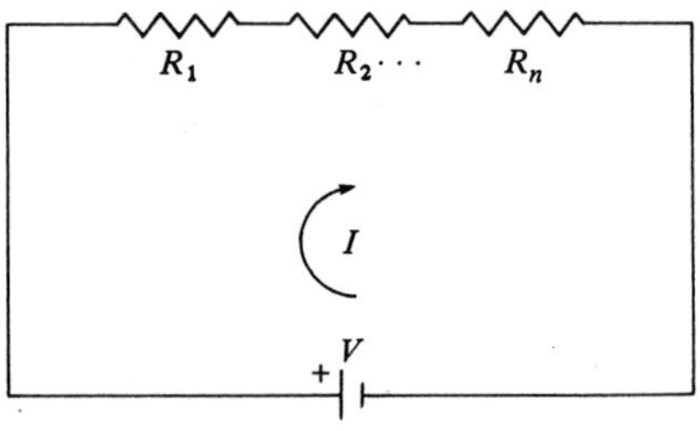

Figure 12.7

equivalent to a single resistor the reciprocal of whose resistance is equal to the sum of the reciprocals of the resistances of the individual resistors.

A group of resistors connected by wires of negligible resistance as shown in figure 12.7 is said to be *connected in series*. Suppose that a battery of voltage V is connected as shown, and that, as a result, a steady current I flows around the circuit. (By Kirchhoff's current law we must have the same current, I, flowing through each branch, because the algebraic sum of the currents at each node must be zero.) Let the resistances of the various resistors be $R_1, R_2, \ldots, R_n$. Since the current I flows through the ith resistor, it follows that the voltage across the ith resistor is $R_i I$. It follows from Kirchhoff's voltage law that the voltage across the battery must equal the sum of the voltage differences across all the resistors. Thus

$$V = (R_1 + R_2 + \cdots + R_n)I.$$

Since we know V, we can solve this equation for I and thus obtain the currents and voltages through each branch. We have completely solved this network problem. Again we have merely reproduced the well-known elementary result which states that if a number of resistors are connected in series they are equivalent to a single resistor whose resistance is the sum of the resistances of the individual resistors.

Let us now consider a slightly more complicated circuit consisting of a battery of voltage V connected to three resistors of resistances R_1, R_2 and R_3, as shown in figure 12.8. A point in the network from which two or more wires run to different elements is called a *node* or a *vertex*. The point A is a node from which wires run to the battery and to one of the resistors. The point B is a node from which wires run to all three resistors. The point C is a node from which there run three wires, one to the battery and two to different resistors. We do not consider the lower right-hand corner as a separate node; since it is connected by a wire of zero resistance to the point C, it must be identified as being the same node as C. Thus the circuit has three nodes, A, B, and C, and four branches, the battery branch joining A to C, the resistor joining A to B, and the two resistors joining B to C. The lengths and shapes of the wires making the interconnections are completely irrelevant. What matters are the branches and the nodes and how the branches and nodes are interconnected. Thus figure 12.9 describes exactly the same circuit as figure 12.8. A network as simple as that shown in figure 12.8 or 12.9 can be analyzed into parallel and series connections,

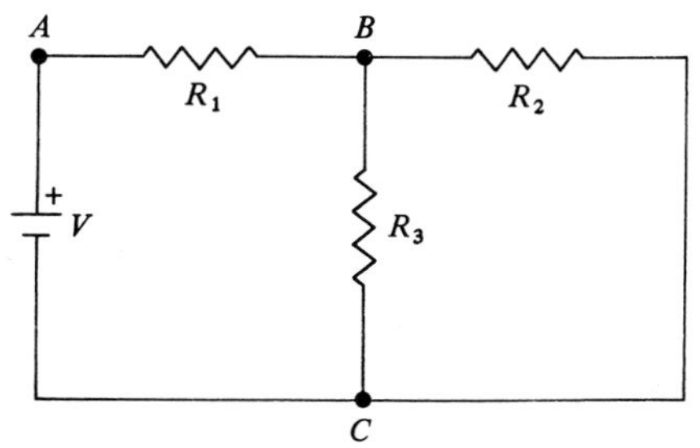

Figure 12.8

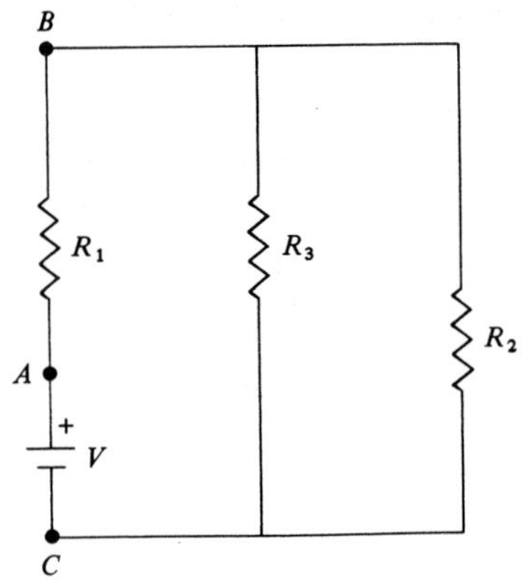

Figure 12.9

and in this way all the information about the network can be calculated. Thus the resistors R_2 and R_3 are in parallel, and hence are equivalent, as far as the rest of the circuit is concerned, to a single resistor whose resistance is the reciprocal of $R_2^{-1} + R_3^{-1}$, i.e., of resistance

$$\frac{R_2 R_3}{R_2 + R_3}.$$

This equivalent resistor is connected in series with the resistor R_1. Thus, as far as the battery is concerned, it is as if there were just a single resistor of resistance

$$R = R_1 + \frac{R_2 R_3}{R_2 + R_3}.$$

Thus the current drawn from the battery is

$$I = V/R.$$

This must also be the current through R_1, so that $I_1 = V/R$ is the current through R_1. The voltage drop across R_1 is then $V^1 = I_1 R_1$. The current I divides between the parallel resistors R_2 and R_3 in proportion to the inverse of their resistances, as we have seen when we discussed the example of resistors in parallel. Thus the currents through R_2 and R_3 are

$$I_2 = \frac{R_3 I}{R_2 + R_3} \quad \text{and} \quad I_3 = \frac{R_2 I}{R_2 + R_3}.$$

From this we see that the common voltage drop across R_2 and R_3 is

$$\frac{R_2 R_3 I}{R_2 + R_3}.$$

In this case we have obtained all the relevant information about the network by considering it as a resistor R_1 in series with a pair of parallel resistors R_2 and R_3. This procedure is frequently the most convenient way of proceeding when the network is not too complicated. A more systematic method is needed for more complicated networks. We will now illustrate the two methods of Maxwell for this same simple network.

Before doing so, we shall draw the network once again, but this time just indicating the branches (with their orientations), the nodes, and how they are joined together; but not indicating the nature of the individual branches. There are three nodes, A, B and C, and four branches, α, β, γ, and δ. The branch α goes from A to B. We shall write this fact symbolically as

$$\partial \alpha = B - A.$$

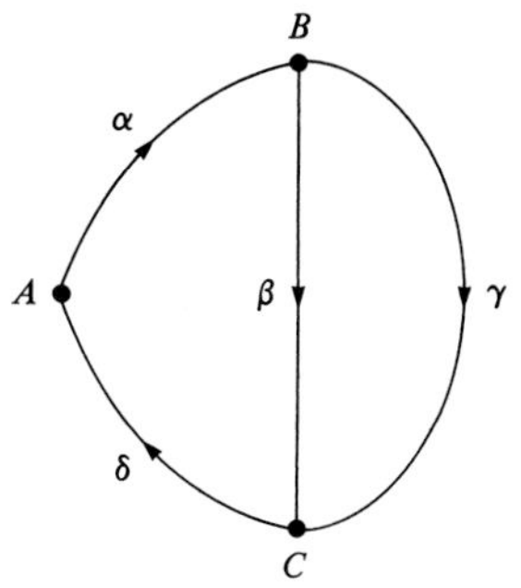

Figure 12.10

The symbol ∂ is called the *boundary operator*, and the above equation is interpreted as saying that the two boundary points of the branch α are B and A; also since the branch leaves A, we count A with a minus sign, and since it goes toward B, we count B with a plus sign. The remaining boundary relations are

$$\partial \beta = C - B, \quad \partial \gamma = C - B, \quad \text{and} \quad \partial \delta = A - C.$$

By a *path* we shall mean a succession of branches, each traversed in its proper direction or backwards, so that the end point of one segment is the origin of the next in the succession. Thus the path $\alpha + \beta$ goes from A to B and then from B to C. The boundary of this path is $\partial(\alpha + \beta) = B - A + C - B = C - A$. The path $\beta - \gamma$ goes from B to C along β and then from C to B along γ. It has no boundary, $\partial(\beta - \gamma) = C - B - (B - C) = 0$. We can think of a path as giving a succession – node, branch, node, branch, ..., branch, node – in which each branch is flanked by the two points of its boundary. The path is said to join the first and the last points in the succession. The path is called *closed* if the first and the last points coincide. In general, we do not suppose that all the branches in a path are distinct; we can pass by the same node or branch several times. If all the elements of the succession are distinct from one another (except for the first and last point if the path is closed), we say that the path is *simple*. A simple closed path is called a *mesh*. Thus $M_1 = \alpha + \beta + \delta$ is a mesh, as is $M_2 = \gamma - \beta$. In a mesh we do not care about the starting point; that is, we regard the mesh $\beta + \delta + \alpha$ as defining the same mesh, M_1, as $\alpha + \beta + \delta$. The direction is important: $-\gamma + \beta = -M_2$ is *not* the same as M_2. We could also consider a third

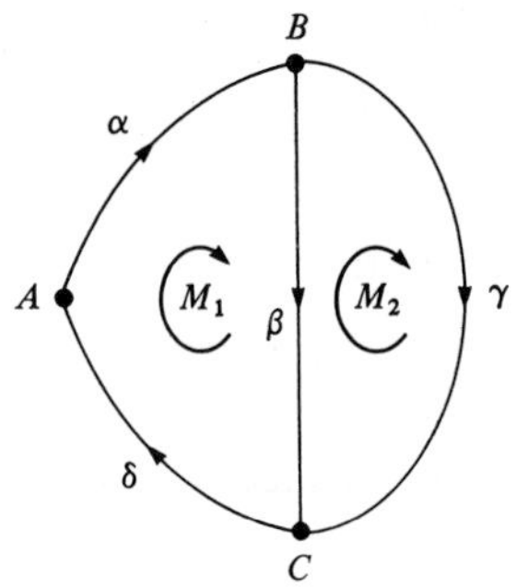

Figure 12.11

mesh, $M_3 = \alpha + \gamma + \delta$. In some sense, however, this mesh is not independent of the other two: if we formally add M_1 and M_2, and allow ourselves the option of applying the commutative law of addition, we get

$$M_1 + M_2 = \alpha + \beta + \delta + \gamma - \beta = \alpha + \gamma + \delta = M_3.$$

We shall spend some time in the next section setting up the mathematics to justify these kinds of manipulations. In Maxwell's mesh-current method we choose an independent set of meshes (precise definition in the next section) such as M_1 and M_2. We introduce unknown currents, J_1 and J_2, flowing around these meshes. Thus J_1 flows through R_1, R_3 and the battery, while J_2 flows downward through R_2 and upward through R_3. If we know the values of J_1 and J_2 we could immediately compute the values of the currents through the various branches. Indeed, since R_1 contributes only to the mesh M_1, the current through M_1 is equal to J_1, and the same goes for the battery. Since R_2 is part of the mesh M_2 only, all the current through it comes from the mesh current J_2 and so the current through R_2 is J_2. The branch containing the resistor R_3 contributes positively to the mesh M_1 and negatively to the mesh M_2, and thus the current through R_3 is $J_1 - J_2$. By Kirchhoff's voltage law, the total change in voltage as we go around any mesh must be zero. Applied to M_1, we get a drop of voltage R_1J_1 across α, a drop of $R_3(J_1 - J_2)$ across β and an

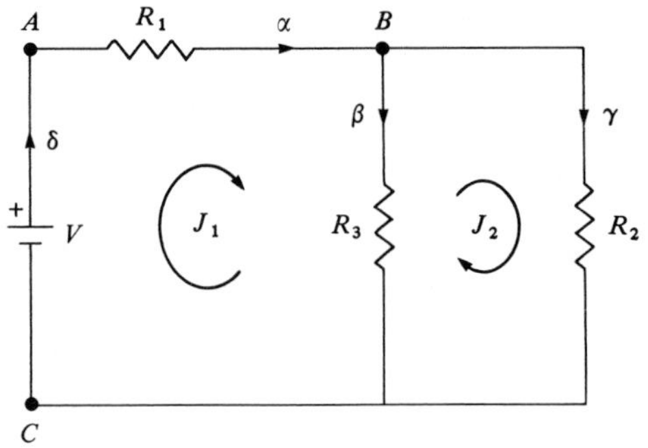

Figure 12.12

increase of V across the battery of δ. Thus

$$R_1J_1 + R_3(J_1 - J_2) - V = 0.$$

Similarly, for M_2 we get

$$R_2J_2 + R_3(J_2 - J_1) = 0.$$

We thus get two equations for the two unknowns, J_1 and J_2; it is easy to check that these can be solved to yield the same results as before.

Thus the gist of the mesh-current method is to choose an independent set of meshes, and assign unknown currents to them. This then determines the currents in all the branches in terms of the mesh currents. We then apply Kirchhoff's voltage law to each mesh in the form which asserts that the total change in voltage around each mesh must vanish. This gives one equation for each mesh, and hence as many equations as there are unknowns. In some sense, the idea of the mesh-current method is to introduce unknown currents in such a fashion that Kirchhoff's current law is automatically satisfied, and then to use the voltage law. We defer the precise definition of 'independent' and the proof of the theorem which asserts that the method works (i.e., that the equations have unique solutions for resistive networks) to the next few sections.

We now explain the node-potential method. Since the electric potential function is determined only up to an additive constant, we may fix the potential at one of the nodes to be zero. (If one of the nodes is connected to ground, it is usually convenient to select this node as having zero potential.) Using our same old network as example, let us set the potential at C equal to zero. The potential at A must then be V, the potential supplied by the battery. The only node whose potential is unknown is B. Let us denote this unknown potential by x. We now apply Kirchhoff's current law to those nodes with unknown potentials. In our case there is only one, node B. We can express the current flowing into node B through each of the three branches connected to it in terms of the resistances and the differences between node potentials. Thus, the current flowing into B through R_1 is $(V - x)/R_1$, the current flowing into B through R_2 is $(0 - x)/R_2$, and the current flowing into

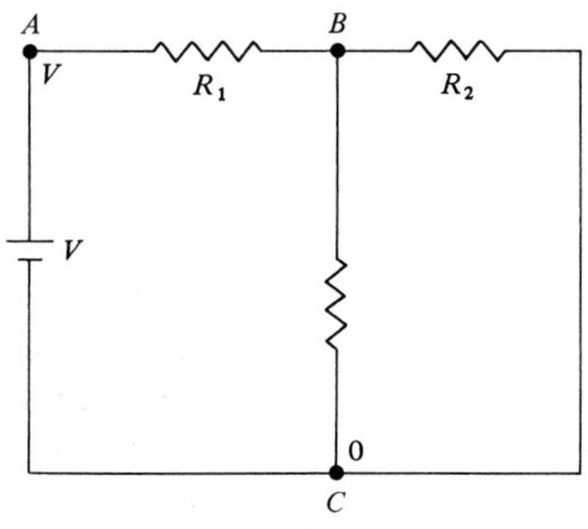

Figure 12.13

B through R_3 is $(0 - x)/R_3$. Kirchhoff's current law states that the sum of these three currents equals zero, so that

$$(V - x)/R_1 - x/R_2 - x/R_3 = 0.$$

We can solve this equation for x, then determine the voltages across and the currents through all branches. The proof of why this method works in general will be deferred to the next section.

Notice that in our example the node-potential method is superior to the mesh method since it involves solving for one unknown rather than two. It is not clear that the node-potential method is superior to our original analysis into parallel and series connections.

In dealing with a general network, it is advisable first to check by inspection whether it can easily be decomposed into parallel and series connections. If not, it is probably not worthwhile to try to figure out such a decomposition. Then check how many independent meshes there are and determine how many unknown mesh currents. Similarly determine how many unknown node potentials there are. (Frequently, symmetry considerations can cut down on the number of unknowns.) Choose the method with the fewer unknowns and use it to solve the network. Although it is not advisable to choose between the latter two methods by casual inspection, a general rule of thumb is that a network with few meshes and many nodes will yield more readily to the mesh-current method, and a network with few nodes and many meshes will yield more readily to the node-potential method. Another relevant factor is how the sources are specified. If the network is energized by sources having specified voltages, this tends to reduce the number of unknown node potentials, and hence favor the node-potential method. If currents are specified, this tends to favour the mesh-current method. To allow for all these considerations, it is usually best to draw two diagrams of the network and mark the number of unknowns of the mesh-current method on one and the number of unknowns of the node-potential method on the other in order to make an intelligent choice between the two methods.

So far we have been considering purely resistive networks. We can also apply these methods to determine the steady-state (oscillating) behavior of linear circuits with inductors and capacitors. (Linear here means that all inductances and capacitances are constant.) If all the generators are sinusoidal with the same frequency, $\omega/2\pi$ i.e., all voltage sources are of the form $V\mathrm{e}^{\mathrm{i}\omega t}$, and all current sources of the form $I\mathrm{e}^{\mathrm{i}\omega t}$, then, as is well-known (and follows trivially from the definitions), an inductor with inductance L acts by the law $V = \mathrm{i}\omega LI$ and a capacitor with capacitance C acts by the law $V = (1/\mathrm{i}C\omega)I$. We can now apply the same rules as before with these *complex resistances* or *impedances*. In this situation, however, solutions need not always exist, due to the phenomenon of resonance. Thus, for example, suppose that an inductor with $L = 1$ is in series with a capacitor with $C = 1$. If we put this series together with a generator with $\omega = 1$ the rule for adding resistances in series gives $R = \mathrm{i} + 1/\mathrm{i} = 0$, i.e., a short circuit drawing an infinite current. In the next section we shall discuss some conditions which avoid this unrealistic situation.

12.2. The topology of one-dimensional complexes

The terms *oriented graph* and *one-dimensional complex* are synonymous. They both refer to a mathematical structure that will represent for us the structure consisting of the various interconnections of the branches of an electrical network when we ignore the nature of the various branches; this structure will allow us to study the nature of Kirchhoff's laws in somewhat abstract form and then see how the properties of the individual branches affect the solution of the network problem.

A *one-dimensional complex* is a collection consisting of two sets: a set of zero-dimensional objects or nodes, $\{A, B, \ldots\} = S_0$ and a set of one-dimensional objects or branches $\{\alpha, \beta, \ldots\} = S_1$, together with a rule which assigns to each branch two distinct nodes, the initial point and the final point of the branch. Thus we are given a map from S_1 to $S_0 \times S_0$.* In what follows we shall assume that the sets S_0 and S_1 are finite.

We define the terms *path*, *mesh*, etc., as in the preceding section. A path determines an integer for each branch, namely the number of times the branch is traversed in the path, with orientation taken into account, so that when the branch is traversed from its initial point to its final point the contribution is $+1$ and when the branch is traversed in the opposite direction the contribution is -1. Thus each path determines a vector $\mathbf{p} = (p_\alpha, p_\beta, \ldots)^{\mathrm{T}}$ with integer coefficients, where the coordinates of the vector $\mathbf{p}$ are labeled by the branches of the graph; with p_α, for example, giving the total number of times the branch α is crossed in the positive direction minus the total number of times it is crossed in the negative direction. We can also think of a current distribution of the network as giving a vector $\mathbf{I} = (I_\alpha, I_\beta, \ldots)^{\mathrm{T}}$ whose coordinates are labeled by the branches, where now I_α, for example, is the real number giving the current, in amperes, through the branch α (if our one-dimensional complex were the complex of branches of an electrical network). This suggests that for any complex we introduce the vector space consisting of all vectors $\mathbf{K} = (K_\alpha, K_\beta, \ldots)^{\mathrm{T}}$ whose components are indexed by the branches. (Unless otherwise specified, we shall assume that these components are real numbers so that we get a real vector space.) We shall denote this vector space by C_1 and call it the space of *one-chains*. We shall identify each branch, κ, with the vector that has 1 in the κ position and zeros elsewhere. Thus $\alpha = (1, 0, 0, \ldots)^{\mathrm{T}}$, $\beta = (0, 1, 0, \ldots)^{\mathrm{T}}$ etc.

Similarly, we construct the real vector space, C_0, consisting of vectors whose components are indexed by the nodes and call this the space of *zero-chains*. Again, we will identify a node A with the vector which is 1 in the Ath position and zero elsewhere, so that a node is now an element of the vector space C_0. In particular, an

* $S_0 \times S_0$ denotes the Cartesian product of the set S_0 with itself. So $S_0 \times S_0$ is just the collection of all ordered pairs of nodes.

expression such as $A - B$ makes sense as an element of the vector space C_0. Notice that $\dim C_1$ is the number of branches and $\dim C_0$ is the number of nodes.

We now define a linear map called the *boundary map*, ∂, from C_1 to C_0. To define the map ∂ it is sufficient to prescribe its values on each of the branches, since the branches form a basis for C_1. Each branch has an initial point and a terminal point and we define ∂ (branch) = (terminal node) − (initial node). Thus, for example, if α is a branch going from A to B, then $\partial\alpha = B - A$.

Let us examine the meaning of the operator ∂. Suppose that* $\mathbf{K} = (K_\alpha, K_\beta, K_\gamma, \ldots)^T$ and $\partial\mathbf{K} = \mathbf{L}$, where $\mathbf{L} = (L_A, L_B, \ldots)^T$. In computing a term such as L_A, we see that we get a sum of certain of the coefficients of $\mathbf{K}$: in fact

$$L_A = (K_{\delta_1} + \cdots + K_{\delta_l}) - (K_{\varepsilon_1} + \cdots + K_{\varepsilon_k})$$

where $\delta_1, \ldots, \delta_l$ are all the branches which go to A and $\varepsilon_1, \ldots, \varepsilon_k$ are all the branches which leave A. Thus, from the example in figure 12.14,

$$L_A = K_\rho - K_\eta - K_\alpha.$$

From this we see that Kirchhoff's current law has a very simple formulation:

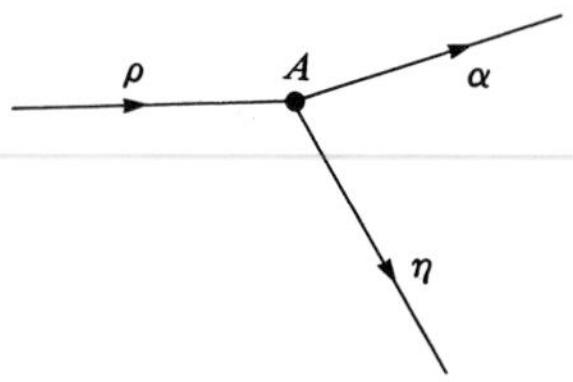

Figure 12.14

> *Kirchhoff's current law*: If $\mathbf{I}$ is the one-chain giving the current distribution of an electric current, then
>
> $$\partial\mathbf{I} = 0.$$

Recall that a simple closed path is called a *mesh*. In figure 12.15, for example, the path $\alpha + \beta + \delta$, represented by the vector $\mathbf{p}_1 = (1, 1, 0, 1)^T$, is a mesh, as is the path $-\beta + \gamma$, represented by the vector $\mathbf{p}_2 = (0, -1, 1, 0)^T$. The sum of these two vectors, $\mathbf{p}_3 = \mathbf{p}_1 + \mathbf{p}_2 = (1, 0, 1, 1)^T$ is another mesh, $\alpha + \gamma + \delta$. Clearly, each of these meshes has no boundary: $\partial\mathbf{p}_1 = \partial\mathbf{p}_2 = \partial\mathbf{p}_3 = 0$.

For any one-dimensional complex we shall denote the subspace of C_1 consisting of those one-chains satisfying $\partial\mathbf{K} = 0$ by Z_1. We express this relationship symbolically as $Z_1 = \ker \partial \subset C_1$. We call the elements of Z_1 *cycles*. Every mesh is a cycle, but

* On the preceding page and in much of what follows we write vectors in $\mathbb{R}^n$ as $(a, b, c, \ldots)^T$ instead of $\begin{pmatrix} a \\ b \\ c \\ \vdots \end{pmatrix}$ in order to save typesetting space.

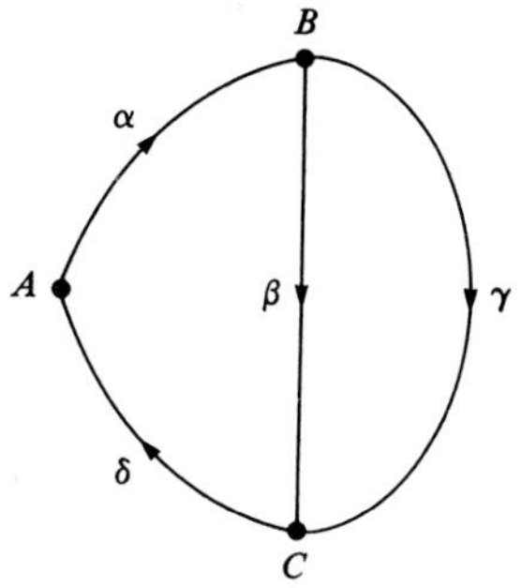

Figure 12.15

not every cycle is a mesh. For example, the vector

$$\mathbf{I}_1 = \begin{pmatrix} \frac{3}{2} \\ 1 \\ \frac{1}{2} \\ \frac{3}{2} \end{pmatrix}$$

satisfies $\partial\mathbf{I}_1 = 0$ but does not describe a mesh since it has entries other than 1, -1, or 0. It does represent a set of currents satisfying Kirchhoff's current laws. In fact, one of the easiest ways to verify that $\mathbf{I}_1$ is a cycle is to write in the appropriate current next to each branch and to check that the algebraic sum of the currents at each node equals zero. See figure 12.16.

Alternatively we could notice that $\mathbf{I}_1 = \frac{3}{2}\mathbf{p}_1 + \frac{1}{2}\mathbf{p}_2$; since $\mathbf{p}_1$ and $\mathbf{p}_2$ are elements of Z_1, so is I_1. In fact, for this simple network, the two meshes $\mathbf{p}_1$ and $\mathbf{p}_2$ form a basis for Z_1.

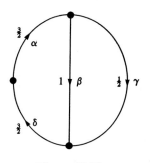

Figure 12.16

Havin defined the subspace $Z_1 \subset C_1$ as being of the *kernel* of the boundary map ∂, we now turn our attention to the *image* of ∂. We denote by B_0 the subspace of C_0 which is the image of ∂. Symbolically, we may write

$$B_0 = \partial C_1 \subset C_0.$$

We call B_0 the space of *boundaries*.

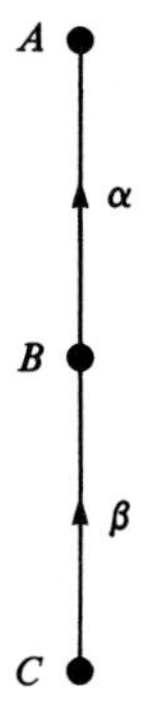

Figure 12.17

The significance of B_0 may be made clearer by reference to figure 12.17. The element of C_0, $A - B = (1, -1, 0)^T$ which is the boundary of the branch α, lies in the subspace B_0; so does $B - C = (0, 1, -1)^T$ which is the boundary of β. The sum of these two vectors, $A - C = (1, 0, -1)^T$, is again an element of C_0, and it is the boundary of the path $\alpha + \beta$.

Not every element of B_0 can be interpreted as the boundary of a path, however. For example, $(2, -1, -1)^T$ is an element of B_0 that corresponds to no single path.

If we consider these elements of C_0 that do *not* lie in the subspace B_0, we find that they do not form a subspace. For the network of figure 12.17, to take a simple example, the vectors $A = (1, 0, 0)^T$, and $B = (0, 1, 0)^T$ are *not* elements of B_0, but their difference $A - B = (1, -1, 0)^T$ *is* an element of B_0. We can, however, form the quotient space $H_0 = C_0/B_0$, whose elements are equivalence classes of elements of C_0 whose difference lies in B_0. The vectors $A = (1, 0, 0)^T$ and $B = (0, 1, 0)^T$ in C_0 correspond to the same vector in H_0 because their difference is the vector $(1, -1, 0)^T$, which lies in B_0. We denote this equivalence class by $\bar{A}$ or by $\overline{(1, 0, 0)}^T$. Similarly, the vectors $(2, 0, 0)^T$ and $(0, 1, 1)^T$ belong to the same equivalence class, $\overline{(2, 0, 0)}^T$, because their difference $(2, -1, -1)^T$ lies in B_0. For the simple network of figure 12.17, in fact, every vector in C_0 lies in the same equivalence class as a vector of the form $(\alpha, 0, 0)^T$, where α is a real number. The quotient space $H_0 = C_0/B_0$ is therefore a one-dimensional vector space, isomorphic with the real numbers.

For an example of a one-complex in which H_0 is two-dimensional, refer to

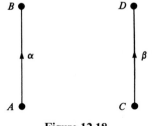

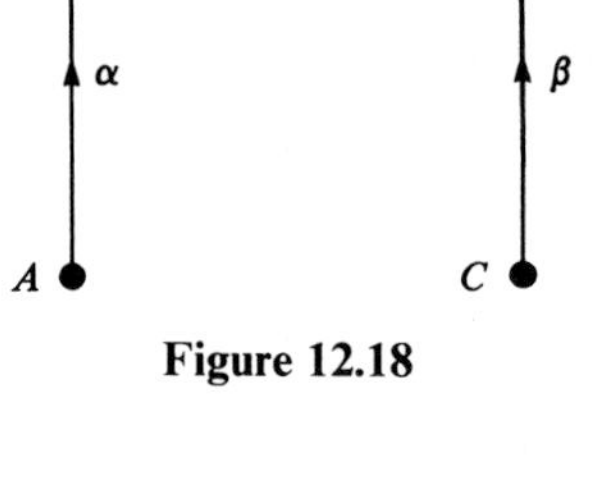

Figure 12.18

figure 12.18. The vectors $(1,0,0,0)^{\mathrm{T}}$ and $(0,0,1,0)^{\mathrm{T}}$ do *not* belong to same equivalence class because their difference $(1,0,-1,0)^{\mathrm{T}}$ is not the boundary of any element of C_1. In that case, the equivalence class of $(1,0,0,0)^{\mathrm{T}}$ and the equivalence class of $(0,0,1,0)^{\mathrm{T}}$ form a basis for the two-dimensional vector space H_0. Notice that, if a branch γ, joining A to C, were added to the complex, then $(1,0,-1,0)^{\mathrm{T}}$ would become an element of B_0 and H_0 would become one-dimensional.

The meaning of H_0 and Z_1

Our immediate goal is to give some geometric interpretation to the spaces H_0 and Z_1; in the process we shall get some understanding of the mesh-current method. We wish to prove the following two facts (whose precise statement we shall give in the course of the discussion):

(i) dim H_0 *is the number of connected components of the complex*;
(ii) *We can find a basis of* Z_1 *consisting of meshes.*

We begin with (i). What do we mean by connected components? We recall that a path joins the node A to the node D if A is the initial point of the first segment of the path and D is the final point of the last segment of the path. If $\mathbf{P}$ is the one-chain with integer coefficients corresponding to this path, then it is clear that $\partial\mathbf{P} = D - A$, since we can compute $\partial\mathbf{P}$ by adding the boundaries of all the individual branches, and all the intermediate nodes will vanish. A complex is called *connected* if every point can be joined to every other point. In this case we can write an arbitrary zero chain

$$\mathbf{L} = \sum_{N=A,B,C,\ldots} L_N N = (L_A, L_B, \ldots)$$

in the form

$$\mathbf{L} = \sum L_N\{(A + (N - A)\} = \sum L_N A + \sum_{N \neq A} L_N(N - A)$$

or, equivalently,

$$\mathbf{L} = (L_A + L_B + \ldots, 0, 0, \ldots)^{\mathrm{T}} + L_B(-1,1,0,0,\ldots)^{\mathrm{T}} + L_C(-1,0,1,0,\ldots)^{\mathrm{T}} + \ldots.$$

For a connected complex, $N - A$ is a boundary for all $N\,(\neq A)$, and

$$\mathbf{L} - \left(\sum_N L_N\right)A = \sum_{N \neq A} L_N(N - A)$$

is an element of B_0. Thus *every* zero-chain $\mathbf{L}$ is in the equivalence class of some multiple of A, and in the quotient space $H_0 = C_0/B_0$ every element is a multiple of $\bar{A}$, the equivalence class of A. On the other hand, A by itself is not a boundary, and so $\bar{A} \neq 0$. We have thus proved that dim $H_0 = 1$ for any connected complex.

Now consider any complex, and let A be some node. Starting with A, we consider all branches which have A as a boundary point, either an initial or a final point. Let us adjoin to A the set of all nodes that are at the other end of these branches. We then adjoin all new branches emanating from these nodes and then all new points at the end of these branches. We continue this way until we have no new branches or nodes to add. We thus obtain a collection $S_0(A)$ of nodes and a collection $S_1(A)$ of branches with the property that the boundary points of any branch in $S_1(A)$ lie in $S_0(A)$. Thus $(S_0(A), S_1(A))$ forms a *subcomplex* of our original complex. If it is not the whole complex, we pick some node B which does not lie in $S_0(A)$ and repeat the process. Proceeding in this way we get a collection of disjoint sets $S_0(A)$, $S_0(B)$, etc., and $S_1(A), S_1(B)$, etc., with each of the subcomplexes $S_0(A), S_1(A); S_0(B), S_1(B)$; etc., connected and none of them connected to any other. We get corresponding vector spaces $C_0(A)$, etc., and direct sum decompositions* $C_0 = C_0(A) \oplus C_0(B) \ldots$ and $C_1 = C_1(A) \oplus C_1(B) \ldots$ with $\partial C_1(A) \subset C_0(A), \partial C_1(B) \subset C_0(B)$, etc. From this we see that $H_0 = H_0(A) \oplus H_0(B) + \cdots$ and so $\dim H_0 =$ the number of summands $=$ the number of connected components.

We now turn to (ii). A mesh is a simple closed path. Hence each mesh determines an element, $\mathbf{M}$, of C_1 whose coordinates are either $+1$, -1 or 0, and furthermore, $\partial\mathbf{M} = 0$. We say that a set of meshes is independent if the corresponding elements of C_1 are linearly independent. (This is the precise formulation of the notion of independence that we used in the preceding section.) We wish to show that we can find a family of meshes so that the corresponding elements $\mathbf{M}_1, \ldots, \mathbf{M}_z$ form a basis of Z_1. In the process of proving this result we shall give a constructive procedure for finding such a family of meshes. Notice that in view of our discussion of (i) it is sufficient to work with a connected complex; if the complex is not connected we simply apply our procedure to each connected component separately. Since the components are completely independent of one another, this will give the result for the full complex.

A connected complex containing no meshes is called a *tree*. In a tree there exists at least one node which is a boundary point of only one branch. (We are uninterested in the trivial case of a complex with one node and no branches.) To prove this fact, simply start at any node. If more than one branch impinges on this node choose one of these branches and move to the other end point of this branch. If this node does not have the desired property, move along a branch, other than the one just used to

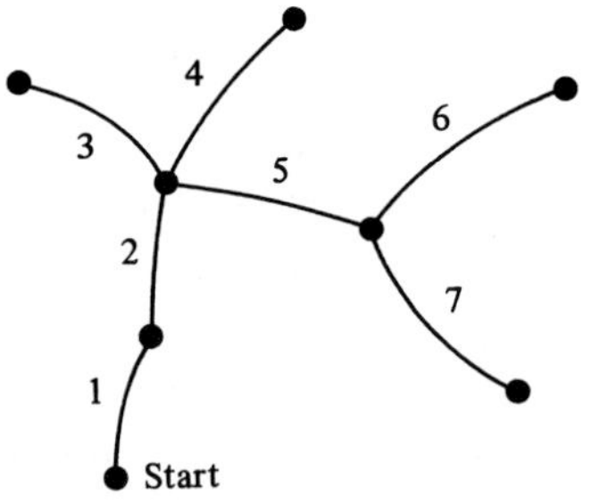

Figure 12.19

* The direct sum of two vector spaces V_1 and V_2 is the Cartesian product $V_1 \times V_2$ made into a vector space in the obvious way: $(v_1, v_2) + (w_1, w_2) = (v_1 + w_1, v_2 + w_2)$ and $r(v_1, v_2) = (rv_1, rv_2)$. If U and W are subspaces of a vector space V so that every $v \in V$ can be written uniquely as $v = u + w$ with $u \in V$ and $v \in W$ then $v \mapsto (u, w)$ gives an isomorphism of V with the direct sum of U and W. In this case we write $V = U \oplus W$, and we have $\dim V = \dim U + \dim W$.

still another node. Continue this procedure. Since there are no meshes, we can never come back to an earlier node. Since there are only a finite number of nodes, eventually we must reach a node which is the boundary of only one branch.

Suppose we have a tree and we start from a node which is at the end of only one branch. From this node we can build up the whole tree by adding one branch at a time. Since there are no meshes, each time we add a new branch we add a new node. Thus, in a tree, the number of nodes is exactly one more than the number of branches: if b denotes the number of branches and n denotes the number of nodes than in any tree

$$n = b + 1.$$

We have proved that, for any connected complex, $\dim H_0 = 1$. Since $H_0 = C_0/B_0$, $\dim H_0 = \dim C_0 - \dim B_0$, and we may conclude that $\dim B_0 = \dim C_0 - \dim H_0 = n - 1 = b$. Therefore B_0, the image of the map ∂, has the same dimension as the space C_1 on which ∂ acts. For any linear map $T: V \to W$ we know that $\dim(\text{im } T) + \dim(\ker T) = \dim V$. Applying this to the map ∂, we conclude that $\dim(\ker \partial) = \dim C_1 - \dim B_0 = b - b = 0$, so that $\ker \partial = \{0\}$; i.e. $Z_1 = \{0\}$. This proves (ii) for the case of a tree: if there are no meshes there are no non-trivial cycles.

Suppose we had any connected complex and built it up as before starting from some arbitrary node. Each time we add a branch we may or may not add a new node. Eventually, when we attach all the branches, we will have added all the nodes since the complex is connected. Thus for a general connected complex we have

$$n \leqslant b + 1.$$

For a connected complex we have already proved that $\dim H_0 = 1$, and we may conclude because $\dim C_0 - \dim B_0 = \dim H_0$ that $n - \dim B_0 = 1$ or $\dim B_0 = n - 1$.

Since B_0 is the image of ∂ and Z_1 is the kernel of ∂, we know that $\dim B_0 + \dim Z_1 = \dim C_1 = b$ or $(n - 1) + \dim Z_1 = b$. Therefore

$$\boxed{\dim Z_1 = b + 1 - n}$$

for any connected complex.

To prove (ii) we must exhibit a family of $b + 1 - n$ independent meshes which form a basis for Z_1. We do this by first dividing the set of branches S_1 into a *maximal tree* T which contains all the nodes of our complex, and a set of branches $\bar{T}$ which are not part of this tree, by the following procedure: Choose any branch which forms part of a mesh, and put it aside as a member of the subset $\bar{T}$. The remaining branches still connect all nodes. Repeat this procedure until no meshes remain. At this point the branches which have *not* been placed in the subset $\bar{T}$ constitute the maximal tree T. This procedure does not determine a unique maximal tree; figure 12.20 shows two different ways in which a given complex can be reduced to a maximal tree (solid lines) by removal of branches (dotted lines) which form part of meshes.

Since T forms a tree which connects all nodes, it contains $n - 1$ branches. There are $b - (n - 1) = b + 1 - n$ branches in $\bar{T}$. Each of these branches in $\bar{T}$ has its ends

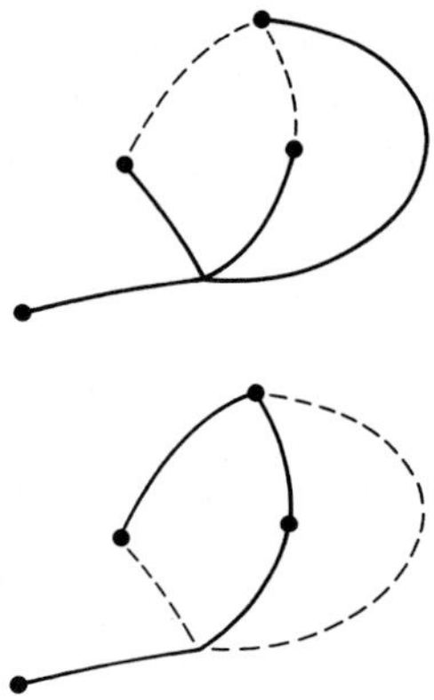

Figure 12.20. Two different maximal trees for the same complex

connected by a set of branches in T, since T connects all nodes; and, furthermore, since T contains no meshes, this connecting path is unique. Each branch in $\bar{T}$, combined with this unique path in T joining its ends, forms a mesh. Since the number of branches in $\bar{T}, b+1-n$, equals the dimension of Z_1, we now have only to prove that the meshes that we have constructed are linearly independent.

To prove linear independence, let α_i denote the ith branch of $\bar{T}$, and let $\mathbf{M}_i$ denote the mesh that includes branch α_i. Consider a linear combination of meshes: $\sum c_i\mathbf{M}_i$. Since α_i occurs *only* in mesh $\mathbf{M}_i$, with a coefficient of $+1$, the coefficient of α_i in the sum must be c_i. Therefore we cannot have $\sum c_i\mathbf{M}_i=0$ unless *all* the $c_i=0$, and we conclude that the meshes $\mathbf{M}_i$ are linearly independent.

Trees and projections

Notice that a choice of maximal tree T determines a projection ρ_T of C_1 onto the subspace Z_1, as follows:

If $\alpha_i \in T$, then $\rho_T(\alpha_i)=0$.

If $\alpha_i \in \bar{T}$, then $\rho_T(\alpha_i)=\mathbf{M}_i$.

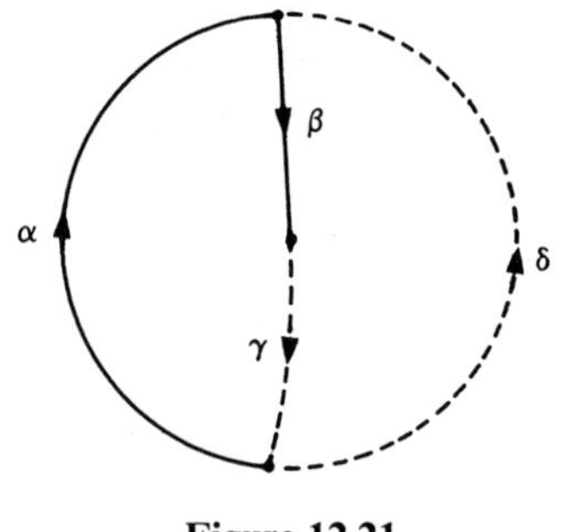

Figure 12.21

For example, if the maximal tree consisting of α and β is chosen in the network of figure 12.4 the projection operator ρ_T is $\rho_T(\alpha)=0$, $\rho_T(\beta)=0$, $\rho_T(\gamma)=\alpha+\beta+\gamma$, $\rho_T(\delta)=-\alpha+\delta$. (Notice that each mesh $\mathbf{M}_i$ includes the branch $\alpha_i \in \bar{T}$ with a coefficient of $+1$, not -1.) The projection ρ_T may be represented by the matrix

$$\begin{pmatrix} 0 & 0 & 1 & -1 \\ 0 & 0 & 1 & 0 \\ 0 & 0 & 1 & 0 \\ 0 & 0 & 0 & 1 \end{pmatrix}.$$

If we are dealing with a complex which is not connected we can complete the proof of (ii) and construct the projection ρ_T by simply choosing a maximal tree in each connected component.

The mesh current method

We now understand the significance of Maxwell's mesh-current method: by choosing an independent family of meshes and assigning *mesh currents* J_i to each mesh, we form an element $J_1\mathbf{M}_1+\cdots+J_m\mathbf{M}_m$ of Z_1, and the most general assignment of currents consistent which Kirchhoff's current law, i.e., the most general element of Z_1, can be obtained in this way.

In the preceding discussion we have made use of the equation

$$1-\dim Z_1 = n-b \tag{12.1}$$

for connected complexes. Applying this to each component of a general complex and adding, we get

$$\dim H_0 - \dim Z_1 = \dim C_0 - \dim C_1. \tag{12.2}$$

Since $\dim H_0 = \dim C_0 - \dim B_0$ and $\dim B_0 = \dim C_1 - \dim Z_1$, (12.2) is an immediate consequence of the definitions of the various spaces associated with ∂.

In Maxwell's mesh-current method, we think of the 'mesh currents as determining the currents in each branch'. We can give a mathematical formulation to this way of thinking as follows: Consider the space of mesh currents just as another copy of Z_1, call it H_1. That is, H_1, the space of mesh currents, is just a copy of Z_1, but thought of as an abstract vector space, not as a subspace of C_1. We shall then consider the operation of 'finding the branch currents determined by the mesh currents' as a linear map, σ, where $\sigma\colon H_1 \to C_1$ is the (identity) map which identifies H_1 with the subspace Z_1 of C_1. For example, in the circuit of figure 12.21, the space H_1 is two-dimensional. The map σ assigns to the mesh M_1 the branch currents $\alpha+\beta+\gamma$, so we write

$$\sigma(M_1)=\alpha+\beta+\gamma$$

and similarly

$$\sigma(M_2)=\gamma-\beta.$$

Thus, if we use M_1 and M_2 as a basis in our abstract space, H_1, of mesh currents, and use $\alpha, \beta, \gamma, \delta$ as a basis of C_1, then the matrix of σ relative to these basis is

$$\begin{pmatrix} 1 & 0 \\ 1 & -1 \\ 0 & 1 \\ 1 & 0 \end{pmatrix}.$$

We shall see that Maxwell's mesh-current method reduces to the problem of inverting a linear map from H_1 to its dual space, which we shall denote by H^1.

Returning to equation (12.2), we can now rewrite it as

$$\dim H_0 - \dim H_1 = \dim C_0 - \dim C_1 \qquad (12.3)$$

Kirchhoff's voltage law states that all branch voltages can be obtained as potential differences from a potential function defined on the nodes. Since only differences of potential are significant, we may arbitrarily assign the potential at one node of each connected component to be zero. The potentials at the remaining nodes then determine all the branch voltages. For example, in figure 12.22, if we assign $\Phi^A = \Phi^D = 0$, then the branch voltages are $V^\alpha = \Phi^C$, $V^\beta = \Phi^B$, $V^\delta = \Phi^C - \Phi^B$, $V^\gamma = \Phi^E$. Since the number of nodes is $\dim C_0$, the number of connected components $\dim H_0$, we see that there are only $\dim C_0 - \dim H_0$ *independent assignments* of branch voltages compatible with Kirchhoff's voltage law. Thus, roughly speaking, the number of conditions imposed by Kirchhoff's voltage law is

$$\dim C_1 - (\dim C_0 - \dim H_0) = \dim H_1.$$

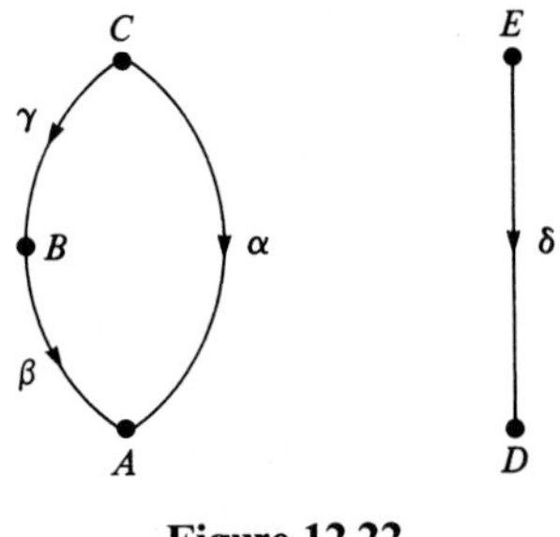

Figure 12.22

Since the number of independent assignments of branch currents compatible with Kirchhoff's current law equals $\dim H_1$, the number of linearly independent meshes, we see that Kirchhoff's current law imposes $\dim C_1 - \dim H_1$ conditions. Thus the two laws together impose $\dim C_1 = b$ conditions. (The laws are independent of each other since one refers to current and the other to voltages.) The equations relating current to voltage in each branch give b equations. Together we get $2b$ equations for the $2b$ unknows consisting of the currents and voltages. *Kirchhoff's theorem* about linear resistive circuits asserts that we can indeed solve these equations. To prove Kirchhoff's theorem we must now turn our attention to Kirchhoff's voltage law.

12.3. Cochains and the d operator

We first observe that voltage is, in a sense, dual to current; the product of the voltage, V^γ across a branch, γ, with the current I_γ through γ gives the power dissipated by γ. Thus, if we want to introduce a *voltage vector* $\mathbf{V} = (V^\alpha, V^\beta, \ldots)$ then the vector $\mathbf{V}$ should lie in the dual space of the space C_1. We therefore introduce this dual space and call it the space of *one-cochains*. We denote it by C^1. Thus C^1 is the space of linear functions on C_1. Similarly, we introduce the space C^0 of linear functions on C_0 and call it the space of *zero-cochains*. We now introduce two bits of notation that will appear somewhat strange at first, but will prove suggestive of some far reaching generalizations later on. Given a $\mathbf{K} = (K_\alpha, K_\beta, \ldots)^{\mathrm{T}} \in C_1$ and a $\mathbf{W} = (W^\alpha, W^\beta, \ldots) \in C^1$ we shall denote the value of the linear function, $\mathbf{W}$, on the vector, $\mathbf{K}$, by $\int_{\mathbf{K}} \mathbf{W}$; thus

$$\int_{\mathbf{K}} \mathbf{W} = W^\alpha K_\alpha + W^\beta K_\beta + \cdots.$$

Similarly, we shall denote the value of a zero-cochain $f = (f^A, f^B, \ldots)^{\mathrm{T}} \in C^0$ on a zero-chain $c = (c_A, c_B, \ldots)^{\mathrm{T}}$ by $\int_c f$; thus

$$\int_c f = f^A c_A + f^B c_B + \cdots.$$

Our second bit of notation has to do with the map ∂. The boundary map, ∂, is a linear map from C_1 to C_0. Its adjoint will be a linear map from C^0 to C^1. We shall denote this adjoint map by d and call it the coboundary operator. Thus, if f is a zero-cochain and $\mathbf{K}$ is a one-chain, the value of df on $\mathbf{K}$ is equal to f evaluated on $\partial\mathbf{K}$. In terms of the notation we have introduced, we can write this as

$$\int_{\mathbf{K}} \mathrm{d}f = \int_{\partial\mathbf{K}} f. \tag{12.4}$$

(Our notation has been chosen so as to make the preceding equation look like the fundamental theorem of the calculus. Of course, in the above equation there are no integrals, just the evaluation of linear functions on vectors in a vector space – i.e., finite sums. However, if we think of f as a differentiable function on the line and df as the expression $f'(x)\mathrm{d}x$, and if we think of K as the interval $A \leqslant x \leqslant B$, then the left-hand side of (12.4) would be interpreted as the integral $\int_K \mathrm{d}f = \int_A^B f'(x)\mathrm{d}x = f(B) - f(A)$, and this is what we would expect the right-hand side to mean. This said, the reader is asked to forget this analogy for the present. We shall return to it in section 15.2.)

If α is a branch and $\partial\alpha = B - A$, then the above equation becomes $\int_\alpha \mathrm{d}f = f(B) - f(A)$. Kirchhoff's voltage law says that there exists a function, Φ, on the nodes, i.e., a zero-cochain, such that $V^\alpha = \Phi(A) - \Phi(B)$ for any branch α with $\partial\alpha = B - A$. Thus Kirchhoff's voltage law can be formulated as

$$\mathbf{V} = -\mathrm{d}\Phi.$$

(To see this we observe that if the equation $\int_{\mathbf{K}} \mathbf{V} = -\int_{\partial\mathbf{K}} \Phi$ holds for all branches then it holds for all one-chains, since the branches form a basis of the space of one-chains. Thus $\mathbf{V}$ and $-\mathrm{d}\Phi$ take on the same value for all one-chains, i.e., $\mathbf{V} = -\mathrm{d}\Phi$.)

An immediate consequence of Kirchhoff's current and voltage laws is the result known as *Tellegen's theorem.* Suppose that, for a given network, $\mathbf{I} \in C_1$ is a distribution of branch currents satisfying Kirchhoff's current law, $\partial\mathbf{I} = 0$. Suppose also that, for the same network, there is a distribution of voltages, $\mathbf{V}$, which satisfies the voltage law $\mathbf{V} = -\mathrm{d}\Phi$. Then the total power dissipated in all branches is

$$P = \sum V^\alpha I_\alpha = \int_{\mathbf{I}} \mathbf{V} = -\int_{\mathbf{I}} \mathrm{d}\Phi.$$

But, by the very definition of the map d,

$$P = -\int_{\mathbf{I}} \mathrm{d}\Phi = -\int_{\partial\mathbf{I}} \Phi = 0,$$

since $\partial\mathbf{I} = 0$. This result shows that energy is conserved in electrical networks – batteries and generators supply energy at the same rate that it is dissipated in resistors. It is characteristic of the power of our notation that no assumptions had to be made about the nature of the branch elements.

In considering currents, we found it useful to introduce two subspaces related to the kernel and image of the boundary map ∂: the subspace of cycles $Z_1 = \ker\partial \subset C_1$ and the subspace of boundaries $B_0 = \operatorname{im}\partial \subset C_0$. Let us now carry out a similar program with the coboundary operator d.

Let $Z^0 \subset C^0$ denote the kernel of the operator d. Thus Z^0 is the subspace of potential functions on the nodes with the property that all branch voltages are zero. For a connected complex, all branch voltages will be zero if and only if the potential is the same at every node. More generally, Z^0 is the subspace of potentials which are constant on each connected component. Adding an element of Z^0 to an element of C^0 has no physical significance; it amounts simply to changing the arbitrary reference level with respect to which any potential is defined.

Suppose that Φ is an element of Z^0, so that $\mathrm{d}\Phi = 0$. Let $\partial\mathbf{K}$ denote an arbitrary element of the space of boundaries B^0. Then

$$\int_{\partial\mathbf{K}} \Phi = \int_{\mathbf{K}} \mathrm{d}\Phi = 0.$$

In other words, any element of Z^0, acting on any element of B_0, gives zero. Conversely, suppose that $\int_{\partial\mathbf{I}}\Phi = 0$ for all $\mathbf{K}$. Then $\int_{\mathbf{I}}\mathrm{d}\Phi = 0$ for all $\mathbf{K}$. So $\mathrm{d}\Phi = 0$, the zero function on C^1, and $\Phi \in Z^0$. For this reason Z^0 is said to be the *annihilator space* of B_0.

Consider now the quotient space $H_0 = C_0/B_0$, whose elements are equivalence classes of elements of C_0 whose difference lies in B_0. Let c_1 and c_2 be elements of C_0 whose difference $c_1 - c_2$ lies in B_0. Then, for any $\Phi \in Z^0$,

$$\int_{c_1} \Phi - \int_{c_2} \Phi = \int_{c_1 - c_2} \Phi = 0$$

so that $\int_{c_1}\Phi = \int_{c_2}\Phi$ if $c_1 = c_2 \pmod{B_0}$. Thus an element of Z^0 has the same value when evaluated on any member of an equivalence class, so that Z^0 may be regarded as the space of linear functions on H_0.

This relationship of Z^0 to C_0, B_0, and H_0 is an example of a general principle which we have encountered: the *kernel* of the *adjoint* of a linear transformation is both the *annihilator* space of the *image* of the transformation. It is also the *dual* space of the *quotient* by the image subspace. Here we have the transformation ∂, acting on C_1, the adjoint transformation d, acting on C^0, and the kernel of the adjoint, Z^0, which is the annihilator space of $B_0 = \operatorname{im}\partial$ and also the dual of $H_0 = C_0/B_0$.

We now consider the quotient space $P^0 = C^0/Z^0$. Since Z^0 consists of functions which are constant on each connected component, the elements of P^0 are equivalence classes of functions which differ only by a constant on each connected component. Thus if we modify a vector in C^0 by adding on a physically insignificant constant, possibly different for each connected component, the resulting vector still corresponds to the same element of P^0. It is P^0 which corresponds to the space of potentials, when we allow for the arbitrary additive constant in each connected component. Given a voltage vector **V** which obeys Kirchhoff's voltage law, the associated potential Φ, for which $\mathbf{V} = -\mathrm{d}\Phi$, is determined uniquely only up to the element of P^0 that Φ defines.

12.4. Bases and dual bases

To choose a basis for P^0, it is convenient to select one node in each connected component as *ground*, then choose a total of $\dim C^0 - \dim Z^0$ basis vectors for which the potential is unity at one node which is not a ground node, zero at all other nodes. For the network of figure 12.23, for instance, we may choose A and D as ground nodes, whereupon the vector (Φ^B, Φ^C, Φ^E) represents an arbitrary element of P^0.

The mapping $\mathrm{d}: C^0 \to C^1$ induces a map $[\mathrm{d}]: P^0 \to C^1$, which we shall call the *restricted coboundary map*. A matrix representation of [d] is obtained from that of d by simply deleting the columns corresponding to ground nodes. For the network of figure 12.23, we cross out columns $1(A)$ and $4(D)$, so

$$\mathrm{d} = \begin{pmatrix} -1 & 1 & 0 & 0 & 0 \\ 0 & -1 & 1 & 0 & 0 \\ 1 & 0 & -1 & 0 & 0 \\ 0 & 0 & 0 & 1 & -1 \end{pmatrix} \quad \text{while} \quad [\mathrm{d}] = \begin{pmatrix} 1 & 0 & 0 \\ -1 & 1 & 0 \\ 0 & -1 & 0 \\ 0 & 0 & -1 \end{pmatrix}$$

Since Z^0 is the annihilator space of B_0, any two elements of C^0 differing by an element of Z^0 give the same result when acting on an element of B_0. Thus the quotient space $P^0 = C^0/Z^0$ may be identified with the dual space of B_0. The adjoint of the map $[\mathrm{d}]: P^0 \to C^1$ is the restricted boundary operator $[\partial]: C_1 \to B_0$. It is convenient to use a basis for B_0 which is dual to that chosen for P^0, so that

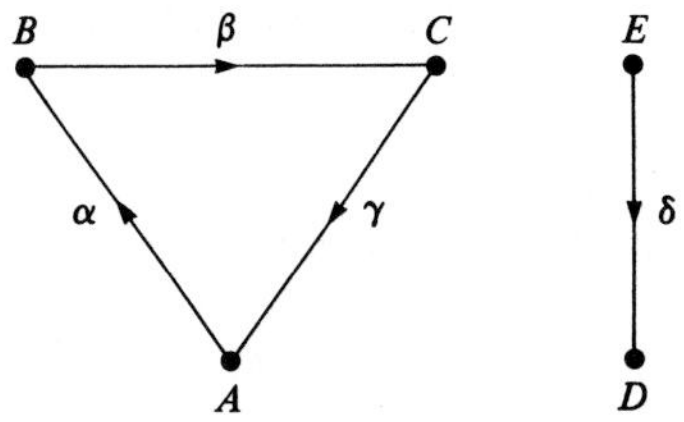

Figure 12.23

the matrix representing $[\partial]$ is just the transpose of the matrix representing $[\mathrm{d}]$. Relative to this basis, a vector in B_0 is represented by simply deleting from its expression as a vector in C_0 the components corresponding to ground nodes. For example, in figure 12.23, with ground nodes A and D, we have

$$\partial\alpha = B - A$$

so that in vector notation

$$\partial\begin{pmatrix}1\\0\\0\\0\\0\end{pmatrix} = \begin{pmatrix}-1\\1\\0\\0\\0\end{pmatrix} \quad \text{and} \quad [\partial]\begin{pmatrix}1\\0\\0\\0\end{pmatrix} = \begin{pmatrix}1\\0\\0\end{pmatrix}$$

and

$$\partial = \begin{pmatrix}-1 & 0 & 1 & 0\\ 1 & -1 & 0 & 0\\ 0 & 1 & -1 & 0\\ 0 & 0 & 0 & 1\\ 0 & 0 & 0 & -1\end{pmatrix} \quad \text{but} \quad [\partial] = \begin{pmatrix}1 & -1 & 0 & 0\\ 0 & 1 & -1 & 0\\ 0 & 0 & 0 & -1\end{pmatrix} = [\mathrm{d}]^{\mathrm{T}}.$$

We now denote by B^1 the subspace of C^1 that is the image of d. Physically, B^1 is the space of branch voltage vectors that obey Kirchhoff's voltage law. Furthermore, by Tellegen's theorem, which we proved earlier, B^1 is the annihilator space of Z_1: any voltage distribution obeying Kirchhoff's voltage law, applied to any current distribution obeying Kirchhoff's current law, gives zero. If we form the quotient space $H^1 = C^1/B^1$, then, it may be identified with the dual space of H_1 (also known as Z_1). The proof is simple. Let $\mathbf{V}_1$ and $\mathbf{V}_2$ be vectors in C^1 whose difference lies in B^1, so that $\mathbf{V}_1 - \mathbf{V}_2 = -\mathrm{d}\Phi$ for some Φ. Then, given an arbitrary current vector $\mathbf{I} \in Z_1$, for which $\partial\mathbf{I} = 0$,

$$\int_{\mathbf{I}}\mathbf{V}_1 - \int_{\mathbf{I}}\mathbf{V}_2 = \int_{\mathbf{I}}(\mathbf{V}_1 - \mathbf{V}_2) = -\int_{\mathbf{I}}\mathrm{d}\Phi = -\int_{\partial\mathbf{I}}\Phi = 0$$

so that $\int_{\mathbf{I}}\mathbf{V}_1 = \int_{\mathbf{I}}\mathbf{V}_2$. Thus $\mathbf{V}_1$ and $\mathbf{V}_2$, which correspond to the same element of H^1, are also equal as elements of the dual space of H_1.

We may summarize the relationship among C^0, C^1, Z^1, B^1, and Z_1 as follows:

The transformation d acts on C^0, and maps $C^0 \to C^1$;

Its adjoint ∂ acts on C_1, mapping $C_1 \to C_0$;

The kernel of ∂ is Z_1, which has as its annihilator space $B^1 = \operatorname{im} d$ and also has as its dual space $H^1 = C^1/B^1$.

12.5. The Maxwell methods

We earlier considered a procedure for constructing a basis of meshes for Z_1 by choosing a maximal tree. We denote by σ the map that identifies each vector in $Z_1 = H_1$ as an element of C_1. We can now choose a basis for H^1 which is dual to the basis chosen for H_1, and denote by s the map that projects the space C^1 onto the quotient space $H^1 = C^1/B^1$. Then s and σ are adjoint transformations, and their matrix representations are transposes of one another.

Constructing the matrices of σ and s is easier than it sounds. Consider the network of figure 12.24, with a maximal tree consisting of α and γ. Then a basis for Z_1 is

$$\mathbf{M}_1 = \alpha + \beta + \gamma = \begin{pmatrix} 1 \\ 1 \\ 1 \\ 0 \end{pmatrix},$$

$$\mathbf{M}_2 = -\gamma + \delta = \begin{pmatrix} 0 \\ 0 \\ -1 \\ +1 \end{pmatrix},$$

so that the matrix of σ is

$$\begin{pmatrix} 1 & 0 \\ 1 & 0 \\ 1 & -1 \\ 0 & +1 \end{pmatrix}.$$

Note that

$$\sigma\left(\begin{pmatrix} J_1 \\ J_2 \end{pmatrix}\right) = \begin{pmatrix} J_1 \\ J_1 \\ J_1 - J_2 \\ J_2 \end{pmatrix}$$

so that σ provides the rule for expressing branch currents in terms of mesh currents.

Now s is just the transpose of σ:

$$s = \begin{pmatrix} 1 & 1 & 1 & 0 \\ 0 & 0 & -1 & 1 \end{pmatrix}.$$

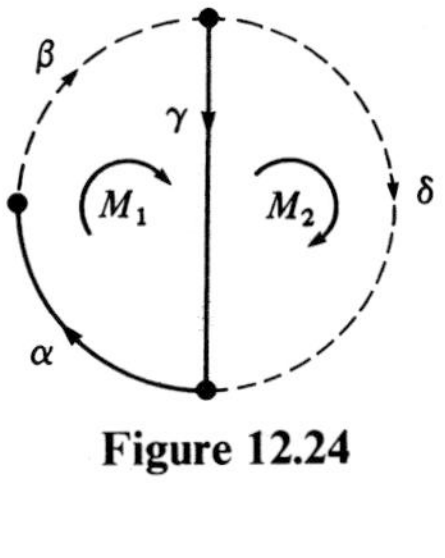

Figure 12.24

Note that

$$s\left(\begin{pmatrix} v^{\alpha} \\ v^{\beta} \\ v^{\gamma} \\ v^{\delta} \end{pmatrix}\right) = \begin{pmatrix} v^{\alpha} + v^{\beta} + v^{\gamma} \\ -v^{\gamma} + v^{\delta} \end{pmatrix}$$

so that the components of $s\mathbf{V}$ simply equal the sum of the branch voltages around the various meshes we choose as a basis for H_1.

We may summarize all the above relations in a single diagram:

$$P^0 \xrightarrow{[\mathrm{d}]} C^1 \xrightarrow{s} H^1$$

$$B_0 \xleftarrow{[\partial]} C_1 \xleftarrow{\sigma} H_1$$

In this diagram, vector spaces in the same column are dual to one another: C^1 and C_1, P^0 and B_0, H^1 and H_1. Furthermore, [d] and $[\partial]$ are adjoint, as are s and σ. Both [d] and σ are injective (have zero kernel) while their adjoints $[\partial]$ and s are surjective (have the entire indicated vector space as their image). Finally, $\operatorname{im}\sigma = \ker[\partial]$ while $\operatorname{im}[\mathrm{d}] = \ker s$. With this scheme we can write both of Kirchhoff's laws in a symmetrical form:

Kirchhoff's current law: $\mathbf{I} = \sigma\mathbf{J}$, $\mathbf{J} \in H_1$; Kirchhoff's voltage law: $\mathbf{V} = -[\mathrm{d}]\Phi$, $\Phi \in P^0$.	Kirchhoff's laws

We are now in a position to discuss Kirchhoff's theorem on resistive circuits. We shall assume that we have an electric circuit made up of resistors for which the operating characteristic of each branch α is a straight line in the (I_α, V^α) plane. We shall also make the realistic (and simplifying) assumption that this line is not

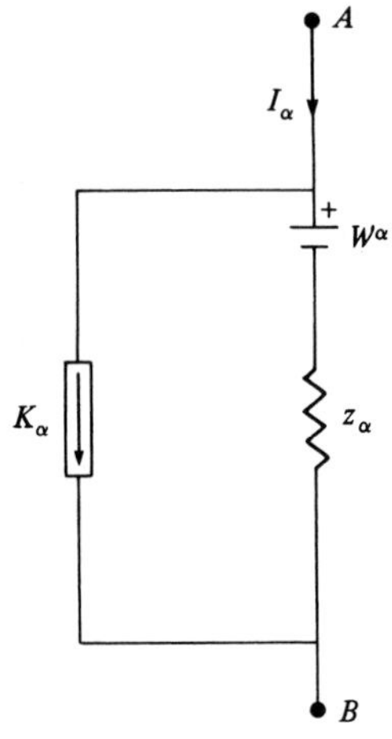

Figure 12.25

parallel to either of the axes. This means that any voltage source is in series with some resistance and any current source is in parallel with some resistance: there is no source that can supply constant voltage no matter what the current drawn and no source that provides constant current no matter what the voltage. We thus assume that each branch α with $\partial\alpha = B - A$ looks like figure 12.25.

The voltage across the resistor is $V^\alpha - W^\alpha$ and the current through the resistor is $I_\alpha - K_\alpha$. Thus the characteristic of the system is given as

$$(V^\alpha - W^\alpha) = z_\alpha(I_\alpha - K_\alpha).$$

Here W^α, K_α (either of which might be zero) are given, as is $z_\alpha \neq 0$. In a purely resistive circuit z_α is a positive real number. We can also consider the case of the steady state of a linear circuit containing capacitors and inductors, in which case z_α is a complex number depending on the frequency. The I_α and V^α are unknowns. We let $Z: C_1 \to C^1$ be the linear transformation whose matrix (in terms of the basis consisting of the branches) is the diagonal matrix with entries z_α. We can then write the above as

$$\mathbf{V} - \mathbf{W} = Z(\mathbf{I} - \mathbf{K}).$$

Combined with Kirchhoff's laws we obtain the equations

$$\boxed{\mathbf{V} - \mathbf{W} = Z(\mathbf{I} - \mathbf{K}), \quad \mathbf{I} = \sigma\mathbf{J}, \quad \mathbf{V} = -[\mathrm{d}]\Phi}$$

for the unknowns $\mathbf{J} \in H_1$ and $\Phi \in P^0$, where $\mathbf{W}$, $\mathbf{K}$ and Z are given. We can try to solve these equations by either of the following two methods.

The mesh current method

(i) Write $\mathbf{I} = \sigma\mathbf{J}$ to insure that Kirchhoff's current law is satisfied, then apply s to the equation $\mathbf{V} - \mathbf{W} = Z(\sigma\mathbf{J} - \mathbf{K})$. Since, by Kirchhoff's voltage law, $s\mathbf{V} = s(-[\mathrm{d}]\Phi) = 0$ we obtain $-s\mathbf{W} = sZ\sigma\mathbf{J} - sZ\mathbf{K}$ or $(sZ\sigma)\mathbf{J} = s(Z\mathbf{K} - \mathbf{W})$. The right-hand side of this equation is given, as is the linear transformation $sZ\sigma$. If $sZ\sigma$ is invertible we obtain

$$\mathbf{J} = (sZ\sigma)^{-1}s(Z\mathbf{K} - \mathbf{W}) \tag{12.5}$$

from which all currents and voltages may be obtained by $\mathbf{I} = \sigma\mathbf{J}$, $\mathbf{V} = \mathbf{W} + Z(\mathbf{I} - \mathbf{K})$.

This is Maxwell's mesh-current method. It depends upon inverting $sZ\sigma$.

The node potential method

(ii) Write $\mathbf{V} = -[\mathrm{d}]\Phi$ to insure that Kirchhoff's voltage law is satisfied. Then invert Z, which is just a diagonal matrix, to obtain $\mathbf{I} - \mathbf{K} = Z^{-1}(-[\mathrm{d}]\Phi - \mathbf{W})$. Now apply $[\partial]$ to both sides. Since, by Kirchhoff's current law, $[\partial]\mathbf{I} = 0$, we obtain

$$-[\partial]\mathbf{K} = -[\partial]Z^{-1}[\mathrm{d}]\Phi - [\partial]Z^{-1}\mathbf{W}$$

or

$$([\partial]Z^{-1}[\mathrm{d}])\Phi = [\partial](\mathbf{K} - Z^{-1}\mathbf{W}).$$

If $[\partial]Z^{-1}[\mathrm{d}]$ is invertible we obtain

$$\mathbf{\Phi} = ([\partial]Z^{-1}[\mathrm{d}])^{-1}[\partial](\mathbf{K} - Z^{-1}\mathbf{W}) \qquad (12.6)$$

and can obtain all voltages and currents from

$$\boxed{\mathbf{V} = -[\mathrm{d}]\mathbf{\Phi}, \quad \mathbf{I} = \mathbf{K} + Z^{-1}(\mathbf{V} - \mathbf{W}). \qquad (12.7)}$$

This is Maxwell's node-potential method. It depends upon inverting $([\partial]Z^{-1}[\mathrm{d}])$.

12.6. Matrix expressions for the operators

Equations (12.5)–(12.7) give the essence of Maxwell's methods. To make them work in practice we need matrix expressions for the operators. This involves appropriate choices of bases and dual bases. In this section we review how these choices are made and work out some examples. The *proof* that the methods always work for resistive networks will be given at the beginning of Section 12.7.

For the spaces B_0, H_0, Z^0, and P^0, a basis follows from a choice of a ground node in each connected component of the network. For example, in the network of figure 12.26, we might choose node A as ground in one connected component, node E in the other. Then a basis for B_0 consists of the boundaries of paths joining ground

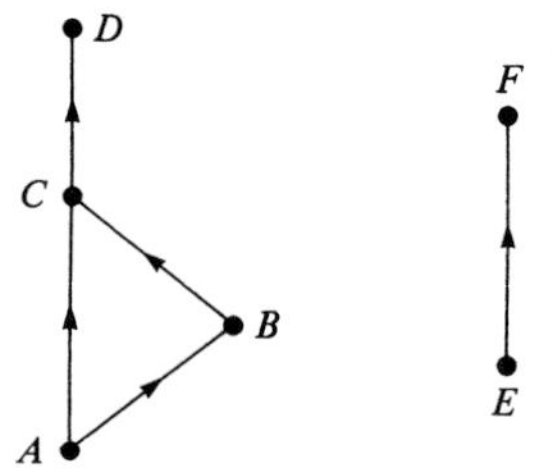

Figure 12.26

nodes to each other node in the same connected component. In the example, these basis elements are

$$\mathbf{b}_1 = \begin{pmatrix} -1 \\ 1 \\ 0 \\ 0 \\ 0 \\ 0 \end{pmatrix}, \quad \mathbf{b}_2 = \begin{pmatrix} -1 \\ 0 \\ 1 \\ 0 \\ 0 \\ 0 \end{pmatrix}, \quad \mathbf{b}_3 = \begin{pmatrix} -1 \\ 0 \\ 0 \\ 1 \\ 0 \\ 0 \end{pmatrix}, \quad \mathbf{b}_4 = \begin{pmatrix} 0 \\ 0 \\ 0 \\ 0 \\ -1 \\ 1 \end{pmatrix}.$$

To construct a basis for the quotient space $H_0 = C_0/B_0$ we extend the basis for B_0 to a basis for C_0 by choosing basis vectors which have 1 in the position of one ground node, zero elsewhere, and we take the equivalence class of each of these. In the example, this gives

$$\mathbf{h}_1 = \overline{\begin{pmatrix} 1 \\ 0 \\ 0 \\ 0 \\ 0 \\ 0 \end{pmatrix}}, \quad \mathbf{h}_2 = \overline{\begin{pmatrix} 0 \\ 0 \\ 0 \\ 0 \\ 1 \\ 0 \end{pmatrix}}.$$

In the space Z^0, the kernel of d, which is dual to H_0, the basis elements are potentials with a constant value of 1 on each connected component in turn. In the example we have

$$\mathbf{z}^1 = \begin{pmatrix} 1 \\ 1 \\ 1 \\ 1 \\ 0 \\ 0 \end{pmatrix}, \quad \mathbf{z}^2 = \begin{pmatrix} 0 \\ 0 \\ 0 \\ 0 \\ 1 \\ 1 \end{pmatrix},$$

which are clearly dual to $\mathbf{h}_1$ and $\mathbf{h}_2$ above. To choose bases for the quotient space $P^0 = C^0/Z^0$, we simply choose potentials which are 1 at each non-ground node in turn, and take the equivalence class of each. In the example, this gives

$$\mathbf{p}^1 = \overline{\begin{pmatrix} 0 \\ 1 \\ 0 \\ 0 \\ 0 \\ 0 \end{pmatrix}}, \quad \mathbf{p}^2 = \overline{\begin{pmatrix} 0 \\ 0 \\ 1 \\ 0 \\ 0 \\ 0 \end{pmatrix}}, \quad \mathbf{p}^3 = \overline{\begin{pmatrix} 0 \\ 0 \\ 0 \\ 1 \\ 0 \\ 0 \end{pmatrix}}, \quad \mathbf{p}^4 = \overline{\begin{pmatrix} 0 \\ 0 \\ 0 \\ 0 \\ 0 \\ 1 \end{pmatrix}}.$$

These equivalence classes are clearly dual to the basis $\{\mathbf{b}_1, \mathbf{b}_2, \mathbf{b}_3, \mathbf{b}_4\}$ for B_0 which we constructed earlier. This is why the matrix representing $[\mathrm{d}]: P^0 \to C^0$ is the transpose of the matrix representing $[\partial]: C_0 \to B_0$. The *columns* of $[\mathrm{d}]$, and the *rows* of $[\partial]$, correspond to the non-ground nodes which are associated with basis vectors.

We turn our attention now to the spaces Z_1, B^1, and H^1 which are associated with the kernel of ∂ and the image of d. For the sake of completeness, we consider

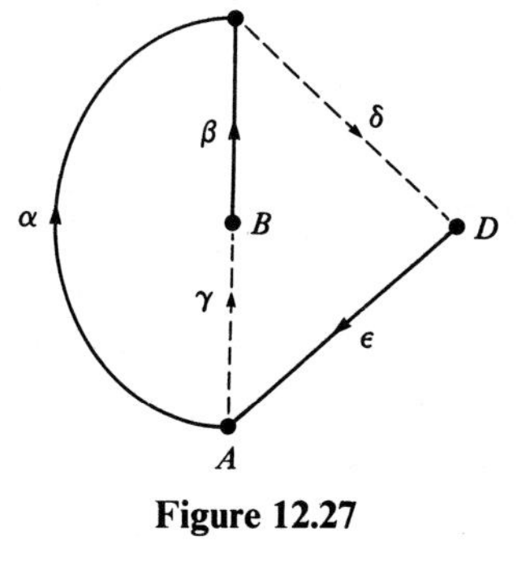

Figure 12.27

also the quotient space $G_1 = C_1/Z_1$, which is dual to B^1. For all these spaces, the choice of basis is governed by a choice of maximal tree, such as branches $\alpha, \beta, \varepsilon$ in figure 12.27.

The basis elements for Z_1, as we have already seen, are the meshes associated with the branches which are *not* in the maximal tree. In the example, these are

$$\mathbf{m}_1 = \begin{pmatrix} -1 \\ 1 \\ 1 \\ 0 \\ 0 \end{pmatrix} \quad \text{and} \quad \mathbf{m}_2 = \begin{pmatrix} 1 \\ 0 \\ 0 \\ 1 \\ 1 \end{pmatrix}$$

associated with branches γ and δ respectively. For the quotient space $G_1 = C_1/Z_1$, which represents violations of Kirchhoff's current law, we may form the equivalence classes of vectors which have 1 in the position of each tree branch in turn. In the example this gives

$$\mathbf{g}_1 = \overline{\begin{pmatrix} 1 \\ 0 \\ 0 \\ 0 \\ 0 \end{pmatrix}}, \quad \mathbf{g}_2 = \overline{\begin{pmatrix} 0 \\ 1 \\ 0 \\ 0 \\ 0 \end{pmatrix}}, \quad \mathbf{g}_3 = \overline{\begin{pmatrix} 0 \\ 0 \\ 0 \\ 0 \\ 1 \end{pmatrix}}.$$

Since these vectors all have 0 in the position of every non-tree branch, no linear combination can be a mesh. This shows that they are independent elements of G_1.

For the subspace B^1 of coboundaries, branch voltages that obey Kirchhoff's voltage law, we may construct a basis by assigning unit voltage to one branch in the maximal tree, zero to the other branches in the maximal tree. This assignment associates a potential (up to a constant) with each node, and these potentials can be used to determine the voltages across the non-tree branches. Equivalently, since each branch which is not in the maximal tree is associated with a unique mesh, containing that branch plus branches that are in the tree, we can use each such mesh in turn to determine the voltages across the non-tree branches. For the example of figure 12.27 this procedure leads to basis vectors associated with $\alpha, \beta, \varepsilon$ respectively:

$$\begin{aligned} \mathbf{b}^1 &= (1 \quad 0 \quad\ \ 1 \quad -1 \quad 0), \\ \mathbf{b}^2 &= (0 \quad 1 \quad -1 \quad\ \ 0 \quad 0), \\ \mathbf{b}^3 &= (0 \quad 0 \quad\ \ 0 \quad -1 \quad 1). \end{aligned}$$

These annihilate the meshes $\mathbf{m}_1$ and $\mathbf{m}_2$, and they are dual to the basis $\mathbf{g}_1, \mathbf{g}_2, \mathbf{g}_3$ for G_1. They do not coincide precisely with the columns of [d], but they span the same subspace of C^1.

Just as the basis vectors for B^1 may be associated with the branches which are in the maximal tree, those for $H^1 = C^1/B^1$ are associated with branches which are not in the tree. They are equivalence classes of voltage vectors which have $+1$ in the position of one non-tree branch, zero elsewhere. In our example, the basis vectors for H_1 are thus

$$\overline{\mathbf{h}^1 = \begin{pmatrix} 0 \\ 0 \\ 1 \\ 0 \\ 0 \end{pmatrix}}, \quad \overline{\mathbf{h}^2 = \begin{pmatrix} 0 \\ 0 \\ 0 \\ 1 \\ 0 \end{pmatrix}},$$

which correspond to *unit violations* of Kirchoff's voltage law in mesh 1 and mesh 2 respectively. Clearly $\mathbf{h}^1$ and $\mathbf{h}^2$ are dual to the meshes $\mathbf{m}_1$ and $\mathbf{m}_2$.Because of this, the matrix representing $s: C^1 \to H^1$ is the transpose of the matrix representing $\sigma: Z_1 \to C_1$.

An alternative method of choosing basis vectors for the space B^1 of voltages obeying Kirchhoff's voltage law is sometimes convenient. We simply choose the potential to be 1 in turn at each of the non-ground nodes, zero at all other nodes, and apply d to the resulting potentials. (Equivalently, we take the potential as -1 in turn at each non-ground node and write down the branch voltages.) In the present example, with node A chosen as ground, this leads to the basis vectors

$$\begin{aligned} \mathbf{a}^1 &= (0 \quad -1 \quad 1 \quad 0 \quad 0), \\ \mathbf{a}^2 &= (1 \quad 1 \quad 0 \quad -1 \quad 0), \\ \mathbf{a}^3 &= (0 \quad 0 \quad 0 \quad 1 \quad -1), \end{aligned}$$

associated with nodes B, C, and D respectively. As dual basis vectors in $G_1 = C_1/Z_1$, we must choose current distributions in which unit current flows from the ground node along the tree to each non-ground node in turn. In the example, this leads to basis vectors

$$\mathbf{k}_1 = \overline{\begin{pmatrix} 1 \\ -1 \\ 0 \\ 0 \\ 0 \end{pmatrix}}, \quad \mathbf{k}_2 = \overline{\begin{pmatrix} 1 \\ 0 \\ 0 \\ 0 \\ 0 \end{pmatrix}}, \quad \mathbf{k}_3 = \overline{\begin{pmatrix} 0 \\ 0 \\ 0 \\ 0 \\ -1 \end{pmatrix}},$$

which correspond respectively to unit violations of Kirchhoff's current law at non-ground nodes B, C, and D respectively.

Let us now apply both of Maxwell's methods to the circuit of figure 12.28, which has the topology which we considered earlier while analyzing figure 12.24. Wemust first make an arbitrary choice of which node will be ground: let us select node A, so that $\Phi^A = 0$. The branch voltages are now expressed in terms of the remaining node potentials as follows:

$$V^\alpha = -\Phi^B, \quad V^\beta = \Phi^B - \Phi^C, \quad V^\gamma = \Phi^C, \quad V^\delta = \Phi^C.$$

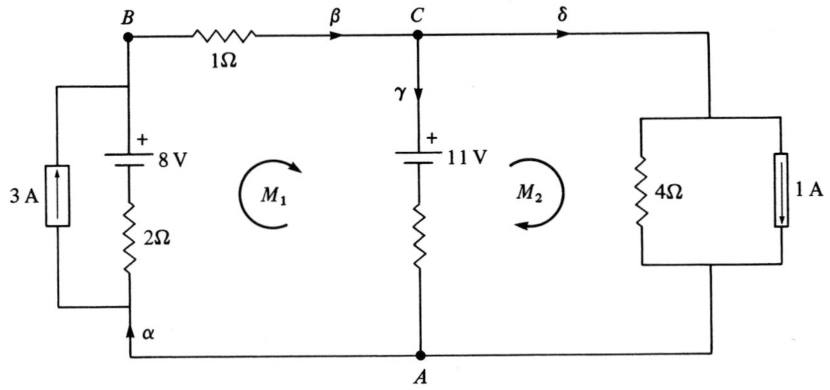

Figure 12.28

Therefore, in order to have $V = -[\mathrm{d}]\Phi$, we have

$$[\mathrm{d}] = \begin{pmatrix} +1 & 0 \\ -1 & +1 \\ 0 & -1 \\ 0 & -1 \end{pmatrix}.$$

As a check on this calculation, we form the matrix ∂. Since $\partial\alpha = B - A$; $\partial\beta = C - B$; $\partial\gamma = A - C$; $\partial\delta = A - C$, we have

$$\partial = \begin{pmatrix} -1 & 0 & +1 & +1 \\ +1 & -1 & 0 & 0 \\ 0 & +1 & -1 & -1 \end{pmatrix}.$$

To form $[\partial]$, we simply delete the row of ∂ which corresponds to the chosen ground node A. With the first row thus deleted we have

$$[\partial] = \begin{pmatrix} +1 & -1 & 0 & 0 \\ 0 & +1 & -1 & -1 \end{pmatrix}.$$

As expected $[\partial]$ is simply the transpose of $[\mathrm{d}]$.

With the meshes $\mathbf{M}_1$ and $\mathbf{M}_2$ chosen as indicated in figure 12.28, we see that the branch currents are expressed in terms of mesh currents J_1 and J_2 as

$$I_\alpha = J_1, \quad I_\beta = J_1, \quad I_\gamma = J_1 - J_2, \quad I_\delta = J_2,$$

so that, since $\mathbf{I} = \sigma\mathbf{J}$, we have

$$\sigma = \begin{pmatrix} 1 & 0 \\ 1 & 0 \\ 1 & -1 \\ 0 & 1 \end{pmatrix}.$$

As a check, we construct s by expressing the voltage drop $\mathscr{E}$ around each mesh in terms of the branch voltages.

Around mesh 1: $\mathscr{E}_1 = V^\alpha + V^\beta + V^\gamma$;

Around mesh 2: $\mathscr{E}_2 = -V^\gamma + V^\delta$.

In order to have $\mathscr{E} = s\mathbf{V}$,

$$s = \begin{pmatrix} 1 & 1 & 1 & 0 \\ 0 & 0 & -1 & 1 \end{pmatrix}.$$

As expected, s is the transpose of σ.

The next step is to write down the vectors **W** and **K** which represent the voltage and current sources in the four branches. The sign convention for **W** is that a battery voltage is considered positive if the battery contributes positively to the voltage drop when the branch is traversed in the sense indicated by the arrow. Referring to figure 12.28, we find

$$W^\alpha = -8, \quad W^\beta = 0, \quad W^\gamma = +11, \quad W^\delta = 0.$$

Similarly, a current source counts as positive if it contributes positively to current flow in the direction of the arrow, so that

$$K_\alpha = +3, \quad K_\beta = 0, \quad K_\gamma = 0, \quad K_\delta = +1.$$

We can therefore describe the sources in the network by the two vectors

$$\mathbf{W} = \begin{pmatrix} -8 \\ 0 \\ 11 \\ 0 \end{pmatrix} \quad \text{and} \quad \mathbf{K} = \begin{pmatrix} 3 \\ 0 \\ 0 \\ 1 \end{pmatrix}$$

Finally, we write down the matrix Z which describes the resistances. Since

$$R_\alpha = 2, \quad R_\beta = 1, \quad R_\gamma = 3, \quad R_\delta = 4,$$

we have

$$Z = \begin{pmatrix} 2 & 0 & 0 & 0 \\ 0 & 1 & 0 & 0 \\ 0 & 0 & 3 & 0 \\ 0 & 0 & 0 & 4 \end{pmatrix}.$$

To apply the mesh-current method we use the equation

$$\mathbf{J} = (sZ\sigma)^{-1} s(Z\mathbf{K} - \mathbf{W}).$$

Matrix multiplication gives

$$sZ\sigma = \begin{pmatrix} 1 & 1 & 1 & 0 \\ 0 & 0 & -1 & 1 \end{pmatrix} \begin{pmatrix} 2 & 0 & 0 & 0 \\ 0 & 1 & 0 & 0 \\ 0 & 0 & 3 & 0 \\ 0 & 0 & 0 & 4 \end{pmatrix} \begin{pmatrix} 1 & 0 \\ 1 & 0 \\ 1 & -1 \\ 0 & 1 \end{pmatrix} = \begin{pmatrix} 6 & -3 \\ -3 & 7 \end{pmatrix},$$

so that

$$(sZ\sigma)^{-1} = \frac{1}{33} \begin{pmatrix} 7 & 3 \\ 3 & 6 \end{pmatrix}.$$

We also find

$$\mathbf{ZK}-\mathbf{W}=\begin{pmatrix}2&0&0&0\\0&1&0&0\\0&0&3&0\\0&0&0&4\end{pmatrix}\begin{pmatrix}3\\0\\0\\1\end{pmatrix}-\begin{pmatrix}-8\\0\\11\\0\end{pmatrix}=\begin{pmatrix}14\\0\\-11\\4\end{pmatrix}$$

so that

$$s(\mathbf{ZK}-\mathbf{W})=\begin{pmatrix}1&1&1&0\\0&0&-1&1\end{pmatrix}\begin{pmatrix}14\\0\\-11\\4\end{pmatrix}=\begin{pmatrix}3\\15\end{pmatrix}.$$

Then, at last,

$$\mathbf{J}=\frac{1}{33}\begin{pmatrix}7&3\\3&6\end{pmatrix}\begin{pmatrix}3\\15\end{pmatrix}=\frac{1}{33}\begin{pmatrix}66\\99\end{pmatrix}=\begin{pmatrix}2\\3\end{pmatrix}$$

so the solution is $J_1=2$, $J_2=3$. Finally,

$$\mathbf{I}=\sigma\mathbf{J}=\begin{pmatrix}1&0\\1&0\\1&-1\\0&1\end{pmatrix}\begin{pmatrix}2\\3\end{pmatrix}=\begin{pmatrix}2\\2\\-1\\3\end{pmatrix}.$$

If you look back at the calculation, you can confirm the following points, which in fact hold for any network.

(1) In $sZ\sigma$, each diagonal entry is the sum of all the resistances in the branches which constitute a mesh. For example, mesh $\mathbf{M}_1$, consisting of branches α, β, and γ, has a total resistance of 6 ohms, the upper left entry of the matrix.

(2) In $sZ\sigma$, each off-diagonal entry is (up to a $\pm$sign) the resistance common to two meshes. In the circuit which we have been considering, meshes 1 and 2 have 3 ohms in common. The $-$ sign arises because γ occurs in M_1 and M_2 with opposite sign.

(3) The components of $(sZ\sigma)\mathbf{J}$ are the voltage drops which would exist around each mesh if there were no sources. With $J_1=2, J_3=3$, you can check, for example, that there would be a net drop of 3 volts around mesh 1.

(4) The components of $\mathbf{ZK}-\mathbf{W}$ are the voltage rises which would exist in the branches if all mesh currents were zero. For example, if J_1 were zero, there would be a current of 3 amperes down through the 2 ohm resistor in branch α and a rise of $(2\cdot 3+8)=14$ volts across α.

(5) The components of $s(\mathbf{ZK}-\mathbf{W})$ are the net voltage rises which would occur around the various meshes if all branch currents were zero. For example, with $J_1=J_2=0$, there would be a total rise of 3 volts around mesh 1.

(6) The equation $(sZ\sigma)\mathbf{J}=s(\mathbf{ZK}-\mathbf{W})$ therefore states that, for each mesh, the sum of the voltage drops caused by the mesh currents flowing through the resistances equals the sum of the voltage rises that would exist if the mesh

currents were all zero, so that the total voltage drop around each mesh, due to both the mesh currents and the sources, is zero as required by Kirchhoff's voltage law.

To apply the node-potential method, we use the equation

$$\Phi = ([\partial]Z^{-1}[d])^{-1}[\partial](\mathbf{K} - Z^{-1}\mathbf{W}).$$

Matrix multiplication gives

$$[\partial]Z^{-1}[d] = \begin{pmatrix} 1 & -1 & 0 & 0 \\ 0 & 1 & -1 & -1 \end{pmatrix} \begin{pmatrix} \frac{1}{2} & 0 & 0 & 0 \\ 0 & 1 & 0 & 0 \\ 0 & 0 & \frac{1}{3} & 0 \\ 0 & 0 & 0 & \frac{1}{4} \end{pmatrix} \begin{pmatrix} 1 & 0 \\ -1 & 1 \\ 0 & -1 \\ 0 & -1 \end{pmatrix}$$

$$= \begin{pmatrix} \frac{3}{2} & -1 \\ -1 & \frac{19}{12} \end{pmatrix}$$

so that

$$([\partial]Z^{-1}[d])^{-1} = \frac{24}{33}\begin{pmatrix} \frac{19}{12} & 1 \\ 1 & \frac{3}{2} \end{pmatrix}.$$

We also find

$$\mathbf{K} - Z^{-1}\mathbf{W} = \begin{pmatrix} 3 \\ 0 \\ 0 \\ 1 \end{pmatrix} - \begin{pmatrix} \frac{1}{2} & 0 & 0 & 0 \\ 0 & 1 & 0 & 0 \\ 0 & 0 & \frac{1}{3} & 0 \\ 0 & 0 & 0 & \frac{1}{4} \end{pmatrix} \begin{pmatrix} -8 \\ 0 \\ 11 \\ 0 \end{pmatrix} = \begin{pmatrix} 7 \\ 0 \\ -\frac{11}{3} \\ 1 \end{pmatrix}$$

so that

$$[\partial](\mathbf{K} - Z^{-1}\mathbf{W}) = \begin{pmatrix} 1 & -1 & 0 & 0 \\ 0 & 1 & -1 & -1 \end{pmatrix} \begin{pmatrix} 7 \\ 0 \\ -\frac{11}{3} \\ 1 \end{pmatrix} = \begin{pmatrix} 7 \\ \frac{8}{3} \end{pmatrix}.$$

Then

$$\Phi = \frac{24}{33}\begin{pmatrix} \frac{19}{12} & 1 \\ 1 & \frac{3}{2} \end{pmatrix}\begin{pmatrix} 7 \\ \frac{8}{3} \end{pmatrix} = \frac{24}{33}\begin{pmatrix} \frac{165}{12} \\ 11 \end{pmatrix} = \begin{pmatrix} 10 \\ 8 \end{pmatrix}$$

so that the unknown node potentials are $\Phi^B = 10, \Phi^C = 8$.

Look back at this calculation to confirm the following points, which again are true in general.

(1) Each diagonal entry in $[\partial]Z^{-1}[d]$ is the sum of the reciprocal resistances (conductivities) of all the resistors connected to a node. For example, node B has resistance of 1 and 2 ohms connected to it, and the upper left entry is just $1 + \frac{1}{2} = \frac{3}{2}$.

(2) Each off-diagonal entry in $[\partial]Z^{-1}[d]$ is (up to a $\pm$ sign) the reciprocal of the resistance in the branch which joins two nodes. Since nodes B and C are joined by a 1 ohm resistor, the off-diagonal entry is -1.

(3) The components of $[\partial]Z^{-1}[d]\Phi$ represent the *net* current which would flow out of each node if there were no sources. For example, with $\Phi^A = 0$, $\Phi^B = 8$ and no sources, there would be 5 amps flowing from B to A through the 2 ohm resistor, and 2 amps flowing from B to C through the 1 ohm resistor a total of 7 amps. You can check that the first component of the vector $[\partial]Z^{-1}[d]\Phi$ is 7.

(4) The components of $\mathbf{K} - Z^{-1}\mathbf{W}$ are the branch currents which would exist if all node potentials were zero. For example, if both Φ^A and Φ^B were zero, then there would be a current of 4 amperes through the 2 ohm resistor in branch α and a total branch current of $4 + 3 = 7$ amps. The first component current of $\mathbf{K} - Z^{-1}\mathbf{W}$ is 7.

(5) The components of $[\partial](\mathbf{K} - Z^{-1}\mathbf{W})$ are the net currents which would flow into each node if all node potentials were equal to zero. For example, in the case $\Phi^A = \Phi^B = \Phi^C$, 7 amps would flow into node B through branch α, no current would flow along branch β to node C, and the net current into B would be 7 amperes.

(6) The equation $[\partial]Z^{-1}[d]\Phi = [\partial](\mathbf{K} - Z^{-1}\mathbf{W})$ states that for each node, the sum of the currents out of each node which flow through the branch resistances as a result of node potential differences equals the sum of the currents which would flow into the node if all node potentials were zero, so that the total current entering each node, due to both the node potentials and the sources, is zero in accordance with Kirchhoff's current law.

If you compare the six statements about the node-potential method with the six about the mesh current, you will notice a remarkable duality. Replace *node* by *mesh*, *current* by *voltage*, and *resistance* by *conductance*, and they are the same.

One feature of the above example that does not hold in general is that the size of the matrix that had to be inverted was the same in both cases. That happened because there were two independent meshes and two non-ground nodes. In general one method may involve inverting a larger matrix than the other.

12.7. Kirchhoff's theorem

The question now arises whether the mesh-current method will work for any network. The issue is whether the mapping $sZ\sigma$ can be inverted. Since $sZ\sigma$ is a map from the space H_1 (mesh currents) to its dual, H^1 (voltage around meshes), so both these spaces have the same dimension, we have only to verify that $sZ\sigma$ is injective; i.e., that its kernel is zero. Consider any non-zero element $\mathbf{J}$ in H_1. Its image $(sZ\sigma)\mathbf{J}$ is an element of the dual space H^1, and we wish to show that it is *not* the zero element. We do this by simply evaluating it on the element $\mathbf{J}$; since s is the adjoint of σ we have $[(sZ\sigma)\mathbf{J}](\mathbf{J}) = [Z\sigma\mathbf{J}](\sigma\mathbf{J})$. But $\sigma\mathbf{J}$ is just the vector of branch currents $\mathbf{I}$, while $Z\sigma\mathbf{J} = Z\mathbf{I}$. Hence

$$[(sZ\sigma)\mathbf{J}]\mathbf{J} = (Z\mathbf{I})\mathbf{I} = \int_{\mathbf{I}} Z\mathbf{I} = \sum_{\gamma} z_\gamma I_\gamma^2$$

where the sum is over all branches. If the matrix Z has only positive diagonal entries, then $\int_I Z\mathbf{I} > 0$ unless $\mathbf{I} = 0$. But, since σ is injective, $\mathbf{I} = \sigma\mathbf{J} = 0$ implies $\mathbf{J} = 0$; that is, the branch currents are all zero only if all mesh currents are zero. Thus, for any non-zero $\mathbf{J}$, $[(sZ\sigma)\mathbf{J}](\mathbf{J}) > 0$. We conclude that $(sZ\sigma)\mathbf{J}$ cannot be the zero element of H^1 if $\mathbf{J} \neq 0$. Therefore $(sZ\sigma)$ has zero kernel and is invertible. This completes the proof that the mesh-current method will succeed for any purely resistive network. A similar argument shows that $[\partial]Z^{-1}[d]$ is invertible in this case, so that the node-potential method also works.

More generally, for a network whose branches are resistors, capacitors, or inductors, the above procedure gives us some interesting information. For a network with b branches, Z will still be given by a $b \times b$ diagonal matrix, whose entries, the impedances of the branches, may be functions of frequency $\omega/2\pi$. If each branch contains only one capacitor or inductor, each impedance will be $i\omega L$ or $-i/\omega C$. If there are m independent meshes, the matrix of $sZ\sigma$ will be an $m \times m$ matrix whose entries are functions of ω, and its determinant will give rise to a polynomial $D(\omega^2)$ whose degree is at most m. The operator $sZ\sigma$ will fail to be invertible if $D(\omega^2) = 0$, which means that there can be at most m values of ω^2 for which $sZ\sigma$ is not invertible. The frequencies $\omega/2\pi$ are the *resonant frequencies* of the system. For each of these resonant frequencies, the equation $(sZ\sigma)\mathbf{J} = 0$ has a non-zero solution even though there are no source terms on the right; that is, currents can flow even though there are no voltage or current sources. The solutions of this equation determine the *normal modes* of the system. A similar argument, applied to the matrix of $[\partial]Z^{-1}[d]$, whose size is $(b - m) \times (b - m)$, shows that there can be at most $(b - m)$ resonant frequencies. So to find the resonant frequencies of a network, we simply form the matrix of $sZ\sigma$ or $[\partial]Z^{-1}[d]$, whichever is smaller, and calculate for what values of ω^2 its determinant is zero.

As an illustration of this method, consider the network of figure 12.29. This has the same topology as figure 12.28, so again

$$s = \begin{pmatrix} 1 & 1 & 1 & 0 \\ 0 & 0 & -1 & 1 \end{pmatrix}, \quad \sigma = \begin{pmatrix} 1 & 0 \\ 1 & 0 \\ 1 & -1 \\ 0 & 1 \end{pmatrix}.$$

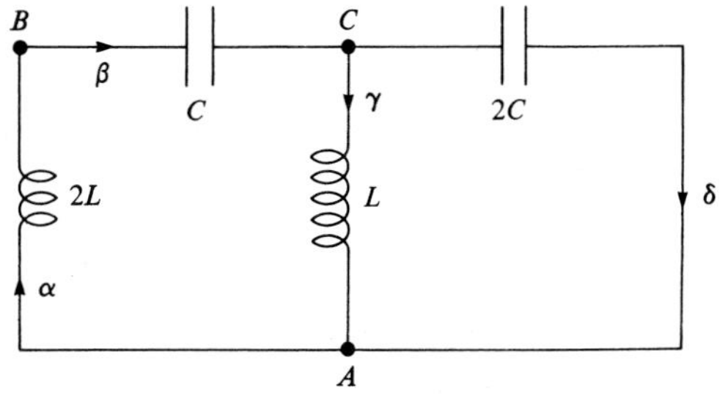

Figure 12.29

The matrix Z now involves the impedances of the various components:

$$z_\alpha = 2i\omega L, \quad z_\beta = -i/\omega C, \quad z_\gamma = i\omega L, \quad z_\delta = -i/2\omega C.$$

Then

$$Z = i\begin{pmatrix} 2\omega L & 0 & 0 & 0 \\ 0 & -(\omega C)^{-1} & 0 & 0 \\ 0 & 0 & \omega L & 0 \\ 0 & 0 & 0 & -(2\omega C)^{-1} \end{pmatrix}.$$

Matrix multiplication yields

$$sZ\sigma = i\begin{pmatrix} 3\omega L - (\omega C)^{-1} & -\omega L \\ -\omega L & \omega L - (2\omega C)^{-1} \end{pmatrix}$$

and we find that

$$\mathrm{Det}\,(sZ\sigma) = -(3\omega L - (\omega C)^{-1})(\omega L - (2\omega C)^{-1}) + \omega^2 L^2$$

or

$$\mathrm{Det}\,(sZ\sigma) = -2\omega^2 L^2 + 5L/2C - 1/2\omega^2 C^2.$$

Setting $\mathrm{Det}\,(sZ\sigma) = 0$, we obtain a quadratic in ω^2:

$$D(\omega^2) = 2L^2\omega^4 - 5L\omega^2/2C + (2C^2)^{-1} = 0.$$

This factors readily:

$$(L\omega^2 - C^{-1})(2L\omega^2 - (2C)^{-1}) = 0.$$

The resonant frequencies of the network are therefore given by

$$\omega^2 = 1/LC \quad \text{and} \quad \omega^2 = 1/4LC.$$

To find the corresponding normal modes, we solve the equation $(sZ\sigma)\mathbf{J} = 0$ for each resonant frequency. If $\omega = \sqrt{(1/LC)}$, for example,

$$sZ\sigma = i\begin{pmatrix} 2\sqrt{(L/C)} & -\sqrt{(L/C)} \\ -\sqrt{(L/C)} & \frac{1}{2}\sqrt{(L/C)} \end{pmatrix}$$

and a solution to $(sZ\sigma)\mathbf{J} = 0$ is $\begin{pmatrix} 1 \\ 2 \end{pmatrix}$, which means that the current in mesh 2 is twice the current in mesh 1. Similarly, setting $\omega = 1/2\sqrt{(LC)}$, we have

$$sZ\sigma = i\begin{pmatrix} -\frac{1}{2}\sqrt{(L/C)} & -\frac{1}{2}\sqrt{(L/C)} \\ -\frac{1}{2}\sqrt{(L/C)} & -\frac{1}{2}\sqrt{(L/C)} \end{pmatrix}$$

so that a solution to $(sZ\sigma)\mathbf{J} = 0$ is $\begin{pmatrix} 1 \\ -1 \end{pmatrix}$, which corresponds to a normal mode of frequency $\omega/2\pi$ where the mesh currents J_1 and J_2 have the same amplitude but opposite phase.

12.8. Steady-state circuits and filters

The concept of impedance, used in the preceding section to determine the resonant frequencies of circuits containing just inductors and capacitors, may be employed also to analyze steady-state alternating-current circuits. We assume that all voltage and current sources have the same frequency $\omega/2\pi$, so that branch currents are of the form

$$\mathbf{I} = \begin{pmatrix} I_\alpha \\ I_\beta \\ \vdots \end{pmatrix} e^{i\omega t}$$

and the branch voltages are of the form

$$\mathbf{V} = \begin{pmatrix} V^\alpha \\ V^\beta \\ \vdots \end{pmatrix} e^{i\omega t}.$$

Here the components $I_\alpha, I_\beta, \ldots$ and $V^\alpha, V^\beta, \ldots$ are *complex* quantities, and it is the real part of each component that represents the true current or voltage. (Recall that $e^{i\omega t} = \cos\omega t + i \sin \omega t$.) Voltage and current sources are represented by

$$\mathbf{W} = \begin{pmatrix} W^\alpha \\ W^\beta \\ \vdots \end{pmatrix} e^{i\omega t} \quad \text{and} \quad \mathbf{K} = \begin{pmatrix} K_\alpha \\ K_\beta \\ \vdots \end{pmatrix} e^{i\omega t}$$

respectively. Then it remains true for each branch that

$$V^\alpha - W^\alpha = z_\alpha(I_\alpha - K_\alpha)$$

where z_α is a diagonal matrix whose entries are now the complex impedances of the various branches. The formalism we have developed for direct current circuits thus carries over without change to steady-state alternating-current circuits, with only a few minor changes:

(1) The true, physical voltage or current in a branch is given by the *real part* of the appropriate component of the complex vector $Ie^{i\omega t}$ or $Ve^{i\omega t}$.

(2) The peak value of the voltage or current in a branch is given by the modulus of the complex quantity that represents the voltage or current; i.e., if the current in branch α is represented by I_α, the peak value is

$$|I_\alpha| = \sqrt{((\operatorname{Re} I_\alpha)^2 + (\operatorname{Im} I_\alpha)^2)}.$$

(3) The average power dissipated in branch α is determined by the average of the product of the *real parts* of the complex quantities V^α and I_α. To be explicit, the average power P is given by

$$\begin{aligned} P_\alpha &= \text{Average of } [\operatorname{Re}(V^\alpha e^{i\omega t}) \operatorname{Re}(I_\alpha e^{i\omega t})] \\ &= \tfrac{1}{4}\text{Average of } [(V^\alpha e^{i\omega t} + V^{\alpha *} e^{-i\omega t})(I_\alpha e^{i\omega t} + I_\alpha^* e^{-i\omega t})] \\ &= \tfrac{1}{4}\text{Average of } (V^\alpha I_\alpha e^{2i\omega t} + V^\alpha I_\alpha^* + V^{\alpha *} I_\alpha + V^{\alpha *} I_\alpha^* e^{-2i\omega t})], \end{aligned}$$

where we have used * to denote complex conjugate.

Since the terms involving $e^{2i\omega t}$ and $e^{-2i\omega t}$ average to zero over time, we have

$$P_\alpha = \tfrac{1}{4}(V^{\alpha *} I_\alpha + V^\alpha I_\alpha^*) = \tfrac{1}{2}\operatorname{Re}(V^\alpha I_\alpha^*).$$

In the case where branch α consists simply of a resistor, so that $V^\alpha = RI_\alpha$, this simplifies to

$$P_\alpha = \tfrac{1}{2}RI_\alpha^2.$$

For a branch consisting simply of a capacitor or inductor, whose impedance z_α is a purely imaginary quantity, the average power dissipated is zero.

As an elementary example of a steady-state circuit, consider the so-called low-pass filter shown in figure 12.30. We may regard this circuit as having two branches. A basis for Z_1 is the mesh $\alpha+\beta$, so that $\sigma = \begin{pmatrix}1\\1\end{pmatrix}$, $s=(1 \quad 1)$. Since $Z = \begin{pmatrix}R & 0\\ 0 & -\mathrm{i}/\omega C\end{pmatrix}$, we have immediately

$$sZ\sigma = (1 \quad 1)\begin{pmatrix}R & 0\\ 0 & -\mathrm{i}/\omega C\end{pmatrix}\begin{pmatrix}1\\1\end{pmatrix} = (R - \mathrm{i}/\omega C).$$

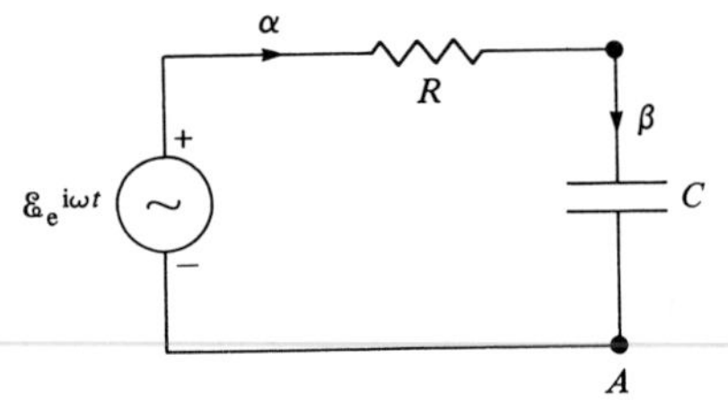

Figure 12.30

The voltage in branch α is $W^\alpha = -\mathscr{E}e^{\mathrm{i}\omega t}$ (with polarity as shown) so that the mesh current is simply

$$J = (sZ\sigma)^{-1}(-sW) = \frac{1}{R - \mathrm{i}/\omega C}\mathscr{E}e^{\mathrm{i}\omega t}.$$

The peak value of this current is the real quantity

$$|J| = \left|\frac{\mathscr{E}}{R - (\mathrm{i}/\omega C)}\right| = \frac{\mathscr{E}}{\sqrt{(R^2 + (\omega^2 C^2)^{-1})}}$$

while the peak value of the voltage across branch β is

$$|V^\beta| = |z_\beta I_\beta| = \left|\frac{-\mathrm{i}\mathscr{E}/\omega C}{R - \mathrm{i}/\omega C}\right| = \frac{\mathscr{E}}{\sqrt{((\omega RC)^2 + 1)}}.$$

This voltage decreases from $\mathscr{E}$ when $\omega = 0$ to zero as $\omega \to \infty$, hence the name *low-pass filter*.

We may make the preceding circuit slightly more complicated by placing an inductor in series with the capacitor. The impedance of branch β then becomes $z_L = \mathrm{i}\omega L - (1/\omega C)$, and the mesh current is $J = \mathscr{E}/(R + \mathrm{i}(\omega L - (\omega C)^{-1}))$. This mesh current clearly has its peak value if $L = 1/\omega C$ or $\omega = \sqrt{(1/LC)}$. This is the so-called *resonant* frequency of the circuit shown in figure 12.31.

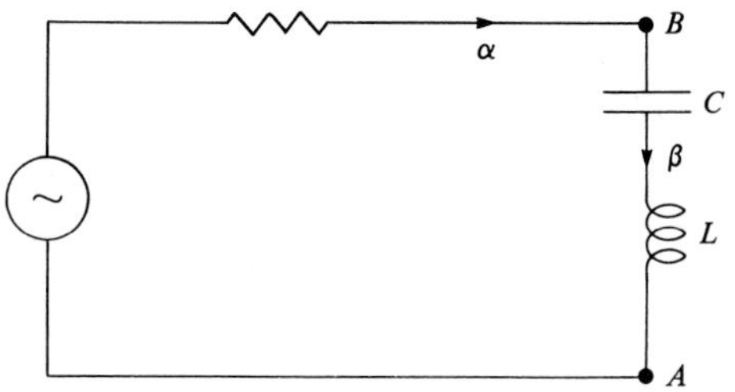

Figure 12.31

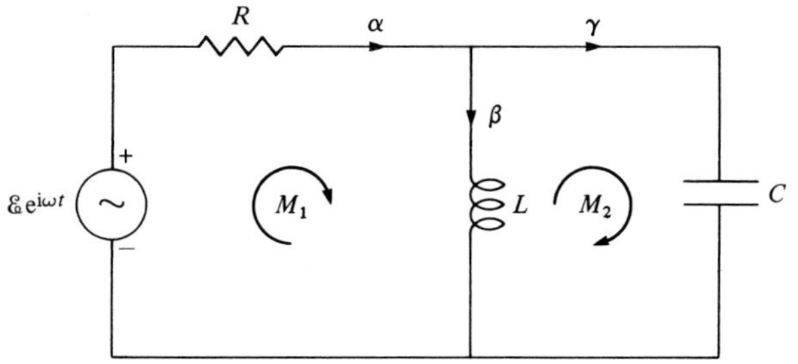

Figure 12.32

As a slightly less trivial example, consider the circuit in figure 12.32. A basis for Z consists of the meshes $M_1 = \alpha + \beta$ and $M_2 = \alpha + \gamma$, so that

$$\sigma = \begin{pmatrix} 1 & 0 \\ 1 & -1 \\ 0 & 1 \end{pmatrix} \quad \text{and} \quad s = \begin{pmatrix} 1 & 1 & 0 \\ 0 & -1 & 1 \end{pmatrix}.$$

Since $Z = \begin{pmatrix} R & 0 & 0 \\ 0 & i\omega L & 0 \\ 0 & 0 & -i/\omega C \end{pmatrix}$, we have

$$sZ\sigma = \begin{pmatrix} R & i\omega L & 0 \\ 0 & -i\omega L & -i/\omega C \end{pmatrix} \begin{pmatrix} 1 & 0 \\ 1 & -1 \\ 0 & 1 \end{pmatrix} = \begin{pmatrix} R + i\omega L & -i\omega L \\ -i\omega L & i\omega L - i/\omega C \end{pmatrix}.$$

This matrix may be inverted by inspection, giving

$$(sZ\sigma)^{-1} = \frac{1}{R(i\omega L - i/\omega C) + LC^{-1}} \begin{pmatrix} i\omega L - i/\omega C & i\omega L \\ i\omega L & R + i\omega L \end{pmatrix}.$$

To determine the mesh currents, we write that

$$\mathbf{W} = \begin{pmatrix} -\mathscr{E} \\ 0 \\ 0 \end{pmatrix}$$

and calculate that

$$J = (sZ\sigma)^{-1}(-sW) = \frac{\mathscr{E}}{R(i\omega L - i/\omega C) + LC^{-1}} \begin{pmatrix} i\omega L - i/\omega C \\ i\omega L \end{pmatrix}.$$

Of interest here is the solution for the resonant frequency $\omega/2\pi$, where $\omega L = -1/\omega C$ and $J = \frac{\mathscr{E}}{L/C}\begin{pmatrix} 0 \\ \mathrm{i}\omega L \end{pmatrix}$ so that there is no current in mesh M, and no power dissipated in the resistor.

Of course our techniques permit us to analyze alternating-current circuits of arbitrary complexity just as easily as direct-current circuits with the same topology. Consider, for example, the following two-stage filter.

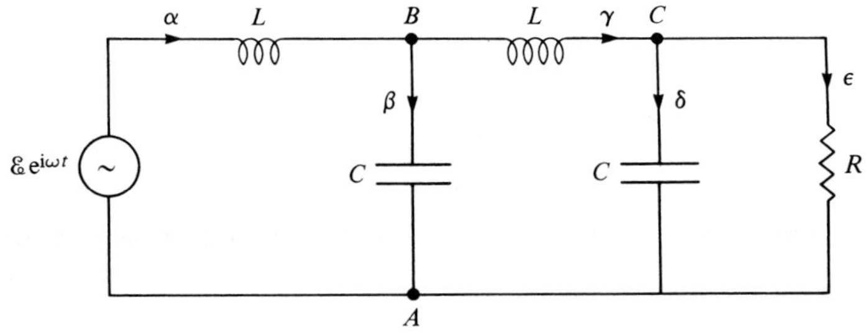

Figure 12.33

Since this circuit has three meshes but only two non-ground nodes, it is easiest to analyze it by the node-potential method. We choose node A as ground node, so that

$$[\partial] = \begin{pmatrix} 1 & -1 & -1 & 0 & 0 \\ 0 & 0 & 1 & -1 & -1 \end{pmatrix} \quad \text{and} \quad [\mathrm{d}] = \begin{pmatrix} 1 & 0 \\ -1 & 0 \\ -1 & 1 \\ 0 & -1 \\ 0 & -1 \end{pmatrix}.$$

Then $[\partial]Z^{-1}[\mathrm{d}] =$

$$\begin{pmatrix} 1 & -1 & -1 & 0 & 0 \\ 0 & 0 & 1 & -1 & -1 \end{pmatrix} \begin{pmatrix} -\mathrm{i}/\omega L & 0 \\ \mathrm{i}\omega C & 0 \\ \mathrm{i}/\omega L & -\mathrm{i}/\omega L \\ 0 & \mathrm{i}\omega C \\ 0 & -1/R \end{pmatrix}$$

$$= \begin{pmatrix} \mathrm{i}\omega C - 2\mathrm{i}/\omega L & \mathrm{i}/\omega L \\ \mathrm{i}/\omega L & \mathrm{i}\omega C - (\mathrm{i}/\omega L) + 1/R \end{pmatrix}.$$

Using the formula

$$\mathbf{\Phi} = ([\partial]Z^{-1}[\mathrm{d}])^{-1}[\partial]Z^{-1}\mathbf{W}$$

it is a straightforward matter to invert this matrix and determine the potentials at nodes B and C. We content ourselves with solving the problem for the resonant frequency at which $\omega C = 1/\omega L$. Then the matrix to be inverted is simply

$$[\partial]Z^{-1}[\mathrm{d}] = \begin{pmatrix} -\mathrm{i}\omega C & \mathrm{i}\omega C \\ \mathrm{i}\omega C & 1/R \end{pmatrix}.$$

By noting that $[\partial]Z^{-1}\mathbf{W} = \begin{pmatrix} -1/i\omega L \\ 0 \end{pmatrix} = \begin{pmatrix} i\mathscr{E}\omega C \\ 0 \end{pmatrix}$ at this frequency, we see immediately

$$\begin{pmatrix} \phi^B \\ \phi^C \end{pmatrix} = \frac{1}{(-i\omega C/R) - \omega^2 C^2} \begin{pmatrix} 1/R & -i\omega C \\ -i\omega C & -i\omega C \end{pmatrix} \begin{pmatrix} i\omega C \\ 0 \end{pmatrix} \mathscr{E}$$

or

$$\begin{pmatrix} \phi^B \\ \phi^C \end{pmatrix} = \frac{\mathscr{E}}{(-i\omega C/R) - \omega^2 C^2} \begin{pmatrix} i\omega C/R \\ \omega^2 C^2 \end{pmatrix}.$$

Summary

A One-complexes

Given a diagram representing a one-complex, you should know how to construct a maximal tree and the associated basis of meshes for Z_1 and write the matrices representing σ, s, ∂, and d.

You should know the definitions of the subspaces B_0, H_0, Z_1, H^0, P^0, and B^1 and be able to construct bases for them.

B Resistive networks

You should be able to write down the relation between V and I for a branch containing a battery, current source, and resistor.

You should be able to derive and apply Maxwell's mesh-current and node-potential formulas for solving electrical networks.

C Alternating-current networks

You should know how to use Maxwell's methods to find steady-state solutions and normal modes for networks involving inductors, capacitors, and sinusoidal voltage or current sources.

Exercises

12.1, 2. For figures 12.34 and 12.35, do the following:

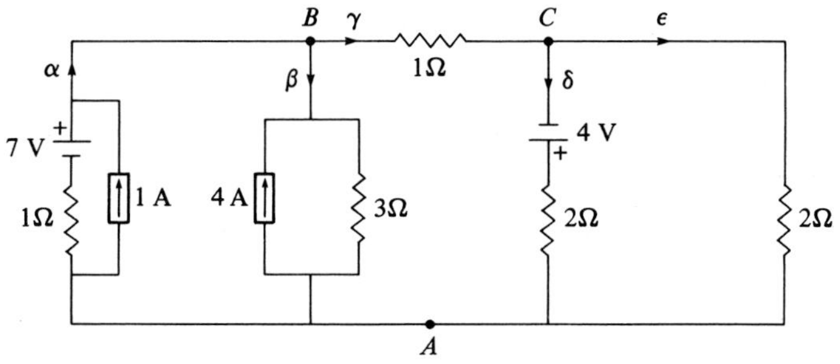

Figure 12.34

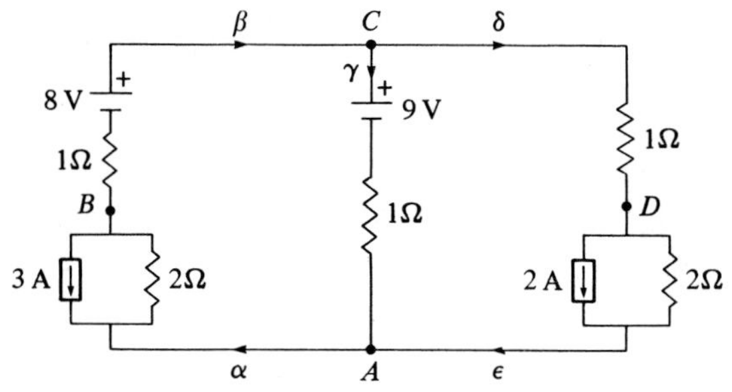

Figure 12.35

(a) Find an independent set of meshes, and write down the element of C_1 corresponding to each mesh. (You may use notation like $M_1 = \alpha + \beta - \gamma$.) Check that the number of independent meshes, m, satisfies $m = b + 1 - n$.

(b) Express the current in each branch in terms of mesh currents, then find expressions for the branch voltages.

(c) By applying Kirchhoff's voltage law to each mesh, construct a set of simultaneous equations in the mesh currents. Solve these equations.

(d) Choose node A as ground and express voltages and currents in terms of the remaining unknown node potentials. By applying Kirchhoff's current law to each node, obtain a set of simultaneous equations for the node potentials, and solve them.

12.3, 4. For figures 12.36 and 12.37, do the following:

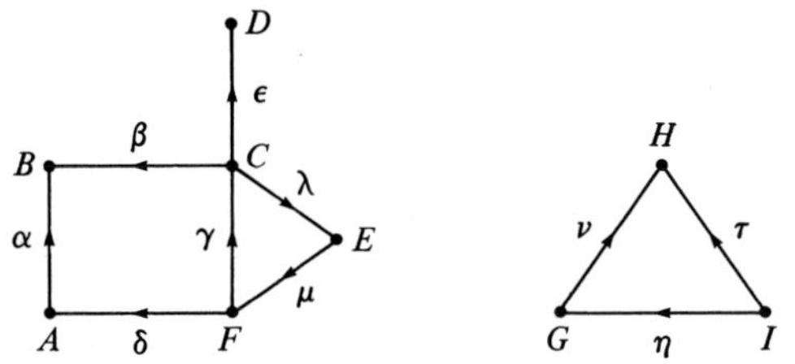

Figure 12.36

(a) Determine the dimension of the spaces C_1, Z_1, C_0, and H_0. Check that $\dim H_0 - \dim Z_1 = \dim C_0 - \dim C_1$.

(b) For each connected component, find a tree T which connects all the nodes. Using these trees, construct a basis for Z_1, and write down explicitly the projection ρ_T in a form such as

$$\rho_T(\alpha) = 0,$$
$$\rho_T(\beta) = \beta + \gamma - \delta, \quad \text{etc.}$$

(c) Determine how many mesh equations and how many node equations would be required to solve a circuit with the given topology.

12.5. For each of the three branches shown in figure 12.38, determine the relation between V and I. In each case, construct a branch with the same relation by using only a suitably chosen battery and resistor (but no

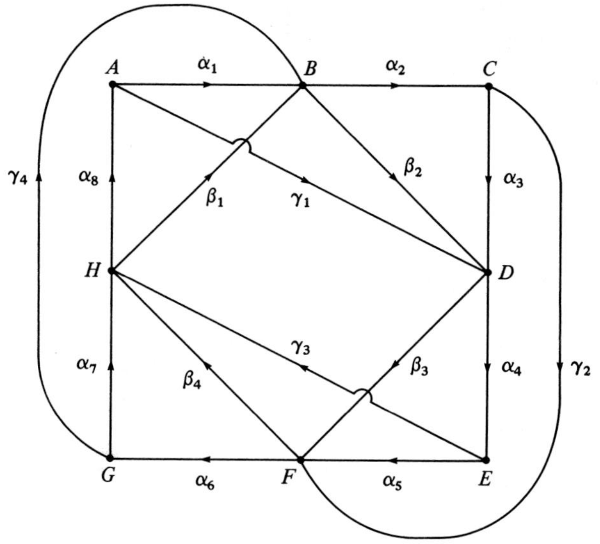

Figure 12.37

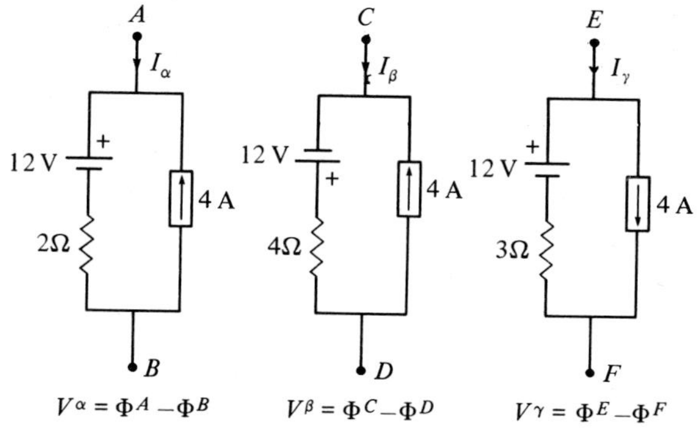

Figure 12.38

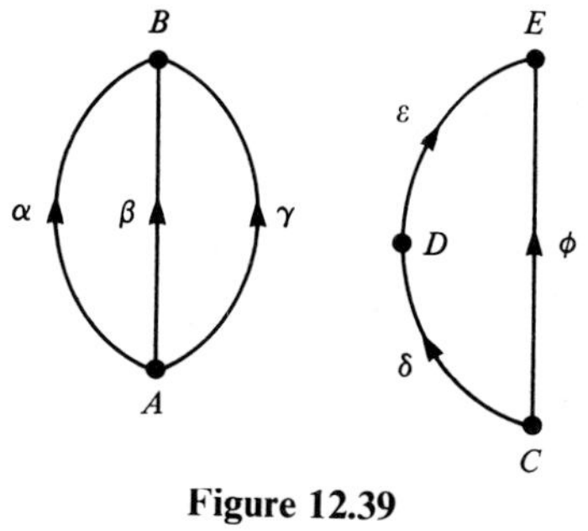

Figure 12.39

current source) and also by using a suitable current source and resistor (but no battery).

12.6. Consider the complex shown in figure 12.39.

(a) Write down the matrices ∂ and d. Take the order of branches to be $\alpha, \beta, \gamma, \delta, \varepsilon, \phi$, even though these are not the first six letters of the Greek alphabet!

(b) Show that $\begin{pmatrix}1\\1\\0\\0\\0\end{pmatrix}$ and $\begin{pmatrix}0\\0\\1\\1\\1\end{pmatrix}$ form a basis for H^0. Select nodes A and C as ground nodes and show that $\overline{\begin{pmatrix}0\\1\\0\\0\\0\end{pmatrix}}$, $\overline{\begin{pmatrix}0\\0\\0\\1\\0\end{pmatrix}}$, and $\overline{\begin{pmatrix}0\\0\\0\\0\\1\end{pmatrix}}$ from a basis for $P^0 = C^0/Z^0$. Construct the matrix which represents the operator $[\mathrm{d}]: P^0 \to C^1$ relative to this basis.

(c) Select a maximal tree in each component by choosing branches α, δ, and ε. The three basis elements of Z_1 are now the meshes formed by combining the remaining branches β, γ, and ϕ with branches in the tree (always choosing the mesh so that β, γ, or δ has coefficient $+1$). Write down the matrix which represents $\sigma: Z_1 \to C_1$ relative to this basis.

(d) Show that the columns of the matrix [d] determine a basis for B^1. Associate vectors of the quotient space $H^1 = C^1/B^1$ with the three branches *not* in the maximal trees, obtaining

$$\overline{\begin{pmatrix}0\\1\\0\\0\\0\\0\end{pmatrix}}, \overline{\begin{pmatrix}0\\0\\1\\0\\0\\0\end{pmatrix}} \text{ and } \overline{\begin{pmatrix}0\\0\\0\\0\\0\\1\end{pmatrix}}.$$

Show that these three elements form a basis for H^1 which is dual to the basis chosen for Z_1.

(e) Relative to this basis for H^1, construct the matrix $s: C^1 \to H^1$. Check that s is the transpose of σ.

(f) Show that the boundaries of the paths joining ground nodes to non-ground nodes form a basis for B_0 which is dual to the basis chosen for P^0.

(g) Show that the equivalence classes of the ground nodes,

$$\overline{\begin{pmatrix}1\\0\\0\\0\\0\end{pmatrix}} \text{ and } \overline{\begin{pmatrix}0\\0\\1\\0\\0\end{pmatrix}}$$

form a basis for $H_0 = C^0/B^0$ which is dual to the basis chosen for Z^0.

12.7. For Exercise 12.1 construct $[\mathrm{d}]$, $[\partial]$, σ, s, Z, $\mathbf{K}$, and $\mathbf{W}$. Set up the mesh-current and node-potential equations by using these matrices and vectors, and check that the equations are the same. Let node A be ground and let β and δ be the maximal tree.

12.8. Solve Exercise 12.2 by the same method, with A as the ground node and α, β, and δ as the maximal tree.

12.9. Use the mesh current method to determine the branch currents $I_\alpha, I_\beta, I_\gamma$ for the network in figure 12.40. Use the meshes defined by choosing β as a maximal tree.

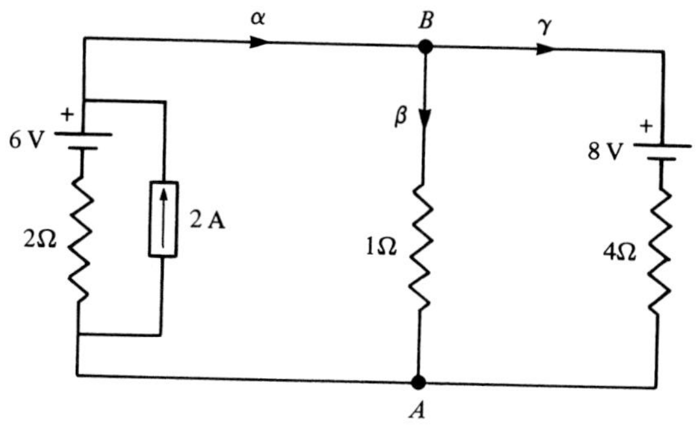

Figure 12.40

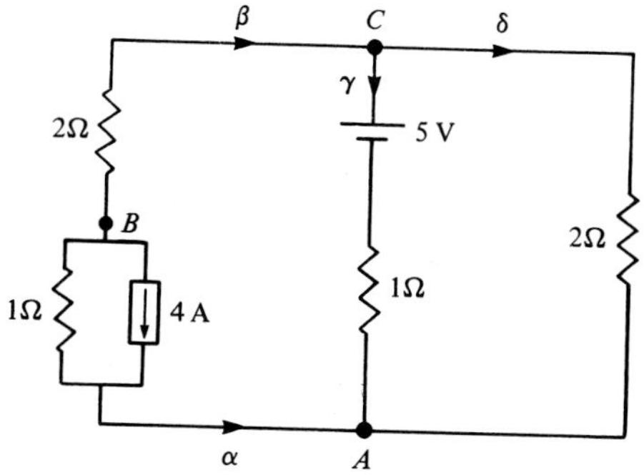

Figure 12.41

12.10. Consider the electrical network in figure 12.41. Suppose that the matrix representation of s is

$$s = \begin{pmatrix} -1 & +1 & 0 & +1 \\ 0 & 0 & +1 & -1 \end{pmatrix}.$$

Indicate clearly what basis for Z_1 is implied by this form of s. What maximal tree would lead to this basis? Suppose that a basis for P^0 has been chosen as follows:

$$\mathbf{p}^1 \text{ is the equivalence class } \overline{\begin{pmatrix} 0 \\ 1 \\ 0 \end{pmatrix}},$$

$$\mathbf{p}^2 \text{ is the equivalence class } \overline{\begin{pmatrix} 0 \\ 0 \\ 1 \end{pmatrix}}.$$

Express $\overline{\begin{pmatrix} 2 \\ 1 \\ 4 \end{pmatrix}}$ in terms of the basis vectors $\mathbf{p}^1$ and $\mathbf{p}^2$, and write down the matrix representations of $[\partial]$ and $[d]$. Using the node-potential method, calculate Φ^B and Φ^C for the given network.

12.11. Consider the electrical network shown in figure 12.42.

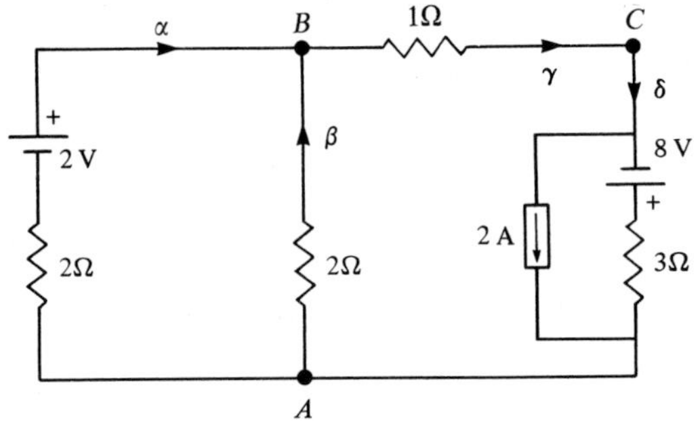

Figure 12.42

Write down the matrices ∂ and d for this network. Branches α and γ constitute a maximal tree for the network. Write down the basis for Z_1 which corresponds to this maximal tree, and construct the matrix σ relative to this basis. Use the mesh-current method to determine all the branch currents for the given network.

12.12.(a) Using the mesh-current method, determine the normal mode frequencies for the network in figure 12.43, i.e., determine the values of $\omega/2\pi$ for which $sZ\sigma$ is singular, and find the solution of $(sZ\sigma)\mathbf{J} = 0$ for these frequencies.

(b) Solve the same network by the node-potential method. With A as a ground node, you still get a 3×3 matrix to invert. Why are there only two normal modes?

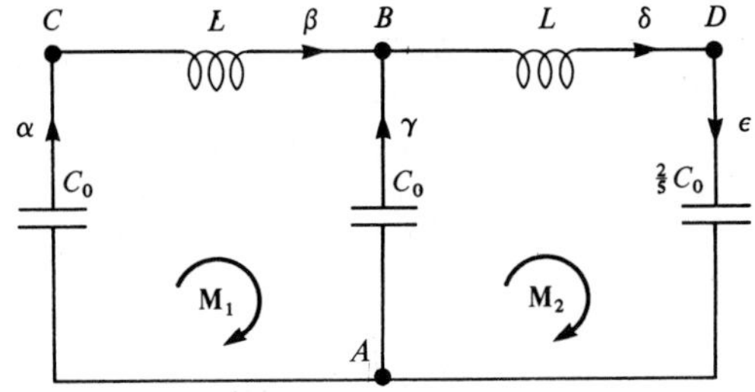

Figure 12.43

(c) Convert the same network to a network with only two nodes by combining α and β in series and δ and ε in series, thereby eliminating nodes C and D. Solve by the node-potential method. This time you only have to look at a 1×1 matrix!

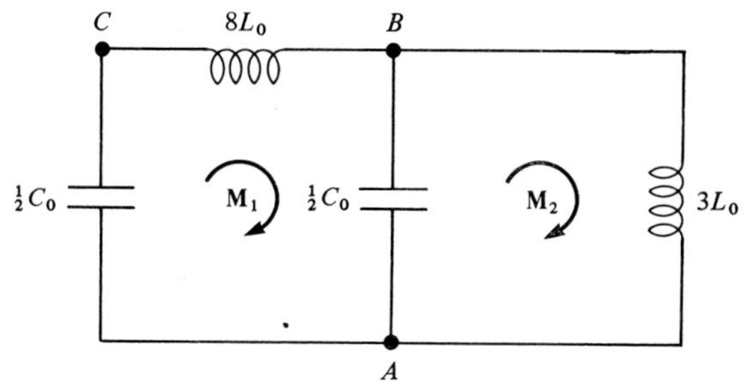

Figure 12.44

12.13.(a) Use the mesh-current method to determine the normal modes of oscillation of the network shown in figure 12.44. For each mode, express the frequency $\omega/2\pi$ in terms of the quantity $\omega_0 = 1\sqrt{(L_0C_0)}$, and determine the ratio, J_2/J_1, of mesh currents.

(b) Use the node-potential method to find the normal mode frequencies in the same network.

12.14.(a) Construct the complex-valued matrix $sZ\sigma$ for the network shown in figure 12.45.

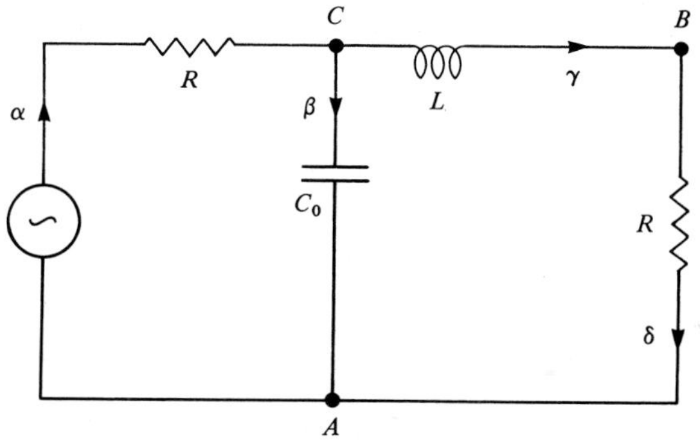

Figure 12.45

(b) Suppose that the generator in branch α supplies a voltage which is the real part of $\mathscr{E}_0 e^{i\omega t}$. Find expressions for the steady-state voltages V^β and V^δ.

12.15. Suppose that a current source that supplies a current $K = \mathrm{Re}(K_0 e^{i\omega_0 t})$, where $\omega_0 = 1/\sqrt{(L_0C_0)}$, is connected in series with the capacitor between nodes A and B of the network of Exercise 12.13. Determine the resulting mesh currents J_1 and J_2.

13

The method of orthogonal projection

In Chapter 13 we continue the study of electrical networks. We examine the boundary-value problems associated with capacitive networks and use these methods to solve some classical problems in electrostatics involving conductors.

13.1. Weyl's method of orthogonal projection

We turn now to two other methods of dealing with the network problem, Weyl's method and Kirchhoff's original method. Both of these methods are restricted to the purely resistive case; i.e., all the z_α are positive real numbers, $z_\alpha = r_a > 0$. However, they are both interesting and worth studying. We begin with Weyl's method.

The equation relating branch voltages $\mathbf{V}$ and branch currents $\mathbf{I}$ is

$$\mathbf{V} - \mathbf{W} = Z(\mathbf{I} - \mathbf{K})$$

which we can rewrite as

$$\mathbf{V} = \mathbf{W} + Z(\mathbf{I} - \mathbf{K})$$

or as

$$\mathbf{V} = Z(Z^{-1}\mathbf{W} - \mathbf{K} + \mathbf{I}).$$

We now use the matrix Z to define a positive-definite scalar product on the space C_1 of branch currents by

$$(\mathbf{I}, \mathbf{I}')_Z = \int_{\mathbf{I}} Z\mathbf{I}' = r_\alpha I_\alpha I'_\alpha + r_\beta I_\beta I'_\beta + \cdots.$$

Suppose now that we have any current distribution $\mathbf{I}'$ satisfying Kirchhoff's current law $\partial \mathbf{I}' = 0$ and any voltage distribution $\mathbf{V}$ satisfying Kirchhoff's voltage law, so that $\mathbf{V} = -d\Phi$. We know then that $\int_{\mathbf{I}'} \mathbf{V} = 0$. Since $\mathbf{V} = Z(Z^{-1}\mathbf{W} + \mathbf{I} - \mathbf{K})$, we may write

$$\int_{\mathbf{I}'} Z(Z^{-1}\mathbf{W} + \mathbf{I} - \mathbf{K}) = 0.$$

Expressed in terms of the scalar product defined above, this equation is

$$(\mathbf{I}', Z^{-1}\mathbf{W} + \mathbf{I} - \mathbf{K})_Z = 0.$$

Thus we wish that the vector $(\mathbf{K} - Z^{-1}\mathbf{W}) - \mathbf{I}$ be orthogonal to all cycles.

We can thus reformulate the resistive network problem as follows. We are given the space C_1 of branch currents, in which lies the subspace Z_1 of cycles. For a network with specified sources $\mathbf{W}$ and $\mathbf{K}$, we can form the vector $\mathbf{K} - Z^{-1}\mathbf{W}$, which describes the branch currents which would exist if all node potentials were set equal to zero. In general this vector is *not* an element of the subspace Z_1. We wish to find a vector $\mathbf{I} \in Z_1$ such that the vector $\mathbf{K} - Z^{-1}\mathbf{W} - \mathbf{I}$ is orthogonal to all cycles. Thus we have reduced the resistive network to the geometrical problem illustrated in figure 13.1: given the vector $\mathbf{K} - Z^{-1}\mathbf{W}$ in the space C_1, we must compute its orthogonal projection onto the subspace Z_1, relative to the scalar product defined by Z. Then we will have expressed $\mathbf{K} - Z^{-1}\mathbf{W}$ in the form

$$\mathbf{K} - Z^{-1}\mathbf{W} = \mathbf{I} + (\mathbf{K} - Z^{-1}\mathbf{W} - \mathbf{I})$$

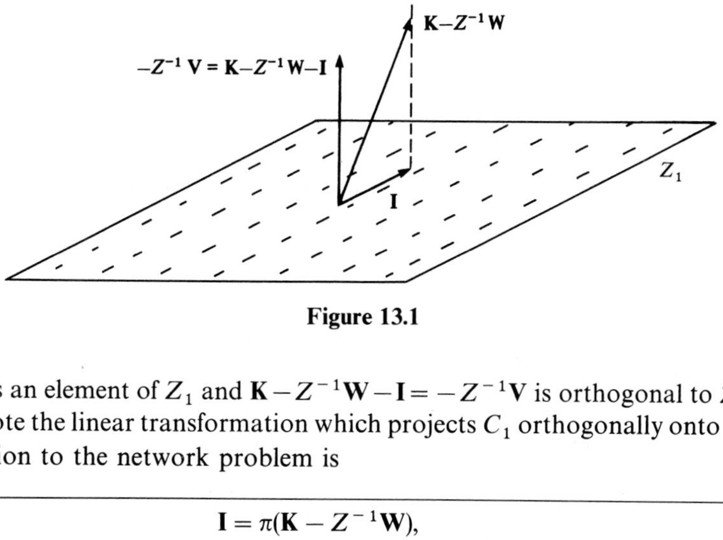

Figure 13.1

where $\mathbf{I}$ is an element of Z_1 and $\mathbf{K} - Z^{-1}\mathbf{W} - \mathbf{I} = -Z^{-1}\mathbf{V}$ is orthogonal to Z_1. If we let π denote the linear transformation which projects C_1 orthogonally onto Z_1, then the solution to the network problem is

$$\mathbf{I} = \pi(\mathbf{K} - Z^{-1}\mathbf{W}),$$
$$\mathbf{V} = Z(\pi - 1)(\mathbf{K} - Z^{-1}\mathbf{W}).$$

It is geometrically clear that a unique projection operator π exists, and that therefore a unique solution to the network problem exists.

One procedure for orthogonal projection from C_1 onto Z_1 makes use of a choice of an orthonormal basis: Let $\{\mathbf{e}_1, \mathbf{e}_2, \ldots, \mathbf{e}_m\}$ denote an orthonormal basis for Z_1.

Then, if $\mathbf{u}$ denotes an element of C_1, its projection onto Z_1 is given by

$$\pi\mathbf{u} = (\mathbf{e}_1, \mathbf{u})\mathbf{e}_1 + (\mathbf{e}_2, \mathbf{u})\mathbf{e}_2 + \cdots + (\mathbf{e}_m, \mathbf{u})\mathbf{e}_m.$$

In practice the computation of the matrix which represents π is a tedious business. We must first choose a basis for the subspace Z_1, say the basis given by an independent family of meshes, and then use the Gram–Schmidt procedure to convert this to a basis which is orthonormal with respect to the scalar product defined by the matrix Z. We must finally take each basis vector of C_1 in turn and calculate its projection onto Z_1 by taking the sum of its projections along the orthonormal basis elements of Z_1. Each such projection determines a column of the matrix π.

As a simple illustration of Weyl's method, we consider the circuit shown in figure 13.2. For the circuit, C_1 is two-dimensional. There is only one mesh, $\alpha + \beta$, so Z_1 is the one-dimensional subspace spanned by the vector $\begin{pmatrix}1\\1\end{pmatrix}$. The matrix Z is $\begin{pmatrix}1 & 0\\0 & 3\end{pmatrix}$.

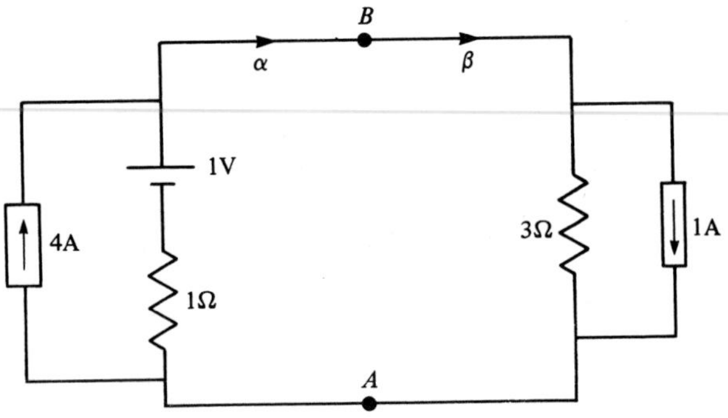

Figure 13.2

To compute the projection matrix π we must project the two basis vectors $\begin{pmatrix}1\\0\end{pmatrix}$ and $\begin{pmatrix}0\\1\end{pmatrix}$ in turn onto the subspace Z_1. We first normalize the basis element of Z_1: the length of $\begin{pmatrix}1\\1\end{pmatrix}$ is $\left(\begin{pmatrix}1\\1\end{pmatrix}, \begin{pmatrix}1 & 0\\0 & 3\end{pmatrix}\begin{pmatrix}1\\1\end{pmatrix}\right) = 4$, so we divide $\begin{pmatrix}1\\1\end{pmatrix}$ by $\sqrt{4} = 2$, obtaining $\begin{pmatrix}\frac{1}{2}\\\frac{1}{2}\end{pmatrix}$, which has unit length. Then the projection of $\begin{pmatrix}1\\0\end{pmatrix}$ is

$$\left(\begin{pmatrix}1\\0\end{pmatrix}, \begin{pmatrix}1 & 0\\0 & 3\end{pmatrix}\begin{pmatrix}\frac{1}{2}\\\frac{1}{2}\end{pmatrix}\right)\begin{pmatrix}\frac{1}{2}\\\frac{1}{2}\end{pmatrix} = \begin{pmatrix}\frac{1}{4}\\\frac{1}{4}\end{pmatrix}$$

while the projection of $\begin{pmatrix}0\\1\end{pmatrix}$ is

$$\left(\begin{pmatrix}0\\1\end{pmatrix}, \begin{pmatrix}1 & 0\\0 & 3\end{pmatrix}\begin{pmatrix}\frac{1}{2}\\\frac{1}{2}\end{pmatrix}\right)\begin{pmatrix}\frac{1}{2}\\\frac{1}{2}\end{pmatrix} = \begin{pmatrix}\frac{3}{4}\\\frac{3}{4}\end{pmatrix}.$$

Therefore $\pi = \begin{pmatrix} \frac{1}{4} & \frac{3}{4} \\ \frac{1}{4} & \frac{3}{4} \end{pmatrix}$. For the network of figure 13.2, $\mathbf{K} = \begin{pmatrix} 4 \\ 1 \end{pmatrix}$ and $\mathbf{W} = \begin{pmatrix} -1 \\ 0 \end{pmatrix}$, so that $\mathbf{K} - Z^{-1}\mathbf{W} = \begin{pmatrix} 4 \\ 1 \end{pmatrix} - \begin{pmatrix} -1 \\ 0 \end{pmatrix} = \begin{pmatrix} 5 \\ 1 \end{pmatrix}$. Applying π to this vector, we find immediately

$$\mathbf{I} = \pi(\mathbf{K} - Z^{-1}\mathbf{W}) = \begin{pmatrix} \frac{1}{4} & \frac{3}{4} \\ \frac{1}{4} & \frac{3}{4} \end{pmatrix}\begin{pmatrix} 5 \\ 1 \end{pmatrix} = \begin{pmatrix} 2 \\ 2 \end{pmatrix}$$

so the solution is $I_\alpha = I_\beta = 2$. Notice that $\mathbf{K} - Z^{-1}\mathbf{W} - \mathbf{I} = \begin{pmatrix} 5 \\ 1 \end{pmatrix} - \begin{pmatrix} 2 \\ 2 \end{pmatrix} = \begin{pmatrix} 3 \\ -1 \end{pmatrix}$, which is orthogonal to $\mathbf{I}$, as promised. The solution to this network problem is displayed graphically in figure 13.3. Because the scalar product is not the ordinary Euclidean scalar product, the subspace Z_1 and its orthogonal complement are not represented by perpendicular lines.

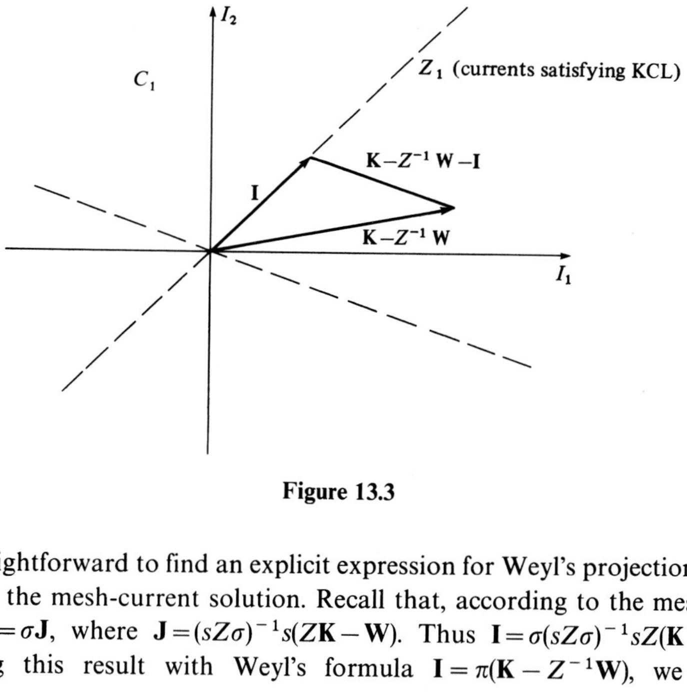

Figure 13.3

It is straightforward to find an explicit expression for Weyl's projection operator π by using the mesh-current solution. Recall that, according to the mesh-current method, $\mathbf{I} = \sigma\mathbf{J}$, where $\mathbf{J} = (sZ\sigma)^{-1}s(Z\mathbf{K} - \mathbf{W})$. Thus $\mathbf{I} = \sigma(sZ\sigma)^{-1}sZ(\mathbf{K} - Z^{-1}\mathbf{W})$. Comparing this result with Weyl's formula $\mathbf{I} = \pi(\mathbf{K} - Z^{-1}\mathbf{W})$, we see that

$$\pi = \sigma(sZ\sigma)^{-1}sZ.$$

It is easy to check that $\pi^2 = \pi$, as must be true for any projection operator.

13.2. Kirchhoff's method

We turn finally to Kirchhoff's method. Although this method was invented many decades before Weyl's, we may regard it from our present point of view as providing an explicit formula for the projection operator π in terms of the trees of the graph.

Suppose that we have a connected complex, and let T denote a maximal tree. Recall that for each such tree we have defined a projection operator ρ_T by

$$\rho_T(\alpha) = \begin{cases} 0 & \text{if } \alpha \in T, \\ \mathbf{M}_\alpha & \text{if } \alpha \notin T, \end{cases}$$

where $\mathbf{M}_\alpha$ is the cycle corresponding to unique mesh containing branch α whose other branches all lie in T. (The cycle is to be chosen so that it contains $+\alpha$, not $-\alpha$.) For example, in figure 13.4, the branch β alone forms a maximal tree. Then

$$\rho_T(\alpha) = \alpha - \beta, \quad \rho_T(\beta) = 0, \quad \rho_T(\gamma) = \gamma + \beta,$$

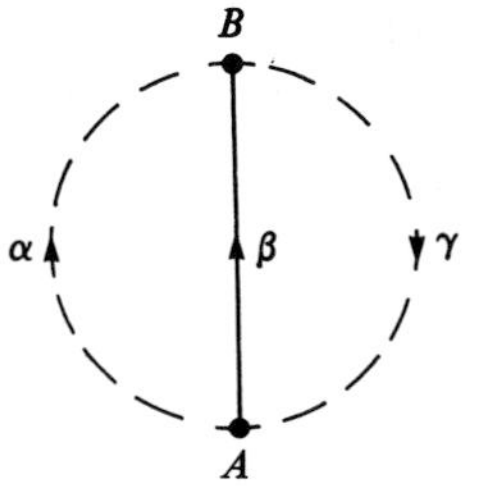

Figure 13.4

so that the matrix representing this projection ρ_T is $\begin{pmatrix} 1 & 0 & 0 \\ -1 & 0 & 1 \\ 0 & 0 & 1 \end{pmatrix}$. Notice that the diagonal entries of ρ_T are all $+1$ or zero.

Any such ρ_T is a projection operator with range Z_1: its image is Z_1, and $\rho_T(\mathbf{M}) = \mathbf{M}$ for $\mathbf{M} \in Z_1$. It is not, however, an *orthogonal* projection operator, because its kernel is not orthogonal (with respect to the scalar product defined by Z) to the subspace Z_1. To put it differently, ρ_T is not self-adjoint: for a pair of branches α and β

$$(\rho_T\alpha, \beta)_Z \neq (\alpha, \rho_T\beta)_Z$$

in general.

While ρ_T for a single maximal tree is not self-adjoint, there are many different maximal trees in a complex. What Kirchhoff discovered was a scheme for taking a weighted average of the various projection operators which *is* self-adjoint. Suppose that for each maximal tree T we have a real number λ_T, with $0 \leqslant \lambda_T \leqslant 1$ and $\sum_T \lambda_T = 1$, where the sum is over all trees. Then

$$\rho_\lambda = \sum_T \lambda_T \rho_T$$

is again a projection onto Z_1: its image is Z_1, and it maps any vector in Z_1 into itself. In general, ρ_λ is not orthogonal, but there is one choice of λ_T for which ρ_λ is orthogonal. Kirchhoff's prescription for this choice is to define, for each maximal tree T,

$$Q_T = \prod_{\beta \notin T} r_\beta$$

the product of the resistances of all the branches *not* in T. The weighting factors λ_T are then made proportional to Q_T, with their sum equal to unity; i.e.,

$$\lambda_T = Q_T/R \quad \text{where} \quad R = \sum_T Q_T.$$

Thus we have an explicit formula for the Weyl projection operator π, namely

$$\boxed{\pi = R^{-1} \sum_T Q_T \rho_T.}$$

We must show that this operator is self-adjoint, or, what amounts to the same thing, that $R\pi = \sum_T Q_T \rho_T$ is self-adjoint. Since the branches form a basis for C_1, it suffices to show that for any pair of branches, α and β, we have

$$\sum_T Q_T(\rho_T \alpha, \beta)_Z = \sum_T Q_T(\alpha, \rho_T \beta)_Z.$$

For fixed α and β, as we sum over all maximal trees, three cases can arise, as shown in figure 13.5.

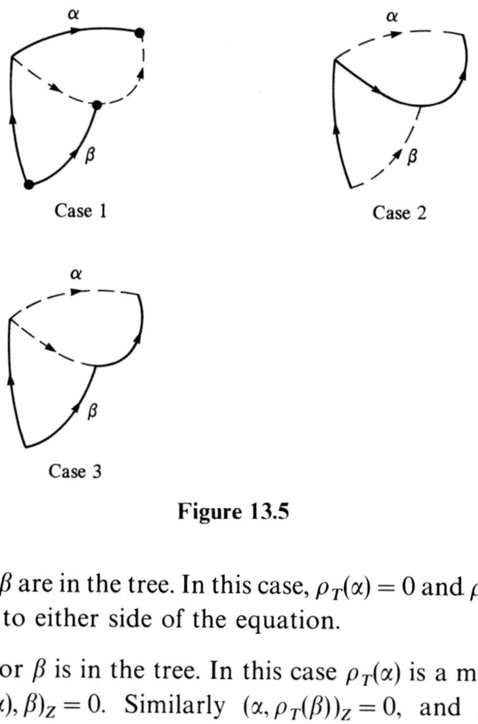

Figure 13.5

Case 1. Both α and β are in the tree. In this case, $\rho_T(\alpha) = 0$ and $\rho_T(\beta) = 0$, so the tree contributes nothing to either side of the equation.

Case 2. Neither α nor β is in the tree. In this case $\rho_T(\alpha)$ is a mesh which does not contain β, so $(\rho_T(\alpha), \beta)_Z = 0$. Similarly $(\alpha, \rho_T(\beta))_Z = 0$, and so the tree again contributes nothing.

Case 3. Branch α is not in the tree, but β is in the tree. In this case, if β is in the mesh $\mathbf{M}_\alpha$ formed by α and branches of the tree, we have

$$(\rho_T \alpha, \beta)_Z = \pm r_\beta$$

where the + sign holds if α and β occur with the same sign in $\mathbf{M}_\alpha$, the − sign if they occur with opposite sign. For example, in figure 13.6(a), $\rho_T(\alpha) = \alpha - \gamma - \beta + \delta$ and $(\rho_T(\alpha), \beta)_Z = -r_\beta$. There is now always a unique maximal tree, T', formed by replacing β by α in the tree T, for which $\rho_{T'}(\beta) = \pm\rho_T(\alpha)$, so that $(\alpha, \rho_{T'}(\beta))_Z = \pm r_\alpha$. Such a tree T' is illustrated in figure 13.6(b).

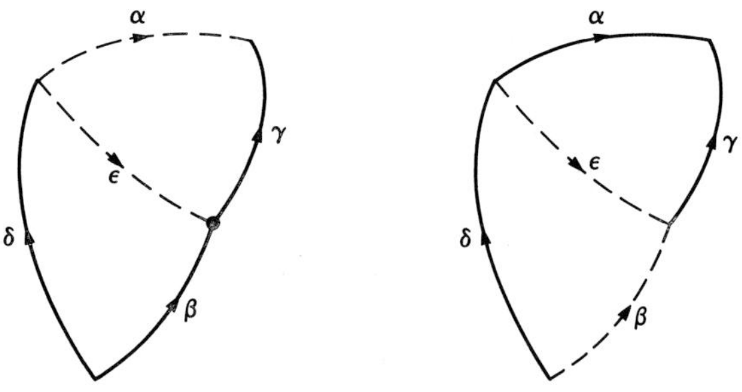

(a) (Tree T) (b) (Tree T')

Figure 13.6

For the pair of trees T and T', $Q_T r_\beta = Q_{T'} r_\alpha$, since deleting branch β from T leaves the same branches as does deleting branch α from T', and both $Q_T r_\beta$ and $Q_{T'} r_\alpha$ are the product of the resistances of these remaining branches. The relationship between T and T' is symmetrical if we interchange the roles of α and β. Thus

$$Q_T(\rho_T\alpha, \beta)_Z = Q_{T'}(\alpha, \rho_{T'}\beta)_Z;$$

and to each T for which the left-hand side is non-zero there corresponds a unique T' for which the equality holds, and vice versa. Thus, if we sum over all T we get the desired result

$$\sum_T Q_T(\rho_T\alpha, \beta)_Z = \sum_T Q_T(\alpha, \rho_T\beta)_Z$$

since we can ignore zero summands. We have thus proved Kirchhoff's formula

$$\pi = R^{-1}\sum_T Q_T\rho_T$$

where

$$Q_T = \prod_{\alpha\notin T} r_\alpha, \quad R = \sum_T Q_T$$

and the sums extend over all maximal trees.

As an illustration of Kirchhoff's method, consider the network of figure 13.7, which has the same topology as figure 13.4. There are three maximal trees, the branch α (tree T_1), the branch β (T_2), and the branch γ (T_3).

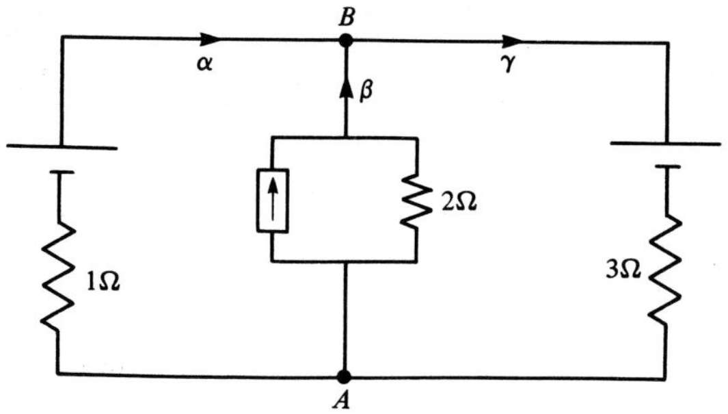

Figure 13.7

For T_1,

$$\rho_{T_1} = \begin{pmatrix} 0 & -1 & 1 \\ 0 & 1 & 0 \\ 0 & 0 & 1 \end{pmatrix} \text{ and } Q_{T_1} = 2\cdot 3 = 6.$$

For T_2,

$$\rho_{T_2} = \begin{pmatrix} 1 & 0 & 0 \\ -1 & 0 & 1 \\ 0 & 0 & 1 \end{pmatrix} \text{ and } Q_{T_2} = 1\cdot 3 = 3.$$

For T_3,

$$\rho_{T_3} = \begin{pmatrix} 1 & 0 & 0 \\ 0 & 1 & 0 \\ 1 & 1 & 0 \end{pmatrix} \text{ and } Q_{T_3} = 1\cdot 2 = 2.$$

Then

$$R = Q_{T_1} + Q_{T_2} + Q_{T_3} = 6 + 3 + 2 = 11.$$

Finally,

$$\pi = \tfrac{1}{11}[6\rho_{T_1} + 3\rho_{T_2} + 2\rho_{T_3})],$$

or

$$\pi = \frac{1}{11}\begin{pmatrix} 5 & -6 & 6 \\ -3 & 8 & 3 \\ 2 & 2 & 9 \end{pmatrix}.$$

This matrix projects onto the subspace Z_1, spanned by the meshes $\begin{pmatrix} 1 \\ -1 \\ 0 \end{pmatrix}$ and $\begin{pmatrix} 0 \\ 1 \\ 1 \end{pmatrix}$; you can check that $\pi\begin{pmatrix} 1 \\ -1 \\ 0 \end{pmatrix} = \begin{pmatrix} 1 \\ -1 \\ 0 \end{pmatrix}$ and $\pi\begin{pmatrix} 0 \\ 1 \\ 1 \end{pmatrix} = \begin{pmatrix} 0 \\ 1 \\ 1 \end{pmatrix}$. The matrix $1 - \pi$ projects onto the orthogonal subspace:

$$1 - \pi = \frac{1}{11}\begin{pmatrix} 6 & 6 & -6 \\ 3 & 3 & -3 \\ -2 & -2 & 2 \end{pmatrix}.$$

Its image, which is also the kernel of π, is the one-dimensional subspace spanned by the vector $\begin{pmatrix} 6 \\ 3 \\ -2 \end{pmatrix}$. This is indeed orthogonal to both the meshes $\begin{pmatrix} 1 \\ -1 \\ 0 \end{pmatrix}$ and $\begin{pmatrix} 0 \\ 1 \\ 1 \end{pmatrix}$. Notice that

$$Z \begin{pmatrix} 6 \\ 3 \\ -2 \end{pmatrix} = \begin{pmatrix} 1 & 0 & 0 \\ 0 & 2 & 0 \\ 0 & 0 & 3 \end{pmatrix} \begin{pmatrix} 6 \\ 3 \\ -2 \end{pmatrix} = \begin{pmatrix} 6 \\ 6 \\ -6 \end{pmatrix}.$$

The vector $\begin{pmatrix} 6 \\ 6 \\ -6 \end{pmatrix}$ describes, as it should, branch voltages which are compatible with Kirchhoff's voltage law.

13.3. Green's reciprocity theorem

We conclude our study of resistive networks with a powerful result called Green's reciprocity theorem, which will appear in a similar form when we study capacitive networks and electrostatics. To prove this theorem we imagine a network in which all but two of the branches contain only resistors, but no voltage or current sources. The remaining two branches, α and β, may contain voltage or current sources in addition to resistors. With a specific choice of sources in branches α and β we will obtain a solution with currents $\mathbf{I}$ satisfying Kirchhoff's current law and voltages $\mathbf{V}$ satisfying Kirchhoff's voltage law. With a different choice of sources we obtain a different solution, again with currents $\hat{\mathbf{I}}$ and voltages $\hat{\mathbf{V}}$ which satisfy both of Kirchhoff's laws. (A typical situation to which the theorem would apply is shown in figure 13.8). Both $\mathbf{I}$ and $\hat{\mathbf{I}}$, since they satisfy Kirchhoff's current law, lie in the space Z_1. Both $\mathbf{V}$ and $\hat{\mathbf{V}}$, since they obey Kirchhoff's voltage law, lie in the space B^1. We know that any element of B^1, acting on any element of Z_1, gives zero, so we may conclude

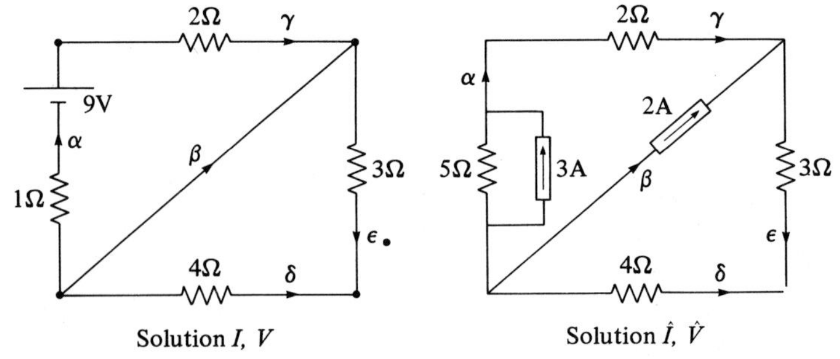

Figure 13.8

that

$$\int_{\mathbf{I}} \hat{\mathbf{V}} = 0 = \int_{\hat{\mathbf{I}}} \mathbf{V}.$$

Thus, in terms of a sum over branches

$$\sum_{\substack{\text{all}\\ \text{branches}}} \hat{V}^\gamma I_\gamma = \sum_{\substack{\text{all}\\ \text{branches}}} V^\gamma \hat{I}_\gamma$$

Now we single out branches α and β:

$$\hat{V}^\alpha I_\alpha + \hat{V}^\beta I_\beta + \sum_{\substack{\text{other}\\ \text{branches}}} \hat{V}^\gamma I_\gamma = V^\alpha \hat{I}_\alpha + V^\beta \hat{I}_\beta + \sum_{\substack{\text{other}\\ \text{branches}}} V^\gamma \hat{I}_\gamma$$

We assumed that all branches except α and β were simply resistors, so that $\hat{V}^\gamma = r_\gamma \hat{I}_\gamma$ and $V^\gamma = r_\gamma I_\gamma$. Thus

$$\sum \hat{V}^\gamma I_\gamma = \sum r_\gamma \hat{I}_\gamma I_\gamma = \sum V^\gamma \hat{I}_\gamma$$

and we may cancel the sum over other branches from both sides, leaving

$$\boxed{\hat{V}^\alpha I_\alpha + \hat{V}^\beta I_\beta = V^\alpha \hat{I}_\alpha + V^\beta \hat{I}_\beta}$$

which is *Green's reciprocity theorem.*

Green's reciprocity theorem may be used to deduce a number of surprising properties of passive resistive networks, ones which contain no sources, only resistors. Given such a network, we can add a new branch in two different ways, as illustrated in figure 13.9. Starting with the purely resistive network of figure 13.9(a), we could connect a new branch α between a pair of existing nodes, as in figure 13.9(b). Such a 'soldering entry' creates no new nodes. The branch α could be a short circuit as shown, or it could include a voltage source, current source, or a resistor. Alternatively, we could make a 'pliers entry' by cutting an existing wire, creating a new node as in figure 13.9(c).

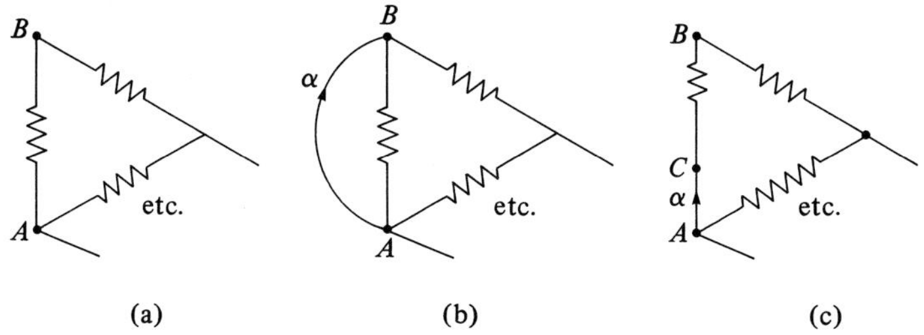

Figure 13.9 (a) Original network. (b) Soldering entry. (c) Pliers entry.

Suppose that we take a linear resistive network, as shown in figure 13.10, and attach two new branches, branch α between nodes A and B and branch β between nodes C and D. Whether we make a soldering entry or pliers entry is immaterial. We can then reach the following conclusions by using the reciprocity theorem.

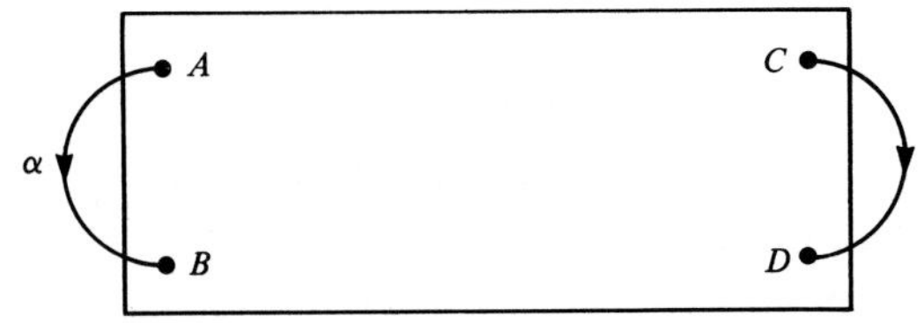

Figure 13.10

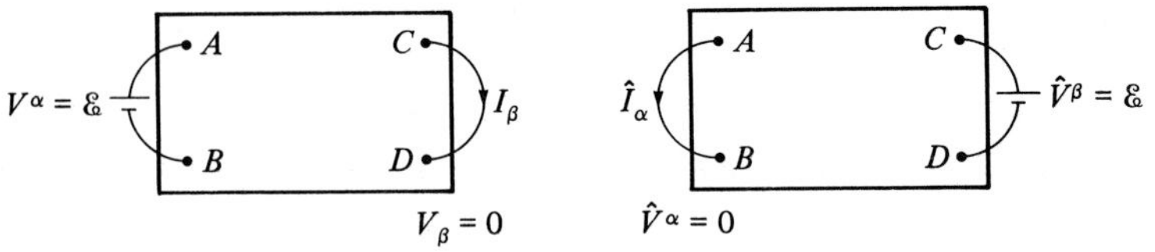

Figure 13.11

Case 1. We insert a battery of voltage $\mathscr{E}$ across α and measure the current I in branch β. We then connect the same battery across β instead, and measure the resulting current through α, as shown in figure 13.11. Since $V^\beta = 0$ and $\hat{V}^\alpha = 0$ (both are short circuits), Green's reciprocity theorem simplifies to

$$\hat{V}^\beta I_\beta = V^\alpha \hat{I}_\alpha.$$

But we used the same battery, so $\hat{V}^\beta = V^\alpha = \mathscr{E}$. Therefore $\mathscr{E} I_\beta = \mathscr{E}\hat{I}_\alpha$ and $I_\beta = \hat{I}_\alpha$. What is remarkable about this result is that, even though the network itself need have no symmetry properties whatever, the relation between applied voltage and resulting current is symmetrical.

Case 2. We insert a current source j across α and measure the resulting voltage across β, which is an open circuit. We then connect the same current source into β and measure the voltage across an open circuit at α, as shown in figure 13.12. Since $I_\beta = 0$ and $\hat{I}_\alpha = 0$ (both are open circuits), the reciprocity theorem simplifies in this case to

$$\hat{V}^\alpha I_\alpha = V^\beta \hat{I}_\beta.$$

But

$$I_\alpha = \hat{I}_\beta = j,$$

so

$$\hat{V}^\alpha j = V^\beta j$$

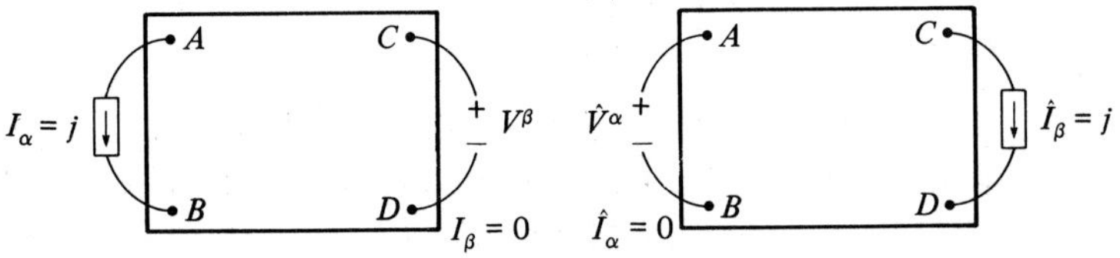

Figure 13.12

and we conclude that

$$\hat{V}^{\alpha} = V^{\beta}.$$

More generally, we can have a *resistive n-port* from which n pairs of wires protrude. If current sources $I_{\alpha}, I_{\beta}, \ldots$ are connected to the various ports, the resulting voltages V^{α}, V^{β} will depend upon the currents according to some relationship

$$\mathbf{V} = R\mathbf{I}$$

where R is an $n \times n$ matrix. The above argument shows that the *matrix R must be symmetric*, because, for any two ports λ and μ, the dependence of V^{λ} on I_{μ} is the same as the dependence of V^{μ} upon I_{λ}.

Green's reciprocity theorem can also be derived as a consequence of the mesh-current solution. In the case where there are no current sources, so that $\mathbf{K} = \mathbf{0}$, we have

$$\mathbf{J} = (sZ\sigma)^{-1}s(-\mathbf{W})$$

so that

$$\mathbf{I} = \sigma\mathbf{J} = -\sigma(sZ\sigma)^{-1}s\mathbf{W}.$$

Because Z is symmetric and s is the transpose of σ, the matrix

$$G = -\sigma(sZ\sigma)^{-1}s$$

is symmetric, so that if we write

$$\mathbf{I} = G\mathbf{W}$$

the matrix G which expresses the branch currents in terms of the imposed battery voltages is symmetric. Similarly we can start from the node-potential solution

$$\Phi = ([\partial]Z^{-1}[\mathrm{d}])^{-1}[\partial](\mathbf{K} - \mathbf{Z}^{-1}\mathbf{W}).$$

If there are no voltage sources, so that $\mathbf{W} = \mathbf{0}$, then

$$\mathbf{V} = -[\mathrm{d}]\Phi = -[\mathrm{d}]([\partial]Z^{-1}[\mathrm{d}])^{-1}[\partial]\mathbf{K}.$$

Again, since $[\mathrm{d}]$ is the transpose of $[\partial]$, the matrix

$$R = [\mathrm{d}]([\partial]Z^{-1}[\mathrm{d}])^{-1}[\partial],$$

which expresses the branch voltages in terms of the imposed currents, is a symmetric matrix.

Incidentally, the dependence of $\mathbf{V}$ upon $\mathbf{W}$, or of $\mathbf{I}$ upon $\mathbf{K}$, is *not* in general described by a symmetric matrix.

13.4. Capacitive networks

A minor modification of the network problem is to have a network of capacitors instead of a network of resistors. Each branch is now allowed to have a battery in series with a capacitor. When the batteries are attached, charge will accumulate on the plates on the capacitors, and eventually a steady state will be reached. In this

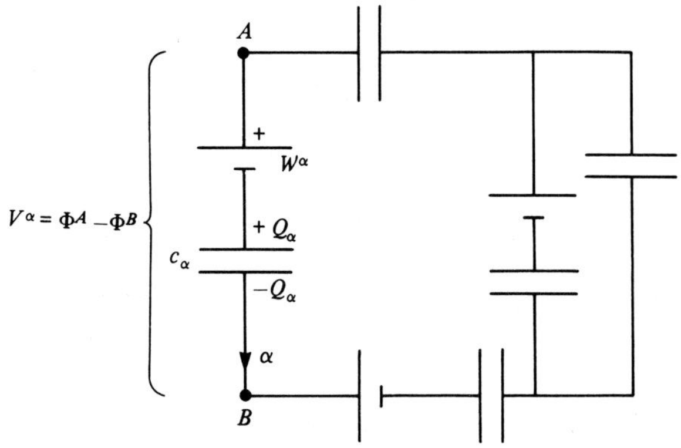

Figure 13.13

steady state no current will be flowing, since charge cannot cross between the plates of the capacitors. We are interested in knowing the charge, Q_α, on the capacitor in branch α and the voltage V^α across the branch. A typical capacitive network, and the sign conventions for a typical branch α, are illustrated in figure 13.13. Notice that, as before, positive V^α and W^α refer to voltage drops when the branch is traversed in the sense defined by the arrow, and that positive Q_α implies that the positively charged plate of the capacitor is encountered first as the branch is traversed. With these conventions, if the capacitance in the branch is c_α, we have the equation

$$V^\alpha - W^\alpha = \frac{Q_\alpha}{C_\alpha}.$$

We may regard the vector $\mathbf{Q} = (Q_\alpha, Q_\beta, \ldots)$ as a one-chain, and the vector $\mathbf{V} = (V^\alpha, V^\beta, \ldots)^{\mathrm{T}}$ as a one-cochain. However, the pairing between voltage and charge now gives energy rather than power. In vector notation, we may write

$$\mathbf{V} - \mathbf{W} = C^{-1}\mathbf{Q}$$

where C is the diagonal matrix with entries C_α. With the replacement of Z by C^{-1} and of power by energy the situation is formally identical to our discussion of resistive circuits. The voltage $\mathbf{V}$ still may be derived from a potential, so $\mathbf{V} = -\mathrm{d}\Phi$. Since charge cannot flow across the capacitors, the total charge on the capacitors connected to any node cannot change. If the capacitors were initially uncharged, this total charge must be zero for each node, and $\partial\mathbf{Q} = 0$, just as $\partial\mathbf{I} = 0$ for resistive networks.

Isolated networks of initially uncharged capacitors, for which $\partial\mathbf{Q} = 0$, may be treated exactly as we have treated networks of resistors. We need only replace $\mathbf{I}$ by $\mathbf{Q}$ and Z by C^{-1} in all our previous results. The batteries are still represented by $\mathbf{W}$, and there is nothing which plays the role of a current source. Then, by analogy with the mesh-current method, we may introduce *mesh charges*, described by a vector $\mathbf{P} \in Z_1$,

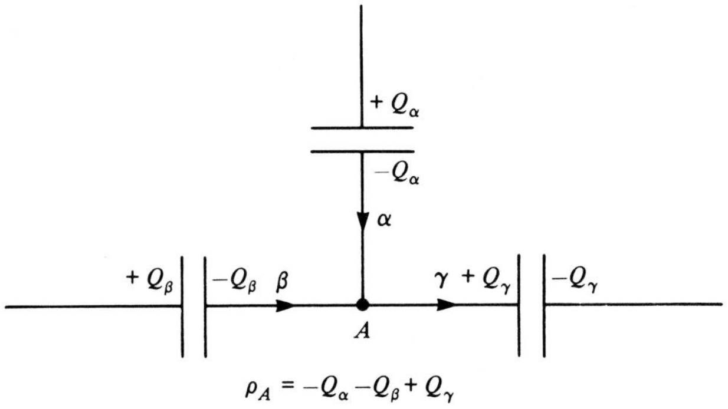

Figure 13.14

so that any charge distribution satisfying $\partial\mathbf{Q} = 0$ may be written as $\mathbf{Q} = \sigma\mathbf{P}$. Then

$$\mathbf{P} = (sC^{-1}\sigma)^{-1}(-s\mathbf{W}).$$

Alternatively, we can replace Z^{-1} by C in the node-potential method to obtain

$$\Phi = -([\partial]C[\mathrm{d}])^{-1}[\partial]C\mathbf{W}.$$

More generally, if the capacitors were initially charged, there may be a fixed total charge ρ_A on the plates connected to each node. The vector $\boldsymbol{\rho} = (\rho_A, \rho_B, \ldots)$ may be regarded as a zero-chain. Since, according to our sign conventions, the *negative* plate of a capacitor is connected to the node at the end of the branch, we write $\partial\mathbf{Q} = -\boldsymbol{\rho}$ in general. (Refer to figure 13.14 in order to convince yourself that the minus sign is correct.) Since the sum of the charges on the two plates of each capacitor is zero, we know that

$$\sum_{\substack{\text{all}\\ \text{nodes}}} \rho_A = 0$$

when the sum is taken over all nodes in a network.

The equation $\partial\mathbf{Q} = -\rho$ is known as *Gauss' law*. It expresses the node charges in terms of the charges across the capacitors in the branches. Looking ahead to our treatment of electrostatics in Chapter 16, we shall see that the analog of the node charges will be the 'charge density' ρ, while the analog of $\mathbf{Q}$ will be the 'dielectric displacement', $\mathbf{D}$. We will explain in Chapter 16 that $\mathbf{D}$ should be thought of as a two-form on three dimensional space and describe how one would perform the physical experiments to measure it. It will then turn out that Gauss' law is nothing other than the three dimensional version of Stokes' theorem. In the 'vector calculus' treatment of electrostatics as presented in most texts, the dielectric displacement $\mathbf{D}$ is thought of as a vector field. The Gauss' law becomes the 'divergence theorem'. But in our present context of the theory of capacitive networks, the equation

$\partial\mathbf{Q} = -\rho$ simply says that we assign to each node the sum of the charges on the plates connected to it.

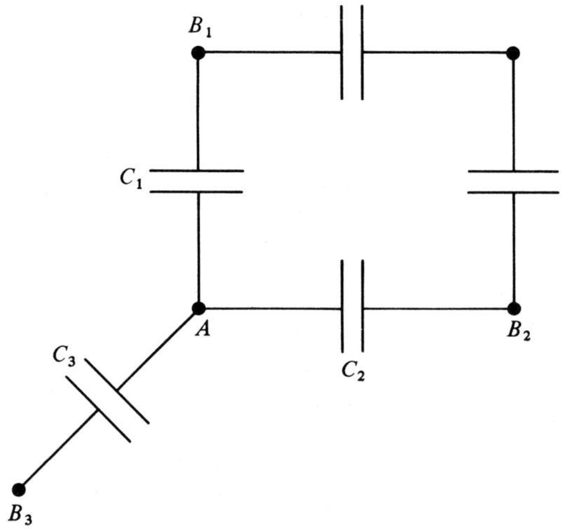

Figure 13.15

For the rest of this chapter we will study capacitive networks *with* initial charges but *without* voltage sources. So we are assuming that $\mathbf{W} = 0$, and our equations become

$$\mathbf{Q} = C\mathbf{V}, \quad \mathbf{V} = -\mathrm{d}\Phi \quad \text{and} \quad \partial\mathbf{Q} = -\rho.$$

We can combine these three equations to obtain

$$-\partial C\mathrm{d}\Phi = -\rho.$$

This equation is known as *Poisson's equation.* It gives the relation between the potential function and the node charges. The operator $-\partial C\mathrm{d}$ occurring on the left hand side of Poisson's equation is called the *Laplacian* and is denoted by Δ. Thus Poisson's equation may be written as

$$\Delta\Phi = -\rho.$$

Now d maps $C^0 \to C^1$, C maps $C^1 \to C_1$ and ∂ maps $C_1 \to C_0$. Therefore the Laplacian, Δ, maps $C^0 \to C_0$. The Φ occurring on the left hand side of Poisson's equation belongs to C^0 while the ρ occurring on the right hand side of Poisson's equation belongs to C_0, as is to be expected.

It is instructive to write out explicitly the form of the Laplacian. Let u be an element of C^0 and A a node. If α is a branch with A one of its boundaries, then

$$\mathrm{d}u(\alpha) = \pm(u(A) - u(B)) \quad \text{if} \quad \partial\alpha = \pm(A - B).$$

So $C\mathrm{d}u$ has coefficient

$$\pm C_\alpha(u(A) - u(B)) \quad \text{at} \quad \alpha.$$

If we now apply ∂ and see what its coefficient is at A, we see that (summing over all α with $\partial\alpha = \pm(A - B)$ for some B)

$$(\Delta u)(A) = \sum_{\alpha:\partial\alpha = \pm(B-A)} C_\alpha(u(B) - u(A)).$$

In particular, u satisfies *Laplace's equation*

$$\Delta u = 0$$

if and only if

$$u(A) = \frac{1}{\sum C_\alpha} \sum C_\alpha u(B),$$

where the sum extends over all nodes B which are the other ends of branches emanating from A, as illustrated in figure 13.15. In other words, Laplace's equation says that the value of u at A is the weighted average of its values at *nearest neighbor* nodes, the weighting being given by the capacitances.

This interpretation should be compared with the partial differential equation

$$\Delta u = \frac{\partial^2 u}{\partial x^2} + \frac{\partial^2 u}{\partial y^2} = 0$$

introduced in Chapter 5. We saw there that this equation says that the value of u at a point P is the average of u over a small circle centered at A. For the concept of circle to be defined, we had to have the Euclidean geometry of the plane. Here to define the nearest neighbor average, we need the matrix C of capacitances. There is an analogy between the choice of C and the choice of Euclidean plane geometry. We shall pursue this deep point in Chapter 16.

Let us now return to our study of Poisson's equation. Notice that the Φ occurring on the left of Poisson's equation is only determined up to a function which is constant on all connected components of the network. Indeed any such function lies in the kernel of d, and hence surely lies in the kernel of $\Delta = -\partial C\mathrm{d}$. Also the ρ which occurs on the right hand side of Poisson's equation lies in the subspace B_0 of C_0 since, by definition, the node charges are determined by applying ∂ to the branch charges. Recall that we introduced the quotient space $P^0 = C^0/Z^0$ and the restricted coboundary operator $[\mathrm{d}]: P^0 \to C^1$, and we also introduced the restricted boundary operator $[\partial]: C_1 \to B_0$. The situation is best summarized by the following diagram of maps:

$$C^0 \longrightarrow P^0 \xrightarrow{[\mathrm{d}]} C^1 \xrightarrow{C} C_1 \xrightarrow{[\partial]} B_0 \longrightarrow C_0$$

$$C^0 \xrightarrow{-\Delta} C_0$$

This suggests that we introduce the *restricted Laplacian* $[\Delta] = -[\partial]C[\mathrm{d}]$ and consider the restricted Poisson equation

$$[\Delta]\Psi = -\rho$$

for $\Psi \in P^0$. (In other words, Ψ is a 'potential determined up to additive constant on each connected component'.) Now since the operator C is given by a positive diagonal matrix in terms of the standard basis of C^1 and C_1 we know from the node potential version of Kirchhoff's theorem that $[\partial]C[\mathrm{d}]$ is invertible and hence the solution of the restricted Poisson equation is given by

$$\Psi = ([\partial]C[\mathrm{d}]^{-1}\rho.$$

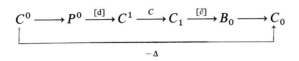

Illustrative example

A simple Poisson equation problem is presented in figure 13.16. Four units of charge are at node B, one unit is at node C, and we wish to find the potential at all nodes. (If you want to specify units, use microfarads for capacitance, microcoulombs for charge, and volts for potential.)

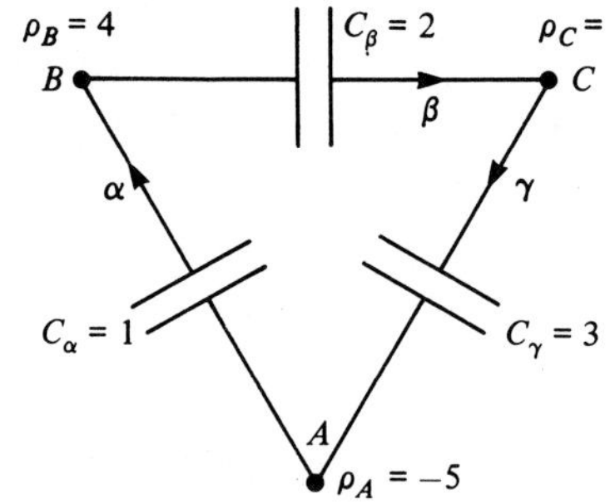

Figure 13.16

We choose A as the ground node. The restricted boundary operator is then

$$[\partial] = \begin{pmatrix} 1 & -1 & 0 \\ 0 & 1 & -1 \end{pmatrix}$$

(remember, the row for node A is deleted). The matrix [d] is the transpose of $[\partial]$:

$$[\mathrm{d}] = \begin{pmatrix} 1 & 0 \\ -1 & 1 \\ 0 & -1 \end{pmatrix}.$$

The matrix C is $\begin{pmatrix} 1 & 0 & 0 \\ 0 & 2 & 0 \\ 0 & 0 & 3 \end{pmatrix}$, so

$$[\partial]C[\mathrm{d}] = \begin{pmatrix} 1 & -1 & 0 \\ 0 & 1 & -1 \end{pmatrix} \begin{pmatrix} 1 & 0 & 0 \\ 0 & 2 & 0 \\ 0 & 0 & 3 \end{pmatrix} \begin{pmatrix} 1 & 0 \\ -1 & 1 \\ 0 & -1 \end{pmatrix} = \begin{pmatrix} 3 & -2 \\ -2 & 5 \end{pmatrix}.$$

The solution to Poisson's equation is

$$([\partial]C[\mathrm{d}])^{-1} = \frac{1}{11}\begin{pmatrix} 5 & 2 \\ 2 & 3 \end{pmatrix}, \quad \Psi = \frac{1}{11}\begin{pmatrix} 5 & 2 \\ 2 & 3 \end{pmatrix}\begin{pmatrix} 4 \\ 1 \end{pmatrix} = \begin{pmatrix} 2 \\ 1 \end{pmatrix}.$$

13.5. Boundary-value problems

The Poisson equation problem that we just solved is the analogue of the electrostatic problem of computing the potential $\Phi(\mathbf{r})$ resulting from a specified isolated charge distribution $\rho(\mathbf{r})$ in vacuum, with no boundary conditions specified. More interesting electrostatic problems are ones in which the boundary condition of zero potential on certain conductors is specified, or in which the potential is specified on

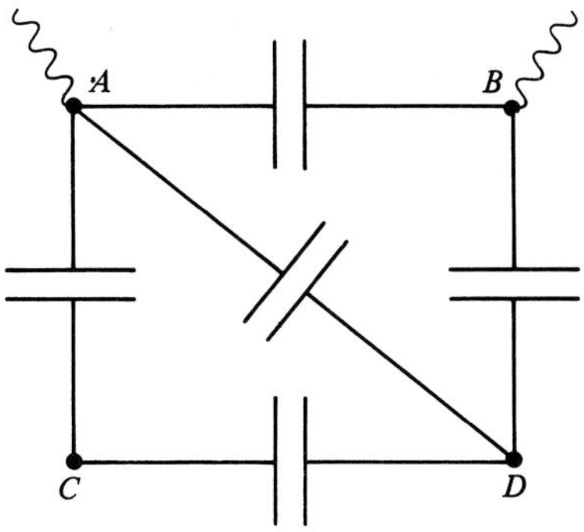

Figure 13.17

the boundary of a region. We shall now consider the discrete analogues of such problems.

We imagine a connected network of capacitors, with no batteries along any of the branches, as shown in figure 13.17. We subdivide the nodes of the network into two classes; *boundary* nodes and *interior* nodes. Boundary nodes, such as A and B, are connected to external sources which maintain their potentials at specified values. Since charge can flow freely along the wires connecting these nodes to the external sources, it will not be reasonable to specify the total charge at a boundary node. Interior nodes, such as C and D, are connected only to other nodes of the network, not to any external sources. The total charge at one of these nodes is constant and may be specified; it is unreasonable, though, to specify the potential at interior nodes.

In the general problem concerning such a network, we would specify the potential at each boundary node and the total charge at each interior node arbitrarily, then try to determine the potential at each interior node and the charge at each boundary node. In fact, such a general problem can be expressed as a superposition of two more restricted problems:

(1) A *Dirichlet problem*, in which the total charge at each interior node is set equal to zero and the potential at each boundary node has a specified (generally non-zero) value.

(2) A *Poisson equation problem*, in which the potential at each boundary node is set equal to zero and the total charge ρ at each interior node has a specified value. The problem posed in figure 13.15 was of this type, with node A functioning as a boundary node.

The solution to the general problem can always be expressed as the superposition of the solution to a Dirichlet problem with the appropriate boundary conditions and a Poisson problem with the appropriate charges at interior nodes.

Corresponding to our decomposition of nodes into two classes, we get a decomposition of the vector spaces C_0 (node charges) and C^0 (potentials). The space

C_0 is the direct sum

$$C_0 = C_0^{\text{bound}} \oplus C_0^{\text{int}},$$

where C_0^{bound} consists of zero-chains having non-zero coefficients only at the boundary nodes and C_0^{int} consists of those zero-chains which have non-zero coefficients only at the interior nodes. Similarly, we have

$$C^0 = C^0_{\text{bound}} \oplus C^0_{\text{int}},$$

where C^0_{bound} consists of those linear functions which vanish on C_0^{int}, while C^0_{int} consists of those functions which vanish on C_0^{bound}.

The potential Φ may be decomposed in a unique way as the sum of an element of each of C^0_{bound} and C^0_{int}. We write

$$\Phi = \Phi_{\text{bound}} + \Phi_{\text{int}}.$$

For the circuit of figure 13.17, with boundary nodes A and B and interior nodes C and D, this decomposition reads simply

$$\begin{pmatrix} \Phi^A \\ \Phi^B \\ \Phi^C \\ \Phi^D \end{pmatrix} = \begin{pmatrix} \Phi^A \\ \Phi^B \\ 0 \\ 0 \end{pmatrix} + \begin{pmatrix} 0 \\ 0 \\ \Phi^C \\ \Phi^D \end{pmatrix}.$$

Similarly, the vector representing node charges may be decomposed as

$$\boldsymbol{\rho} = \boldsymbol{\rho}^{\text{bound}} + \boldsymbol{\rho}^{\text{int}}$$

so that, in the circuit of figure 13.17

$$\begin{pmatrix} \rho_A \\ \rho_B \\ \rho_C \\ \rho_D \end{pmatrix} = \begin{pmatrix} \rho_A \\ \rho_B \\ 0 \\ 0 \end{pmatrix} + \begin{pmatrix} 0 \\ 0 \\ \rho_C \\ \rho_D \end{pmatrix}.$$

In terms of these vector space decompositions, we can formulate our problems as follows:

(1) **Dirichlet problem:** Since the potential is specified at each boundary node, Φ_{bound} is a known vector. Furthermore, since there is no charge at any interior node, the vector $\boldsymbol{\rho}^{\text{int}} = 0$. We must determine the vector Φ_{int} so that

$$\Delta(\Phi_{\text{int}} + \Phi_{\text{bound}}) = 0$$

at all interior nodes. At the boundary nodes, we can calculate

$$\Delta(\Phi_{\text{int}} + \Phi_{\text{bound}}) = -\boldsymbol{\rho}^{\text{bound}}$$

and will then know the potential and charge at all nodes.

(2) **Poisson equation problem:** Since the charge is specified at each interior node, $\boldsymbol{\rho}^{\text{int}}$ is a known vector. Furthermore, since the potential is specified as zero on each boundary node, $\Phi_{\text{bound}} = 0$. We must determine the vector Φ_{int} so that

$$\Delta(\Phi_{\text{bound}} + \Phi_{\text{int}}) = -\boldsymbol{\rho}^{\text{int}}$$

at all interior nodes. Again, the charge distribution at the boundary nodes is determined once the potential is known, by the equation

$$\rho^{\text{bound}} = -\Delta(\Phi_{\text{bound}} + \Phi_{\text{int}}).$$

13.6. Solution of the boundary-value problem by Weyl's method of orthogonal projection

We study a connected network with n_i interior nodes and n_0 boundary nodes. We may solve both these problems by using Weyl's method of orthogonal projection. We use the capacitance matrix $C: C^1 \to C_1$ to define a scalar product on the space C^1 of branch voltages by the formula

$$(\mathbf{V}, \hat{\mathbf{V}})_C = \int_{CV} \hat{\mathbf{V}} = \sum_{\substack{\text{all} \\ \text{branches}}} C_\alpha V^\alpha \hat{V}^\alpha.$$

Since C is a diagonal matrix whose entries are all positive, this is a positive-definite scalar product. Not surprisingly, it has a physical interpretation in terms of energy: for any V the expression

$$\tfrac{1}{2}(\mathbf{V}, \mathbf{V})_C = \sum \tfrac{1}{2} C_\alpha V^{\alpha^2}$$

represents the total energy stored in all the capacitors.

Under this scalar product, the space C^1 decomposes into the orthogonal direct sum

$$C^1 = \mathrm{d}(C^0) \oplus C^{-1} Z_1.$$

Indeed, we know that $\mathrm{d}(C^0)$ is the annihilator space of the subspace Z_1 of C_1. But taking the scalar product of an element V of C_1 on an element of the form $C^{-1}I$ is the same as evaluating V on I:

$$(V, C^{-1}I)_C = \int_I V.$$

Let us decompose the space $\mathrm{d}(C^0)$ further: Since $\mathrm{d}C^0_{\text{int}}$ is a subspace of C^0, its orthogonal complement inside $\mathrm{d}C^0$ is a subspace of $\mathrm{d}C^0$; call it D^1. So we have the direct sum decomposition

$$\mathrm{d}C^0 = D^1 \oplus \mathrm{d}C^0_{\text{int}}$$

so therefore

$$\boxed{C^1 = D^1 \oplus \mathrm{d}C^0_{\text{int}} \oplus C^{-1} Z_1} \qquad (13.1)$$

is a decomposition of C^1 into three mutually orthogonal subspaces. Notice that an element of D^1 is of the form $\mathrm{d}\Phi$ since $D^1 \subset \mathrm{d}C^0$. Also if $V \in D^1$ so $V \perp \mathrm{d}C^0_{\text{int}}$, then

$$0 = (V, \mathrm{d}\Phi_{\text{int}})_C = \int_{CV} \mathrm{d}\Phi_{\text{int}} = \int_{\partial CV} \Phi_{\text{int}}.$$

At any interior node, we can find a Φ_{int} which does not vanish at that node, and vanishes at all other nodes. Thus ∂CV must have vanishing components at all interior nodes. I.e., $\partial CV \in C_0^{\text{bound}}$. But $V = -\mathrm{d}\Phi$ so

$$\Delta\Phi = 0 \quad \text{at all interior nodes.}$$

In other words, D^1 *consists of* $\mathrm{d}\Phi$ *where* Φ *is a solution of the Dirichlet problem.* Now $\dim C^{-1}Z_1 = \dim Z_1 = m$. If there is at least one boundary node $\mathrm{d}\colon C^0_{\text{int}} \to C^1$ is injective, so $\dim \mathrm{d}C^0_{\text{int}} = n_{\text{i}}$. Also, $\dim C^1 = n + m - 1 = n_{\text{b}} - 1 + n_{\text{i}} + m$. It thus follows from (13.1) that

$$\dim D^1 = n_{\text{b}} - 1.$$

As there are $n_{\text{b}} - 1$ boundary nodes left over if we exclude ground, we see that we can arbitrarily assign the value of Φ at the remaining boundary nodes. Thus we can always solve the Dirichlet problem.

As an explicit example of the decomposition (13.1), consider the network of figure 13.18. Here A, B, and C are boundary nodes; D is an interior node. Since there are four branches, $\dim C^1 = 4$. The orthogonal subspaces are as follows.

(1) $C^{-1}Z_1$ is a one-dimensional subspace with basis $\begin{pmatrix} 1 \\ \frac{1}{2} \\ -\frac{1}{2} \\ 0 \end{pmatrix}$. This basis element corresponds to a charge distribution, $C\begin{pmatrix} 1 \\ \frac{1}{2} \\ -\frac{1}{2} \\ 0 \end{pmatrix} = \begin{pmatrix} 1 \\ 1 \\ -1 \\ 0 \end{pmatrix}$, for which $\rho = 0$ at all

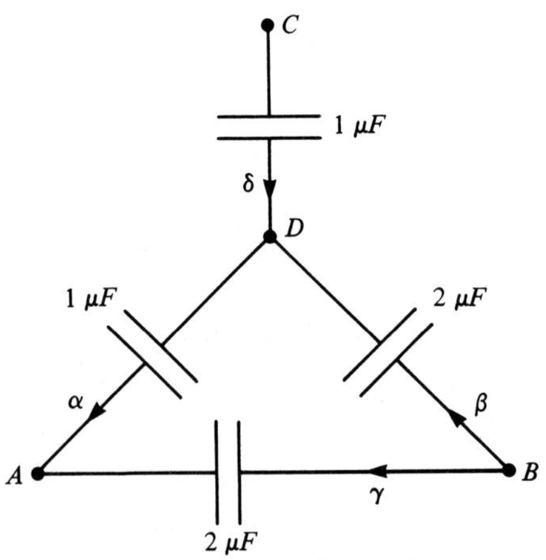

Figure 13.18

nodes. It violates Kirchhoff's voltage law, of course, and so is not derivable from a potential.

(2) dC^0_{int} is a one-dimensional subspace with basis $\begin{pmatrix} 1 \\ -1 \\ 0 \\ -1 \end{pmatrix}$. This basis element describes branch voltages arising from a potential

$$\Phi^A = \Phi^B = \Phi^C = 0, \Phi^D = 1.$$

(3) D^1 is a two-dimensional subspace with basis elements

$$\begin{pmatrix} \frac{2}{3} \\ \frac{1}{3} \\ 1 \\ 0 \end{pmatrix} \text{ and } \begin{pmatrix} \frac{1}{3} \\ -\frac{1}{6} \\ \frac{1}{6} \\ \frac{2}{3} \end{pmatrix}.$$

These describe branch voltages arising from

$$\Phi^A = \Phi^C = 0, \quad \Phi^B = 1 \quad \text{and} \quad \Phi^A = \Phi^B = 0, \quad \Phi^C = 1$$

respectively, with $\rho_D = 0$ in either case. (In general, it is not a straightforward matter to write down a basis for D^1 by inspection.)

Solution of Poisson's equation

We are now in a position to solve Poisson's equation by Weyl's method of orthogonal projection. Suppose the charge ρ is specified at all interior nodes. We construct *any* voltage distribution $\hat{\mathbf{V}}$ for which $-\partial C\hat{\mathbf{V}} = \boldsymbol{\rho}$ has the specified values at all interior nodes. ($\hat{\mathbf{V}}$ may even violate Kirchhoff's voltage law.) Then there is a unique decomposition

$$\hat{\mathbf{V}} = \mathbf{V} + \mathbf{W}$$

where $\mathbf{V}$ is an element of dC^0_{int} and $\mathbf{W}$ is orthogonal to dC^0_{int}. Being an element of dC^0_{int}, $\mathbf{V}$ is derivable from a potential with $\Phi = 0$ on the boundary. Since $\mathbf{W}$ is orthogonal to $\mathbf{V}$, it lies in $D^1 \oplus C^{-1}Z_1$, and it therefore satisfies $\partial C\mathbf{W} = 0$ at all interior nodes. Thus $-\partial C\mathbf{V} = -\partial C\hat{\mathbf{V}} = \boldsymbol{\rho}$ at interior nodes, and $\mathbf{V}$ is the solution to the problem. If we denote by π the orthogonal projection of C^1 onto dC^0_{int}, we may express the solution as

$$\mathbf{V} = \pi\hat{\mathbf{V}}.$$

Solution of Dirichlet's problem

Similarly, we may solve Dirichlet's problem by orthogonal projection. Let $\hat{\boldsymbol{\Phi}}$ denote a potential which has the specified values at boundary nodes and which is zero at all interior nodes. Let $\hat{\mathbf{V}} = -d\hat{\boldsymbol{\Phi}}$. Then

$$\hat{\mathbf{V}} = \pi\hat{\mathbf{V}} + (1 - \pi)\hat{\mathbf{V}}.$$

Since $\pi\hat{\mathbf{V}} \in dC^0_{int}$, we may write $\pi\hat{\mathbf{V}} = -d\psi$, where $\psi = 0$ at all boundary nodes. If we write

$$(1-\pi)\hat{V} = -\mathrm{d}\Phi,$$

we know, since $(1-\pi)\hat{V}\in D^1$, that $\Delta\Phi = 0$ at all interior nodes. Furthermore, since

$$-\mathrm{d}\hat{\Phi} = -\mathrm{d}\psi - \mathrm{d}\Phi$$

and $\psi = 0$ at all boundary nodes, $\Phi = \hat{\Phi}$ on the boundary. Thus Φ is the desired solution to Dirichlet's problem.

Examples

Both of these methods may be illustrated for the network of figure 13.19. For that network, a basis for $\mathrm{d}C^0_{\text{int}}$ was

$$\mathbf{U} = \begin{pmatrix} 1 \\ -1 \\ 0 \\ -1 \end{pmatrix},$$

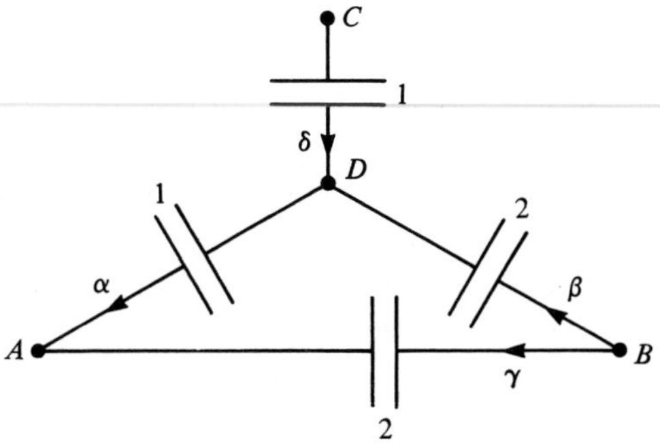

Figure 13.19

a vector whose (length)^2 (relative to the scalar product defined by C), is $1+2+1=4$. Then, for any $\mathbf{V}\in C^1$,

$$\pi\mathbf{V} = \tfrac{1}{4}(\mathbf{V}, \mathbf{U})_C\mathbf{U}.$$

Applying π to the basis vectors $\begin{pmatrix} 1 \\ 0 \\ 0 \\ 0 \end{pmatrix}, \begin{pmatrix} 0 \\ 1 \\ 0 \\ 0 \end{pmatrix}, \ldots,$ we find the matrices

$$\pi = \frac{1}{4}\begin{pmatrix} 1 & -2 & 0 & -1 \\ -1 & 2 & 0 & 1 \\ 0 & 0 & 0 & 0 \\ -1 & 2 & 0 & 1 \end{pmatrix} \quad \text{and} \quad (1-\pi) = \frac{1}{4}\begin{pmatrix} 3 & 2 & 0 & 1 \\ 1 & 2 & 0 & -1 \\ 0 & 0 & 4 & 0 \\ 1 & -2 & 0 & 3 \end{pmatrix}.$$

Suppose now that we wish to solve Poisson's equation for $\rho_D = 1$. One possible way of having one unit of charge at D is to choose $\hat{Q}_\alpha = \hat{Q}_\beta = \hat{Q}_\gamma = 0$, $\hat{Q}_\delta = -1$, so that $\hat{\mathbf{V}} = \begin{pmatrix} 0 \\ 0 \\ 0 \\ -1 \end{pmatrix}$. Then $\pi\hat{\mathbf{V}} = \frac{1}{4}\begin{pmatrix} +1 \\ -1 \\ 0 \\ -1 \end{pmatrix}$ gives the solution to the problem.

Notice that it is not necessary for $\hat{V}$ to satisfy Kirchhoff's voltage law. Another charge distribution for which $\rho_D = 1$ is $\hat{Q}_\alpha = \frac{1}{3}$, $\hat{Q}_\beta = -\frac{1}{3}$, $\hat{Q}_\gamma = 1$, $\hat{Q}_\delta = -\frac{1}{3}$. Then

$$\hat{\mathbf{V}} = C^{-1}\hat{\mathbf{Q}} = \begin{pmatrix} \frac{1}{3} \\ -\frac{1}{6} \\ 2 \\ -\frac{1}{3} \end{pmatrix},$$

which fails to satisfy Kirchhoff's voltage law because the sum of the voltage drops around the mesh $\beta + \alpha - \gamma$ is not zero. Still

$$\pi\hat{\mathbf{V}} = \frac{1}{4}\begin{pmatrix} 1 & -2 & 0 & -1 \\ -1 & 2 & 0 & 1 \\ 0 & 0 & 0 & 0 \\ -1 & 2 & 0 & 1 \end{pmatrix}\begin{pmatrix} \frac{1}{3} \\ -\frac{1}{6} \\ 2 \\ -\frac{1}{3} \end{pmatrix} = \frac{1}{4}\begin{pmatrix} 1 \\ -1 \\ 0 \\ -1 \end{pmatrix}$$

and we again obtain the correct solution.

In constructing $\hat{\mathbf{V}}$, it is simplest, whenever possible, to assign all the charge on each interior node to a single branch which connects that branch to a boundary node. In complicated networks where there are interior nodes connected only to other interior nodes, it is advisable to begin with such nodes and work from the inside out. For example, in figure 13.20, where A and E are boundary nodes, it is best to begin

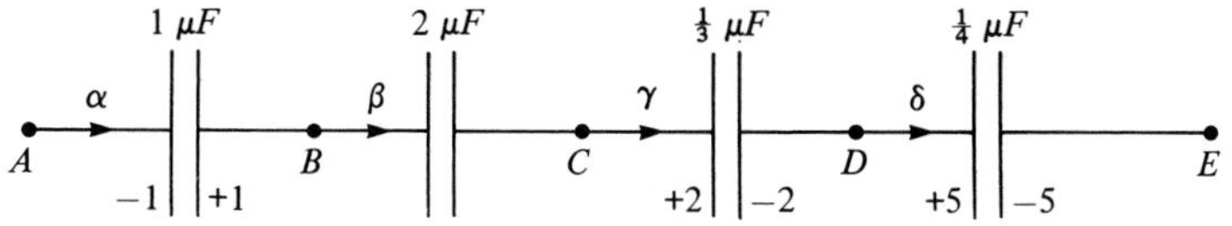

Figure 13.20

with node C. If $\rho_B = 1$, $\rho_C = 2$, $\rho_D = 3$, choose $\hat{Q}_\gamma = 2$, then $\hat{Q}_\delta = 5$, then $\hat{Q}_\alpha = -1$ so that

$$\hat{\mathbf{Q}} = \begin{pmatrix} -1 \\ 0 \\ 2 \\ 5 \end{pmatrix} \quad \text{and} \quad \hat{\mathbf{V}} = \begin{pmatrix} -1 \\ 0 \\ 6 \\ 20 \end{pmatrix}.$$

Returning to figure 13.19, we solve Dirichlet's problem for $\Phi^A = 0$, $\Phi^B = 5$, $\Phi^C = 6$.

We set $\hat{\Phi} = \Phi$ on the boundary and $\hat{\Phi}^D = 0$, so that $\hat{\mathbf{V}} = -\mathrm{d}\hat{\Phi} = \begin{pmatrix} 0 \\ 5 \\ 5 \\ 6 \end{pmatrix}$. Then

$$\mathbf{V}=(1-\pi)\hat{\mathbf{V}}=\frac{1}{4}\begin{pmatrix}3 & 2 & 0 & 1\\ 1 & 2 & 0 & -1\\ 0 & 0 & 4 & 0\\ 1 & -2 & 0 & 3\end{pmatrix}\begin{pmatrix}0\\5\\5\\6\end{pmatrix}=\frac{1}{4}\begin{pmatrix}16\\4\\20\\8\end{pmatrix}=\begin{pmatrix}4\\1\\5\\2\end{pmatrix}$$

which describes the correct solution to the problem, with $\Phi^D=4$. The same solution would have been obtained with any other arbitrarily chosen value for $\hat{\Phi}^D$.

13.7. Green's functions

The map which assigns to the charge distribution ρ the corresponding potential u which solves the Poisson equation is called the *Green's operator*. The matrix entries of this operator $\mathbf{G}$ relative to the basis of C_0 consisting of the nodes will be denoted by $G(A,B)$, where A and B are nodes. We use the notation $G(A,B)$ instead of the more familiar G_{AB} to make the notation look more like the corresponding formulas that will appear later when we treat continuous charge distributions. Thus

$G(A,B)=$ potential at B due to solving the Poisson equation with unit charge at A.

The function G that assigns to the nodes A and B the matrix element $G(A,B)$ is called the *Green's function.* So G is a function of two variables. When considered as a function of the second variable, with the first variable fixed, it is an element of C^0_{int}. We let Δ_2 denote the operator Δ applied to the element $G(A,\cdot)$ of C^0_{int}. Thus

$$-\Delta_2 G(A,\cdot)=\delta_A$$
$$G(A,B)=0 \quad \text{for } B \text{ on the boundary.}$$

To repeat,

$$\Delta\colon C^0_{\text{int}}\to C_0^{\text{int}}.$$

The inverse of $-\Delta$ is G so

$$-\Delta u=\rho \quad \text{iff} \quad \mathbf{G}\rho=u.$$

Because Δ is symmetric, so is G. In terms of the basis given by the nodes, we can write

$$\boxed{u(B)=\sum\rho(A)G(A,B) \quad \text{for all } B}$$

where the sum extends over all interior A.

We can also use the Green's function to solve the Dirichlet problem. For this it is convenient to make use of a summation formula known as *Green's formula.* Notice that, if u and v are arbitrary elements of C^0, we have

$$\sum_{-\text{all }A}u(A)\Delta v(A)=\sum_{\text{all }A}u(A)\partial C\,\mathrm{d}v(A)=\int_{\partial C\mathrm{d}v}u=\int_{C\mathrm{d}v}\mathrm{d}u=(\mathrm{d}u,\mathrm{d}v),$$

where $(,)$ denotes the symmetric scalar product on C^1. Thus

$$\sum_{\text{all } A} u(A)\Delta v(A) = \sum_{\text{all } A} \Delta u(A)v(A).$$

If we split the sum up into two parts, one over the boundary and the other over the interior, we get Green's formula:

$$\sum_{\text{interior } A} (u(A)\Delta v(A) - \Delta u(A)v(A)) = - \sum_{\text{boundary } B} (u(B)\Delta v(B) - \Delta u(B)v(B)). \quad (13.2)$$

Let us take two interior points A_1 and A_2, and set $u = G(A_1, \cdot)$ and $v = G(A_2, \cdot)$ in Green's formula; the right-hand side in (13.2) vanishes since $G(A_1, \cdot)$ and $G(A_2, \cdot)$ vanish on the boundary. At interior points $u(A) = 0$ if $A \neq A_1$ and $u(A_1) = 1$. Similarly for v and A_2. Thus the vanishing of the left-hand side (13.2) says that

$$G(A_1, A_2) = G(A_2, A_1).$$

If we write $u = \Phi$, $\Delta u = -\rho$ and $v = \hat{\Phi}$, $\Delta v = -\hat{\rho}$, (13.2) becomes

$$\sum_{\substack{\text{interior} \\ \text{nodes } C}} (\rho_C \hat{\Phi}^C - \hat{\rho}_C \Phi^C) = \sum_{\substack{\text{boundary} \\ \text{nodes } B}} (\hat{\rho}_B \Phi^B - \rho_B \hat{\Phi}^B). \quad (13.3)$$

We see that Green's formula is just Green's reciprocity theorem for capacitive networks.

We now suppose that $\hat{\Phi}$ is a solution to the Dirichlet problem with specified values on the boundary, so $\hat{\rho} = 0$ in the interior. Take $\Phi = G(A, \cdot)$ in (13.3). Then we have

$$\sum_{\substack{\text{interior} \\ \text{nodes } C}} (-\Delta G(A, C)\hat{\Phi}^C - \hat{\rho}_C G(A, C)) = \sum_{\substack{\text{boundary} \\ \text{nodes } B}} (\hat{\rho}_B G(A, B) + \Delta G(A, B)\hat{\Phi}^B).$$

Recall now that $-\Delta G(A, C) = 1$ if $C = A$ and 0 otherwise, because $G(A, C)$ is the solution to Poisson's equation with one unit of charge at A, zero at other interior nodes. Furthermore, we have assumed that $\hat{\rho}_C = 0$ at interior nodes. So only one term, $\hat{\Phi}^A$, survives on the left-hand side of the equation. On the right, $G(A, B) = 0$, so we have finally

$$\hat{\Phi}^A = \sum_{\substack{\text{boundary} \\ \text{nodes}}} \Delta G(A, B)\hat{\Phi}^B.$$

That is, $\hat{\Phi}^A$, at an interior node A, is a weighted average of the values of $\hat{\Phi}^B$ at boundary nodes, with the weighting factor $\Delta G(A, B)$ being the quantity of negative charge which is induced at boundary node B when unit positive charge is placed at A with the entire boundary held at zero potential. The matrix $(\Delta G(A, B))$, which represents a linear transformation from C^0_{bound} to C^0_{int}, is known as the *Poisson kernel*.

It is a simple matter to apply the Green's function approach to the network of figure 13.18. Since there is only one interior node, D, we need only to solve Poisson's equation for the case $\rho_D = 1$. We have already done this, obtaining voltages

$$\mathbf{V} = \frac{1}{4}\begin{pmatrix} 1 \\ -1 \\ 0 \\ -1 \end{pmatrix}.$$

Therefore, with A, B, and C grounded, the potential at D is $\frac{1}{4}$, and $G(D, D) = \frac{1}{4}$. The charges at the boundary nodes under the circumstances are $\rho_A = -\frac{1}{4}$, $\rho_B = -\frac{1}{2}$, $\rho_C = -\frac{1}{4}$; so we have $\Delta G(D, A) = \frac{1}{4}$, $\Delta G(D, B) = \frac{1}{2}$, $\Delta G(D, C) = \frac{1}{4}$. Armed with this Green's function, we can easily solve Dirichlet's problem for the network. If $\rho_D = 0$, then Φ^D is given by a weighted average of the potentials on the boundary, with ΔG supplying the weights:

$$\Phi^D = \Delta G(D, A)\Phi^A + \Delta G(D, B)\Phi^B + \Delta G(D, C)\Phi^C$$

or

$$\Phi^D = \tfrac{1}{4}\Phi^A + \tfrac{1}{2}\Phi^B + \tfrac{1}{4}\Phi^C.$$

There is another version of Green's formula (called Green's second formula) which is sometimes useful. Let us now look at a boundary point, A. In the formula

$$\Delta u(A) = \sum_{\alpha \text{ with } \pm\partial\alpha = (B - A)} C_\alpha(u(A) - u(B))$$

let us divide the sum on the right into two parts according to whether the branch α joins two points on the boundary or joins the boundary point to an interior point. Thus

$$\Delta u(A) = \sum_{\substack{\alpha \text{ with } \pm\partial\alpha = B - A \\ B \text{ an interior point}}} C_\alpha(u(A) - u(B)) + \sum_{\substack{\alpha \text{ with } \pm\partial\alpha = B - A \\ B \text{ a boundary point}}} C_\alpha(u(A) - u(B)).$$

Let us denote the second sum by $\Delta^{\text{bound}}u(A)$. We can think of the boundary, together with the branches of the network which join two boundary points, as forming a network in its own right. Then Δ^{bound} would be the Laplace operator on this subnetwork, determined by the capacitances of these boundary-to-boundary branches. Therefore,

$$\sum_{\text{boundary } A} (u(A)\Delta^{\text{bound}}v(A) - \Delta^{\text{bound}}u(A)v(A)) = 0$$

and so

$$\sum_{A \text{ in boundary}} (u(A)\Delta v(A) - \Delta u(A)v(A))$$
$$= \sum_{\substack{A \text{ in boundary} \\ \partial\alpha = \pm(B - A) \\ B \text{ in interior}}} (u(A)C_\alpha(v(A) - v(B)) - C_\alpha(u(A) - u(B))v(A)).$$

The terms involving $u(A)v(A)$ on the right cancel and the right-hand side simplifies to

$$\sum_{\substack{A \text{ in boundary} \\ \partial\alpha = \pm(B-A) \\ B \text{ in interior}}} c_\alpha(u(B)v(A) - u(A)v(B)).$$

Substituting into Green's formula, we get *Green's second formula*:

$$\boxed{\sum_{A \text{ in interior}} (u(A)\Delta v(A) - \Delta u(A)v(A)) = \sum_{\substack{A \text{ in boundary} \\ \partial\alpha = \pm(B-A) \\ B \text{ in interior}}} c_\alpha(u(A)v(B) - u(B)v(A)).}$$

13.8. The Poisson kernel and random walk

We continue our study of the Dirichlet problem. Let $Q = \Delta G$ be the Poisson kernel. Thus, if $\phi \in C^0_{\text{bound}}$ gives the specified boundary potential, then u, given by

$$u(A) = \sum_{B \text{ on boundary}} Q(A, B)\phi(B),$$

gives the solution of the Dirichlet problem with boundary values ϕ. In the preceding section we saw how to find G, and hence Q, by the method of orthogonal projection. In this section we shall give a probabilistic construction of Q. While this method is not usually amenable to direct computation, it does give new insights into the problem and also provides an iterative scheme for computing Q.

We first recall exactly what the Dirichlet problem says. Let A be any interior node. Let B be a nearest neighbor node to A. That is, B is a node at the other end of some branch β with end point $\pm A$. (B can be interior or boundary.) Let

$$P_{B,A} = C_\beta\left(\sum_{\partial\alpha = \pm(E-A)} c_\alpha\right)^{-1}.$$

Then the condition

$$(\Delta u)(A) = 0$$

is the same as

$$u(A) = \sum u(B)P_{B,A}$$

summed over all nearest neighbors. Notice that by construction

$$P_{A,B} > 0, \quad \sum_B P_{B,A} = 1.$$

Let us set

$$P_{A,A} = 1, \quad P_{A',A} = 0, \quad A' \neq A$$

at all boundary nodes. Then let P denote the $n \times n$ matrix whose entries are $P_{B,A}$. To fix the notation, let us write the boundary nodes first, so that upper left-hand corner of P will be a $k \times k$ identity matrix, where k is the number of boundary nodes. The column corresponding to an interior node will have $P_{B,A}$ at the B'th node, if B is a nearest neighbor of A, and 0 otherwise. The condition

$$\Delta u = 0 \text{ at interior } A$$

becomes

$$u = uP$$

if we think of $u = (u(B_1), u(B_2), \ldots)$ as a row vector. If the first k components of u are given by ϕ, then so are the first k components of uP, since the first k columns of P have 1 in the kth position and 0 elsewhere. Thus, the Dirichlet problem can be formulated as

$$uP = u \text{ and the first } k \text{ entries of } u \text{ are given by } \phi.$$

On the other hand, the matrix P has the structure of a matrix of *transition probabilities*. Indeed, suppose we had a particle which is situated at some node of the network at each instant of discrete time $t = 1, 2, \ldots$. We suppose that it moves according to the following rule. If at some time t it is located at a boundary point, it stays put. If it is located at an interior point, then it must move to a nearest neighbor. The probability of its moving to a nearest neighbor B is $P_{B,A}$.

Now the matrix P^N represents the transition probability after N steps, and it is clear what the limiting behavior of the matrix P^N will be like from the probability interpretation: Starting at any interior point, there will, after a sufficiently large number of steps (the number being independent of the point if large enough), be a non-zero probability of hitting the boundary, and thus a probability less than one of remaining in the interior; and therefore, as the number of steps increases, the probability of remaining in the interior goes to zero. Thus, in the limit, the matrix P^N tends to a matrix of the form

$$H = \begin{pmatrix} I & Q \\ 0 & 0 \end{pmatrix},$$

where I is the $k \times k$ identity matrix and Q is a matrix with k rows and $n - k$ columns, where $n - k$ is the number of interior nodes. The entry $Q(A, B)$ represents the probability that the particle starting at the interior point A will eventually end up at the particular boundary point B. (It must eventually end at *some* boundary point.) Now $P^N \to H$ and hence $P^{N+1} = P^N P \to H$ and thus $HP = H$. Therefore, if $\mathbf{v}$ is any potential, $\mathbf{v}H$ satisfies

$$(\mathbf{v}H)P = \mathbf{v}H$$

and hence

$$\mathbf{u} = \mathbf{v}H$$

is a solution of the Dirichlet problem, whose boundary values are those given by the first k entries in $\mathbf{v}$. If we choose $\mathbf{v}$ to be a vector whose first k components are given by ϕ, i.e., if we choose $\mathbf{v}$ of the form

$$\mathbf{v} = (\phi, \mathbf{w}) \quad \mathbf{w} \text{ any } n - k \text{ row vector},$$

then $\mathbf{v}H$ is a solution of the Dirichlet problem. From the form of the matrix H

we see that, at any interior point A, we have

$$\mathbf{u}(A) = (\mathbf{v}H)(A) = \sum_{B \text{ in boundary}} \phi(A)Q(A, B)$$

so that Q is indeed the Poisson kernel.

The form of H as the limit of P^N can be given an interesting geometrical interpretation. The entry $P^N(A_2, A_1)$ is the probability of getting from A_1 to A_2 in N steps. This probability is the sum of the probabilities of all N-step paths leading from A_1 to A_2. (If A_2 is a boundary node, then some of these paths may really be of smaller length – that is, the last r 'steps' in the path may consist of the particle standing still at A_2.) The probability of any path is simply the product of the $P_{B,A}$ over all the branches α traversed in the path. If A is an interior point and B is a boundary point, it is clear that as N increases, we are simply adding longer and longer paths in the computation of $P^N(B, A)$. Thus

$$Q(B, A) = \sum_{\text{all paths joining } A \text{ to } B} (\text{probability of path}).$$

13.9. Green's reciprocity theorem in electrostatics

As a slight generalization of capacitive networks, we may consider a system of charged conductors, each of which has a well-defined charge ρ and a well-defined potential ϕ. To describe the analogue of a branch for this sort of system requires the introduction of the dielectric displacement field and must be deferred until Chapter 16, where we consider electrostatics. The description in terms of charges and potentials, however, has much in common with capacitive networks.

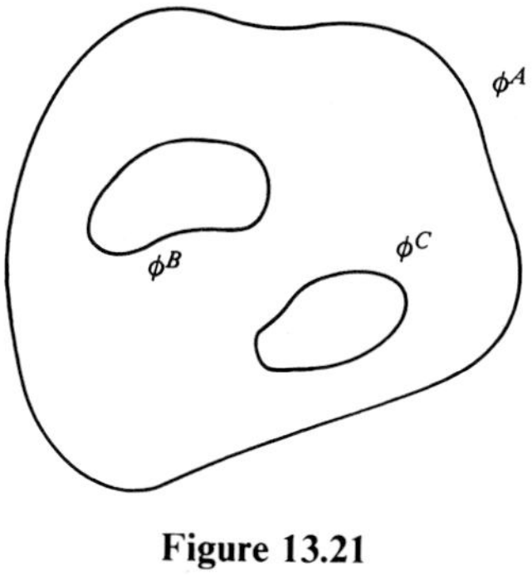

Figure 13.21

The total charges on each of the various conductors may be described in terms of a vector $\boldsymbol{\rho} = \begin{pmatrix} \rho_A \\ \rho_B \\ \vdots \end{pmatrix}$ in a space we may call C_0, while the potentials form a vector $\boldsymbol{\phi} = (\phi^A, \phi^B, \ldots)$ in its dual space C^0. As with capacitive networks, the evaluation of an element of C^0 on an element of C_0 gives the electrostatic energy of the system of conductors: to be precise, the energy is $\frac{1}{2}\int_\rho \boldsymbol{\phi} = \frac{1}{2}\sum \rho_A \phi^A$.

(In Chapter 16 we will see that this expression will be equal to a certain integral

$$\int (\mathrm{d}\phi, \mathrm{d}\phi) \mathrm{d}x \, \mathrm{d}y \, \mathrm{d}z$$

over the space between the conductors, where ϕ is the *electrostatic potential.* For now, we must content ourselves with a description of what happens at the conductors.)

The total conductor charges $\boldsymbol{\rho}$ may be expressed in terms of the potentials $\boldsymbol{\phi}$ by a Laplace operator Δ, so that

$$\boldsymbol{\rho} = -\Delta \boldsymbol{\phi}.$$

The operator Δ depends upon the shapes of the various conductors, on their distribution in space, and on fundamental constants of electrostatics. Except in certain cases involving parallel planes or concentric spheres or cylinders, calculation of Δ is extremely difficult. However, as we shall learn in Chapter 16, a Green's reciprocity theorem holds for systems of conductors just as it does for capacitive networks. To be specific, if $\boldsymbol{\rho}$ and $\boldsymbol{\phi}$ denote one possible distribution of total charges and potentials for a system of conductors, while $\boldsymbol{\rho}'$ and $\boldsymbol{\phi}'$ denote another possible distribution, then Green's reciprocity theorem states that

$$\int_{\rho'} \boldsymbol{\phi} = \int_{\rho} \boldsymbol{\phi}'.$$

It follows immediately from this theorem that Δ is a self-adjoint operator, represented by a symmetric matrix. To see this, let

$$\boldsymbol{\rho} = -\Delta \boldsymbol{\phi}, \quad \boldsymbol{\rho}' = -\Delta \boldsymbol{\phi}'.$$

Then

$$\int_{-\Delta\boldsymbol{\phi}'} \boldsymbol{\phi} = \int_{-\Delta\phi} \boldsymbol{\phi}'$$

i.e.,

$$(\Delta \boldsymbol{\phi}', \boldsymbol{\phi}) = (\boldsymbol{\phi}', \Delta \boldsymbol{\phi})$$

from which it follows that Δ is self-adjoint. In the physics literature, $-\Delta$ is usually called the *matrix of capacitance coefficients.*

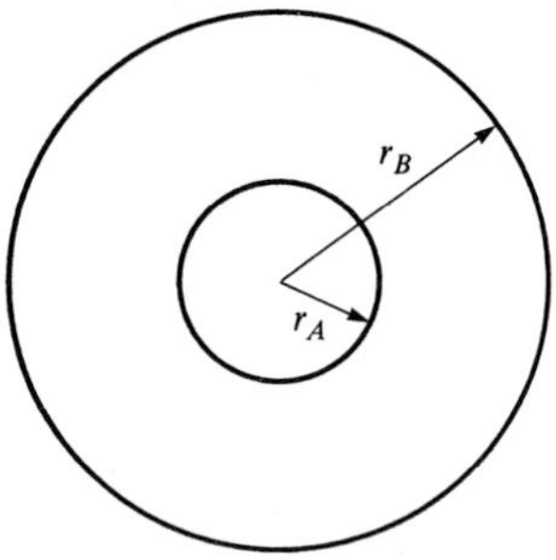

Figure 13.22

The inverse of the matrix $-\Delta$ is called the matrix of potential coefficients. In certain simple cases we can write this matrix down in terms of well-known formulas of elementary electrostatics. Consider, for example, a system of two concentric spheres, with radii r_A and r_B. For a sphere of radius r_0 bearing unit positive charge, the electric potential, expressed in Gaussian units, is

$$\phi(r) = \begin{cases} 1/r_0 & (r \leqslant r_0), \\ 1/r & (r > r_0). \end{cases}$$

(See, for example, Purcell, *Electricity and Magnetism*, section 1.11.) For an alternative treatment, see later in sections 16.1 and 16.2 where we derive this result from first principles.

If $\rho_A = 1$, $\rho_B = 0$, the potentials are therefore $\phi^A = 1/r_A$, $\phi^B = 1/r_B$, while, if $\rho_A = 0$, $\rho_B = 1$, both potentials are equal: $\phi^A = \phi^B = 1/r_B$. It follows that

$$-\Delta^{-1} = \begin{pmatrix} 1/r_A & 1/r_B \\ 1/r_B & 1/r_B \end{pmatrix}.$$

This matrix permits us to calculate the potential of the two spheres for an arbitrary charge distribution. Its inverse gives the Laplace operator:

$$-\Delta = \left(\frac{1}{r_B}\frac{1}{r_A} - \frac{1}{r_B^2}\right)^{-1} \begin{pmatrix} 1/r_B & -1/r_B \\ -1/r_B & 1/r_A \end{pmatrix}$$

or

$$-\Delta = \frac{r_A r_B}{r_B - r_A} \begin{pmatrix} 1 & -1 \\ -1 & r_B/r_A \end{pmatrix} = \frac{r_B}{r_B - r_A} \begin{pmatrix} r_A & -r_A \\ -r_A & r_B \end{pmatrix}.$$

This matrix determines the charges on the two spheres for specified potentials. For the case $\phi_A = 1$, $\phi_B = 0$ it gives

$$\begin{pmatrix} \rho_A \\ \rho_B \end{pmatrix} = \frac{r_A r_B}{r_B - r_A} \begin{pmatrix} 1 \\ -1 \end{pmatrix},$$

i.e., there are equal and opposite charges of magnitude $r_A r_B/(r_B - r_A)$ on the two spheres. This quantity $r_A r_B/(r_B - r_A)$ is called the capacitance of the pair of spheres.

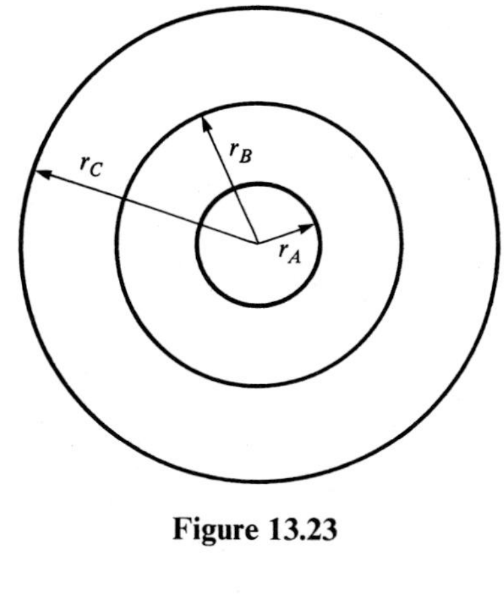

Figure 13.23

The matrix $-\Delta^{-1}$ is equally easy to write down for any number of concentric spheres. For example, with three spheres as shown in figure 13.23 it is

$$-\Delta=\begin{pmatrix}1/r_A & 1/r_B & 1/r_C\\ 1/r_B & 1/r_B & 1/r_C\\ 1/r_C & 1/r_C & 1/r_C\end{pmatrix}.$$

Let us now, as for capacitive networks, imagine that some of the conductors are *boundary conductors* whose potential may be established by connecting them to batteries, while others are *interior conductors* whose charge may be specified. In a Poisson equation problem, we are given $\phi'=0$ on all boundary conductors, while ρ' is specified on all interior conductors, and we are required to determine ϕ' for the interior conductors, ρ' for the boundary conductors. In a Dirichlet problem, on the other hand, we are given $\rho=0$ on interior conductors, while ϕ is specified on all boundary conductors, and we must determine ϕ for interior conductors, ρ for boundary conductors.

The two types of problems are related by Green's reciprocity theorem. In general

$$\sum_{\text{boundary}} \rho'_A\phi^A+\sum_{\text{interior}} \rho'_B\phi^B=\sum_{\text{boundary}} \rho_A\phi'^A+\sum_{\text{interior}} \rho_B\phi'^B.$$

But given that $\phi'=0$ on the boundary (Poisson) and $\rho=0$ in the interior (Dirichlet), we get

$$\sum_{\text{interior}} \rho'_B\phi^B=\sum_{\text{boundary}} \rho'_A\phi^A.$$

Suppose we denote by $Q(C,A)$ the charge ρ'_A on boundary conductor A for the Poisson equation problem in which unit charge is placed on interior conductor C, while the charge is zero for all other interior conductors. Then only one term survives on the left-hand side, and we have

$$\phi^C=-\sum Q(C,A)\phi^A,\quad \text{where } Q(C,A)=-\Delta G(C,A)$$

This is the Green's function solution to the Dirichlet problem.

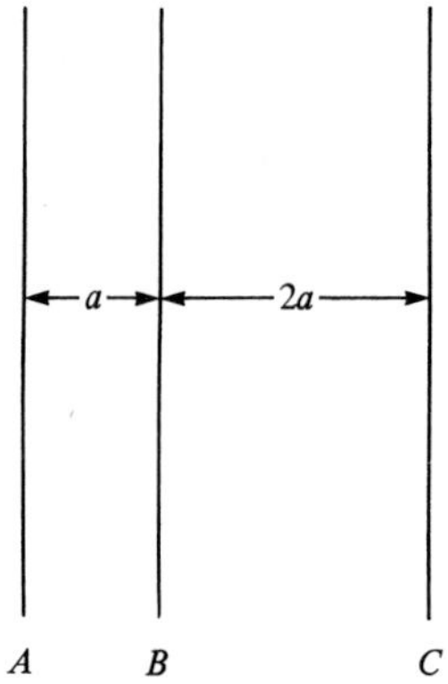

Figure 13.24

In practice, it is frequently easy, for systems of conductors, to solve Dirichlet's problem, by inspection, then to use Green's reciprocity theorem to solve Poisson's equation. Consider, for example, the system of three large parallel conducting planes shown in figure 13.24. We regard A and C as boundary conductors, B as an interior conductor. Since potential is a linear function of position between the planes for this geometry, it is apparent that

$$\phi^B = \phi^A + \tfrac{1}{3}(\phi^C - \phi^A)$$

or

$$\phi^B = \tfrac{2}{3}\phi^A + \tfrac{1}{3}\phi^C.$$

This is a solution to Dirichlet's problem.

Now consider Poisson's equation, with charge ρ'_B on the middle plane, $\phi'^A = \phi'^C = 0$. By the reciprocity theorem,

$$\rho'_B\phi^B = -\rho'_A\phi^A - \rho'_C\phi^C.$$

But $\phi^B = \frac{2}{3}\phi^A + \frac{1}{3}\phi^C$ for an arbitrary Dirichlet problem, so

$$\rho'_B \cdot \tfrac{2}{3}\phi^A + \rho'_B \cdot \tfrac{1}{3}\phi^C = -\rho'_A\phi^A - \rho'_C\phi^C.$$

It follows that

$$\rho'_A = -\tfrac{2}{3}\rho'_B, \quad \rho'_C = -\tfrac{1}{3}\rho'_B$$

so that we have learned how the induced charge is distributed between the two boundary planes.

More generally, Green's reciprocity theorem may be applied to problems involving both conductors and specified distributions of charge in the region bounded by these conductors. For example, if a charge $+q$ is placed near an infinitely large conducting plane, the electric potential everywhere on the same side of the plane of the charge may be determined by the *method of images*,

$$\phi(P) = \frac{q}{r} - \frac{q}{r'},$$

see section 16.1. From this potential function it is possible to determine the distribution of negative charge over the conducting plane. This in turn makes possible

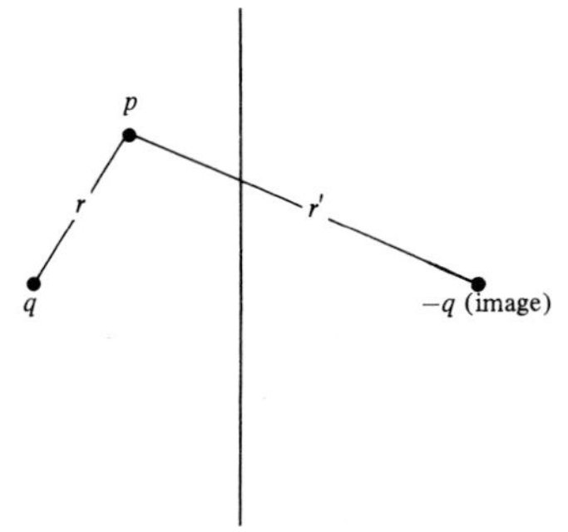

Figure 13.25

the solution of the Dirichlet problem in which the potential is specified everywhere on one plane and it is required to determine the potential throughout all space. In a similar manner, one can construct the Green's function for the Poisson equation problem of a point charge inside a grounded conducting sphere and use it so solve the Dirichlet problem in which the potential is specified on the surface of a sphere, and it is required to determine the potential within the charge-free region bounded by the sphere.

Summary

A Orthogonal projection

You should be able to state the properties of the self-adjoint projection operator π for resistive networks and to use π to solve network problems.

You should be able to construct π by using the Gram–Schmidt process or by Kirchhoff's method.

B Capacitive networks

You should be able to solve capacitive network problems by using Maxwell's methods or by inverting the Laplace operator.

You should be able to solve Poisson's equation or the Dirichlet problem for a capacitive network by means of orthogonal projection and to describe the associated decomposition of C^1 into orthogonal subspaces.

C Green's functions

You should know how to construct the Green's function for a capacitive network and to use it in solving Poisson's equation or the Dirichlet problem.

Exercises

13.1. Consider the network shown in figure 13.26.

(a) Write down a basis for Z_1. Explain the meaning of the statement 'π represents a projection from C_1 onto the subspace Z_1'. By carrying out

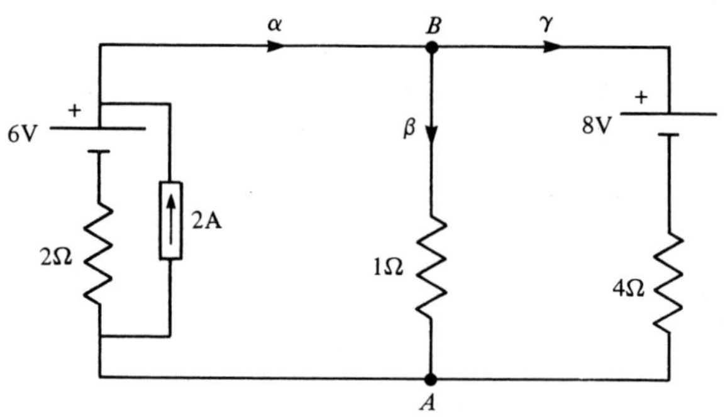

Figure 13.26

appropriate computations verify that the matrix

$$\pi = \frac{1}{7}\begin{pmatrix} 5 & 2 & 2 \\ 4 & 3 & -4 \\ 1 & -1 & 6 \end{pmatrix}$$

represents a projection from C_1 onto Z_1.

(b) Use π to determine the branch currents in the above network. Make a sketch which shows the image and kernel of π and indicates how it solves the problem.

(c) What linear transformation is represented by π^{T}, the *transpose* of π? Name, and characterize in physical terms, the space which is the kernel of this transformation. Support your answer either by invoking a general theorem or by citing appropriate properties of the matrix π^{T}.

13.2.(a) Construct the 2×2 matrix that represents $\pi: C_1 \to Z$, for the network shown in figure 13.27.

(b) Construct a sketch showing the image and kernel of π. On the sketch display the vectors $\mathbf{K}$ and $\mathbf{I} = \pi\mathbf{K}$.

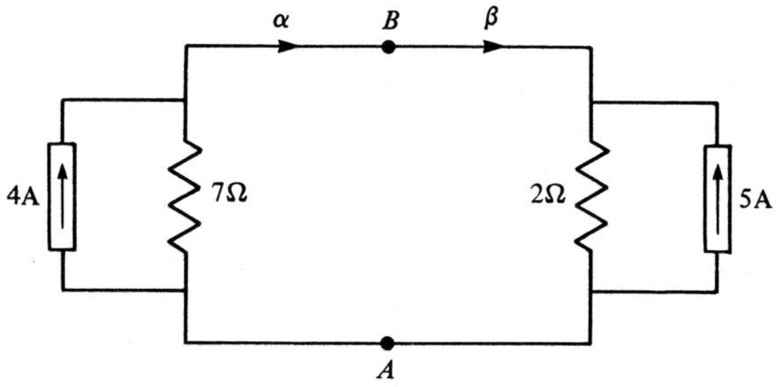

Figure 13.27

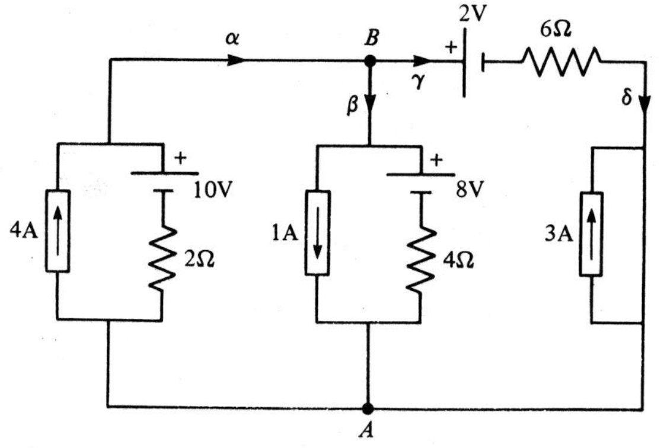

Figure 13.28

13.3. Consider the network in figure 13.28.

(a) Using the Gram–Schmidt process, convert the basis

$$\mathbf{M}_1 = \begin{pmatrix} 1 \\ 1 \\ 0 \\ 0 \end{pmatrix}, \quad \mathbf{M}_2 = \begin{pmatrix} 0 \\ -1 \\ 1 \\ 1 \end{pmatrix}$$

to an orthonormal basis $\{\mathbf{E}_1, \mathbf{E}_2\}$ for Z_1. By applying $\pi\mathbf{I} = (\mathbf{E}_1, \mathbf{I})_Z\mathbf{E}_1 + (\mathbf{E}_2, \mathbf{I})_Z\mathbf{E}_2$ in turn to each of the four basis vectors of C_1, construct the matrix representing π, and use it to solve the above network.

(b) Verify the formula $\pi = \sigma(sZ\sigma)^{-1}sZ$ for the above network.

(c) Construct the matrix $Z(1 - \pi)$, and show that its image is the space B^1.

$$\text{Answer:} \quad \pi = \frac{1}{50}\begin{pmatrix} 22 & 28 & 24 & 4 \\ 14 & 36 & -12 & -2 \\ 8 & -8 & 36 & 6 \\ 8 & -8 & 36 & 6 \end{pmatrix}.$$

13.4.(a) For the network of Exercise 13.3, construct a matrix to represent

$$\tau: C^1 \to C^1$$

which projects C^1 onto B^1, orthogonally with respect to the scalar product defined by Z^{-1}, i.e., $(\mathbf{V}, \mathbf{V}')_{Z^{-1}} = \int_{Z^{-1}\mathbf{V}'}\mathbf{V}$. To construct a basis for B^1, choose the potential to be 0 at A (ground) and -1 in turn at B and at C.

(b) Show that $\mathbf{V} = \tau(\mathbf{W} - Z\mathbf{K})$, and use this result to solve the network of Exercise 13.3.

13.5.(a) Find an explicit expression for the matrix τ of Exercise 13.4 in terms of $[\mathrm{d}]$, Z^{-1}, and $[\partial]$. Use it to show in general that $\tau^2 = \tau$, that the image of τ is B^1, and that τ is self-adjoint with respect to the scalar product defined by Z^{-1}

(b) Show that for any network, the matrix

$$Z^{-1}[\mathrm{d}]([\partial]Z^{-1}[\mathrm{d}])^{-1}[\partial] + \sigma(sZ\sigma)^{-1}sZ$$

which represents a linear transformation from C_1 to C_1, is the identity matrix.

13.6. Apply Kirchhoff's method to Exercise 13.3, as follows:

(a) There are five different maximal trees. Find the projection operator ρ_T(a 4×4 matrix) for each tree. Note that each pair of branches except $\{\alpha, \beta\}$ defines a tree. (Watch the signs – all diagonal entries in ρ_T must be $\geqslant 0$.)

(b) Compute Q_T for each tree and find $R = \sum Q_T$. (We get $R = 50$, the same as $\mathrm{Det}(sZ\sigma)$. Is that true in general?)

(c) Use Kirchhoff's formula

$$\pi = \frac{1}{R}\sum_T Q_T\rho_T$$

to obtain the projection operator π of the Weyl method.

13.7. Invent a variation of Kirchhoff's method which constructs the self-adjoint projection operator $\tau: C^1 \to C^1$, with the property that $\mathbf{V} = \tau(\mathbf{W} - Z\mathbf{K})$, as a weighted average of projections from C^1 onto Z^1 which are not self-adjoint. One way to proceed is to start from Kirchhoff's method in C_1 and multiply both sides by Z.

13.8. Prove the formula $\pi = \sigma(sZ\sigma)^{-1}sZ$ for the Weyl projection operator by verifying that π has all the required properties:

(a) π is a projection: $\pi^2 = \pi$.

(b) The image of π is Z_1.

(c) π is self-adjoint relative to $(\ ,\ )_Z$; i.e., $(\pi\mathbf{I}, \mathbf{I}')_Z = (\mathbf{I}, \pi\mathbf{I}')_Z$.

13.9.(a) Construct the matrix π which represents the orthogonal projection of C_1 onto the subspace Z_1 for the network shown in figure 13.29. Use this

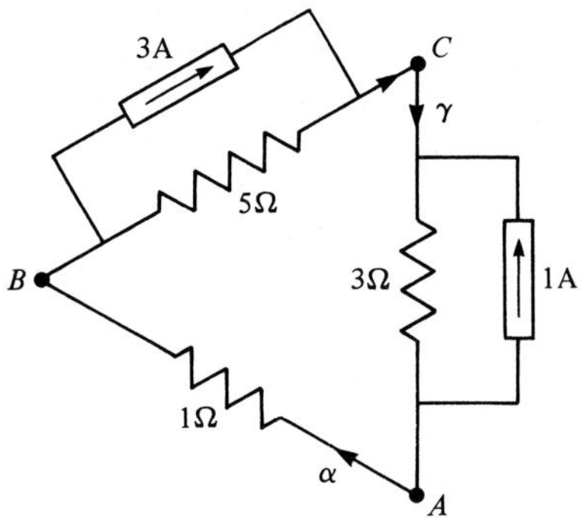

Figure 13.29

projection operator to calculate the branch currents **I** in the network. (They are all integers.)

(b) Construct a basis for the kernel of π for the network, and use it to show explicitly that the kernel of π is orthogonal to its image.

13.10. As an illustration of Green's reciprocity theorem, consider the network shown in figure 13.30.

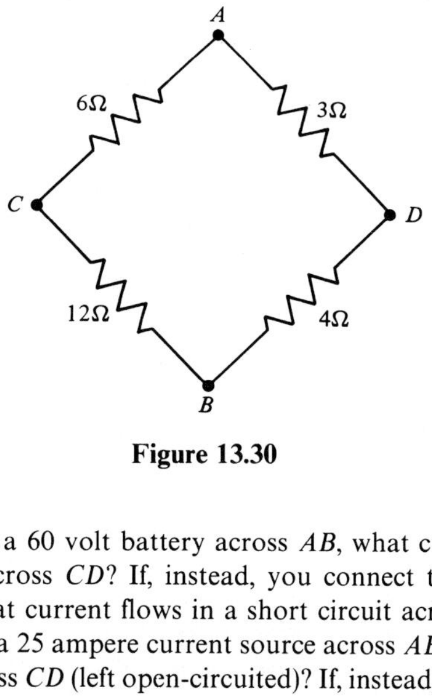

Figure 13.30

(a) If you connect a 60 volt battery across AB, what current flows in a short circuit across CD? If, instead, you connect the same battery across CD, what current flows in a short circuit across AB?

(b) If you connect a 25 ampere current source across AB, what voltage is developed across CD (left open-circuited)? If, instead, you connect the same current source across CD, what voltage is developed across AB?

13.11.(a) For the network shown in figure 13.31, construct the symmetric matrix $Z\pi$. Use it to determine the branch voltages **V** if an 8 ampere current source is inserted in branch α. Also determine **V** for the case of an 8 ampere current source in branch γ.

(b) For the same network, construct the symmetric matrix πZ^{-1}. Use it to determine the branch currents **I** for the case of a 16 volt battery inserted in branch α or as in branch γ.

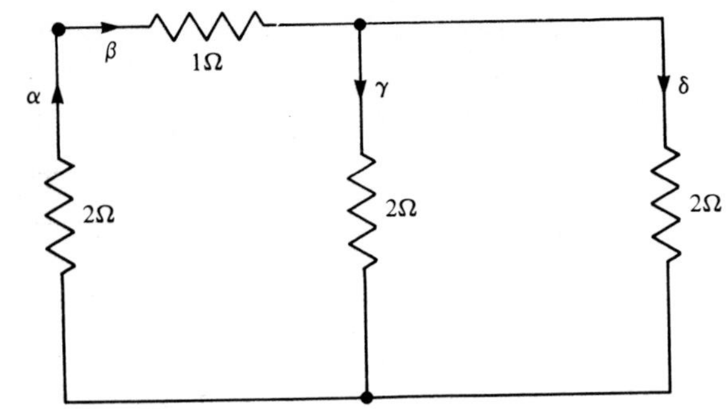

Figure 13.31

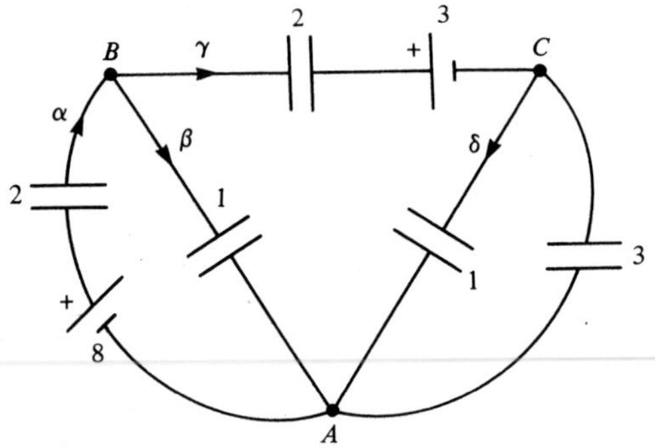

Figure 13.32

13.12. Determine the charge on each capacitor in the network in figure 13.32, by using both the mesh-charge method and the node-potential method. Assume that $[\partial]\mathbf{Q} = 0$, i.e., that all the capacitors were uncharged before the batteries were connected. (If you insist on units, V is in volts, C in microfarads, Q in microcoulombs.)

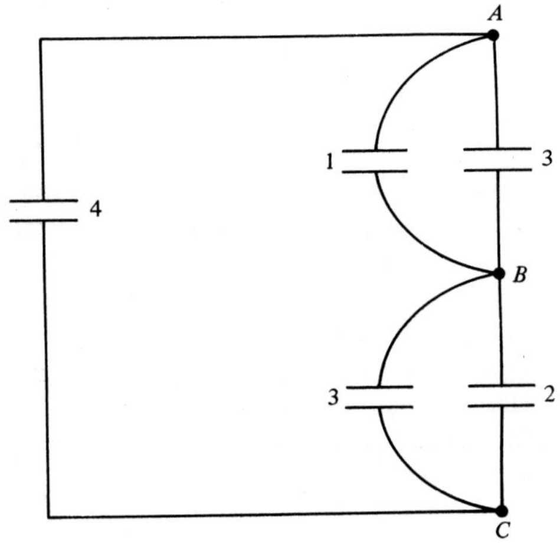

Figure 13.33

13.13. In the network shown in figure 13.33, $\rho_B = 13$ and $\rho_C = -1$ (so $\rho_A = -12$). Solve Poisson's equation $\Delta\Phi = -\rho$ for this network to determine Φ^B and Φ^C. Calculate the branch charges $\mathbf{Q}$ and check that $\partial\mathbf{Q} = -\begin{pmatrix} -12 \\ 13 \\ -1 \end{pmatrix}$.

13.14. Write down the matrices d and ∂ for the capacitive network shown in figure 13.34. Given that $\rho_B = 10$, $\rho_c = -2$, solve Poisson's equation to determine the node potentials Φ^B and Φ^C.

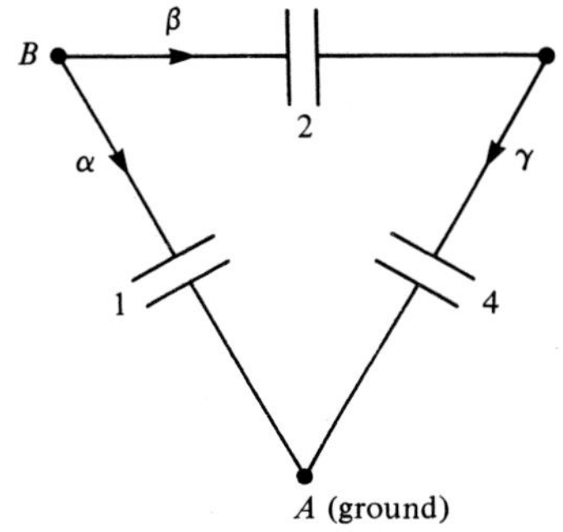

Figure 13.34

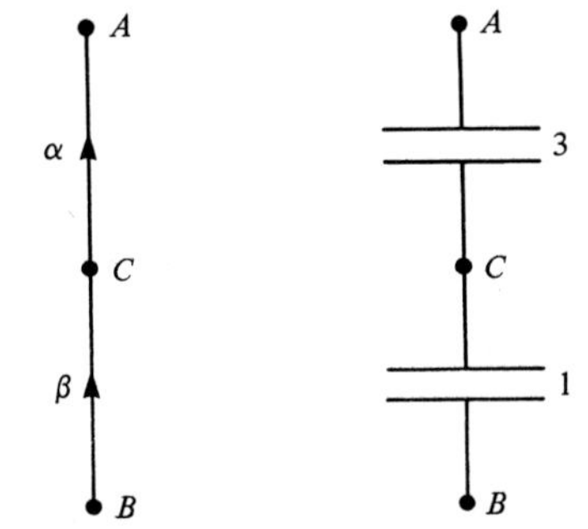

Figure 13.35

13.15. Consider the network of capacitors shown in figure 13.35. Nodes A and B are boundary nodes; node C is an interior node.
(a) Construct the 2×2 matrix π which is the orthogonal projection (relative to the scalar product defined by C) of C' onto $\mathrm{d}C^0_{\mathrm{int}}$.
(b) The vector $\mathbf{W} = \begin{pmatrix} 0 \\ -4 \end{pmatrix}$ corresponds to a situation in which there are four units of charge at node C. Use π to determine how this charge will distribute itself if nodes A and B are both grounded.
(c) Suppose the potential at A is $\Phi^A = 0$, the potential at B is $\Phi^B = 4$, and there is zero charge at node C. Use π to determine the branch voltages V^α and V^β.

13.16. Consider the network of capacitors shown in figure 13.36. A and D are boundary nodes; B and C are interior nodes.

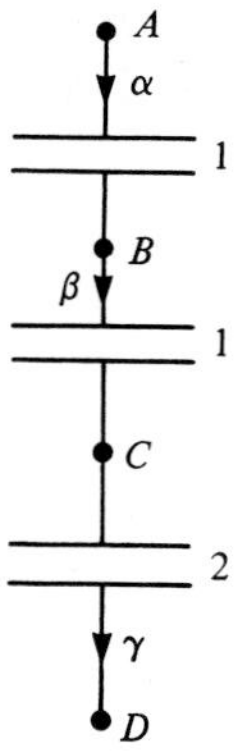

Figure 13.36

(a) Write down the matrices ∂ and d for this network.

(b) Let $\phi_1 = \begin{pmatrix} 0 \\ 0 \\ 1 \\ 0 \end{pmatrix}$ and $\phi_2 = \begin{pmatrix} 0 \\ 3 \\ 1 \\ 0 \end{pmatrix}$ be vectors in C^0. Calculate $\mathrm{d}\phi_1$ and $\mathrm{d}\phi_2$, and verify that they are orthogonal with respect to the scalar product defined by the capacitance matrix C.

(c) Let π represent the orthogonal projection from C^1 onto the subspace spanned by $\mathrm{d}\phi_1$ and $\mathrm{d}\phi_2$. Evaluate $\pi\begin{pmatrix} 1 \\ 0 \\ 0 \end{pmatrix}$.

(d) Suppose that there is one unit of charge at node B, no charge at node C, and nodes A and D are grounded. Use orthogonal projection to calculate the potential at B and at C. Find the charge at A and D.

(e) Suppose that $\Phi^A = 5$, $\Phi^D = 0$, and there is no charge at nodes B and C. Use the method of orthogonal projection to determine the branch voltages $\mathbf{V}$.

(f) For the problem posed in (e) show how to determine Φ^B by using the Green's function of part (d).

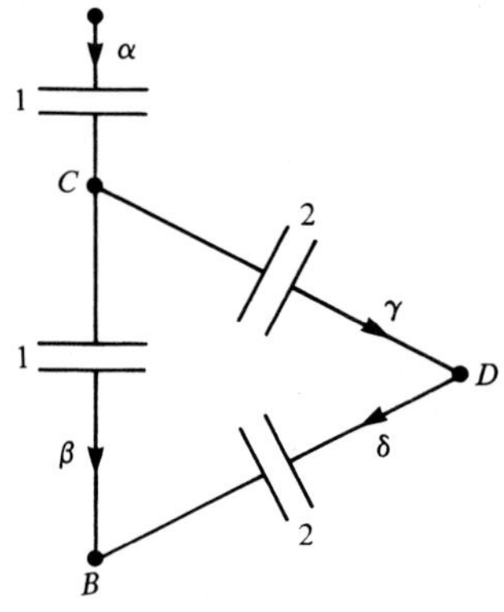

Figure 13.37

13.17. Consider the network of capacitors shown in figure 13.37. Nodes A and B are boundary nodes; nodes C and D are interior nodes.

(a) Consider the vectors

$$\psi_1 = \begin{pmatrix} 0 \\ 0 \\ 0 \\ 1 \end{pmatrix} \quad \text{and} \quad \psi_2 = \begin{pmatrix} 0 \\ 0 \\ 2 \\ 1 \end{pmatrix},$$

both elements of the space C^0. Name, and characterize in physical terms, the subspace which they span. Determine $d\psi_1$ and $d\psi_2$, and verify that they form an orthogonal basis for a subspace of C^1, relative to the scalar product defined by C.

(b) Let π denote the orthogonal projection of C^1 onto the subspace for which you have just found a basis. Characterize in physical terms the vectors which lie in the *kernel* of π. Determine $\pi\begin{pmatrix} 0 \\ 0 \\ 0 \\ 1 \end{pmatrix}$.

(c) Suppose there is 1 unit of positive charge at D, none at C, and nodes A and B are grounded. Using the result of part (b), find all the branch voltages and charges.

(d) Suppose there are 2 units of charge at D, no charge at C, $\Phi^A = 4$ and $\Phi^B = 1$. Using the Green's function which you have just constructed, determine the potential at D.

(Hint: The problem is a superposition of Poisson and Dirichlet.)

13.18. Consider the network in figure 13.38. Regard A, B, and C as boundary nodes, D and E as interior nodes.

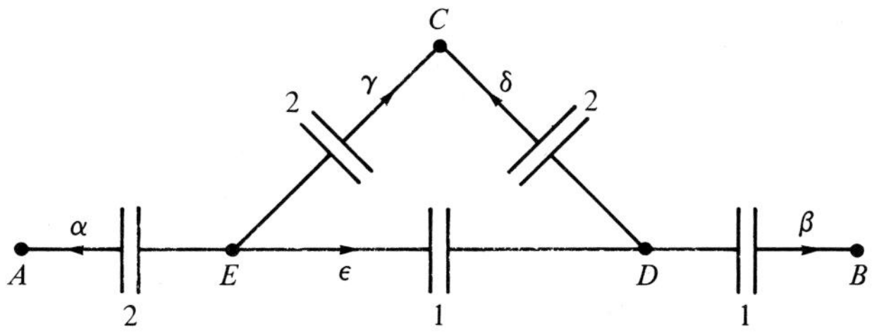

Figure 13.38

(a) Find a basis for the subspace $C^0_{\text{int}} \subset C^0$ and for the subspace $dC^0_{\text{int}} \subset C^1$. Verify that the matrix

$$\pi = \frac{1}{19}\begin{pmatrix} 8 & 1 & 8 & 2 & 3 \\ 2 & 5 & 2 & 10 & -4 \\ 8 & 1 & 8 & 2 & 3 \\ 2 & 5 & 2 & 10 & -4 \\ 6 & -4 & 6 & -8 & 7 \end{pmatrix}$$

projects C^1 onto this subspace.

(b) Construct $1 - \pi$, and verify that its image is orthogonal to the image of π.

(c) Verify that H^1 (the image of Z_1 under the action of C^{-1}) is in the kernel of π.

(d) Construct the matrix $(1-\pi)\mathrm{d}$, which represents a transformation from C^0 to C^1. Verify explicitly that if $\hat{\Phi}$ is any basis vector in C^0_{bound} then $V = -(1-\pi)\mathrm{d}\hat{\Phi}$ describes a set of branch voltages for which $\partial C\mathbf{V} = 0$ at interior nodes D and E.

13.19. For the network of exercise 13.18 use the matrix $(1-\pi)\mathrm{d}$ to solve Dirichlet's problem for potentials $\Phi^A = 3$, $\Phi^B = -5$, $\Phi^C = 0$, with $\rho_D = \rho_E = 0$. Determine Φ^D, Φ^E, ρ_A, ρ_B and ρ_C.

13.20. In the preceding exercise, solve the Poisson equation for the case where

$$\rho_E = 1, \quad \rho_D = 0, \quad \Phi^A = \Phi^B = \Phi^C = 0,$$

as follows:

(a) Choose any $\hat{Q}$ compatible with the conditions $\rho_D = 0$, $\rho_E = 1$. Now form $\hat{\mathbf{V}} = C^{-1}\hat{Q}$, the associated branch voltages.

(b) Calculate $\mathbf{V} = \pi\hat{\mathbf{V}}$, and determine a potential such that $\Phi = 0$ on the boundary and $-\mathrm{d}\Phi = \mathbf{V}$.

(c) Repeat steps (a) and (b) with a different $\hat{Q}$. You should get the same answer.

(d) Solve Poisson's equation for the case $\rho_D = 1$, $\rho_E = 0$. You have now constructed the entire Green's function for this network. Use it to solve Poisson's equation for $\rho_D = 10$, $\rho_E = 7$.

(e) Using the Green's function, construct the Poisson kernel $\Delta G(A, B)$. This is a matrix with three columns and two rows which represents a mapping of

$$\begin{pmatrix} \Phi^A \\ \Phi^B \\ \Phi^C \end{pmatrix} \quad \text{into} \quad \begin{pmatrix} \Phi^D \\ \Phi^E \end{pmatrix}.$$

Its transpose is the matrix which maps

$$\begin{pmatrix} \rho_D \\ \rho_E \end{pmatrix} \quad \text{into} \quad -\begin{pmatrix} \rho_A \\ \rho_B \\ \rho_C \end{pmatrix}.$$

(f) Check that the Poisson kernel gives the correct solution to the Dirichlet problem posed in Exercise 13.18(c).

13.21. A capacitor consists of three long coaxial cylinders, each of length l, whose radii are $r_A = a$, $r_B = 2a$, $r_C = 8a$.

(a) Solve the Dirichlet problem in which cylinder B is uncharged and the potential difference between cylinders A and C is Φ^0.

(b) Using Green's reciprocity theorem, solve the Poisson equation problem in which cylinders A and C are grounded and charge Q is placed on cylinder B.

13.22. Three concentric conducting spheres have radii $r_A = 2$, $r_B = 3$, $r_C = 4$ respectively.

(a) Write down the matrix $-\Delta^{-1}$ for this system, and invert it to construct the Laplace operator $-\Delta$.

(b) Suppose the potentials are $\phi^A = 0$, $\phi^B = 1$, $\phi^C = 0$. Determine the charge on each sphere.

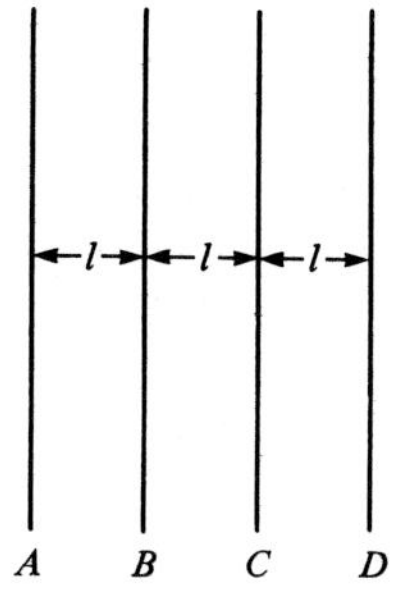

Figure 13.39

13.23. Four large parallel conducting planes, each of area A_0, are separated by a distance l.

(a) Solve by inspection the Dirichlet problem in which ϕ^A and ϕ^D are specified, while planes B and C are uncharged.

(b) Using Green's reciprocity theorem, solve the Poisson equation problem in which planes A and D are grounded while charges Q_B and Q_C are placed on the interior planes. Determine the potentials ϕ^B and ϕ^C and the charges Q_A and Q_D.

14

Higher-dimensional complexes

In Chapter 14 we conclude our introduction to algebraic topology by sketching how the one-dimensional results of Chapters 12 and 13 generalize to higher dimensions

14.1. Complexes and homology

We now interrupt our study of electrical circuits in order to study some extensions of the mathematical methods of the preceding two sections. We have defined a one-dimensional complex as a system made of zero-dimensional objects called nodes and one-dimensional objects called branches, with the branches joined together at the nodes. Of particular interest in the study of such a complex are the notion of connectivity, which reflects itself in the space H_0, and the number of independent meshes, which equals the dimension of the space H_1. We now want to generalize to higher dimensions by considering complexes which include elements of two, three, or more dimensions. A two-dimensional complex or *two-complex* will include, in addition to nodes and branches, two-dimensional elements (bits of surface), which we shall call *two-cells* or *faces*. As indicated in figure 14.1, these faces can attach to

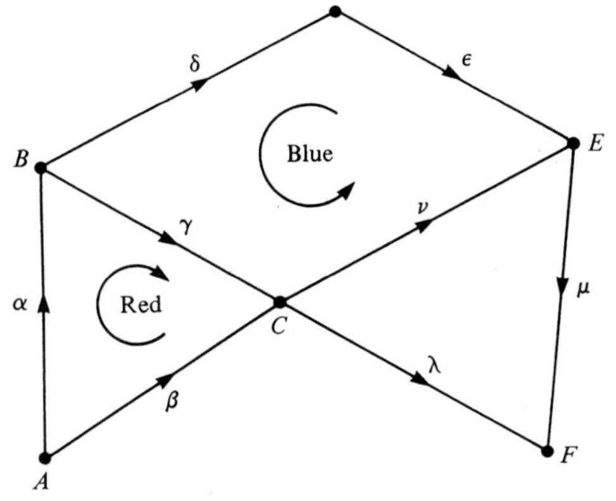

Figure 14.1

one another along the branches. The boundary of a face will consist of a collection of branches, each with a plus or minus sign. Similarly, a three-dimensional complex will include three-dimensional elements (bits of volume), called three-cells, which can attach to one another along faces. The boundary of a three-cell will consist of a collection of faces, again each with a plus or minus sign.

More generally, we define an *n-complex* to consist of $n+1$ (finite) sets $S_0, S_1, S_2, \ldots, S_n$, related by some conditions which we shall describe below after we have introduced some notation. The set S_0 will consist of the zero-dimensional elements, or nodes, of the complex, the set S_1 of the one-dimensional elements, or branches, the set S_2 of the two-dimensional elements, or faces, and so on. Since we do not have enough alphabets handy to assign a different alphabet to each dimension, we shall sometimes denote the elements of S_0 by $\mathbf{0}_1, \mathbf{0}_2, \mathbf{0}_3, \ldots$, the elements of S_1 by $\mathbf{1}_1, \mathbf{1}_2, \mathbf{1}_3, \ldots$, the elements of S_2 by $\mathbf{2}_1, \mathbf{2}_2, \mathbf{2}_3, \ldots$, and so on. Less formally, we shall continue in examples to use Latin letters to denote elements of S_0, Greek letters to denote elements of S_1, and names (e.g., Red, Bill, Canada) to denote elements of S_2, S_3, etc.

For each set S_k we now introduce the vector space of k-chains, C_k, consisting of vectors whose components are indexed by the elements of S_k. The spaces C_0 and C_1 are already familiar from our study of electric networks. In general, the dimension of the space C_k equals the number of elements of S_k. As before, we shall identify an element of S_k with the vector which is one in the position corresponding to that element, zero in all other positions. Thus, for example, in the complex of figure 14.1, we have vectors in C_0: $\mathbf{0}_1 = A = (1,0,0,\ldots)^T$, $\mathbf{0}_2 = B = (0,1,0,\ldots)^T$, etc.; vectors in C_1: $\mathbf{1}_1 = \alpha = (1,0,0,\ldots)^T$, $\mathbf{1}_2 = \beta = (0,1,0,\ldots)^T$, etc.: vectors in C_2: $\mathbf{2}_1 = \text{Red} = (1,0)^T$ and $\mathbf{2}_2 = \text{Blue} = (0,1)^T$.

To complete the description of a complex, we must specify the boundary operator, actually a sequence of boundary operators, which we shall denote all by the same symbol: $\partial: C_k \to C_{k-1}$ for each k. We think of ∂ as assigning to each element of S_k some combination of elements of S_{k-1} (with appropriate signs) which constitute its boundary: this defines ∂ on each basis element of C_k and hence determines a linear map of C_k into C_{k-1}. (For $k = 0$ we take $\partial = 0$, since there are no elements of negative dimension.)

For the two-complex of figure 14.1 we have

$$\partial\alpha = B - A, \partial\beta = C - A, \quad \text{etc.}$$

This determines the boundary map from C_1 to C_0, which we could represent by an 8×6 matrix. To write down this operator, it was essential to have specified the orientation of each branch by an arrow on the diagram. Similarly, in order to specify the boundary operator from C_2 to C_1, we must have specified an orientation for each face, either clockwise or counterclockwise as shown on the diagram. The boundary of a face is the closed path around its perimeter, traversed in the direction determined by the orientation of the face. In figure 14.1, for example, ∂ (Red) $= \alpha + \gamma - \beta$ while ∂ (Blue) $= \gamma + \nu - \varepsilon - \delta$. As a map from C_2 to C_1, the boundary operator is represented by the matrix

$$\partial = \begin{pmatrix} 1 & 0 \\ -1 & 0 \\ 1 & 1 \\ 0 & -1 \\ 0 & -1 \\ 0 & 0 \\ 0 & 0 \\ 0 & 1 \end{pmatrix}.$$

with respect to the bases {Red, Blue} of C_2 and $\{\alpha, \beta, \gamma, \varepsilon, \lambda, v\}$ of C_1.

In our study of one-complexes we found it worthwhile to consider the subspaces that were the kernel and image of the operator ∂. We now generalize this idea by defining the space of k-cycles, Z_k, to be subspace of C_k consisting of elements with zero boundary, i.e., the *kernel* of $\partial: C_k \to C_{k-1}$. In the complex of figure 14.1, for example:

> Z_2 is the zero subspace, because no non-trivial linear combination of Red and Blue has zero boundary;
>
> Z_1 is the three-dimensional subspace spanned by the meshes, for which a basis is
>
> $$\mathbf{M}_1 = \alpha + \gamma - \beta, \mathbf{M}_2 = \gamma + v - \varepsilon - \delta, \mathbf{M}_3 = \lambda - \mu - v;$$
>
> Z_0 is the entire space C_0, since by definition $\partial c = 0$ for any zero-chain c.

Similarly, we define the space B_k of k-boundaries to be the subspace of C_k consisting of elements which are boundaries of elements of C_{k+1}; i.e. the *image* of $\partial: C_{k+1} \to C_k$. For the complex of figure 14.1;

> B_2 is the zero subspace, because there are no three-dimensional elements in a two-complex (similarly, in the one-complexes which we have considered, the subspace B_1 was always the zero subspace);
>
> B_1 is the two-dimensional space spanned by ∂ (Red) $= \alpha + \gamma - \beta$ and ∂ (Blue) $= \gamma + v - \varepsilon - \delta$;
>
> B_0 is a five-dimensional subspace, exactly as if the complex were a one-complex rather than a two-complex.

In this example, each space B_k is a subspace of the corresponding Z_k; i.e., *every boundary is a cycle.* A moment's thought will convince you that this property must hold for any two-complex defined by a diagram like figure 14.1. Since B_2 is empty, it is trivially a subspace of Z_2. Since the boundary of any polygon is a closed path, B_1 is a subspace of Z_1. Finally, since Z_0 is the entire space C_0, B_0 is trivially a subspace of Z_0. The condition $B_k \subset Z_k$ can be stated somewhat differently. Since B_k is the image of $\partial: C_{k+1} \to C_k$ and Z_k is the kernel of $\partial: C_k \to C_{k-1}$, $B_k \subset Z_k$ means that the composition $\partial \circ \partial: C_{k+1} \to C_{k-1}$ is zero. To put it in words: the boundary of a boundary is zero. In the examples which follow, it will become clearer why this is so for any complex.

Homology

Because B_k is a subspace of Z_k, we can form the quotient space $H_k = Z_k/B_k$. H_k, called the kth *homology space* of the complex, will turn out to be of more fundamental significance than either Z_k or B_k individually. While Z_k and B_k relate to properties of a specific complex, H_k, as we shall see, relates to properties of the space built up from the elements of the complex.

We have already met the spaces H_0 and H_1 in our study of networks. Recall that the space H_0 has a dimension equal to the number of connected components, and that its dual space H^0 can be interpreted as the space of potentials which are constant on each connected component. Since the space B_1 of boundaries is empty for a one-complex (there are no two-dimensional elements), $H_1 = Z_1/B_1 = Z_1$. Thus,

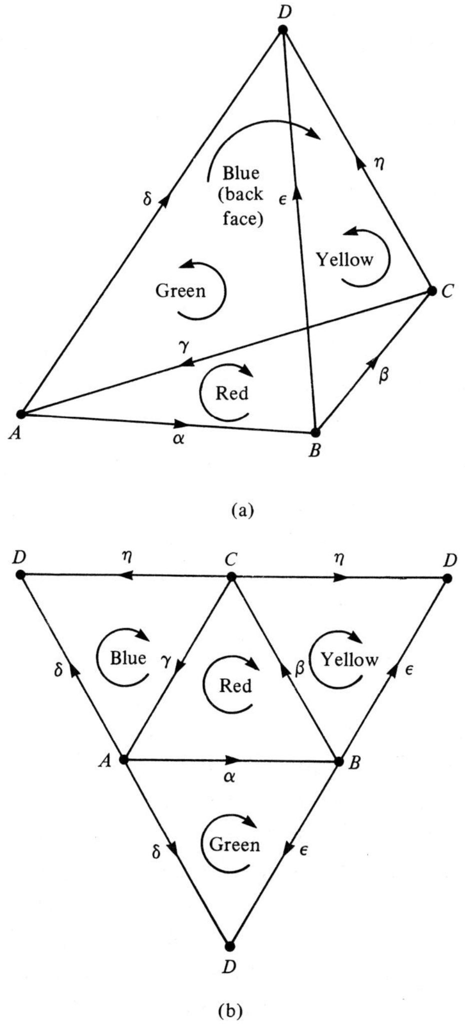

Figure 14.2

for any one-complex, H_1 is just the space Z_1 of cycles which played such an important role in Maxwell's mesh-current method.

Let us now consider the spaces H_k for the two-complex of figure 14.1. Since there are no two-cycles, Z_2 is $\{0\}$, and so is H_2. Since Z_1 is three-dimensional, while B_1 is only two-dimensional, the quotient space $H_1 = Z_1/B_1$ is one-dimensional. Looking at the figure, we see why: the path $\lambda - \mu - \nu$ is not the boundary of any element of C_2. If the triangle CEF were added to the complex as a third basis element of C_2, then $\lambda - \mu - \nu$ would become a boundary, and H_1 would have dimension zero. Finally, because the complex is connected, the space H_0 is one-dimensional.

As an example of a three-complex, consider a solid tetrahedron with all its faces, edges (branches), and vertices (nodes), as shown in figure 14.2(a). The surface of the tetrahedron, unfolded, is shown in figure 14.2(b). There are four vertices, A, B, C, D, and six edges, $\alpha, \beta, \gamma, \delta, \varepsilon, \eta$. There are four faces, labelled Red, Yellow, Green, and Blue in the figure. Each face has been assigned an orientation, described by the circulating arrows in the figure. These orientations have all been chosen so that the arrows circulate counterclockwise when the tetrahedron is viewed from the outside (see Yellow and Green in figure 14.2(a)) and clockwise when the tetrahedron is viewed from the inside (see figure 14.2(b)).

From the figure it is easy to read off the boundary operator from C_2 to C_1, for example:

$$\partial\,(\text{Red}) = -\alpha - \beta - \gamma,$$
$$\partial\,(\text{Green}) = \alpha + \varepsilon - \delta.$$

Since the boundary of each face is a cycle, each such boundary itself has zero boundary, for example:

$$\partial[\partial(\text{Green})] = \partial(\alpha + \varepsilon - \delta) = (B - A) + (D - B) - (D - A) = 0.$$

Finally, there is one three-dimensional element in the complex, the solid tetrahedron itself. In order to write down the boundary operator from C_3 to C_2, we must assign this element an orientation. It is no longer an easy matter, though, to depict this orientation on a diagram: we must do it more abstractly. We will choose the so-called *right-handed* orientation, which means that, when the element is viewed from outside, all of the faces which comprise its boundary appear with a counterclockwise orientation. The reason for the term right-handed is that if you were to place your *right* hand on the surface of the tetrahedron, with the thumb pointing outward, then the fingers of the right hand would circulate in the sense appropriate to a face which appears with a *plus* signs in the boundary. Of course, this is stated much more succinctly in algebraic terms by writing down the boundary operator from C_3 to C_2:

$$\partial(\text{Tetrahedron}) = \text{Red} + \text{Yellow} + \text{Green} + \text{Blue}.$$

If a left-hand orientation had been chosen for the tetrahedron, then all four faces would have appeared with *negative* signs in the expression for the boundary of the tetrahedron. Likewise, reversing the orientation of any face would change its sign in the expression for the boundary.

It is obvious that the boundary of the tetrahedron, a closed surface, has zero boundary. This may readily be confirmed algebraically by using the expressions for the boundary operator from C_3 to C_2 and from C_2 to C_1:

$$\begin{aligned}\partial[\partial(\text{Tetrahedron})] &= \partial[\text{Red} + \text{Yellow} + \text{Green} + \text{Blue}]\\ &= (-\alpha - \beta - \gamma) + (\beta + \eta - \varepsilon) + (\alpha + \varepsilon - \delta) + (\gamma + \delta - \eta) = 0.\end{aligned}$$

This result in no way depended on our choice of orientation for the tetrahedron or for any of its faces or edges. (You might try reversing the orientation of one face and a couple of edges just to appreciate why this is so.)

So far our complexes have all been defined initially by figures, and we have translated the figures into an algebraic specification of the boundary operator ∂. It is apparent that the boundary operator determines the complex as completely as the figure does, and it is tempting to try to specify a complex without drawing a diagram, simply by listing the elements and writing down the boundary operator. When we do so, we shall insist that for any element c of the complex $\partial(\partial c) = 0$.

Let us see why the condition $\partial(\partial c) = 0$ is a reasonable one to impose on any complex, even, for example, a surface which cannot be assembled in three-dimensional space or a complex whose elements include four-dimensional regions of spacetime. To verify that $\partial(\partial c) = 0$ holds for all c, it is sufficient to verify that it holds for the generators, i.e., for the elements of S_k. Now for any such element, we can construct the *subcomplex* consisting of this element, all the elements of S_{k-1} which enter non-trivially into the expression for its boundary, all their boundaries, and so on. The boundary relations between these elements are really properties of the single component we started with. So, although the interconnections between the various building blocks might be complicated, we expect that the individual elements out of which we are building our complex are reasonably familiar geometrical objects. For example, we might want to assume that every element of S_3 looks like some convex polyhedron in three-dimensional space (at least as far as its relations with its boundary are concerned – we will allow for all kinds of continuous geometrical distortion in the actual shape of object, much as we did not care about the shape of the wires in studying circuit theory). Now the boundary of such a polyhedron will consist of a finite number of polygonal faces, each of which can be oriented counterclockwise when we look at the polyhedron from the outside. When we compute the sum of the boundaries of these polygonal faces, each edge is shared by exactly two faces and they induce opposite orientations, so the sum of the boundaries is zero. Thus the boundary of every element of S_3 is a cycle in this case.

Figure 14.3

Similarly, we might be willing to assume that every element of S_2 looks like some convex polygon in the plane. It is then clear that the boundary of any element of S_2 is a cycle (in fact it corresponds to a mesh).

Let us investigate the homology spaces of the complex of the tetrahedron.

As far as H_0 is concerned, we can ignore C_2 and C_3 and consider the one-complex determined by the vertices and edges. This is a connected one-complex and so $\dim H_0 = 1$.

To determine H_1, we may consider the two-complex determined by the vertices, edges, and faces. We can analyze the space Z_1 by the method already familiar from our study of electric networks. Referring to figure 14.2, we see that branches δ, ε, and η form a maximal tree. Each of the remaining branches α, β, γ determines an independent cycle and so $\dim Z_1 = 3$. Each of these cycles is the boundary of a face and so $\dim B_1 = 3$. Hence $\dim H_1 = \dim Z_1 - \dim B_1 = 0$, so $H_1 = \{0\}$ is the trivial zero-dimensional vector space.

Turning our attention next to H_2, we first consider the space Z_2 of two-cycles, the kernel of the operator ∂ acting on C_2. We have already seen that the image of ∂, the space $B_1 \subset C_2$, is three-dimensional. Since C_2 is four-dimensional, the kernel of ∂ is one-dimensional. A basis for the space Z_2 is the element Red + Yellow + Green + Blue, the surface of the tetrahedron, which clearly has zero boundary. Of course this element belongs also to B_2, since it is the boundary of the solid tetrahedron. Hence $\dim B_2 = 1$, $\dim Z_2 = 1$ and $\dim H_2 = 0$.

Turning our attention finally to H_3, we note that there are no three-cycles so that $\dim Z_3 = 0$ and $\dim H_3 = 0$.

Euler's Theorem

In our study of electrical networks, we proved that

$$\dim H_0 - \dim H_1 = \dim C_0 - \dim C_1.$$

The generalization of this result to n-complexes is Euler's theorem:

$$\begin{aligned}\dim H_0 - \dim H_1 + \dim H_2 - \cdots \pm \dim H_n \\ = \dim C_0 - \dim C_1 + \dim C_2 - \cdots \pm \dim C_n.\end{aligned}$$

The number given by either side of this equation is called the *Euler characteristic* of the complex. We shall prove this result shortly, but let us check it immediately for the complex of the tetrahedron. Since $\dim H_0 = 1$, while the spaces H_1, H_2 and H_3 are $\{0\}$,

$$\dim H_0 - \dim H_1 + \dim H_2 - \dim H_3 = 1.$$

On the other hand,

$$\dim C_0 - \dim C_1 + \dim C_2 - \dim C_3 = 4 - 6 + 4 - 1 = 1.$$

Suppose, instead of the solid tetrahedron, we consider the two-complex consisting of the surface of the tetrahedron. The only change is that there is now no three-dimensional element in the complex. As before, $\dim H_0 = 1$ and $\dim H_1 = 0$. It is still true that $\dim Z_2 = 1$, since the surface of the tetrahedron has zero boundary, but now $\dim B_2 = 0$, since there is no three-dimensional element to give rise to a two-dimensional element which is a boundary. Therefore $\dim H_2 = \dim Z_2 - \dim B_2 = 1$ in this case. For this complex, then, $\dim H_0 - \dim H_1 + \dim H_2 = 1 - 0 + 1 = 2$ while $\dim C_0 - \dim C_1 + \dim C_2 = 4 - 6 + 4 = 2$ also.

Let us now do a similar computation for the solid cube, shown in perspective in figure 14.4(a), with its surface shown unfolded in figure 14.4(b). As before, $\dim H_0 = 1$

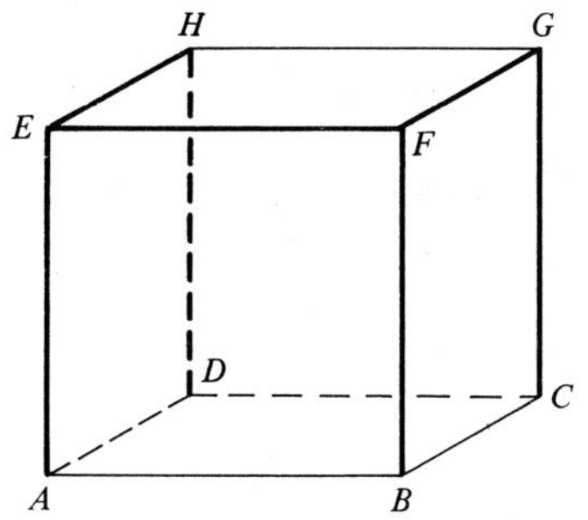

(a)

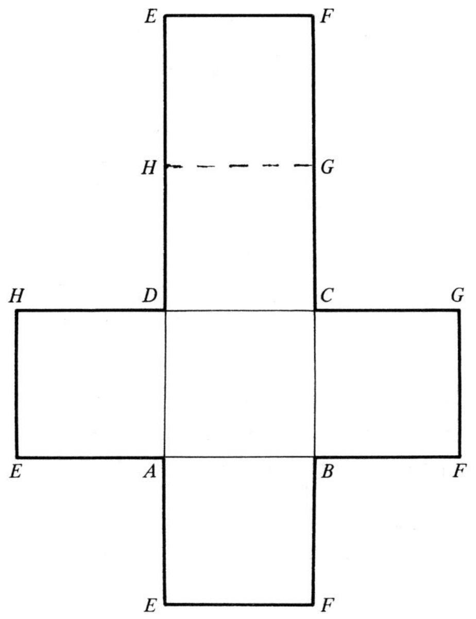

(b)

Figure 14.4

because the complex is connected, and $\dim H_3 = 0$ because there are no three-cycles.

To determine the dimension of Z_1, consider the maximal tree of seven branches which is shown in figure 14.4 (thick lines). Each of the remaining five branches determines a cycle. Hence $\dim Z_1 = 5$, but $\dim B_1 = 5$, and so $\dim H_1 = 0$.

Turning our attention to H_2, we proceed as for the tetrahedron. Since $\dim C_2 = 6$ and $\dim B_1 = 5$, we conclude that $\dim Z_2 = 1$. A basis for Z_2 is the entire surface of the cube. But this surface is the boundary of the cube, so $\dim B_2 = 1$ and $\dim H_2 = \dim Z_2 - \dim B_2 = 0$.

To homology spaces of the cube have turned out to be the same as those of the tetrahedron. Again $\dim H_0 - \dim H_1 + \dim H_2 - \dim H_3 = 1$, and we can check that

$$\dim C_0 - \dim C_1 + \dim C_2 - \dim C_3 = 8 - 12 + 6 - 1 = 1 \text{ also.}$$

Invariance of homology

From the point of view of topology there is a fundamental reason why the complex of the solid cube and the solid tetrahedron have the same homology spaces.

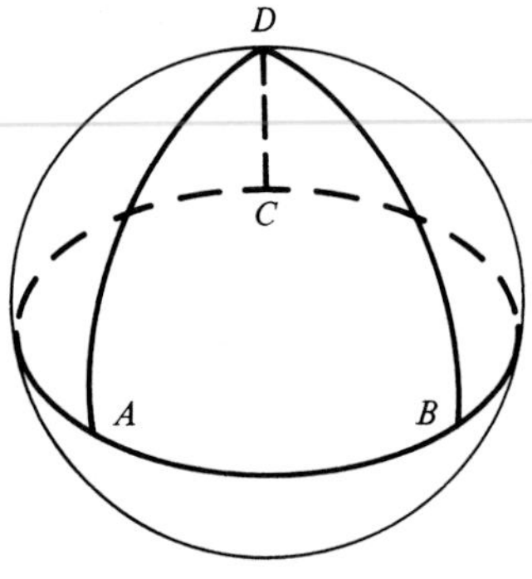

(a)

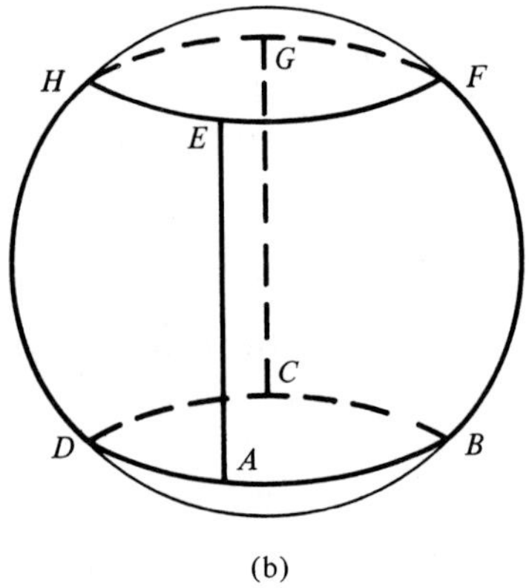

(b)

Figure 14.5

The *space* consisting of the solid cube is the same as the space consisting of the solid tetrahedron, which in turn is the same as the space consisting of the solid ball – they can all be continuously deformed into one another. Similarly, the surface of the tetrahedron, cube, and sphere are all the same space. Figure 14.5(a) illustrates how the surface of the sphere may be broken up as a tetrahedron, while figure 14.5(b) shows how it may be broken up as a cube. Of course the *complex* of the tetrahedron and the *complex* of the cube are different, but they represent different ways of breaking up the *same* space.

There is a basic theorem of topology which says roughly the following. Suppose we can build up a space X by attaching various cells to one another along their boundary cells of one lower dimension. For example, we could build up the surface of the sphere by attaching four two-cells together along their edges as in figure 14.5(a), or by attaching six two-cells as in figure 14.5(b). Alternatively, we could build up a three-dimensional cluster of polyhedra joined along their faces, like a cluster of soap bubbles. Then, says the theorem, *the homology spaces of the resulting complex depend only on the space of X and not upon how we built it up*; if we can decompose X into cells in some other way, to get some other complex, the spaces H_k of the two complexes will be isomorphic for all k.

To continue with the example of the sphere, we can find a decomposition which is simpler than either the tetrahedral or the cubical decomposition, as illustrated in figure 14.7. The surface is broken into two triangles, corresponding to the north and south hemispheres respectively, which are joined along the equator.

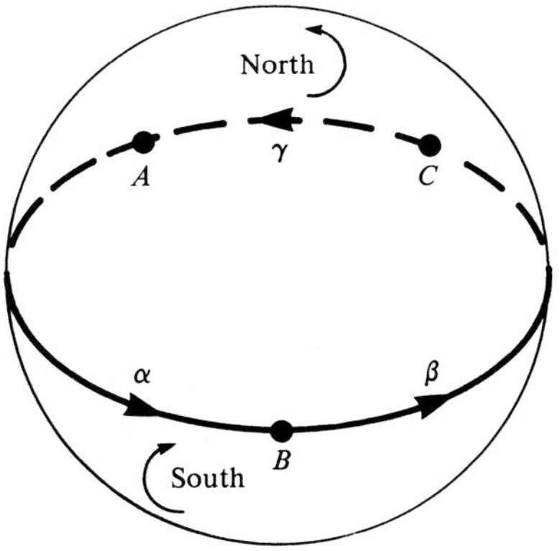

Figure 14.6

For this two-complex there is only one mesh, $\alpha + \beta + \gamma$, and this mesh is clearly the boundary of either hemispherical face. Hence $\dim Z_1 = \dim B_1 = 1$ and again $\dim H_1 = 0$. The sum of the two faces has zero boundary, since

$$\partial(\text{North}) = \alpha + \beta + \gamma$$

while

$$\partial(\text{South}) = -\alpha - \beta - \gamma.$$

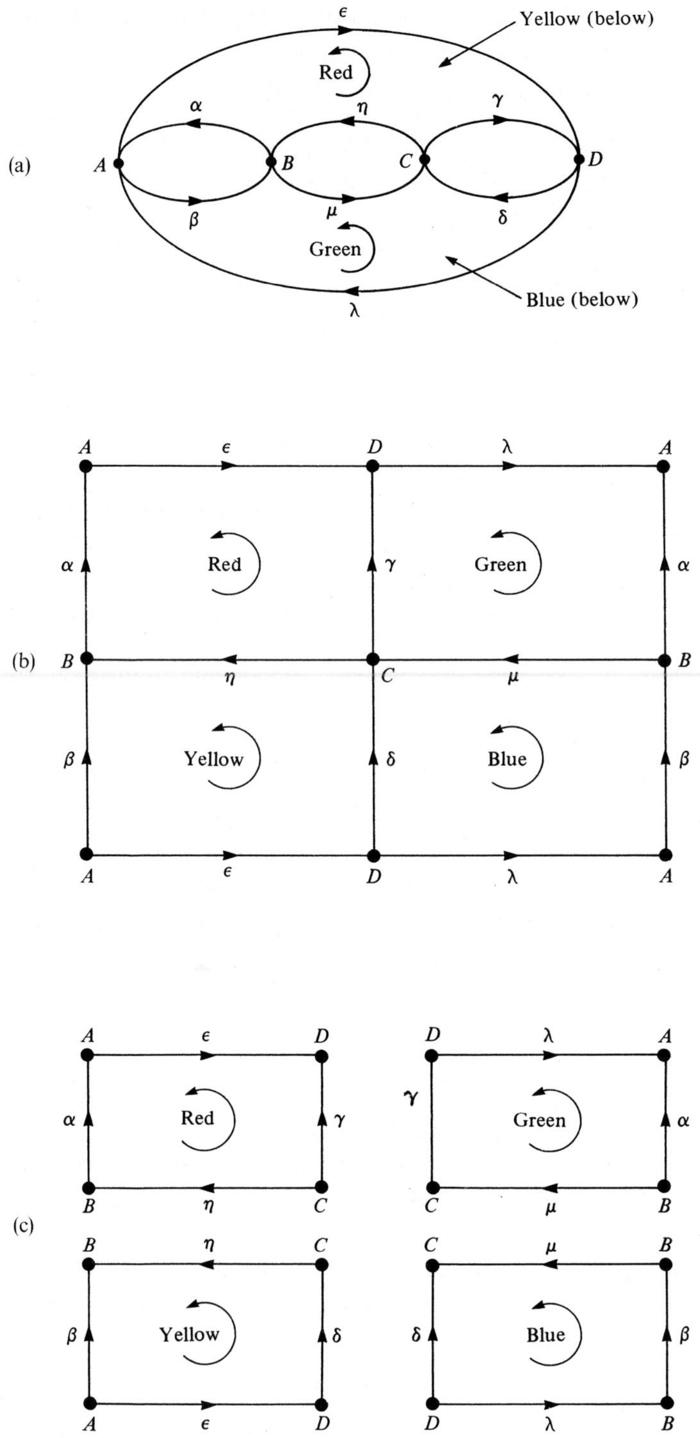
(a)
ε
Yellow (below)
Red
α
η
γ
A
B
C
D
β
μ
δ
Green
Blue (below)
λ
(b)
A
ε
D
λ
A
α
Red
γ
Green
α
B
η
C
μ
B
β
Yellow
δ
Blue
β
A
ε
D
λ
A
(c)
A
ε
D
D
λ
A
α
Red
γ
γ
Green
α
B
η
C
C
μ
B
B
η
C
C
μ
B
β
Yellow
δ
δ
Blue
β
A
ε
D
D
λ
B

Figure 14.7

So again $\dim H_2 = \dim Z_2 = 1$ as for the surface of the cube or tetrahedron. Again we notice, for this two-complex, that

$$\dim C_0 - \dim C_1 + \dim C_2 = 3 - 3 + 2 = 2 = \dim H_0 - \dim H_1 + \dim H_2.$$

We are not in a position to prove the theorem sketched above or even to state it precisely. To do so we would have to define exactly what we mean by *space*, *deform*, *attach*, and so on. Once these notions have been defined properly, it is not too difficult to state the theorem precisely and to prove it, but it would take us too far afield to set up all the necessary concepts in a mathematically correct form. Nevertheless, let us work with the theorem in the back of our minds, no matter how imprecisely formulated, while we do some more computations in order to gain more familiarity with the spaces H_k.

The torus

To obtain an example of a two-complex that has different homology spaces from those which we have been considering, we consider a decomposition of the two-dimensional torus; i.e., the surface of a donut. (Instead of cutting up a beach ball, we are now cutting up an inner tube!) Figure 14.7(a) shows the torus broken up into four regions; figure 14.7(b) shows these regions flattened out, but still attached; figure 14.7(c) shows the regions disassembled, but still labeled so that it is possible to reassemble them. We might think of figure 14.7(c) as depicting a 'torus kit' and try to discover in what fundamental ways it differs from the 'sphere kit' which would be obtained by cutting apart figure 14.7(a).

Counting vertices, edges, and faces, we find that $\dim C_0 = 4$, $\dim C_1 = 8$, and $\dim C_2 = 4$, so that $\dim C_0 - \dim C_1 + \dim C_2 = 4 - 8 + 4 = 0$. (Recall that, for a two-complex formed from the surface of the sphere, this quantity was always 2.)

We now consider the homology spaces of the complex of the torus. Since the complex is connected, we know that H_0 is one-dimensional. Since the entire surface of the torus has zero boundary, $\dim Z_2 = 1$. (Check for yourself that, with the orientations shown, $\partial(\text{Red} + \text{Yellow} + \text{Green} + \text{Blue}) = 0$.) Since we are considering a two-complex, there can be no elements of C_2 which are boundaries, so $\dim B_2 = 0$ and $\dim H_2 = \dim Z_2 - \dim B_2 = 1$. (More generally, it is clear that, for any two-complex which is obtained by subdividing a smooth, connected surface which is the boundary of a region in $\mathbb{R}^3$, $\dim H_2 = 1$ for the same reasons.)

We turn finally to the space H_1. Figure 14.8 shows a maximal tree consisting of only three edges. Take the time to convince yourself that any of the other five edges, when added to this maximal tree, completes a cycle! For example, $\alpha + \beta$ is a cycle, as is $\varepsilon + \lambda$ or $\mu + \eta$. Therefore $\dim Z_1 = 5$ for this complex. On the other hand, $\dim C_2 = 4$. The kernel of ∂ acting on C_2, the space Z_2, is one-dimensional, so the image of ∂, the space B_1, has dimension 3. Therefore

$$\dim H_1 = \dim Z_1 - \dim B_1 = 5 - 3 = 2.$$

It is instructive to proceed further and construct bases for the spaces B_1 and H_1. A simple choice of basis elements for B_1 consists of the meshes which are boundaries of

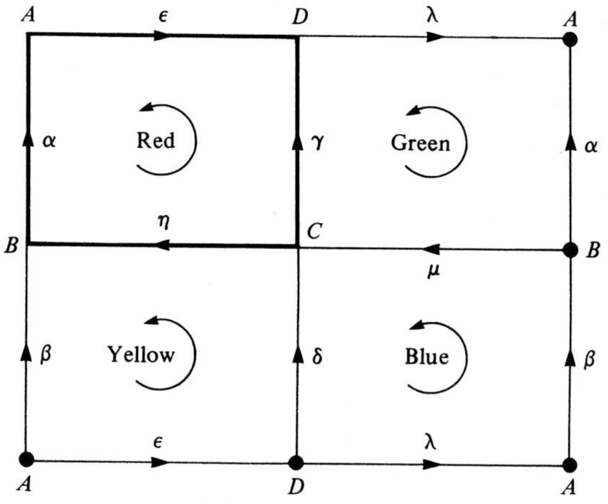

Figure 14.8

three of the four faces, for example:

$$\partial(\text{Yellow}) = \eta - \beta + \varepsilon + \delta,$$

$$\partial(\text{Green}) = \alpha - \lambda - \gamma - \mu,$$

$$\partial(\text{Blue}) = \lambda + \beta + \mu - \delta.$$

Since H_1 is the quotient space Z_1/B_1, its elements are equivalence classes *modulo* B_1. We must therefore find two cycles which are not boundaries, with the additional property that no linear combination of them is a boundary. Referring to figure 14.7(a), we can readily find two such cycles: the cycle $\alpha + \beta$ which goes around the torus in one direction, and the cycle $\varepsilon + \lambda$ which goes around the torus in another independent direction. A basis for H_1 therefore consists of the equivalence classes $\overline{\alpha + \beta}$ and $\overline{\varepsilon + \lambda}$. Of course, there are many other cycles which are not boundaries, but they all fall into the equivalence class of some linear combination of $\alpha + \beta$ and $\varepsilon + \lambda$. For example, the cycle $\gamma + \delta$ lies in the equivalence class of $\alpha + \beta$ because

$$\gamma + \delta = \alpha + \beta + \partial(\text{Red}) + \partial(\text{Yellow}).$$

The cycle $\alpha + \varepsilon + \delta - \mu$ may be expressed

$$(\alpha + \beta) + (\varepsilon + \lambda) - \partial(\text{Blue})$$

and so it lies in the same equivalence class as

$$\alpha + \beta + \varepsilon + \lambda.$$

The meaning of H_1

We are now in a position to understand the general significance of dim H_1: it is, roughly speaking, the number of independent closed curves which do not bound any region. For the sphere, any closed curve is necessarily the boundary of some region and so dim $H_1 = 0$. For a torus, on the other hand, there are two distinct types of closed curves which do not bound any region, as illustrated by figure 14.9, and so

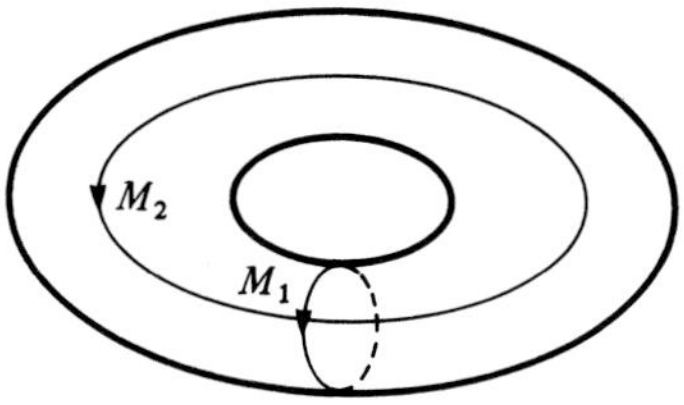

Figure 14.9

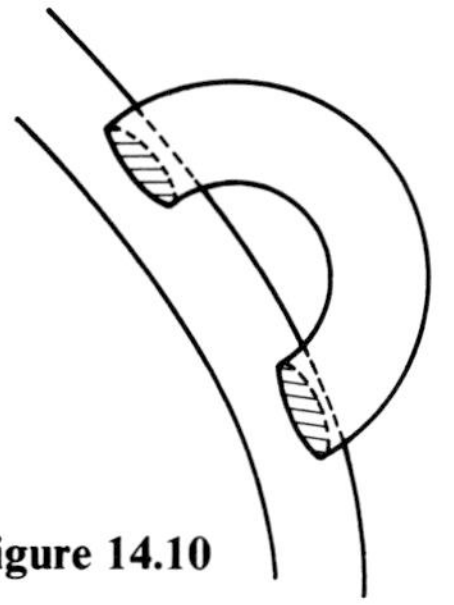

Figure 14.10

dim $H_1 = 2$. Notice that for the complex which we constructed for the torus we had dim C_0 − dim C_1 + dim $C_2 = 4 - 8 + 4 =$ dim H_0 − dim H_1 + dim H_2. Let us examine how the spaces H_1 change when we modify a surface by attaching a 'handle'. By this we mean that we cut two small disks out of the surface and attach a (curved) cylinder, open at the top and the bottom, to the surface by attaching the top of the cylinder to the boundary of one of the disks and the bottom of the cylinder to the boundary of the other disk. Let us choose a decomposition of our original surface so that each of the disks is a cell in our decomposition, say a square, and we think of the cylinder as being made up of four sides. Thus in attaching the cylinder we add no new nodes, we add four new branches (the edges of the square cylinder), remove two former elements of S_2 (the interiors of the disks) and add four new elements of S_2 (the four sides of the cylinder). For the new surface we still have that H_0 and H_2 are one-dimensional. In computing Z_1 for the new complex we can clearly choose a maximal tree which does not contain any of the four new branches (since each of them clearly lies on a mesh – recall our procedure for constructing maximal trees in our study of electrical networks). Thus we can construct a maximal tree consisting entirely of branches of the complex of the old surfaces. Each branch of the cylinder then gives a new independent mesh, and we have increased the dimension of Z_1 by 4. On the other hand, we have increased the dimension of C_2, and hence of B_1, by 2, and so we have increased the dimension of H_1 by 2. Thus we have shown that, each time we add a handle, we increase the dimension of H_1 by 2. A sphere with k handles attached has dim $H_0 = 1$, dim $H_1 = 2k$ and dim $H_2 = 1$. For example, we can think of the torus as a sphere with one handle attached, giving us another proof of our preceding results for the torus.

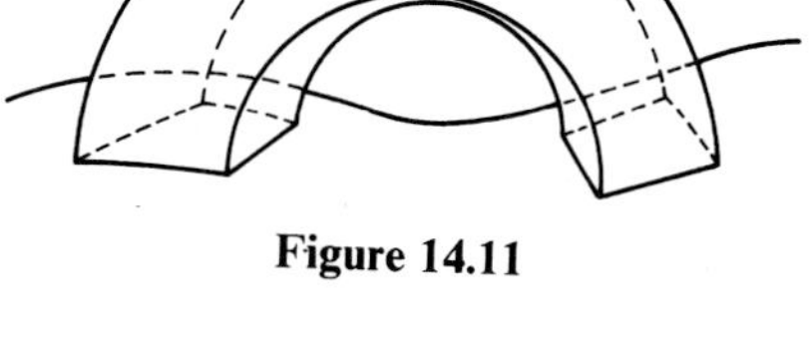

Figure 14.11

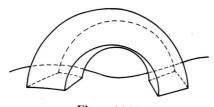

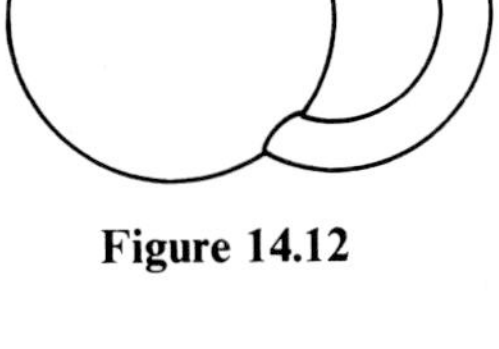

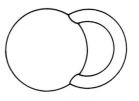

Figure 14.12

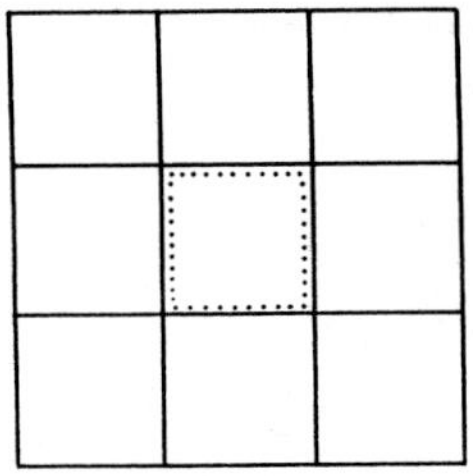

Figure 14.13

We can also investigate the effect of cutting a hole out of a surface. Consider, for example, a complex in the plane in the form of a square together with its interior. For such a complex it is clear that $\dim H_0 = 1$ while $\dim H_1 = 0$ and $\dim H_2 = 0$. Suppose we cut out a hole from the interior of the square. If we had a decomposition of our original square for which the hole we removed was one of the elements of S_2, we see that the process of removing the hole does not change Z_1, but does decrease the dimension of C_2, and hence of B_1, by 1; and hence increases the dimension of H_1 by 1. Thus, if we have a region in the plane with k holes cut out, $\dim H_1 = k$, while $\dim H_0 = 1$ and $\dim H_2 = 0$. A similar argument in three dimensions shows that if we have a region in three-dimensional space with k (three-dimensional) holes cut out then $\dim H_0 = 1$, $\dim H_1 = 0$, $\dim H_2 = k$ and $\dim H_3 = 0$.

The Klein bottle

You may have become aware, in your study of electrical networks, that there are networks, like the one shown in figure 14.14, which cannot be assembled in the plane

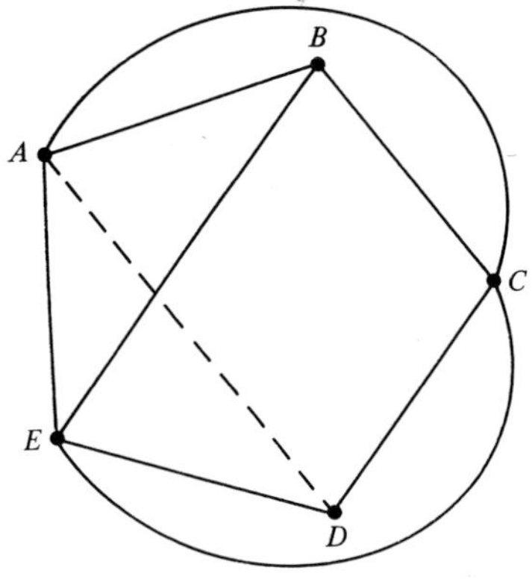

Figure 14.14

without having wires cross. Such non-planar networks can, however, always be assembled in three-dimensional space. Similarly there exist two-complexes which cannot be assembled in three-dimensional space. A well-known example is the so-called *Klein bottle*, a surface which gives rise to the complex of figure 14.15(a). In spite of its close resemblance to the torus of figure 14.7(b), this surface cannot be

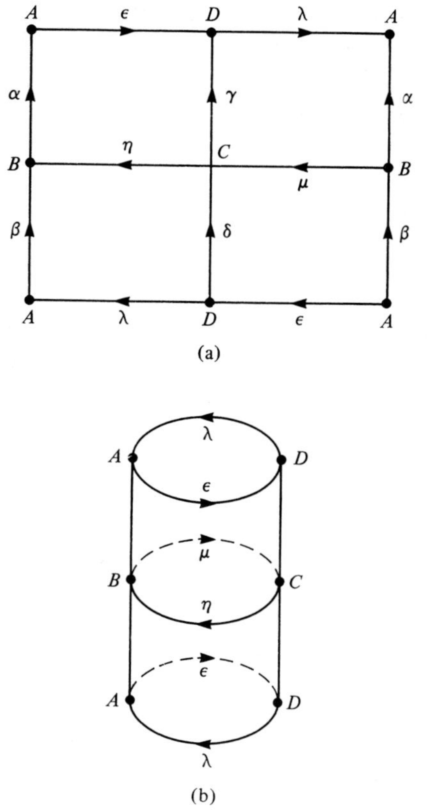

Figure 14.15

constructed in three-dimensional space. The flattened out complex can be formed into a cylinder as shown in figure 14.15(b), but there is then no way of fastening edges ε and λ together with their correct orientations. For the complex of the Klein bottle, the homology spaces are somewhat different than for the torus. Of course, $\dim H_0 = 1$ because the complex is connected. However, $\dim H_2 = 0$, because there is no element of C_2 which has zero boundary. (The boundary of the entire flattened out complex of figure 14.15(a) is $2\varepsilon + 2\lambda$, and you can readily convince yourself that changing the sign of one or more two-cells will still not yield an element of C_2 whose boundary is zero.) As for the complex of figure 14.7, $\dim Z_1 = 5$, but, for the Klein bottle complex, $\dim B_1 = 4$ and $\dim H_1 = 5 - 4 = 1$. This means that there is only one independent cycle which is not a boundary. The cycle $\alpha + \beta$ is not a boundary in either figure 14.7(b) or figure 14.15(a), but the cycle $\varepsilon + \lambda$, which was not an element of B_1 in the case of the torus, is an element of B_1 in the case of the Klein bottle: it is half the boundary of the entire complex of figure 14.15(a).

For both the torus complex and the Klein bottle complex which we have considered, $\dim C_0 = 4$, $\dim C_1 = 8$, and $\dim C_2 = 4$, so that

$$\dim C_0 - \dim C_1 + \dim C_2 = 4 - 8 + 4 = 0.$$

It is reassuring to notice that, for the Klein bottle case,

$$\dim H_0 - \dim H_1 + \dim H_2 = 1 - 1 + 0 = 0 \text{ also.}$$

Proof of Euler's theorem

We now prove the general result

$$\dim H_0 - \dim H_1 + \dim H_2 - \cdots \pm \dim H_n = \dim C_0 - \dim C_1 + \cdots \pm \dim C_n$$

which we have checked in many special cases. The ingredients of the proof are simple.

(1) Since $H_k = Z_k/B_k$, $\dim H_k = \dim Z_k - \dim B_k$.

(2) As a consequence of the rank–nullity theorem applied to $\partial: C_{k+1} \to C_k$, $\dim C_{k+1} = \dim Z_{k+1} + \dim B_k$.

(3) For an n-complex, $\dim B_n = 0$ because there are no $(n+1)$-dimensional objects.

(4) $\dim Z_0 = \dim C_0$ because the boundary of any zero-dimensional cell is zero by definition, there being no objects of negative dimension the complex.

Combining (1) and (2) we find

$$\dim H_k = \dim Z_k - \dim C_{k+1} + \dim Z_{k+1}$$

for $1 \leqslant k < n$. For $k = 0$ we have, by virtue of (4),

$$\dim H_0 = \dim C_0 - \dim C_1 + \dim Z_1$$

while for $k = n$ we have, by virtue of (3),

$$\dim H_n = \dim Z_n.$$

Thus

$$\begin{aligned}
&\dim H_0 - \dim H_1 + \dim H_2 + \cdots \pm \dim H_n \\
&\quad = (\dim C_0 - \dim C_1 + \dim Z_1) - (\dim Z_1 - \dim C_2 + \dim Z_2) \\
&\qquad + (\dim Z_2 - \dim C_3 + \dim Z_3) - \cdots \pm \dim Z_n \\
&\quad = \dim C_0 - \dim C_1 + \dim C_2 - \cdots \pm \dim C_n.
\end{aligned}$$

As an application of this result, we consider a convex polyhedron, which we regard as a complex formed by cutting up the surface of the sphere into cells. If we believe the theorem quoted earlier to the effect that the dimensions of the spaces H_0, H_1, and H_2 are the same for any complex formed by decomposing a given surface, we can deduce that $\dim H_0 - \dim H_1 + \dim H_2 = 1 - 0 + 1 = 2$ as we found earlier for the cube and tetrahedron. It follows that, for any convex polyhedron,

$$\dim C_0 - \dim C_1 + \dim C_2 = 2$$

or

$$\text{number of vertices} - \text{number of edges} + \text{number of faces} = 2.$$

Orientation and the ∂ operator

As we have said above, we have not given a precise definition of the notion of the underlying space of a complex. But here is how we would like you to think about the situation. Each k-cell, that is, each element of S_k, should be thought of as a bounded convex polyhedron in $\mathbb{R}^k$ with a definite orientation. (Remember that we can choose one of two orientations on $\mathbb{R}^k$, and hence on any open subset, by choosing an (ordered) basis.) Now the boundary of this convex polyhedron will be a union of finitely many convex polyhedra of dimension $k-1$, lying in some affine subspace of dimension $k-1$. By a choice of origin and basis we may identify each such subspace with $\mathbb{R}^{k-1}$. We assume that, after we have made such identifications, each of the corresponding $(k-1)$-dimensional polyhedra corresponds to a $(k-1)$-cell in our complex, that is, to an element of S_{k-1}. But the elements of S_{k-1} come with orientations. So we must see how the orientation of a k-cell relates to the orientation of the $(k-1)$-cells on its boundary. This information, of course, is coded into the ∂ operator.

Let $\mathbf{v}_1, \ldots, \mathbf{v}_k$ be a basis of $\mathbb{R}^k$ chosen so that $\mathbf{v}_2, \ldots, \mathbf{v}_k$ is tangent to a specified $(k-1)$-dimensional face of our k-cell C, and so that $\mathbf{v}_1$ points out of the k-cell. Suppose that the orientation of $\mathbb{R}^k$ determined by $\mathbf{v}_1, \ldots, \mathbf{v}_k$ coincides with the orientation of C. Then $\mathbf{v}_2, \ldots, \mathbf{v}_k$ determines an orientation of the particular $k-1$ dimensional face corresponding to, say, $F \subset S_{k-1}$. Now this orientation may or may not coincide with the orientation of F. If it does, we will assign a + sign to F, otherwise, a − sign. Thus

$$\partial C = \sum \pm F$$

where the sum is over those F corresponding to the boundary $k-1$ faces of C. If the basis $\mathbf{v}_1, \ldots, \mathbf{v}_k$ corresponded to the opposite orientation of C, the signs would all be reversed, of course.

If $k \geqslant 2$ we could replace $\mathbf{v}_k$ by $-\mathbf{v}_k$ if necessary to get a 'good' basis. For $k = 1$ the space $\mathbb{R}^{k-1}$ is just the zero vector space. An 'orientation' here is just a choice of + or − sign. Then our rule says to pick the + sign if the vector $\mathbf{v}_1$ giving the orientation points out and the − sign otherwise:

•---->---•

$-$ $\quad \mathbf{v}_1 \quad$ $+$

Of course, each $(k-1)$-cell might occur in the boundary of several k-cells. But since the k-cells form a basis of C, defining as above for each k-cell determines the linear map $\partial\colon C_k \to C_{k-1}$.

To check that $\partial \circ \partial = 0$ it is sufficient to check that $\partial(\partial C) = 0$ for each k-cell. Now every $(k-2)$-cell E that can occur in the expression for $\partial(\partial C)$ occurs because it is the boundary of exactly two $(k-1)$-dimensional faces, say F_1 and F_2, of C. You should now check your understanding of the notion of orientation and our definition of ∂ to see that $\partial(\partial C) = 0$.

Having realized each element of S_k as a convex polyhedron, we can define a smooth map of the complex as a whole into $\mathbb{R}^N$. This will be a rule f that assigns to each point of each k-cell a point of $\mathbb{R}^N$. Whenever we have made one of our identifications of a k-cell C as a convex polyhedron in $\mathbb{R}^N$, the map f, restricted to C, can be thought of as a map from a convex polyhedron in $\mathbb{R}^k$ to $\mathbb{R}^N$. We want this map to be smooth. In fact, we want it to be the restriction of a map g defined in a slightly larger region (so as to include C and all its boundary in the interior) where g is smooth. We will study this in more detail in Chapter 15.

14.2. Dual spaces and cohomology

We turn our attention now to the collection of dual spaces to the spaces C_k. We denote these spaces by C^k. Thus, an element of C^k, which we call a k-cochain, is a linear function of the elements of C_k, which we call k-chains. If $\boldsymbol{\sigma}$ is a k-cochain and $\mathbf{c}$ is a k-chain, we shall denote the value of the cochain $\boldsymbol{\sigma}$ on the chain $\mathbf{c}$ by $\int_{\mathbf{c}}\boldsymbol{\sigma}$, just as we have already done in the cases $k=0$ and $k=1$ while considering electrical networks.

We can now consider the transpose, d, of the boundary operator $\partial: C_{k+1} \to C_k$, which we shall call the coboundary operator. Thus $\mathrm{d}: C^k \to C^{k+1}$ is defined by the formula

$$\int_{\mathbf{c}} \mathrm{d}\boldsymbol{\sigma} = \int_{\partial\mathbf{c}} \boldsymbol{\sigma} \quad \text{where} \quad \boldsymbol{\sigma} \in C^k \quad \text{and} \quad \mathbf{c} \in C_{k+1}.$$

It follows directly from the fact that the boundary of a boundary is zero that $\mathrm{d} \circ \mathrm{d}$ is zero. (Of course, this is the composition of $\mathrm{d}: C^k \to C^{k+1}$ with $\mathrm{d}: C^{k+1} \to C^{k+2}$.) In the notation just introduced,

$$\int_{\mathbf{c}} \mathrm{d}(\mathrm{d}\boldsymbol{\sigma}) = \int_{\partial\mathbf{c}} \mathrm{d}\boldsymbol{\sigma} = \int_{\partial(\partial\mathbf{c})} \boldsymbol{\sigma} = 0 \quad \text{since } \partial(\partial\mathbf{c}) = 0.$$

In order to gain a feeling for the meaning of a cochain and the significance of the coboundary operator d we consider the two-complex of figure 14.16. An element of C^0, a zero-cochain Φ, is defined completely by specifying its value at each node, for example

$$\Phi^A = 1, \Phi^B = 3, \Phi^C = 2, \Phi^D = 5.$$

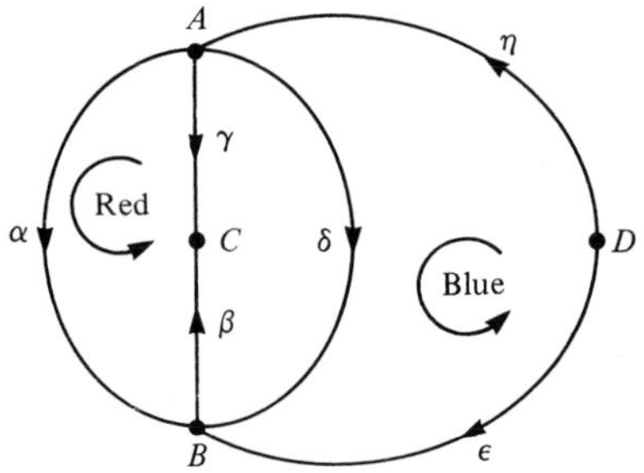

Figure 14.16

The potential functions which figured prominently in electrical network theory are examples of zero-cochains.

The action of the coboundary operator d on a zero-cochain yields a one-cochain $\mathrm{d}\Phi$. To compute the value of $\mathrm{d}\Phi$ on a branch, we use the fact that d is the adjoint of ∂. For example, since $\partial\alpha = B - A$, $\int_\alpha \mathrm{d}\Phi = \int_{\partial\alpha} \Phi = \Phi^B - \Phi^A$. As is already familiar, the operator $\mathrm{d}: C^0 \to C^1$ converts a potential function, which is a zero-cochain or linear function defined on nodes, into a voltage rise function, a one-cochain defined as a linear function on branches.

We can now proceed to investigate the significance of $d: C^1 \to C^2$. An example of a one-cochain might be a linear function $\mathbf{W}$ with values such as

$$W^\alpha = 1,\ W^\beta = 2,\ W^\gamma = 1,\ W^\delta = 2,\ W^\varepsilon = -3,\ W^\eta = 2.$$

The battery voltages in an electric network defined such a one-cochain.

Now the effect of d on a one-cochain $\mathbf{W}$ is to yield a two-cochain $d\mathbf{W}$ which is a linear function on two-chains. We can compute the value of $d\mathbf{W}$ on the two cell Red, for example, by using the definition

$$\int_{\text{Red}} d\mathbf{W} = \int_{\partial(\text{Red})} \mathbf{W}.$$

Since $\partial(\text{Red}) = \alpha + \beta - \gamma$, we have

$$\int_{\text{Red}} d\mathbf{W} = W^\alpha + W^\beta - W^\gamma = 1 + 2 - 1 = 2.$$

Similarly

$$\int_{\text{Blue}} d\mathbf{W} = W^\delta - W^\varepsilon + W^\eta = 2 + 3 + 2 = 7.$$

Now that the values of $d\mathbf{W}$ on a basis for C_2 are known, its values on any element of C_2 may be found by linearity.

As a final example, consider the three-complex of the tetrahedron shown in figure 14.17. Let R, Y, G, and B denote the four faces of the tetrahedron, all oriented

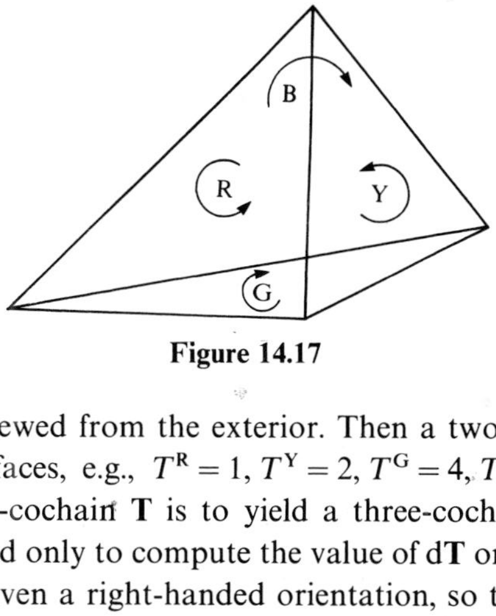

Figure 14.17

counterclockwise as viewed from the exterior. Then a two-cochain $\mathbf{T}$ is specified by its values on the faces, e.g., $T^{\text{R}} = 1$, $T^{\text{Y}} = 2$, $T^{\text{G}} = 4$, $T^B = -1$. The effect of $d: C^2 \to C^3$ on the two-cochain $\mathbf{T}$ is to yield a three-cochain $d\mathbf{T}$ defined on the elements of C_3. We need only to compute the value of $d\mathbf{T}$ on the solid tetrahedron. If the tetrahedron is given a right-handed orientation, so that

$$\partial(\text{Tetrahedron}) = \text{R} + \text{Y} + \text{B} + \text{G},$$

then

$$\int_{\text{Tetrahedron}} d\mathbf{T} = T^{\text{R}} + T^{\text{Y}} + T^{\text{G}} + T^{\text{B}} = 1 + 2 + 4 - 1 = 6.$$

The reader who is familiar with electromagnetic theory may already have noticed the similarity between the coboundary operator and certain operations of significance in physics. For example, the two-cochain **T** might have specified the flux of the electric field through the various faces of a tetrahedron, in which case the three-cochain d**T** would be, by virtue of Gauss's Law, proportional to the total electric charge within the solid tetrahedron. In the earlier example, if the one-cochain **W** had represented voltages induced by a changing magnetic field, then the two-cochain d**W** would be, by virtue of Faraday's Law, proportional to the rate of change of magnetic flux through each of the various two-cells. Indeed, the coboundary operator d is going to evolve into a differential operator that will permit a concise statement of all of Maxwell's equations.

We can now consider the image and kernel of the coboundary operator in order to construct subspaces of the spaces C^k. To be specific, the image of $\mathrm{d}: C^{k-1} \to C^k$ is called B^k, the subspace of *coboundaries*, while the kernel of $\mathrm{d}: C^k \to C^{k+1}$ is called Z^k, the subspace of *cocycles*. Since $\mathrm{d} \circ \mathrm{d} = 0$, any coboundary is necessarily a cocycle, and B^k is therefore a subspace of Z^k. We form the quotient space $H^k = Z^k/B^k$ called the *kth cohomology space* of the complex.

Many aspects of the construction of these spaces have appeared already in our study of electrical networks. In the case $k = 0$, for example, the space Z^0 (zero-cocycles) was the space of physically meaningful potentials that were constant on each connected component of a network. The space B^0 was empty, since there is no way for a zero-cochain to lie in the image of d. As a result, the quotient space $H^0 = Z^0/B^0 = Z^0$. (We used the notation H^0 exclusively in discussing this space earlier.)

For a one-complex, the space Z^1 is the entire space C^1, for the simple reason that, with no two-cochains available, every one-cochain must lie in the kernel of d. The space B^1 of coboundaries contains those voltage distributions which are derivable from a potential. The quotient space $H^1 = Z^1/B^1$ is, in this case, also the space C^1/B^1. You will recall that the space H^1 is dual to the space H_1 (for a one-complex, the same as Z_1).

The simplest example in which Z^k, B^k, and H^k are all non-trivial subspaces is the case $k = 1$ for a two-complex. Consider the complex shown in figure 14.18.

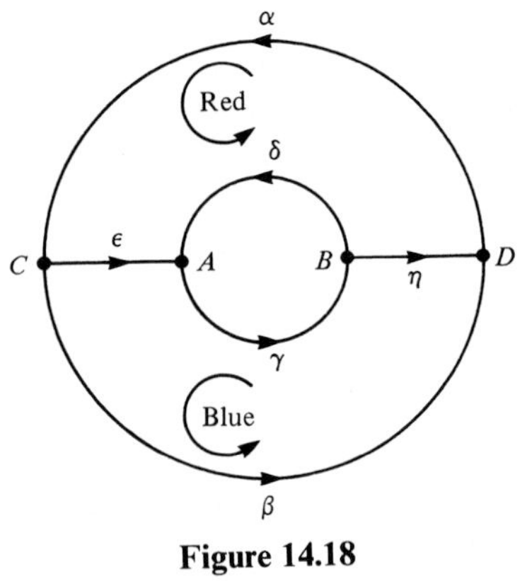

Figure 14.18

Here, since there are six branches, the space C^1 is six-dimensional. To determine $\dim Z^1$, we must look at the image and kernel of $d: C^1 \to C^2$.

Consider first the image of $d: C^1 \to C^2$, the space B^2. Since C^2 is two-dimensional, B^2 cannot have dimension greater than 2. To show that its dimension is 2, it is sufficient to exhibit a basis for C^2 which lies in the image of d. The one-cochain $\mathbf{V}$ which assigns 1 to branch α and 0 to all other branches has the property that

$$d\mathbf{V}(\text{Red}) = 1, d\mathbf{V}(\text{Blue}) = 0.$$

Similarly, the one-cochain $\mathbf{W}$ which assigns 1 to branch β and 0 to all other branches satisfies

$$d\mathbf{W}(\text{Red}) = 0, d\mathbf{W}(\text{Blue}) = 1.$$

Therefore B^2, the image of d, spanned by $d\mathbf{V}$ and $d\mathbf{W}$, has dimension 2, and by the rank–nullity theorem the kernel of d, the space Z^1, satisfies

$$\dim Z^1 = \dim C^1 - \dim B^2 = 6 - 2 = 4.$$

Turning our attention now to B^1, we consider $d: C^0 \to C^1$. The kernel of this operator, the space Z^0 (which is the same as H^0) has dimension 1, because the complex is connected. Therefore

$$\dim B^1 = \dim C^0 - \dim Z^0 = 4 - 1 = 3.$$

That is, there are three independent one-cochains which are derivable from potentials.

We see now that the space Z^1 of one-cocycles has dimension 4, while the space B^1 of one-coboundaries has dimension 3. Therefore $\dim H^1 = \dim Z^1 - \dim B^1 = 4 - 3 = 1$. This means that there exist 'interesting' cocycles which lie in the kernel of d even though they cannot be derived from a potential. The equivalence class of one such cocycle will provide a basis for the one-dimensional quotient space Z^1. Let us choose, for example, the cocycle $\mathbf{V}$ for which $V^\alpha = \frac{1}{2}, V^\beta = \frac{1}{2}, V^\gamma = \frac{1}{2}, V^\delta = \frac{1}{2}, V^\varepsilon = 0, V^\eta = 0$. This is clearly *not* derivable from a potential, because the sum of the voltage drops around a cycle such as $\alpha + \beta$ or $\gamma + \delta$ is not zero. On the other hand, $\mathbf{V}$ lies in the kernel of $d: C^1 \to C^2$, since $d\mathbf{V}(\text{Red}) = V^\alpha - V^\delta = \frac{1}{2} - \frac{1}{2} = 0$ and $d\mathbf{V}(\text{Blue}) = V^\beta - V^\gamma = \frac{1}{2} - \frac{1}{2} = 0$.

It is clear from the above analysis that the existence of an 'interesting' cocycle is intimately related to the presence of a *hole* in the complex. If the disk bounded by γ and δ were added to the complex (call it Yellow) then the only cocycles would be the coboundaries; the cochain $\mathbf{V}$ considered above would have $d\mathbf{V}(\text{Yellow}) = 1$, for example, and would not be a cocycle. In this way the analysis of the cohomology spaces of the complex reveals the presence of the hole just as the analysis of the homology spaces would have.

In this example, the space H^1 is readily seen to be dual to H_1 in the sense that the basis element $\mathbf{V}$ assigns 1 to any cycle which encircles the hole once in a counter-clockwise sense. Consider, for example, the cycle $\mathbf{I} = \alpha + \varepsilon - \delta + \eta$. We see that $\int_\mathbf{I} \mathbf{V} = V^\alpha + V^\varepsilon - V^\delta + V^\eta = \frac{1}{2} - \frac{1}{2} = 0$. But all cycles which encircle the hole once counterclockwise constitute an equivalence class which can serve as a basis for H_1. Furthermore, the specific choice of the element **V** as a representative of its equivalence class did not matter. Suppose we had used $\mathbf{W} = \mathbf{V} + \mathrm{d}\mathbf{\Phi}$ instead. Then for any cycle **M**

$$\int_{\mathbf{M}} \mathbf{W} = \int_{\mathbf{M}} \mathbf{V} + \int_{\mathbf{M}} \mathrm{d}\mathbf{\Phi} = \int_{\mathbf{M}} \mathbf{V} + \int_{\partial\mathbf{M}} \mathbf{\Phi} = \int_{\mathbf{M}} \mathbf{V}$$

since $\partial\mathbf{M} = 0$ if **M** is a cycle. In this sense, the equivalence class $\bar{\mathbf{V}}$ (a basis for H^1) determines a linear function on the equivalence class $\bar{\mathbf{I}}$ (a basis for H_1). To evaluate the function, we evaluate $\int_{\mathbf{I}}\mathbf{V}$, using *any* member of either equivalence class.

Let us now proceed to prove that H^k can be identified in general with the dual space of H_k. We shall first establish that $\dim H^k = \dim H_k$, then show that any element of H^k defines a linear function on the space H_k.

Everything will follow from the now-familiar result: the kernel of the adjoint annihilates the image; the image of the adjoint annihilates the kernel. Consider the following diagram:

$$C^{k-1} \xrightarrow{\mathrm{d}} C^k \xrightarrow{\mathrm{d}} C^{k+1}$$

$$C_{k-1} \xleftarrow{\partial} C_k \xleftarrow{\partial} C_{k+1}$$

We look first at $\partial: C_{k+1} \to C_k$ and its adjoint $\mathrm{d}: C^k \to C^{k+1}$. The *kernel* of the adjoint d is the space Z^k of cocyles; it annihilates the *image* of ∂, which is the space B_k of boundaries. We turn our attention next to $\partial: C_k \to C_{k-1}$ and its adjoint $\mathrm{d}: C^{k-1} \to C^k$. The *image* of the adjoint is the space B^k of coboundaries; it annihilates the *kernel* of ∂, which is the space Z_k of cycles.

But we know that, for any subspace of C_k, the dimension of the subspace plus the dimension of its annihilator space must equal the dimension of C_k. Hence

$$\dim B_k + \dim Z^k = \dim C_k = \dim Z_k + \dim B^k.$$

It follows that

$$\dim Z^k - \dim B^k = \dim Z_k - \dim B_k,$$

i.e., that

$$\dim H^k = \dim H_k$$

since

$$H^k = Z^k/B^k \quad \text{and} \quad H_k = Z_k/B_k.$$

A typical element of H^k is the equivalence class $\bar{\mathbf{V}}$ consisting of cocycles of the form $\mathbf{V} + \mathrm{d}\mathbf{\Phi}$. If $\bar{\mathbf{V}}$ is different from the zero element, then **V** is *not* a coboundary. Similarly, a typical element of H_k is the equivalence class $\bar{\mathbf{I}}$ consisting of cycles of the form $\mathbf{I} + \partial\tau$. If $\bar{\mathbf{I}}$ is different from the zero element, then **I** is not a boundary. We *define* the value of $\bar{\mathbf{V}}$ acting on $\bar{\mathbf{I}}$, $\int_{\mathbf{I}}\bar{\mathbf{V}}$, by letting any element of the class $\bar{\mathbf{V}}$ act on any element of the class $\bar{\mathbf{I}}$. Thus

$$\int_{\mathbf{I}} \bar{\mathbf{V}} = \int_{\mathbf{I}+\partial\tau} \mathbf{V} + \mathrm{d}\Phi = \int_{\mathbf{I}} \mathbf{V} + \int_{\mathbf{I}} \mathrm{d}\Phi + \int_{\partial\tau} \mathbf{V} + \int_{\partial\tau} \mathrm{d}\Phi$$

$$= \int_{\mathbf{I}} \mathbf{V} + \int_{\partial\mathbf{I}} \Phi + \int_{\tau} \mathrm{d}\mathbf{V} + \int_{\tau} \mathrm{d}(\mathrm{d}\Phi).$$

But $\partial\mathbf{I} = 0$ because $\mathbf{I}$ is a cycle, $\mathrm{d}\mathbf{V} = 0$ because $\mathbf{V}$ is a cocycle, and of course $\mathrm{d}(\mathrm{d}\Phi) = 0$. Hence

$$\int_{\mathbf{I}} \bar{\mathbf{V}} = \int_{\mathbf{I}} \mathbf{V}.$$

In other words, it makes no difference which element of the equivalence classes we choose, and so we have a well-defined action of the elements of H^k as linear functions on H_k. This completes the identification of H^k with the dual of H_k.

It was mentioned previously that the homology spaces H_k associated with a complex depend only upon the underlying space and not upon how it was cut up to form the complex. It follows now that the same must hold true for the cohomology spaces H^k. Because the cochains which give rise to the H^k are closely related to quantities of physical significance, these spaces will eventually reveal the impact of topology on electromagnetic theory.

Before leaving the subject of complexes, it is worthwhile to summarize everything which we have studied in a single diagram, figure 14.19. While drawn for a three-complex, it suggests all the relationships among the spaces that we have considered.

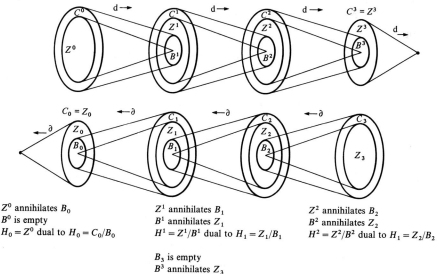

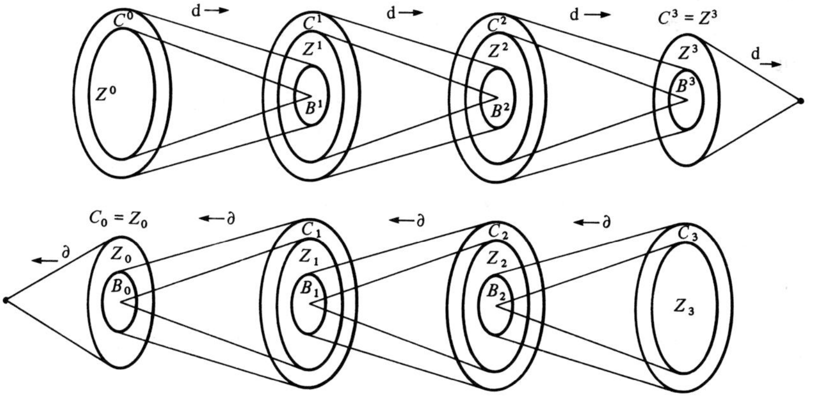

Z^0 annihilates B_0
B^0 is empty
$H_0 = Z^0$ dual to $H_0 = C_0/B_0$

Z^1 annihilates B_1
B^1 annihilates Z_1
$H^1 = Z^1/B^1$ dual to $H_1 = Z_1/B_1$

Z^2 annihilates B_2
B^2 annihilates Z_2
$H^2 = Z^2/B^2$ dual to $H_1 = Z_2/B_2$

B_3 is empty
B^3 annihilates Z_3
$H^3 = C^3/B^3$ dual to $H_3 = Z_3$

Figure 14.19

Summary

A Higher dimensional complexes

You should know the definitions of the spaces Z_k, B_k, and H_k for an n-complex and be able to construct bases for these spaces.

Given diagrams that represent the cells of a two-complex, you should be able to determine whether the complex corresponds to a sphere, a torus, or something else.

B Cohomology and the d operator

For an n-complex, you should be able to define the d operator and the spaces Z^k, B^k, and H^k and be able to construct bases for these spaces.

You should understand how to prove and apply the duality of H^k and H_k.

Exercises

14.1. The five two-complexes in figure 14.20 represent (in some order):
(a) the curved surface of a cylinder;
(b) a Möbius strip (a cylinder with a twist in it);
(c) a sphere;
(d) a torus;
(e) a Klein bottle.
Identify them, and calculate $\dim H_1$ and $\dim H_2$ in each case. Find bases for Z_1, B_1, and H_1 in each case, and check that $\dim C_0 - \dim C_1 + \dim C_2 = \dim H_0 - \dim H_1 + \dim H_2$.

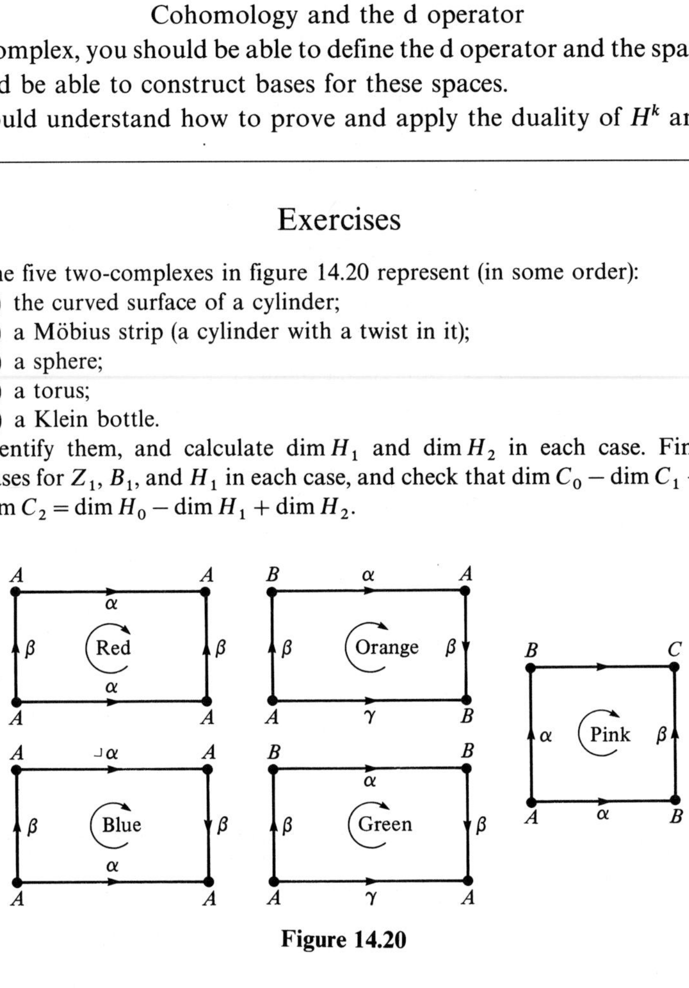

Figure 14.20

14.2. For each complex of Exercise 14.1, construct bases for Z^1, B^1, and H^1.

14.3. An inflatable rubber swimming-pool toy has been divided into three colored regions. Figure 14.21 shows two of the regions.
(a) Determine ∂(Red) and ∂(Blue).
(b) Determine the dimension of Z_2, H_2, and H_0.
(c) Find bases for Z_1, B_1, and H_1.

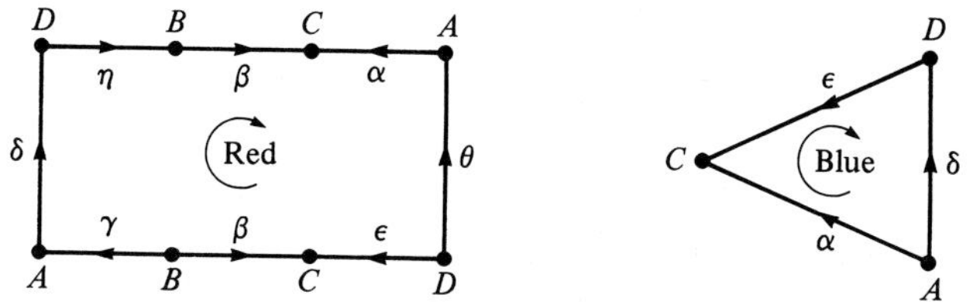

Figure 14.21

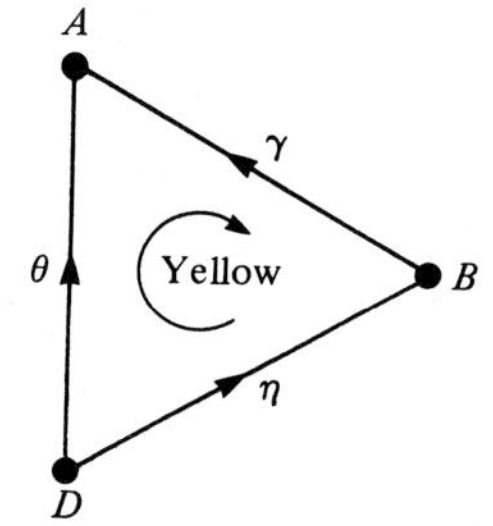

Figure 14.22

(d) Figure 14.22 shows the third region:
After the complete toy has been assembled, what are the dimensions of C_0, C_1, C_2, H_1, H_2? Is the assembled toy a beach ball ($\dim H_0 = 1$, $\dim H_1 = 0$, $\dim H_2 = 1$), a life ring ($\dim H_0 = 1$, $\dim H_1 = 2$, $\dim H_2 = 1$), or something else?

14.4. Three employees of an automobile-tire manufacturing company have divided the surface of an inner tube into three oriented regions, and each employee has attached his name to one region. The resulting two-complex is shown in Figure 14.23.

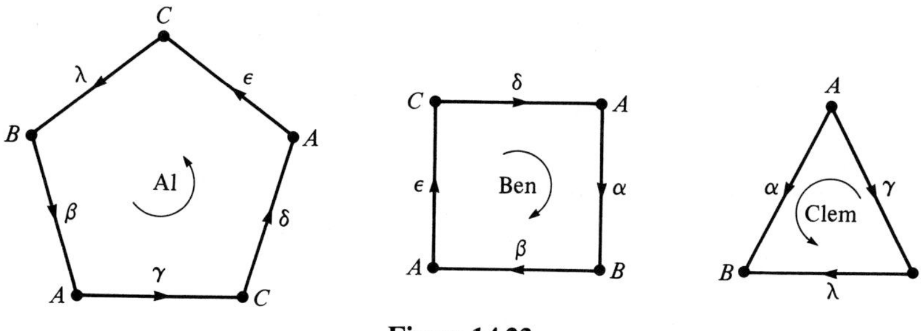

Figure 14.23

(a) What is the dimension of H_0?
(b) Determine the dimension of H_2, and write down a basis for this space. (The two-cells are named Al, Ben, and Clem; just form an appropriate linear combination, but be careful with orientation.) What does your result imply about the assembled 'kit'?
(c) Determine $\dim C_0$, $\dim C_1$, and $\dim C_2$, then calculate $\dim H_1$.
(d) Write down a basis for Z_1, a basis for B_1 and a basis for the quotient space $H_1 = Z_1/B_1$.

(e) Here are three one-cochains, elements of C^1:

$$\begin{array}{lll}
\sigma(\alpha)=1 & \tau(\alpha)=1 & \omega(\alpha)=1 \\
\sigma(\beta)=1 & \tau(\beta)=-1 & \omega(\beta)=-1 \\
\sigma(\gamma)=2 & \tau(\gamma)=2 & \omega(\gamma)=2 \\
\sigma(\delta)=-2 & \tau(\delta)=2 & \omega(\delta)=-2 \\
\sigma(\varepsilon)=-2 & \tau(\varepsilon)=-2 & \omega(\varepsilon)=2 \\
\sigma(\lambda)=1 & \tau(\lambda)=-1 & \omega(\lambda)=-1
\end{array}$$

Identify the cochain which is not a cocycle and determine the two-cochain which is its coboundary (e.g., $d\Omega(\mathrm{Al})=1$, $d\Omega(\mathrm{Ben})=3$, $d\Omega(\mathrm{Clem})=2$).

(f) Identify the cochain which is a coboundary and write down a zero-cochain f whose coboundary it is.

(g) Show that the remaining cochain is a cocycle but not a coboundary. Evaluate it on your two basis elements for H_1.

(h) Construct a second cocycle which is not a coboundary and which will serve as a second basis element for H^1.

(i) Write down a basis for B^1 and a basis for Z^1.

14.5. G.J. Caesar, special envoy of Galactic Rome, was sent to survey the space station New Gaul. His report, which began 'Nova Gallia omnis est divisa in partes tres', included the three maps shown in figure 14.24, which show the portions of the space station controlled by the Aquitani (Aqu), Belgae (Bel) and Celts (Cel) respectively. These warring tribes have been unable to agree on a consistent orientation for the entire surface of the station.

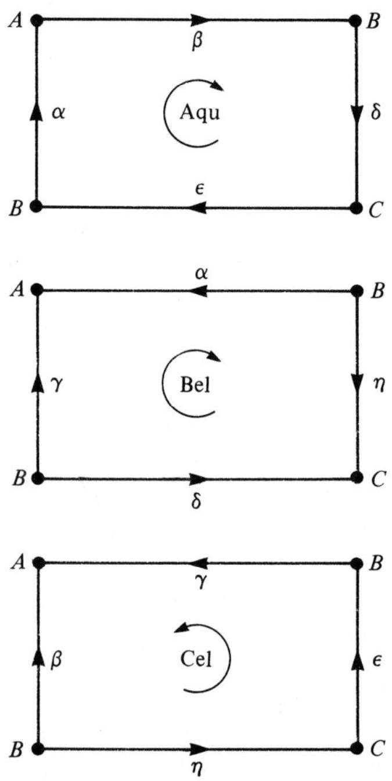

Figure 14.24

Answer the following questions about the New Gaul complex:

(a) Calculate the boundary of each region; ∂(Aqu), etc.

(b) Find a maximal tree. Using this tree, calculate $\dim Z_1$, $\dim B_1$, and $\dim H_1$, and construct a basis for H_1. Express $\alpha - \gamma + \delta - \eta$ and $\beta + \gamma$ in terms of this basis.

(c) If ω is a one-coboundary (element of B^1) satisfying $\omega^\alpha = 2, \omega^\delta = 1$, then what are the values ω^γ and ω^η?

(d) If τ is a one-cochain whose value on each branch is 1, then what are the values $d\tau$(Aqu) and $d\tau$(Bel)?

(e) Calculate $\dim Z^1$. If σ is a one-cocycle (element of Z_1) with $\sigma^\alpha = 3$, $\sigma^\beta = 1$, $\sigma^\delta = 2$, $\sigma^\eta = 4$, what are the values σ^γ and σ^ε?

(f) Construct a basis of H^1 which is dual to the basis for H_1 from part (b). Express the equivalence class $\bar{\sigma}$ in terms of this basis.

(g) Could the surface of New Gaul be a sphere? A cylinder? A torus?

14.6. A colony on a small planet is divided into two provinces, shown as Red and Green in figure 14.25.

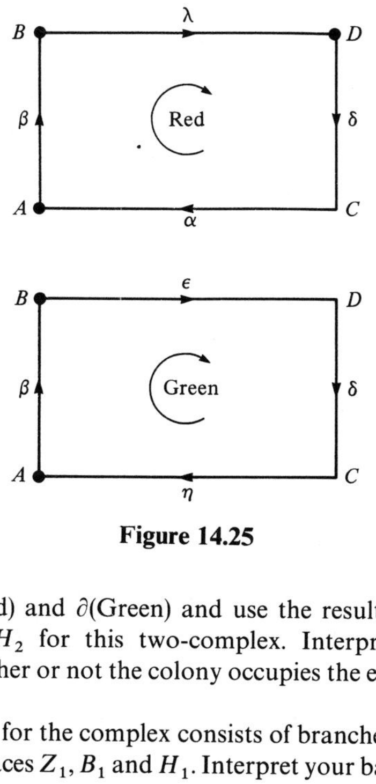

Figure 14.25

(a) Calculate ∂(Red) and ∂(Green) and use the result to determine the dimension of H_2 for this two-complex. Interpret your result to determine whether or not the colony occupies the entire surface of the planet.

(b) A maximal tree for the complex consists of branches α, β, and δ. Find bases for the spaces Z_1, B_1 and H_1. Interpret your basis element for H_1 geographically.

(c) Define the space B^1. Of what space is it the annihilator space? Prove it. For the remainder of the problem, consider a one-cochain $\mathbf{W}$ for which $W^\alpha = 2$, $W^\beta = -1$, $W^\delta = 1$.

(d) Find values of W^γ, W^ε, W^η so that $\mathbf{W}$ is a coboundary.

(e) Suppose that $W^\gamma = 3$, $W^\varepsilon = 2$, $W^\eta = 1$. Calculate $d\mathbf{W}$.

(f) Find values of W^γ, W^ε, W^η such that $\mathbf{W}$ is a cocycle but not a coboundary. Evaluate this cochain on your basis for H_1.

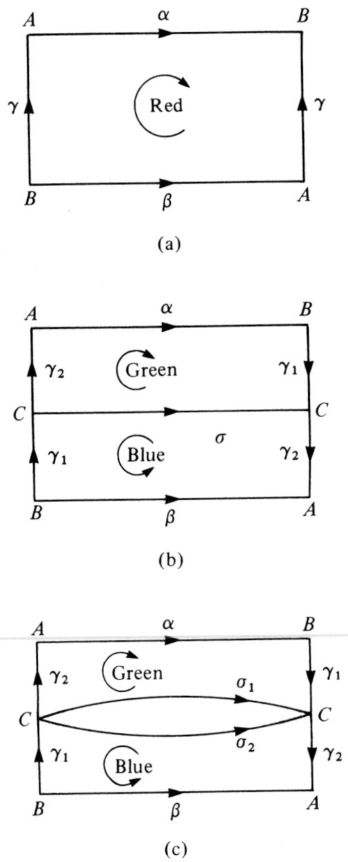

Figure 14.26

14.7.(a) Find the dimension of the spaces C_1, C_2, Z_1, B_1, and H_1 for the complex illustrated in figure 14.26(a).

(b) Find the dimension of H_0 and H_2 and verify

$$\dim C_0 - \dim C_1 + \dim C_2 = \dim H_0 - \dim H_1 + \dim H_2. \quad (14.1)$$

(c) Let a new branch σ and a new node C be added as shown in figure 14.27(b). Write down the matrices $\partial: C_1 \to C_0$, $\partial: C_2 \to C_1$ and compute the matrix $\partial \circ \partial: C_2 \to C_0$ explicitly.

(d) Now a cut is made along branch σ separating it into σ_1 (on the Green side) and σ_2 (on the Blue side). Write down $\partial: C_1 \to C_0$. Find the dimension of H_0 by constructing a basis. What does this say about the network? Construct a basis of H_1 and verify (14.1).

(e) Let τ be a one-cochain with the value 1 on all branches between distinct points, zero otherwise. What are $d\tau$(Green) and $d\tau$(Blue)?

14.8. The two pieces shown in figure 14.27 define a two-complex.

(a) Calculate ∂(Front) and ∂(Back), and use the result to determine $\dim Z_2$.

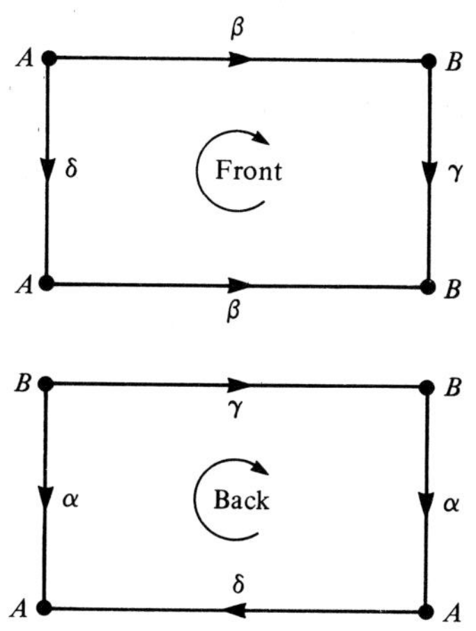

Figure 14.27

(b) Find a maximal tree for the complex, and use it to determine $\dim Z_1$, $\dim B_1$, and $\dim H_1$.
(c) Find a basis for H_1 and express $\bar{\gamma}$ and $\overline{\alpha + \delta + \beta}$ in terms of the basis.
(d) Let ω be a one-coboundary satisfying $\omega^{\beta} = 2$. Calculate ω^{α}.
(e) Let τ be the one-cochain whose value on each branch is 1. Calculate $d\tau$.
(f) Determine $\dim B^2$ for this complex.

15

Complexes situated in $\mathbb{R}^n$

Chapters 15–18 develop the exterior differential calculus as a continuous version of the discrete theory of complexes. In Chapter 15 the basic facts of the exterior calculus are presented: exterior algebra, k-forms, pullback, exterior derivative and Stokes' theorem.

Introduction

In our previous consideration of complexes we have avoided making use of the fact that a complex might be situated in $\mathbb{R}^n$. We ignored the shapes of the wires in electrical networks, and even the question of whether a given network could be constructed in the plane. Now we turn our attention to the special case of a complex situated in $\mathbb{R}^n$ in order to see how cochains can arise naturally from physical or geometrical considerations.

Think first of a one-complex, such as figure 15.1, situated in $\mathbb{R}^3$. The nodes are points in $\mathbb{R}^3$; the branches are differentiable curves in $\mathbb{R}^3$. Since a zero-cochain is determined once we know its value on each node of the complex, we can regard it as a function on the nodes. One obvious way of getting such a function is to have a function defined on $\mathbb{R}^3$. Each such function $\phi(x, y, z)$ determines a zero-cochain Φ by the rule $\Phi^A = \phi(x^A, y^A, z^A)$ where x^A, y^A, z^A are the coordinates of

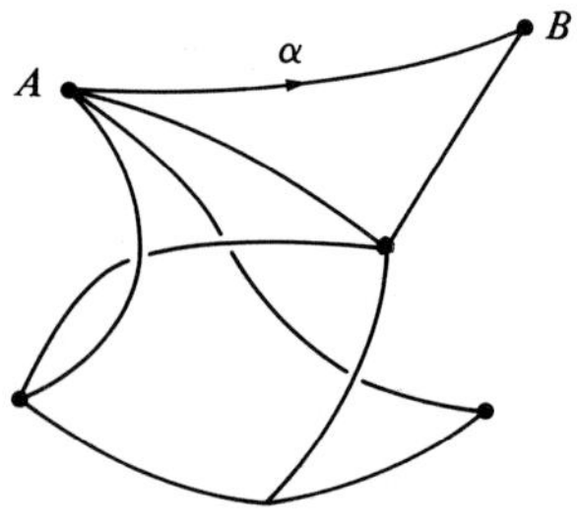

Figure 15.1

node A. The function ϕ may be regarded as an *incipient zero-cochain* in the sense that it defines a zero-cochain on any complex which may be situated in the space.

In order to obtain a one-cochain in a similar manner we need a rule that assigns a number to any branch of a complex. Such a rule can arise from a differential one-form, ω, defined on $\mathbb{R}^3$. Then, if each branch of the complex is a differentiable arc, we can construct a one-cochain by integrating ω along each branch. For example, if branch α is described by the parameterized curve $\alpha(t)$, the one-cochain determined by ω will have the value $\int_\alpha \omega$ on branch α. In this way, a one-form ω can be regarded as an *incipient one-cochain*; it determines a one-cochain on any complex which may be situated in $\mathbb{R}^3$.

Now suppose that ϕ is a continuously differentiable function and we form its differential, the one-form $\omega = \mathrm{d}\phi$. Suppose that α is a branch with $\partial\alpha = B - A$, so that α has a curve $\begin{pmatrix} x(t) \\ y(t) \\ z(t) \end{pmatrix}$ with

$$\begin{pmatrix} x(0) \\ y(0) \\ z(0) \end{pmatrix} = \begin{pmatrix} x_A \\ y_A \\ z_A \end{pmatrix} \quad \text{and} \quad \begin{pmatrix} x(1) \\ y(1) \\ z(1) \end{pmatrix} = \begin{pmatrix} x_B \\ y_B \\ z_B \end{pmatrix}.$$

Then

$$\int_\alpha \omega = \int_\alpha \mathrm{d}\phi = \int_0^1 \alpha^* \mathrm{d}\phi = \int_0^1 \mathrm{d}(\alpha^*\phi) = \alpha^*\phi(1) - \alpha^*\phi(0) = \phi(B) - \phi(A).$$

But the function ϕ determines a zero-cochain Φ, with $\Phi^A = \phi(A)$ and $\Phi^B = \phi(B)$. From this zero-cochain we can form the one-cochain $\mathrm{d}\Phi$, whose value on α is

$$\int_\alpha \mathrm{d}\Phi = \int_{\partial\alpha} \Phi = \Phi^B - \Phi^A = \phi(B) - \phi(A).$$

We have been using the notation d in two quite different ways: as a linear operator which acts on a zero-cochain Φ to produce a one-cochain $\mathrm{d}\Phi$, and as a differential operator which acts on a function ϕ to produce a differential form $\mathrm{d}\phi$. It is now clear that this use of d is consistent in the sense that if the function ϕ gives rise to the zero-cochain Φ, then the one-form $\mathrm{d}\phi$ gives rise to the one-cochain $\mathrm{d}\Phi$. If we denote the space of smooth functions on $\mathbb{R}^3$ by $\Omega^0(\mathbb{R}^3)$ and the space of smooth one-forms by $\Omega^1(\mathbb{R}^3)$, we can summarize the consistency of d by the diagram

$$\begin{array}{ccc} C^0 & \xrightarrow{\mathrm{d}} & C^1 \\ \uparrow & & \uparrow \\ \Omega^0(\mathbb{R}^3) & \xrightarrow[\mathrm{d}]{} & \Omega^1(\mathbb{R}^3) \end{array}$$

which is *commutative* in the sense that we can proceed from a function ϕ to a one-cochain either via C^0 or via Ω^1.

Our aim will be to do something similar for two-cochains and three-cochains. We will introduce, in the next section, objects called differential two-forms and

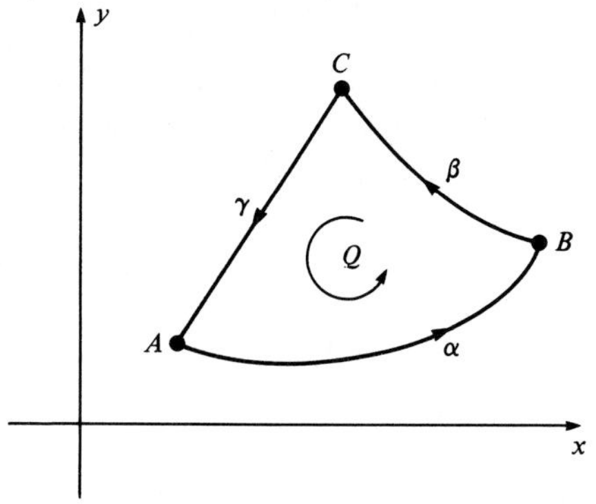

Figure 15.2

three-forms, which will give rise, by suitable processes of integration, to two-cochains and three-cochains respectively. The problem is then to define a differential operator d that converts one-forms into two-forms and two-forms into three-forms, in a manner consistent with the action of d on the corresponding cochains.

For the very special case of a two-complex situated in the plane we can see how this program might work out. An *incipient two-chain* will have to assign a number to each oriented region on the plane; the value assigned to each cell Q in a complex will then define a two-cochain. Now we can proceed from a one-form (incipient one-cochain) ω to a two-cochain $\mathbf{T}$ in two different ways, as suggested by figure 15.2: consider, for example, the path $\alpha + \beta + \gamma$ which is the boundary of the cell Q. One approach is to form a one-cochain $\mathbf{W}$ from ω by evaluating its path integral along each branch of the complex, then apply the operator $\mathrm{d}: C^1 \to C^2$. This yields

$$\int_Q \mathbf{T} = \int_Q \mathrm{d}\mathbf{W} = \int_{\partial Q} \mathbf{W} = W^\alpha + W^\beta + W^\gamma = \int_\alpha \omega + \int_\beta \omega + \int_\gamma \omega.$$

Alternatively, we could hope to apply a differential operator d to ω to obtain a two-form τ, which could then be integrated over the region Q to yield the value of the two-cochain T on the cell Q.

In this case Green's theorem in the plane tells us what form the operator d must take. Recall that, according to Green's theorem, if an oriented region Q in the plane has boundary ∂Q, then the path integral

$$\int_{\partial Q} A\,\mathrm{d}x + B\,\mathrm{d}y = \int_Q \left(\frac{\partial B}{\partial x} - \frac{\partial A}{\partial y}\right)\mathrm{d}x \wedge \mathrm{d}y.$$

Thus it seems promising, in this special case, to introduce two-forms of the form $\tau = f(x, y)\,\mathrm{d}x \wedge \mathrm{d}y$. Such a two-form gives rise to a two-cochain by the process of iterated integration, and a two-form can be obtained from a one-form by the rule

$$\mathrm{d}(A\,\mathrm{d}x + B\,\mathrm{d}y) = \left(\frac{\partial B}{\partial x} - \frac{\partial A}{\partial y}\right)\mathrm{d}x \wedge \mathrm{d}y.$$

15.1. Exterior algebra

The one case in which we already have a differential operator that is perfectly consistent with the coboundary operator is the case of zero-cochains and one-cochains. An incipient zero-cochain is a differentiable function ϕ, which assigns a real number to every point. An incipient one-chain is a one-form ω, which assigns to every point an element of the dual space V^* of whatever underlying vector space V we are using. A function ϕ belongs to the space $\Omega^0(V)$; a one-form ω belongs to the space $\Omega^1(V)$, and we have found an operator $\mathrm{d}: \Omega^0(V) \to \Omega^1(V)$ that is consistent with the coboundary operator.

We shall now construct new spaces $\Omega^2(V), \Omega^3(V), \ldots, \Omega^k(V)$, which will permit the differential operator d to be generalized. An element ω of $\Omega^k(V)$, called a *differential k-form*, will assign to every point in the domain of ω an element of a space called $\Lambda^k(V^*)$, usually pronounced 'wedge k of V^*', as a generalization of the dual space V^*. A k-form is what we shall integrate to determine a k-cochain. We will then be able to define a differential operator $\mathrm{d}: \Omega^k(V) \to \Omega^{k+1}(V)$ that will be consistent with $\mathrm{d}: C^k \to C^{k+1}$ for any complex. As a first step in this direction, we introduce the space $\Lambda^k(V^*)$.

Let V be a vector space of dimension n, and let V^* be its dual space.

The dual space, V^*, is the space of linear functions from V to $\mathbb{R}$: $\mathscr{L}(V; \mathbb{R})$. Let us now consider the space $\mathscr{B}(V, V; \mathbb{R})$ of *bilinear* functions from $V \times V$ to $\mathbb{R}$. To say that such a function, $\tau(\mathbf{v}_1, \mathbf{v}_2)$, is bilinear means that, for fixed $\mathbf{v}_2$, $\tau(\mathbf{v}_1, \mathbf{v}_2)$ is a linear function of its first argument, i.e., $\tau(\alpha\mathbf{v}_1 + \beta\mathbf{w}_1, \mathbf{v}_2) = \alpha\tau(\mathbf{v}_1, \mathbf{v}_2) + \beta\tau(\mathbf{w}_1, \mathbf{v}_2)$. Similarly, for fixed $\mathbf{v}_1$, τ is a linear function of its second argument. The space $\mathscr{B}(V, V; \mathbb{R})$ is a vector space under the usual rule for addition of functions or multiplying a function by a constant.

Within the space of functions $\mathscr{B}(V, V; \mathbb{R})$ is the subspace, denoted $\Lambda^2(V^*)$, of *alternating* bilinear functions. These satisfy the condition $\tau(\mathbf{v}_1, \mathbf{v}_2) = -\tau(\mathbf{v}_2, \mathbf{v}_1)$. From any bilinear function $\sigma(\mathbf{v}_1, \mathbf{v}_2)$ we can create an alternating function by the construction $\tau(\mathbf{v}_1, \mathbf{v}_2) = \sigma(\mathbf{v}_1, \mathbf{v}_2) - \sigma(\mathbf{v}_2, \mathbf{v}_1)$.

It is not difficult to construct elements of $\Lambda^2(V^*)$ from elements of V^*. Given two elements $\omega^i, \omega^j \in V^*$ we can define a function $\omega^i \wedge \omega^j$ by

$$\omega^i \wedge \omega^j(\mathbf{v}_1, \mathbf{v}_2) = \omega^i(\mathbf{v}_1)\omega^j(\mathbf{v}_2) - \omega^i(\mathbf{v}_2)\omega^j(\mathbf{v}_1).$$

Since the functions ω^i and ω^j are linear, $\omega^i \wedge \omega^j$ is bilinear. Furthermore, $\omega^i \wedge \omega^j(\mathbf{v}_2, \mathbf{v}_1) = \omega^i(\mathbf{v}_2)\omega^j(\mathbf{v}_1) - \omega^i(\mathbf{v}_1)\omega^j(\mathbf{v}_2) = -\omega^i \wedge \omega^j(\mathbf{v}_1, \mathbf{v}_2)$ so that $\omega^i \wedge \omega^j$ is alternating.

It follows from the definition of the wedge product that it is distributive with respect to addition:

$$(\omega^1 + \omega^2) \wedge \omega^3 = \omega^1 \wedge \omega^3 + \omega^2 \wedge \omega^3.$$

To prove this, let $\mathbf{v}_1$ and $\mathbf{v}_2$ be arbitrary vectors. Then:

$$\begin{aligned}
[(\omega^1 + \omega^2) \wedge \omega^3](\mathbf{v}_1, \mathbf{v}_2) &= [(\omega^1 + \omega^2)(\mathbf{v}_1)]\omega^3(\mathbf{v}_2) - [(\omega^1 + \omega^2)(\mathbf{v}_2)]\omega^3(\mathbf{v}_1) \\
&= [\omega^1(\mathbf{v}_1) + \omega^2(\mathbf{v}_1)]\omega^3(\mathbf{v}_2) - [\omega^1(\mathbf{v}_2) + \omega^2(\mathbf{v}_2)]\omega^3(\mathbf{v}_1) \\
&= \omega^1 \wedge \omega^3(\mathbf{v}_1, \mathbf{v}_2) + \omega^2 \wedge \omega^3(\mathbf{v}_1, \mathbf{v}_2).
\end{aligned}$$

Suppose that we have chosen dual bases: a basis $\{\mathbf{e}_1, \mathbf{e}_2, \ldots, \mathbf{e}_n\}$ for V, and a basis $\{\varepsilon^1, \varepsilon^2, \ldots, \varepsilon^n\}$ for V^*, such that

$$\varepsilon^i(\mathbf{e}_j) = \begin{cases} 1 & \text{if } i = j, \\ 0 & \text{if } i \neq j. \end{cases}$$

If we express ω^1 and ω^2 in terms of a basis for V^*, then $\omega^1 \wedge \omega^2 = (a_1\varepsilon^1 + a_2\varepsilon^2 + \cdots) \wedge (b_1\varepsilon^1 + b_2\varepsilon^2 + \cdots)$ is a sum of terms of the form $\varepsilon^i \wedge \varepsilon^j$. However, $\varepsilon^j \wedge \varepsilon^i = -\varepsilon^i \wedge \varepsilon^j$. So every element of the form $\omega^1 \wedge \omega^2$ can be expressed in terms of the $\varepsilon^i \wedge \varepsilon^j$ with $i < j$. More generally, let τ be any element of $\Lambda^2(V^*)$. Then the bilinearity and antisymmetry of τ show that τ is determined by the values $\tau(\mathbf{e}_i, \mathbf{e}_j)$, $i < j$. Let us call $\tau(\mathbf{e}_i, \mathbf{e}_j) = b_{ij}$. Then $\sum b_{ij}\varepsilon^i \wedge \varepsilon^j$ takes on the same values as τ on all $\mathbf{e}_k, \mathbf{e}_l$. Hence, $\tau = \sum b_{ij}\varepsilon^i \wedge \varepsilon^j$. Thus the $\varepsilon^i \wedge \varepsilon^j$ span $\Lambda^2(V^*)$. On the other hand, it is easy to check that they are linearly independent. Thus, from the n basis elements of V^*, we can construct $\frac{1}{2}n(n-1)$ independent elements of $\Lambda^2(V^*)$: the elements $\varepsilon^i \wedge \varepsilon^j$, with $i < j$, form a basis for $\Lambda^2(V)$.

In the case where V is the space $\mathbb{R}^3$, it is convenient to name the basis elements of V^* as $\mathrm{d}x$, $\mathrm{d}y$, and $\mathrm{d}z$. Then a basis for $\Lambda^2(\mathbb{R}^{3*})$ consists of the three elements $\mathrm{d}x \wedge \mathrm{d}y$, $\mathrm{d}x \wedge \mathrm{d}z$, and $\mathrm{d}y \wedge \mathrm{d}z$.

Given two elements of V^*, we can use the wedge operator and the rule $\omega^j \wedge \omega^i = -\omega^i \wedge \omega^j$ to multiply and simplify, e.g., $(\mathrm{d}x + \mathrm{d}y) \wedge (\mathrm{d}x + 2\mathrm{d}y - 3\mathrm{d}z) = 2\mathrm{d}x \wedge \mathrm{d}y - 3\mathrm{d}x \wedge \mathrm{d}z + \mathrm{d}y \wedge \mathrm{d}x - 3\mathrm{d}y \wedge \mathrm{d}z = \mathrm{d}x \wedge \mathrm{d}y - 3\mathrm{d}x \wedge \mathrm{d}z - 3\mathrm{d}y \wedge \mathrm{d}z$.

A convenient way to express the action of an element of $\Lambda^2(V^*)$ is by use of determinants:

$$\omega^1 \wedge \omega^2(\mathbf{v}_1, \mathbf{v}_2) = \mathrm{Det}\begin{pmatrix} \omega^1(\mathbf{v}_1) & \omega^1(\mathbf{v}_2) \\ \omega^2(\mathbf{v}_1) & \omega^2(\mathbf{v}_2) \end{pmatrix}.$$

We shall use this notation to generalize the wedge product to the case of more than two factors.

A function f of three vectors in V is called *trilinear* if $f(\mathbf{v}_1, \mathbf{v}_2, \mathbf{v}_3)$ depends linearly on $\mathbf{v}_i$ ($i = 1, 2$ or 3) when the other two vectors are kept fixed. A trilinear function f is called *antisymmetric* if

$$f(\mathbf{v}_2, \mathbf{v}_1, \mathbf{v}_3) = -f(\mathbf{v}_1, \mathbf{v}_2, \mathbf{v}_3)$$

and

$$f(\mathbf{v}_1, \mathbf{v}_3, \mathbf{v}_2) = -f(\mathbf{v}_1, \mathbf{v}_2, \mathbf{v}_3).$$

From these equations it follows, for example, that

$$f(\mathbf{v}_3, \mathbf{v}_2, \mathbf{v}_1) = -f(\mathbf{v}_1, \mathbf{v}_2, \mathbf{v}_3).$$

For instance, the determinant of a 3×3 matrix is a trilinear antisymmetric function of the columns.

We turn now to the construction of the space $\Lambda^3(V^*)$, the space of alternating trilinear functions from V to $\mathbb{R}$. We can construct elements for this space by using the wedge notation and an extension of the definition in terms of determinants, as follows:

$$\omega^i \wedge \omega^j \wedge \omega^k(\mathbf{v}_1, \mathbf{v}_2, \mathbf{v}_3) = \mathrm{Det}\begin{pmatrix} \omega^i(\mathbf{v}_1) & \omega^i(\mathbf{v}_2) & \omega^i(\mathbf{v}_3) \\ \omega^j(\mathbf{v}_1) & \omega^j(\mathbf{v}_2) & \omega^j(\mathbf{v}_3) \\ \omega^k(\mathbf{v}_1) & \omega^k(\mathbf{v}_2) & \omega^k(\mathbf{v}_3) \end{pmatrix}.$$

The following properties of the function $\omega^i \wedge \omega^j \wedge \omega^k$ are immediate consequences of the linearity of $\omega^i(\mathbf{v})$ and elementary properties of determinants.

(a) For fixed $\mathbf{v}_2, \mathbf{v}_3, \omega^i \wedge \omega^j \wedge \omega^k$ is a linear function of $\mathbf{v}_1$, because the entries of the first column are linear in $\mathbf{v}_1$, and the determinant is a linear function of the entries in its first column.

(b) $\omega^i \wedge \omega^j \wedge \omega^k$ is also linear in $\mathbf{v}_2$ and $\mathbf{v}_3$; i.e. it is trilinear.

(c) $\omega^i \wedge \omega^j \wedge \omega^k$ changes sign if two adjacent arguments (e.g. $\mathbf{v}_1$ and $\mathbf{v}_2$, or $\mathbf{v}_2$ and $\mathbf{v}_3$) are interchanged, because a determinant changes sign if adjacent columns are interchanged.

(d) $\omega^i \wedge \omega^j \wedge \omega^k$ changes sign if two adjacent 'factors' are interchanged (e.g. $\varepsilon^j \wedge \varepsilon^i \wedge \varepsilon^k = -\varepsilon^i \wedge \varepsilon^j \wedge \varepsilon^k$), because a determinant changes sign if adjacent rows are interchanged.

(e) $\omega^i \wedge \omega^i \wedge \omega^j = 0$, because a determinant with two equal rows is zero.

(f) The triple wedge product is distributive with respect to addition in each factor, because a determinant is distributive with respect to addition in each row.

(g) If $\mathbf{v}_1, \mathbf{v}_2$, and $\mathbf{v}_3$ are linearly dependent, then $\omega^i \wedge \omega^j \wedge \omega^k\ (\mathbf{v}_1, \mathbf{v}_2, \mathbf{v}_3) = 0$ because a determinant is zero if its columns are linearly dependent.

(h) If ω^i, ω^j, and ω^k are linearly dependent elements of V^*, then $\omega^i \wedge \omega^j \wedge \omega^k = 0$, because a determinant is zero if its rows are linearly dependent.

As above, it is straightforward to prove that the elements $\varepsilon^i \wedge \varepsilon^j \wedge \varepsilon^k$, $i<j<k$, form a basis for $\Lambda^3(V^*)$ where $\varepsilon^i, \varepsilon^j, \varepsilon^k$ are basis elements of V^*. There are n choices for $i, n-1$ for j, $n-2$ for k, because, if any two indices are equal, $\varepsilon^i \wedge \varepsilon^j \wedge \varepsilon^k = 0$. But now each combination of three indices has appeared in $3 \cdot 2 \cdot 1 = 6$ permutations, which correspond to only one element with $i<j<k$. Hence $\dim \Lambda^3(V^*) = \frac{1}{6}n(n-1)(n-2) = n!/(3!(n-3)!)$.

It is now straightforward to extend the preceding definition to the space $\Lambda^k(V^*)$. This is the subspace of the space of multilinear functions $\mathscr{L}(V, V, \ldots, V; \mathbb{R})$ which is alternating in the sense that it changes sign if any two adjacent arguments are interchanged. An element of $\Lambda^k(V^*)$, called a k-form, can be constructed from k elements of V^* by the definition:

$$\omega^{i_1} \wedge \omega^{i_2} \wedge \cdots \wedge \omega^{i_k}(\mathbf{v}_1, \mathbf{v}_2, \ldots \mathbf{v}_k) = \mathrm{Det}\begin{pmatrix} \omega^{i_1}(\mathbf{v}_1) & \omega^{i_1}(\mathbf{v}_2) & \cdots & \omega^{i_1}(\mathbf{v}_k) \\ \omega^{i_2}(\mathbf{v}_1) & \omega^{i_2}(\mathbf{v}_2) & \cdots & \omega^{i_2}(\mathbf{v}_k) \\ \vdots & & & \\ \omega^{i_k}(\mathbf{v}_1) & \omega^{i_k}(\mathbf{v}_2) & \cdots & \omega^{i_k}(\mathbf{v}_k) \end{pmatrix}.$$

The dimension of $\Lambda^k(V^*)$ is $n!/(k!\,(n-k)!)$ and a basis consists of all elements $\varepsilon^{i_1} \wedge \varepsilon^{i_2} \wedge \cdots \varepsilon^{i_k}$ with $i_1 < i_2 < \cdots < i_k$. Notice that $\Lambda^n(V^*)$ is one-dimensional, with the single basis element $\varepsilon^1 \wedge \varepsilon^2 \wedge \cdots \varepsilon^n$, and $\Lambda^k(V^*) = \{0\}$ if $k > n$.

There is one further algebraic operation that will be of importance to us – 'exterior' (sometimes called 'wedge') multiplication of an element of $\Lambda^p(V^*)$ with an element of $\Lambda^q(V^*)$ to obtain an element of $\Lambda^{p+q}(V^*)$. On basis elements it is defined by

$$(\varepsilon^{i_1} \wedge \varepsilon^{i_2} \wedge \cdots \wedge \varepsilon^{i_p}) \wedge (\varepsilon^{j_1} \wedge \cdots \wedge \varepsilon^{j_q}) = \varepsilon^{i_1} \wedge \cdots \wedge \varepsilon^{i_p} \wedge \varepsilon^{j_1} \cdots \wedge \varepsilon^{j_q}.$$

It is then extended to linear combinations of basis elements so as to be linear in each factor separately, i.e. the distributive law for this multiplication should hold. For example, if

$$\omega = 5\varepsilon^1 \wedge \varepsilon^7 + 3\varepsilon^2 \wedge \varepsilon^4 \quad \text{and} \quad \sigma = 6\varepsilon^1 \wedge \varepsilon^5 + 2\varepsilon^3 \wedge \varepsilon^5 - 9\varepsilon^3 \wedge \varepsilon^6$$

are both elements of $\Lambda^2(V^*)$ then the element $\omega \wedge \sigma$ of $\Lambda^4(V^*)$ is computed as

$$\begin{aligned}\omega \wedge \sigma &= 5(\varepsilon^1 \wedge \varepsilon^7) \wedge \sigma + 3(\varepsilon^2 \wedge \varepsilon^4) \wedge \sigma \\ &= 30\varepsilon^1 \wedge \varepsilon^7 \wedge \varepsilon^1 \wedge \varepsilon^5 + 10\varepsilon^1 \wedge \varepsilon^7 \wedge \varepsilon^3 \wedge \varepsilon^5 - 45\varepsilon^1 \wedge \varepsilon^7 \wedge \varepsilon^3 \wedge \varepsilon^6 \\ &\quad + 18\varepsilon^2 \wedge \varepsilon^4 \wedge \varepsilon^1 \wedge \varepsilon^5 + 6\varepsilon^2 \wedge \varepsilon^4 \wedge \varepsilon^3 \wedge \varepsilon^5 - 27\varepsilon^2 \wedge \varepsilon^4 \wedge \varepsilon^3 \wedge \varepsilon^6.\end{aligned}$$

The first summand vanishes because of the repeated factor of ε^1. The remaining terms need some rearranging to make them into basis elements, and this may introduce some sign changes. For example,

$$\varepsilon^1 \wedge \varepsilon^7 \wedge \varepsilon^3 \wedge \varepsilon^5 = \varepsilon^1 \wedge \varepsilon^3 \wedge \varepsilon^5 \wedge \varepsilon^7$$

since the ε^7 has to be moved past two factors to get it to its proper position while

$$\varepsilon^2 \wedge \varepsilon^4 \wedge \varepsilon^3 \wedge \varepsilon^6 = -\varepsilon^2 \wedge \varepsilon^3 \wedge \varepsilon^4 \wedge \varepsilon^6.$$

Thus

$$\begin{aligned}\omega \wedge \sigma = {} & 10\varepsilon^1 \wedge \varepsilon^3 \wedge \varepsilon^5 \wedge \varepsilon^7 - 45\varepsilon^1 \wedge \varepsilon^3 \wedge \varepsilon^6 \wedge \varepsilon^7 + 18\varepsilon^1 \wedge \varepsilon^2 \wedge \varepsilon^4 \wedge \varepsilon^5 \\ & - 6\varepsilon^2 \wedge \varepsilon^3 \wedge \varepsilon^4 \wedge \varepsilon^5 + 27\varepsilon^2 \wedge \varepsilon^3 \wedge \varepsilon^4 \wedge \varepsilon^6.\end{aligned}$$

At this point we strongly urge you to write down many examples of exterior multiplication so as to get the hang of how it works. It will be one of the basic computational tools for the rest of the book so the effort invested now in gaining this computational skill is very worthwhile. By working out many examples you should convince yourself of the truth of the following rules:

The associative law: $(\omega \wedge \sigma) \wedge \tau = \omega \wedge (\sigma \wedge \tau)$.

The distributive law: $\omega \wedge (\sigma + \tau) = \omega \wedge \sigma + \omega \wedge \tau$.

Anticommutativity (nowadays called supercommutativity):

$$\omega \wedge \tau = (-1)^{pq}\tau \wedge \omega \quad \text{if} \quad \omega \varepsilon \Lambda^p(V^*) \quad \text{and} \quad \tau \varepsilon \Lambda^q(V^*).$$

These rules, in turn, facilitate the computation of exterior multiplication.

From a logical point of view, our definition is somewhat unsatisfactory in that it seems to depend on the choice of basis. Strictly speaking we should prove that, if we choose a different basis of V, and, correspondingly, different bases of $\Lambda^p(V^*)$, the actual multiplication of elements does not change. This can be done directly, but is a bit tedious. A better way is to give a more abstract definition of multiplication which automatically satisfies the above rules so that the formula we wrote down for multiplication becomes a consequence of the definition. We do this in the appendix to Chapter 18, and if you are so inclined, you can read that appendix right now. It can be read independently of any other material. But the important point is to gain some computational familiarity with exterior multiplication.

As an example, let the vector space V be four-dimensional. With eventual application to spacetime in mind, we shall name the basis elements of V^* as $\mathrm{d}t$, $\mathrm{d}x$, $\mathrm{d}y$, and $\mathrm{d}z$. The complete collection of spaces $\Lambda^k(V^*)$ is then as follows:

$\Lambda^0(V^*)$ is one-dimensional: if we need to name the basis element, we call it 1.

$\Lambda^1(V^*)$ is four-dimensional: it is the space V^*. A basis is $\{\mathrm{d}t, \mathrm{d}x, \mathrm{d}y, \mathrm{d}z\}$.

$\Lambda^2(V^*)$ is six-dimensional, with basis elements $\mathrm{d}t \wedge \mathrm{d}x, \mathrm{d}t \wedge \mathrm{d}y$, $\mathrm{d}t \wedge \mathrm{d}z$, $\mathrm{d}x \wedge \mathrm{d}y$, $\mathrm{d}x \wedge \mathrm{d}z$, $\mathrm{d}y \wedge \mathrm{d}z$.

$\Lambda^3(V^*)$ is four-dimensional, with basis elements $\mathrm{d}t \wedge \mathrm{d}x \wedge \mathrm{d}y, \mathrm{d}t \wedge \mathrm{d}x \wedge \mathrm{d}z$, $\mathrm{d}t \wedge \mathrm{d}y \wedge \mathrm{d}z$, $\mathrm{d}x \wedge \mathrm{d}y \wedge \mathrm{d}z$.

$\Lambda^4(V^*)$ is one-dimensional, with basis element $\mathrm{d}t \wedge \mathrm{d}x \wedge \mathrm{d}y \wedge \mathrm{d}z$.

15.2. *k*-forms and the d operator

To proceed from the space $\Lambda^k(V^*)$ to the space $\Omega^k(V)$, we now make the same extension as in going from the dual space V^* (also called $\Lambda^1(V^*)$) to the space $\Omega^1(V)$. An element of $\Omega^k(V)$ is a function which assigns an element of $\Lambda^k(V^*)$ to each point; that is, the coefficients of the basis elements of $\Lambda^k(V^*)$ are now real-valued functions. The general element of $\Omega^2(\mathbb{R}^3)$, for example, is

$$a(x, y, z)\mathrm{d}x \wedge \mathrm{d}y + b(x, y, z)\mathrm{d}x \wedge \mathrm{d}z + c(x, y, z)\mathrm{d}y \wedge \mathrm{d}z$$

where a, b, c are real valued *functions*. As far as *algebraic* operations involving addition and the wedge multiplication are concerned, elements of $\Omega^k(V)$ behave exactly as elements of $\Lambda^k(V^*)$ do. These elements of $\Omega^k(V)$, called *differential k-forms* or more commonly just *k-forms*, will serve as our incipient k-cochains. We must now learn how to differentiate and integrate them. We take $V = \mathbb{R}^n$ with $\mathrm{d}x^1, \ldots, \mathrm{d}x^n$ a basis for V^*. (We also use $\mathrm{d}x$, $\mathrm{d}y$, etc.)

The differential operator d which assigns a one-form to a zero-form is already familiar. If $f(x, y, z)$ is a differentiable function (a zero-form), then

$$\mathrm{d}f = \frac{\partial f}{\partial x}\mathrm{d}x + \frac{\partial f}{\partial y}\mathrm{d}y + \frac{\partial f}{\partial z}\mathrm{d}z$$

Eventually, we want to consider a k-form as an object which we integrate over a 'k-dimensional hypersurface' and want to prove Stokes' theorem which says that

$$\int_S \mathrm{d}\omega = \int_{\partial S} \omega,$$

where S is a 'k-dimensional hypersurface' and ∂S its 'boundary'. In Chapter 8 we saw that this equation, applied to infinitesimal parallelograms, forced the definition of the d operator as applied to one forms. We saw there that

$$\mathrm{d}(f\,\mathrm{d}g) = df \wedge \mathrm{d}g.$$

Since every one-form is a sum of such expressions, this determined the definition of d on one-forms. Perhaps now is a useful time to go back and review the basic formulas of Chapter 8 as summarized on page 305.

Let us now proceed more algebraically and follow the example provided by one-forms. We would like to have

$$\mathrm{d}(f\,\mathrm{d}g \wedge \mathrm{d}h \wedge \cdots) = \mathrm{d}f \wedge \mathrm{d}g \wedge \mathrm{d}h \wedge \cdots. \tag{*}$$

This forces us to define the operator d acting on k-forms as follows:

$$\begin{aligned}&\text{If } \tau = f\,\mathrm{d}x^{i_1} \wedge \cdots \wedge \mathrm{d}x^{i_k} + g\,\mathrm{d}x^{j_1} \wedge \cdots \wedge \mathrm{d}x^{j_k} + \cdots \\ &\text{then } \mathrm{d}\tau = \mathrm{d}f \wedge \mathrm{d}x^{i_1} \cdots \wedge \mathrm{d}x^{i_k} + \mathrm{d}g \wedge \mathrm{d}x^{j_1} \cdots \wedge \mathrm{d}x^{j_k}\end{aligned} \tag{15.1}$$

After the operator d has been applied, some rearrangement of terms may be necessary in order to collect the coefficients of each basis element.

As an example of the action of d on a two-form in $\mathbb{R}^3$, let $\tau = B_x\mathrm{d}y \wedge \mathrm{d}z - B_y\mathrm{d}x \wedge \mathrm{d}z + B_z\mathrm{d}x \wedge \mathrm{d}y$ where B_x, B_y, B_z are differentiable functions of x, y, z which may be regarded as the components of a vector field. Then $\mathrm{d}\tau = \mathrm{d}B_x \wedge \mathrm{d}y \wedge \mathrm{d}z - \mathrm{d}B_y \wedge \mathrm{d}x \wedge \mathrm{d}z + \mathrm{d}B_z \wedge \mathrm{d}x \wedge \mathrm{d}y$. Expanding the differentials, and making use of the fact that a wedge product is zero if any two factors are the same, we have:

$$\mathrm{d}\tau = \left(\frac{\partial B_x}{\partial x} + \frac{\partial B_y}{\partial y} + \frac{\partial B_z}{\partial z}\right)\mathrm{d}x \wedge \mathrm{d}y \wedge \mathrm{d}z.$$

For those of you familiar with vector analysis, notice the similarity between $\mathrm{d}:\Omega^2(\mathbb{R}^3) \to \Omega^3(\mathbb{R}^3)$ and the operation div.

There are several properties of the operator d that are useful and worth recording. In every case, it will be sufficient to verify the property for the case of a k-form $\Omega = f\,\mathrm{d}x \wedge \mathrm{d}y \wedge \cdots$. The general case then follows immediately by the linearity of d.

The first property concerns the result of applying d to the product of a function and a k-form. Let f and y be differentiable functions; let $\Omega = g\mathrm{d}x \wedge \mathrm{d}y \wedge \cdots$. Then $\mathrm{d}(f\Omega) = \mathrm{d}(fg\mathrm{d}x \wedge \mathrm{d}y \wedge \cdots) = \mathrm{d}(fg) \wedge \mathrm{d}x \wedge \mathrm{d}y$. But we have the product rule $\mathrm{d}(fg) = f\,\mathrm{d}g + g\mathrm{d}f$ for differentials. It follows that $\mathrm{d}(f\Omega) = f\mathrm{d}g \wedge \mathrm{d}x \wedge \mathrm{d}y \wedge \cdots + g\mathrm{d}f \wedge \mathrm{d}x \wedge \mathrm{d}y \wedge \cdots$ or $\mathrm{d}(f\Omega) = f\mathrm{d}(g\mathrm{d}x \wedge \mathrm{d}y \wedge \cdots) + \mathrm{d}f \wedge (g\mathrm{d}x \wedge \mathrm{d}y \wedge)\cdots$. We conclude that

$$\mathrm{d}(f\Omega) = f\mathrm{d}\Omega + \mathrm{d}f \wedge \Omega.$$

The second property is a more general product rule which applies to the product of a p-form ω and a q-form Ω. Let $\omega = f\mathrm{d}x^1 \wedge \cdots \wedge \mathrm{d}x^p$ and $\Omega = g\mathrm{d}y^1 \wedge \cdots \wedge \mathrm{d}y^q$ where f and g are differentiable functions. $\mathrm{d}(\omega \wedge \Omega) = \mathrm{d}(fg\mathrm{d}x^1 \wedge \cdots \wedge \mathrm{d}x^p \wedge \mathrm{d}y^1 \wedge \cdots \wedge \mathrm{d}y^q)$. So $\mathrm{d}(\omega \wedge \Omega) = f(\mathrm{d}g \wedge \mathrm{d}x^1 \wedge \cdots \wedge \mathrm{d}x^p \wedge \mathrm{d}y^1 \wedge \cdots \wedge \mathrm{d}y^q) + g\mathrm{d}f \wedge \mathrm{d}x^1 \wedge \cdots \wedge \mathrm{d}x^p \wedge \mathrm{d}y^1 \wedge \cdots \wedge \mathrm{d}y^q$. In the first term, we now interchange $\mathrm{d}g$ with the factor $\mathrm{d}x^1 \wedge \cdots \wedge \mathrm{d}x^p$, which introduces a sign $(-1)^p$. Then:

$$\begin{aligned}\mathrm{d}(\omega \wedge \Omega) = {} & (-1)^p(f\mathrm{d}x^1 \wedge \cdots \wedge \mathrm{d}x^p) \wedge (\mathrm{d}g \wedge \mathrm{d}y^1 \wedge \cdots \wedge \mathrm{d}y^p) \\ & + (\mathrm{d}f \wedge \mathrm{d}x^1 \wedge \cdots \wedge \mathrm{d}x^p) \wedge (g\mathrm{d}y^1 \wedge \cdots \wedge \mathrm{d}y^p).\end{aligned}$$

As every p-form is a sum of terms of the form $f\mathrm{d}x^1 \wedge \cdots \wedge \mathrm{d}x^p$ and every q-form is a sum of terms of the form $g\mathrm{d}y^1 \wedge \cdots \wedge \mathrm{d}y^p$, we conclude that

$$\mathrm{d}(\omega \wedge \Omega) = (-1)^p \omega \wedge \mathrm{d}\Omega + \mathrm{d}\omega \wedge \Omega. \tag{15.2}$$

The third property concerns the application of d twice in succession. We consider first $\mathrm{d}(\mathrm{d}f)$ where f is a twice-differentiable function $f(x, y, \ldots)$. In this case,

$$\mathrm{d}(\mathrm{d}f) = \mathrm{d}\left(\frac{\partial f}{\partial x}\mathrm{d}x + \frac{\partial f}{\partial y}\mathrm{d}y + \cdots\right) = \frac{\partial^2 f}{\partial y \partial x}\mathrm{d}y \wedge \mathrm{d}x + \frac{\partial^2 f}{\partial x \partial y}\mathrm{d}x \wedge \mathrm{d}y + \cdots.$$

From the equality of mixed partial derivatives and the relation $\mathrm{d}y \wedge \mathrm{d}x = -\mathrm{d}x \wedge \mathrm{d}y$, it follows that $\mathrm{d}(\mathrm{d}f) = 0$. Now consider the more general case of a k-form $\omega = f\,\mathrm{d}x \wedge \mathrm{d}y \wedge \cdots$. In this case, $\mathrm{d}\omega = \mathrm{d}f \wedge (\mathrm{d}x \wedge \mathrm{d}y \wedge \cdots)$ and $\mathrm{d}(\mathrm{d}\omega) = \mathrm{d}f \wedge \mathrm{d}(\mathrm{d}x \wedge \mathrm{d}y \wedge \cdots) + \mathrm{d}(\mathrm{d}f) \wedge \mathrm{d}x \wedge \mathrm{d}y \wedge \cdots$. But $\mathrm{d}(\mathrm{d}x \wedge \mathrm{d}y \wedge \cdots) = 0$ and $\mathrm{d}(\mathrm{d}f) = 0$, so we conclude that, in general,

$$\mathrm{d}(\mathrm{d}\omega) = 0. \tag{15.3}$$

This is consistent with the property

$$\boxed{\mathrm{d} \circ \mathrm{d} = 0}$$

of the coboundary operator.

Of course equation (*) now follows from (15.2) and (15.3)

If we think of $\mathrm{d}x$ as being the result of applying the operator d to the coordinate function x, then the definition (15.1) becomes a consequence of (15.2) and (15.3). So once you have convinced yourself that the operator d actually exists, all that you have to remember about d are the following four properties:

(1) $\mathrm{d}f = (\partial f/\partial x)\mathrm{d}x + (\partial f/\partial y)\mathrm{d}y + \cdots$, where $x, y, \ldots$ are the coordinate functions,
(2) d respects addition, i.e. $\mathrm{d}(\omega + \sigma) = \mathrm{d}\omega + \mathrm{d}\sigma$,
(3) how d acts on a product, i.e. (15.2), and
(4) $\mathrm{d} \circ \mathrm{d} = 0$.

From these four rules you can compute d of any form.

15.3. Integration of k-forms

To complete the identification of k-forms as *incipient k-cochains* we must now explain how a k-form Ω assigns a k-cochain to every complex with differentiable cells situated in an n-dimensional space. Let us first review the case $k = 1$. Given a smooth curve in $\mathbb{R}^n$ which joins point A to point B, and a one-form ω, we wish to evaluate ω on the curve. To achieve this, we choose any smooth parameterization of the curve, any smooth mapping $\alpha\colon \mathbb{R} \to \mathbb{R}^n$ which maps the interval $[0, 1]$ into the desired curve, with $\alpha(0) = A$ and $\alpha(1) = B$, as shown in figure 15.3. We then *pull back* the one-form ω to obtain $\alpha^*\omega$, an expression of the form $g(t)\mathrm{d}t$, and we define the path integral of the form ω over the curve α as $\int_\alpha \omega = \int_0^1 (\alpha^*\omega)$. The

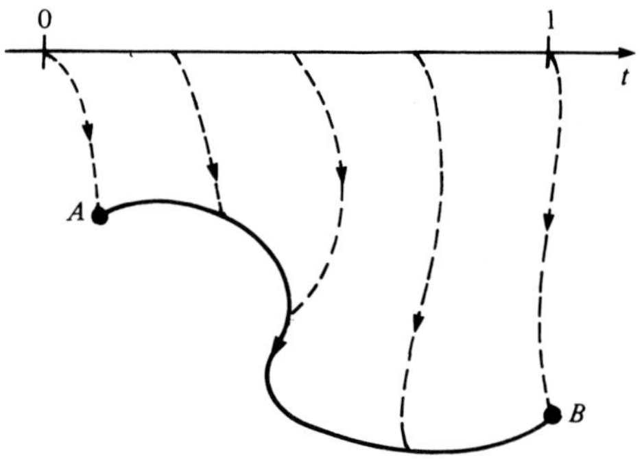

Figure 15.3

usefulness of this definition lies in the fact that the result is independent of the particular parameterization, α, chosen for the curve. The strategy in this case was to reduce the integral over a curve to an integral over a much simpler region (the interval [0, 1]). The crucial construction was the pullback procedure for a one-form: if $\omega = f\mathrm{d}x + g\mathrm{d}y + \cdots$ then $\alpha^*\omega = (\alpha^*f)\mathrm{d}(\alpha^*x) + (\alpha^*g)\mathrm{d}(\alpha^*y) + \cdots$. What had to be proved (from the chain rule) was that the result was independent of the parameterization.

The crucial ingredient in this definition, and in the corresponding theory of two-dimensional integration that we discussed in Chapter 8, is the notion of pullback. So our first order of business is to generalize the notion of pullback: Let $\phi:\mathbb{R}^k \to \mathbb{R}^n$ be a differentiable map. We wish to define an operation, ϕ^*, called pullback, which assigns to any differential form ω (of any degree) on $\mathbb{R}^n$ a differential form $\phi^*\omega$ (of the same degree) on $\mathbb{R}^k$. We would like ϕ^* to preserve addition and multiplication of forms, that is, we would like to have

$$\phi^*(\omega_1 + \omega_2) = \phi^*\omega_1 + \phi^*\omega_2 \tag{15.4}$$

and

$$\phi^*(\omega_1 \wedge \omega_2) = \phi^*\omega_1 \wedge \phi^*\omega_2. \tag{15.5}$$

We would also like ϕ^* to be our old pullback when applied to 'zero-forms', that is to functions, and when applied to linear differential forms. More precisely, if $f \in \Omega^0(\mathbb{R}^n)$ is a function, we want

$$\phi^* f = f \circ \phi, \tag{15.6}$$

and

$$\phi^*\mathrm{d}f = \mathrm{d}(\phi^* f). \tag{15.7}$$

The requirements (15.4)–(15.7) completely force our hand. For example, consider the map $\phi: \mathbb{R}^3 \to \mathbb{R}^3$ given by

$$\phi\begin{pmatrix} r \\ \psi \\ \theta \end{pmatrix} = \begin{pmatrix} x \\ y \\ z \end{pmatrix} \quad \text{where} \quad \begin{aligned} x &= r\sin\theta\cos\psi, \\ y &= r\sin\theta\sin\psi, \\ z &= r\cos\theta. \end{aligned}$$

Thus

$$\phi^* x = r \sin\theta \cos\psi$$

etc. by (15.6). Suppose we want to compute $\phi(\mathrm{d}x \wedge \mathrm{d}y \wedge \mathrm{d}z)$. Then

$$\begin{aligned}\phi^*(\mathrm{d}x \wedge \mathrm{d}y \wedge \mathrm{d}z) &= \phi^* \mathrm{d}x \wedge \phi^* \mathrm{d}y \wedge \phi^* \mathrm{d}z && \text{by (15.5)}\\ &= (\mathrm{d}\phi^* x) \wedge (\mathrm{d}\phi^* y) \wedge (\mathrm{d}\phi^* z) && \text{by (15.7)}\\ &= [\mathrm{d}(r \sin\theta \cos\psi)] \wedge [\mathrm{d}(r \sin\theta \sin\psi)] \wedge [\mathrm{d}(r\cos\theta)] && \text{by (15.6).}\end{aligned}$$

Within each factor of this triple product we apply the usual rule for computing the differential of a function, so that $\mathrm{d}(r \sin\theta \cos\psi) = \sin\theta \cos\psi \, \mathrm{d}r - r \sin\theta \sin\psi \, \mathrm{d}\psi + r \cos\theta \cos\psi \, \mathrm{d}\theta$, etc. We then use the rules of exterior multiplication to express the triple product in terms of $\mathrm{d}r \wedge \mathrm{d}\theta \wedge \mathrm{d}\psi$. We find that

$$\phi^*(\mathrm{d}x \wedge \mathrm{d}y \wedge \mathrm{d}z) = r^2 \sin\theta \, \mathrm{d}r \wedge \mathrm{d}\theta \wedge \mathrm{d}\psi.$$

(This formula is usually called the 'expression of Euclidean volume in polar coordinates'.)

In general, requirements (15.4)–(15.7) force us to define

$$\phi^*(f \, \mathrm{d}x^i \wedge \mathrm{d}x^i \wedge \cdots \wedge \mathrm{d}x^i) = (f \circ \phi) \mathrm{d}(x^i \circ \phi) \wedge \cdots \wedge \mathrm{d}(x^i \circ \phi), \tag{15.8}$$

where the x^i ($i = 1, \ldots, n$) are coordinates on $\mathbf{R}^n$. We must then define ϕ^* on the most general form, which is a sum of such expressions, by (15.4). If we take (15.8) and its extension to sums as the definition of pullback, then we must go back and check that (15.4)–(15.7) actually hold. This is straightforward and will be left as an exercise.

From the definition of d and the properties of pullback it follows that

$$\begin{aligned}\phi^*[\mathrm{d}(f \, \mathrm{d}x^i \wedge \mathrm{d}x^j \wedge \cdots)] &= \phi^*[\mathrm{d}f \wedge \mathrm{d}x^i \wedge \cdots]\\ &= \mathrm{d}(\phi^* f) \wedge \mathrm{d}\phi^* x^i \wedge \cdots\\ &= \mathrm{d}[(\phi^* f) \mathrm{d}\phi^* x^i \wedge \cdots]\\ &= \mathrm{d}[\phi^*(f \, \mathrm{d}x^i \wedge \cdots)].\end{aligned}$$

Applying this computation to sums of expressions of the above form we get the basic formula

$$\phi^* \mathrm{d}\omega = \mathrm{d}\phi^* \omega. \tag{15.9}$$

This result is of central importance. It says that first applying d and then pulling back is the same as first pulling back and then applying d. It can be thought of as a version of the chain rule.

There is one final formula about pullbacks that we should record. Suppose that $\psi : \mathbb{R}^p \to \mathbb{R}^k$ so that we can form the composite map $\phi \circ \psi : \mathbb{R}^p \to \mathbb{R}^n$. Then for any function f we know that

$$(\phi \circ \psi)^* f = f \circ (\phi \circ \psi) = (f \circ \phi) \circ \psi = \psi^* \phi^* f.$$

But then it follows from our rules that

$$(\psi \circ \phi)^* \omega = \phi^*(\psi^* \omega) \tag{15.10}$$

for any differential form ω.

Let us collect the various rules of the exterior differential calculus.

Algebraic operations: addition and multiplication of differential forms.

the differential forms (of any fixed degree) constitute a vector space under addition and multiplication by scalars.

the product of a form of degree p by a form of degree q is a form of degree $p+q$

distributive law: $(\omega_1 + \omega_2) \wedge \sigma = \omega_1 \wedge \sigma + \omega_2 \wedge \sigma$.

associative law: $(\omega_1 \wedge \omega_2) \wedge \omega_3 = \omega_1 \wedge (\omega_2 \wedge \omega_3)$.

anticommutativity: $\omega_1 \wedge \omega_2 = (-1)^{pq}\omega_2 \wedge \omega_1$ if $\deg \omega_2 = p$ and $\deg \omega_2 = q$.

The operator d.

$$\text{if } \omega \text{ is a form of degree } p \text{ then } \mathrm{d}\omega \text{ is a form of degree } p+1,$$

where

$$\mathrm{d}f = (\partial f/\partial x)\,\mathrm{d}x + (\partial f/\partial y)\,\mathrm{d}y + \cdots, \text{ in terms of coordinates } (x, y, \ldots).$$

Also

$$\mathrm{d}(\omega_1 + \omega_2) = \mathrm{d}\omega_1 + \mathrm{d}\omega_2,$$

$$\mathrm{d}(\omega \wedge \sigma) = \mathrm{d}\omega \wedge \sigma + (-1)^p \omega \wedge \mathrm{d}\sigma \quad \text{if } \omega \text{ is of degree } p,$$

and

$$\mathrm{d}(\mathrm{d}\omega) = 0.$$

Pullback

$$\phi^* f = f \circ \phi \quad \text{for a function } f,$$

$$\phi^*(\omega_1 + \omega_2) = \phi^*\omega_1 + \phi^*\omega_2,$$

$$\phi^*(\omega \wedge \sigma) = \phi^*\omega \wedge \phi^*\sigma,$$

$$\phi^*\mathrm{d} = \mathrm{d}\phi^*,$$

and

$$(\psi \circ \phi)^* = \phi^* \circ \psi^*.$$

In the above discussion it was assumed that the maps ϕ, ψ etc. and the forms ω, σ etc. were defined over the entire space. Of course this is not needed; we can define the various operations for forms, maps and so on which are defined on various open sets. We just must make sure that the domains of definition match. Thus, if ϕ is defined only on some open set U, it is enough that ω be defined on some open set O which contains $\phi(U)$ in order to be able to define the pullback $\phi^*\omega$. If ψ is only defined on some open set O then we must assume that $\psi(O) \subset U$ if we want to define $\phi \circ \psi$.

Definition of integration over regions in $\mathbb{R}^k$. We now want to define the integral of a k-form over an oriented region, U, in $\mathbb{R}^k$. That is, we are considering the situation where the degree of the form equals the dimension of the space. We will assume that U is a bounded region and is 'nice' in the following sense. Suppose we put a mesh of size ε on $\mathbb{R}^k$ so that $\mathbb{R}^k$ is a union of non-overlapping cubes of edge size ε. Let $\mathrm{Inn}_\varepsilon(U)$ denote the total volume of all the cubes completely contained in U. Let $\mathrm{Out}_\varepsilon(U)$ denote the total volume of all the cubes which have non-empty intersection with U.

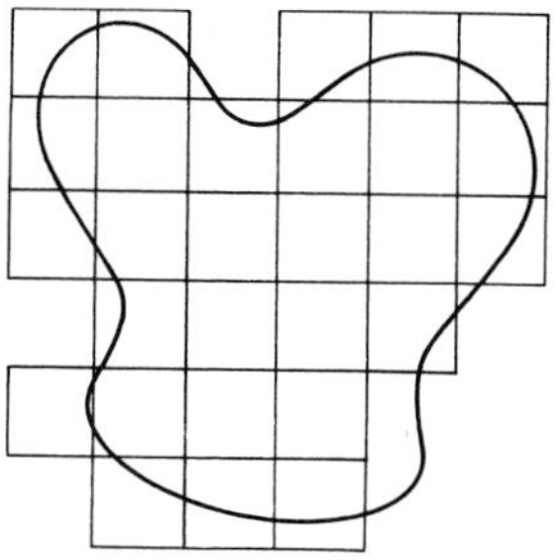

Figure 15.4

(See Fig. 15.4.) Then we assume that we can made the difference

$$\mathrm{Out}_\varepsilon(U) - \mathrm{Inn}_\varepsilon(U)$$

as small as we like by choosing ε sufficiently small. This means that in computing Riemann approximating sums we don't have to worry about 'cubes along the boundary'. For example, it is clear that any bounded polyhedron is nice in this sense. Also, if U is nice and if ψ is a differentiable map defined in a slightly larger region than U, then the mean value theorem implies that $\psi(U)$ is also nice. (You can prove this yourself, or, if you prefer, read Chapter 8 of Loomis–Sternberg *Advanced Calculus* where the basic facts about the Riemann integral in n dimensions are explained in detail.) We will also assume that orientation on U is the standard orientation (given by the standard basis and associated coordinates $x^1, \ldots, x^k$).

Now let Ω be a k-form on $\mathbb{R}^k$. Then we can write

$$\Omega = f\,\mathrm{d}x^1 \wedge \cdots \wedge \mathrm{d}x^k,$$

where f is a differentiable function. For any ε mesh we can construct the Riemann approximating sum

$$\sum f(\mathbf{p}_i)\,\mathrm{vol}(\Box_i) = \varepsilon^k \sum f(\mathbf{p}_i)$$

where $\mathbf{p}_i \in \Box_i$ and the $\Box_i$ range over all the cubes in the mesh contained in (or having non-empty intersection with) U. The (uniform) continuity of f implies that this sum approaches a limit as $\varepsilon \to 0$ independent of the choice of the $\mathbf{p}_i$ or of the mesh. This limit will be denoted by

$$\int_U \Omega.$$

It is important to remember that in this definition U has the standard orientation of $\mathbb{R}^k$. The following are obvious properties of the integral. Suppose that $U = U_1 \cup \cdots \cup U_p$ is a finite union of nice subregions, all with the standard orientation, then

$$\int_U \Omega = \int_{U_1} \Omega + \int_{U_2} \Omega + \cdots + \int_{U_p} \Omega.$$

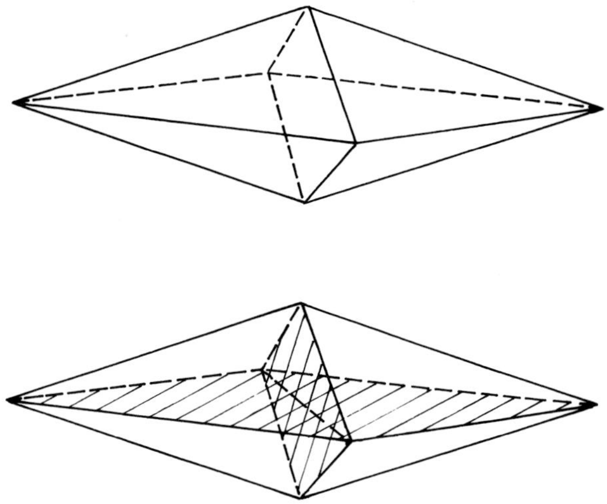

Figure 15.5

Irrelevance of lower dimensional pieces. Let us say that a set S has zero content if, for any positive number δ, we have $\mathrm{Out}_{\varepsilon(S)} < \delta$ if ε is chosen sufficiently small. (In other words, S has content zero if the total volume of the cubes in a mesh which intersect S can be made as small as we like by choosing the mesh size small enough.) For example, if S lies in a linear subspace of dimension $< k$ (or any translate thereof) then S clearly has zero content. Then sets of zero content are irrelevant as far as integration is concerned. That is, if $U = U' \cup S$ where S has zero content then $\int_U \Omega = \int_{U'} \Omega$. This implies, for example, that if we divide a polyhedron up into subpolyhedra, we don't have to worry about the lower dimensional faces in computing the integral.

Also,

$$\text{if } U \text{ is the unit cube, } \{0 \leqslant x_i \leqslant 1,\ i = 1, \ldots, k\} \text{ then}$$

$$\int_U \Omega = k\text{-fold iterated integral of } f = \int_0^1 \cdots \int_0^1 f \, dx_1 \cdots dx_k.$$

The proof is the same as in two dimensions, cf. p. 279 of Chapter 8. Of course, just as in two dimensions, any region that lends itself to iterated integration will do, not just a cube.

The change of variables formula. We now come to a basic fact, the **change of variables formula**. It says:

Suppose that $\phi: \mathbb{R}^k \to \mathbb{R}^k$ is a one-to-one differentiable *orientation-preserving* map with differentiable inverse. Then for any k-form ω defined on the image space, and for any nice region U in the domain space, we have

(As usual, ϕ need only be defined on some region slightly larger than U and ω need only be defined on some region slightly larger than $\phi(U)$.)

Proof of the change of variables formula

(i) Let $A:\mathbb{R}^k \to \mathbb{R}^k$ be a linear map. Then for any bounded nice region $(\text{vol}(AU)/\text{vol}(U)) = |\det A|$.

One begins by proving that the left-hand side of the equation in (i) is independent of U. This goes just as in the two-dimensional case, cf. section 1.9. Let us call the quotient on the left side of the equation $\text{vol}(A)$. So $\text{vol}(A)$ measures the proportional change in volume effected by A. It follows from the definition that

$$\text{vol}(AB) = \text{vol}(A)\cdot\text{vol}(B),$$

and we know from Chapter 11 that

$$\det(AB) = \det A\cdot\det B.$$

We wish to prove that

$$\text{vol}(A) = |\det A| \tag{$*$}$$

for all k by k matrices. From the preceding two equations we can conclude that if (*) is true for A and for B then it is true for AB. If A is a diagonal matrix then (*) is true by inspection. If the matrix A has zero entries below the diagonal, all 1 on the diagonal, and only one nonzero entry above the diagonal, then $\text{vol}(A) = 1$ (this is really the two-dimensional assertion about shear transformations). From this it follows that if $A = U$ is an upper triangular matrix with 1 on the diagonal (and all zeros below the diagonal) then $\text{vol}(U) = \det U = 1$. Similarly for lower triangular matrices. We conclude that (*) is true for all matrices A which can be written as

$$A = LDU$$

with L lower triangular, D diagonal, and U upper triangular. When does a matrix A have such a decomposition? An examination of the row reduction procedure of section 10.8 shows that this will happen if and only if the row reduction of A needs no transpositions, and this will occur if and only if none of the principal minors vanish (The principal minors are the determinants of the square matrices coming from taking the first r row and columns of A.) Thus the conditions are

$$a_{11} \neq 0, \quad (a_{11}a_{22} - a_{21}a_{12}) \neq 0, \ldots, \det A \neq 0.$$

But for any matrix A we can arrange that all these inequalities hold by making arbitrarily small changes in the matrix entries. Since both $\text{vol}(A)$ and $\det A$ are continuous functions of A we conclude that (*) holds for all matrices.

Here is an alternative proof of (*) based on a different way of decomposing non-singular matrices into products. For any matrix A, the matrix A^*A is self-adjoint, since

$$(A^*A)^* = A^*A^{**} = A^*A.$$

Furthermore, if A is non-singular, then the eigenvalues of A^*A are all positive, since $(A^*Ax, x) = (Ax, Ax) > 0$ for any non-zero vector x. Therefore A^*A has a positive definite square root, i.e. a matrix S which is symmetric and positive definite and which satisfies $S^2 = A^*A$. (Indeed, we may write $A^*A = OLO^*$ where L is a diagonal matrix with positive entries on the diagonal, and O is an orthogonal matrix. Then $S = OMO^*$ where M is the diagonal matrix whose entries are the square roots of the entries of L.) We claim that the matrix AS^{-1} is orthogonal. Indeed

$$(AS^{-1}x, AS^{-1}y) = (S^{-1}x, A^*AS^{-1}y) = (S^{-1}x, S^2S^{-1}y) = (x, y)$$

since S^{-1} is self-adjoint. Thus we may write

$$A = SK$$

where $K = AS^{-1}$ is orthogonal. But $S = OMO^*$. So

$$A = OMO^*K$$

where O (and $O^* = O^{-1}$) and K are orthogonal and M is diagonal, with positive entries. For M we have $\mathrm{vol}(M) = \det M$. Orthogonal matrices preserve length and therefore preserve volume, so $\mathrm{vol}(O) = \mathrm{vol}(K) = 1$. On the other hand, $\det O = \pm 1$ (since $\det O \cdot \det O^* = \det OO^* = 1$ and $\det O = \det O^*$). This gives an alternative proof of (*).

(ii) Let $\phi = A: \mathbb{R}^k \to \mathbb{R}^k$ be a linear map. Then

$$\phi^*(\mathrm{d}x^1 \wedge \cdots \wedge \mathrm{d}x^k) = (\det A)\,\mathrm{d}x^1 \wedge \cdots \wedge \mathrm{d}x^k. \qquad (**)$$

This follows from the definition of pullback and exterior product: $\phi^*x^1 = a_{11}x^1 + a_{12}x^2 + \cdots + a_{1k}x^k$ and since the a_{ij} are constants

$$\phi^*\,\mathrm{d}x^1 = a_{11}\,\mathrm{d}x^1 + \cdots + a_{1k}\,\mathrm{d}x^k,$$
$$\vdots$$
$$\phi^*\,\mathrm{d}x^k = a_{k1}\,\mathrm{d}x + \cdots + a_{kk}\,\mathrm{d}x^k.$$

Now $\phi^*\,\mathrm{d}x^1 \wedge \cdots \wedge \phi^*\,\mathrm{d}x^k$ will be some multiple of $\mathrm{d}x^1 \wedge \cdots \wedge \mathrm{d}x^k$, and this multiple, call it $f(A)$, is some numerical function of the matrix A which satisfies the axioms for the determinant function. Hence by the uniqueness properties of the determinant function proved in Chapter 11, we conclude that (**) holds.

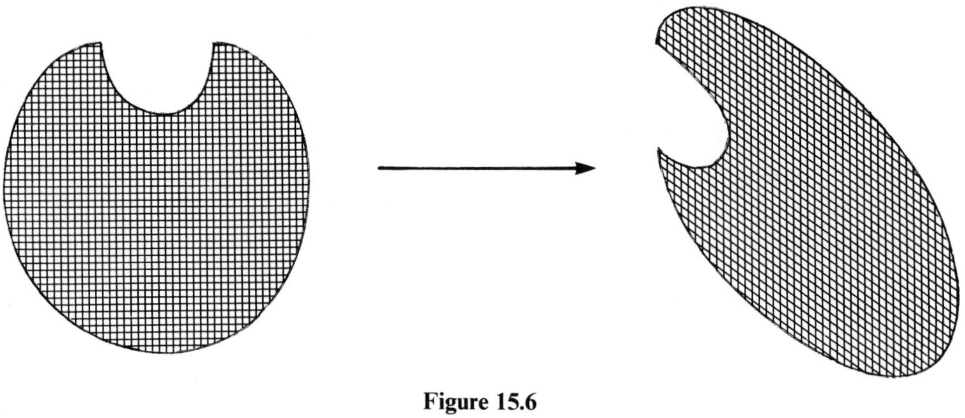

Figure 15.6

(iii) Steps (i) and (ii) imply the change of variables formula for the case that ϕ is a linear map, i.e. $\phi = A$ with $\det A > 0$. Here is the proof: Write $\omega = g\,\mathrm{d}x^1 \wedge \cdots \wedge \mathrm{d}x^k$, where g is a function. Then by (ii),

$$\phi^*\omega = h \det A\,\mathrm{d}x^1 \wedge \cdots \wedge \mathrm{d}x^k \quad \text{where} \quad h(x) = g(Ax).$$

Cover the region U by a mesh of small cubes, see Fig. 15.6. This has the effect of covering the region $\phi(U)$ by a mesh of small parallelepipeds, and the volume of each parallelepiped differs from the volume of the corresponding cube by a factor of $\det A$.

Given any $\varepsilon > 0$, we can choose the mesh size sufficiently small so that the function g does not vary by more that ε on each parallelepiped:

$$|g(\mathbf{p}) - g(\mathbf{q})| < \varepsilon,$$

whenever $\mathbf{p}$ and $\mathbf{q}$ lie in the same parallelepiped. This implies, that if $\square$ denotes any one of the internal cubes, and so $\phi(\square)$ the corresponding parallelepiped,

$$\left| \int_{\phi(\square)} \omega - g(\mathbf{p}) \operatorname{vol}(\phi(\square)) \right| < \varepsilon, \tag{***}$$

where $\mathbf{p}$ is any point in $\phi(\square)$. But

$$\operatorname{vol}(\phi(\square)) = \det A \operatorname{vol}(\square) \quad \text{by (i).}$$

So

$$\left| \int_{\phi(\square)} \omega - g(\mathbf{p}) \det A \operatorname{vol}(\square) \right| < \varepsilon. \tag{****}$$

Let U' denote the union of all the internal cube of the mesh. Then summing the inequalities (****) over all the internal cubes implies that

$$\left| \int_{\phi(\square)} \omega - \sum g(A\mathbf{r}) \det A \operatorname{vol}(\square) \right| < \varepsilon,$$

where $\mathbf{r}$ is any point in the cube $\square$, and the sum extends over all internal cubes. As the mesh gets finer and finer, the cubes along the boundary can be ignored and

$$\int_{\phi(U')} \omega \to \int_{\phi(U)} \omega$$

while

$$\sum g(A\mathbf{r}) \det A \operatorname{vol}(\square) \to \int \phi^*\omega.$$

(iv) Now let us see what has to be modified in the preceeding argument when ϕ is no longer assumed to be linear. First of all, $\phi^*\omega = h \det(\partial\phi/\partial x)\,\mathrm{d}x^1 \wedge \cdots \wedge \mathrm{d}x^k$. So the constant $\det A$ is replaced by a function, $\det(\partial\phi/\partial x)$ where $(\partial\phi/\partial x)$ denotes the matrix-valued function

$$(\partial\phi/\partial x) = \begin{pmatrix} \partial\phi^1/\partial x^1 \cdots \partial\phi^1/\partial x^k \\ \partial\phi^k/\partial x^1 \cdots \partial\phi^k/\partial x^k \end{pmatrix}.$$

The Riemann approximating sum now becomes

$$\sum g(A\mathbf{r}) \det(\partial\phi/\partial x)(\mathbf{r}) \operatorname{vol}(\square).$$

The inequality (***) will still hold if the mesh is fine enough, but $\phi(\square)$ is no longer a parallelepiped, and we will not have the exact equality

$$\mathrm{vol}\,(\phi(\square)) = \det\,(\partial\phi/\partial x)(\mathbf{r})\,\mathrm{vol}\,(\square).$$

But it would be enough to know that this holds approximately in the sense that

$$|\mathrm{vol}\,(\phi(\square) - \det\,(\partial\phi/\partial x)(\mathbf{r})\,\mathrm{vol}\,(\square)| < \varepsilon\,\mathrm{vol}\,(\square).$$

This inequality is a consequence of the mean value theorem. The details of the proof are not too complicated and can be found in Loomis–Sternberg *Advanced Calculus* Chapter 8, pp. 343–4. The reader is referred once again to that whole chapter for a comprehensive treatment of the theory of integration in $\mathbb{R}^k$. This completes our discussion of the change of variables formula.

Integration of *k*-forms over *k*-chains in a complex. Now suppose that ω is a k-form on $\mathbb{R}^n$. Let $\phi:\mathbb{R}^k \to \mathbb{R}^n$ be a smooth map. Let U be a region in $\mathbb{R}^k$. Then we can consider the integral

$$\int_U \phi^*\omega.$$

Let K be a complex. Let ϕ be a map of this complex into $\mathbb{R}^n$. Recall what this means. It is a map of each cell of the complex into $\mathbb{R}^n$ with the property that

the restriction ϕ to each cell C of the complex is a differentiable map.

This makes sense because each cell C is a convex polyhedron in some Euclidean space. In fact, we require that the map ϕ satisfies a technically slightly stronger condition: For each k-cell C of the complex, there is a neighborhood U of C in $\mathbb{R}^k$ and a differentiable map $\psi: U \to \mathbb{R}^n$ such that ϕ and ψ restrict to the same map on C. Each k-cell, C, of the complex is oriented. We can therefore consider the integral

$$\int_C \phi^*\omega$$

over the cell C. Let $\mathbf{c}$ be a k-chain. So $\mathbf{c}$ is a linear combination of k-cells:

$$\mathbf{c} = \sum r_j C_j$$

where the C_j are k-cells and the r_j are real numbers. Then we define the integral of $\phi^*\omega$ over the chain $\mathbf{c}$ by the formula

$$\int_{\mathbf{c}} \phi^*\omega = \sum r_j \int_{C_j} \phi^*\omega.$$

It is clear that this expression depends linearly on the chain $\mathbf{c}$. Thus

the k-form ω together with the map ϕ defines a k-cochain on the complex K.

For example, suppose that our complex K consists of the faces, edges and vertices of a tetrahedron. For simplicity let us assume that the faces are oriented 'consistently'. By this we mean that if we take

$$\mathbf{c} = C_1 + C_2 + C_3 + C_4 \quad \text{then} \quad \partial\mathbf{c} = 0.$$

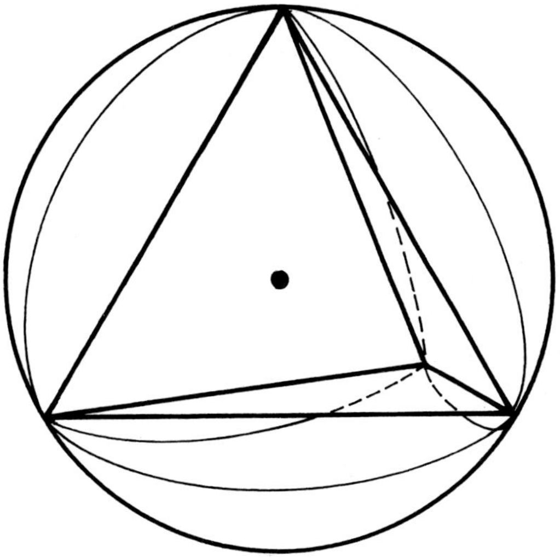

Figure 15.7

To fix the ideas, let us imagine that the tetrahedron is drawn with its center at the origin in $\mathbb{R}^3$.

Independence of cellular subdivision. Let us now consider projection from the center of the tetrahedron onto the surface of the unit sphere, S, and call this map ϕ.

It is easy to see that ϕ is a smooth map of the tetrahedron into $\mathbb{R}^3$. Each face of the tetrahedron is mapped onto a portion of the sphere and we can think of ϕ as mapping the chain $\mathbf{c}$ onto the surface of the sphere (with a definite choice of orientation). If ω is a two-form in $\mathbb{R}^3$ then we can think of $\int_{\mathbf{c}} \phi^* \omega$ as the integral of ω over the sphere (with a definite choice of orientation) and write this as

$$\int_S \omega.$$

This notation implicitly assumes a number of justifications all of which are correct. First of all, it assumes that the boundary curves of the triangular regions into which we have divided the sphere are irrelevant. This is because in computing a two-dimensional integral, the definition via Riemann sums implies that (smooth) curves make no contribution. A more important assumption implicit in the notation is that the subdivision into triangular regions – indeed the complex K and the map ϕ – are irrelevant. All that matters is the surface of the sphere (covered once) with a definite orientation. This is a consequence of the change of variables theorem. Let us explain. Suppose we consider a different complex, say the complex L consisting of all the faces, edges, and vertices of a cube which we draw centered at the origin, and consider the map μ of L onto the surface of the sphere given by projecting from the center. We also assume that the faces of L are consistently oriented so that

$$\mathbf{e} = F_1 + F_2 + F_3 + F_4 + F_5 + F_6 \quad \text{satisfies} \quad \partial \mathbf{e} = 0$$

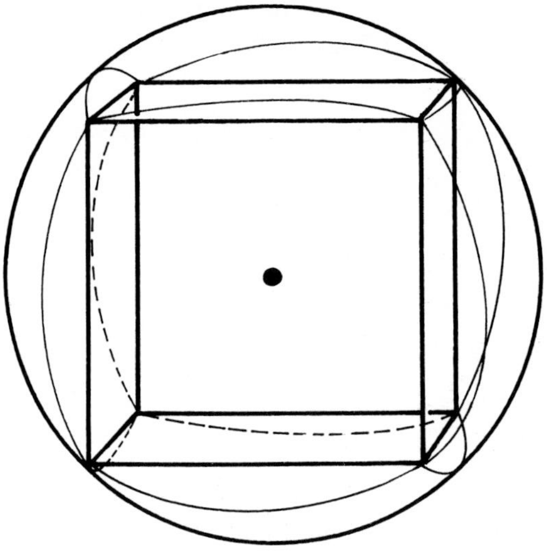

Figure 15.8

where the F_i ($i = 1, \ldots, 6$) are the faces of the cube.

We claim that

$$\int_{\mathbf{e}} \mu^* \omega = \pm \int_{\mathbf{c}} \phi^* \omega \tag{15.11}$$

with the + sign if the orientations of **e** and **c** are the 'same'. (We shall explain what this means in the course of proving the above assertion.) Indeed, consider the curves on the sphere which are the images of the edges of K (under ϕ) and of L (under μ). These curves subdivide the surface of the sphere into a large number of subregions.

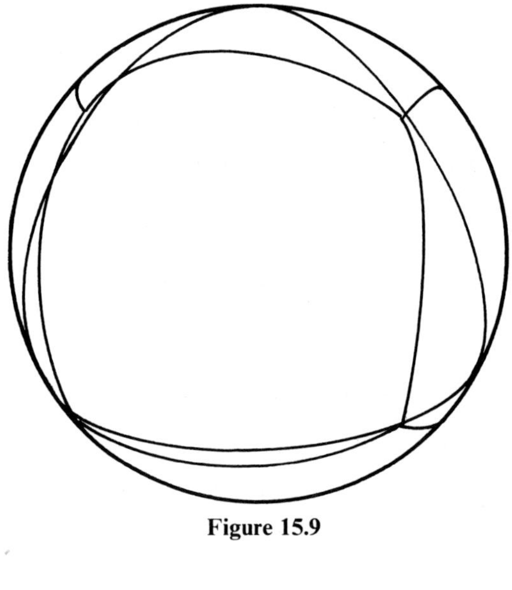

Figure 15.9

Each triangular region on the sphere is a union of several such subregions and so is each 'square region' coming from the cube. Let W be one such subregion. Then we can write

$$W = \phi(U) \text{ where } U \text{ is a subregion of one of the faces of } K$$

and

$$W = \mu(V) \text{ where } V \text{ is a subregion of one of the faces of } L.$$

Consider the map $\psi: U \to V$ given by $\psi = \mu^{-1} \circ \phi$. Then

$$\phi = \mu \circ \psi$$

so

$$\phi^*\omega = \psi^*(\mu^*\omega)$$

and

$$V = \psi(U).$$

(It is easy to check that ψ is a differentiable map.)

The change of variables theorem says that

$$\int_V \mu^*\omega = \pm \int_U \phi^*\omega$$

with the plus sign if and only if ψ is orientation preserving. It is easy to see that the map ψ that we have constructed for the subregion W will be orientation preserving if and only if the corresponding maps for all the others will also be orientation preserving. If this happens we say that **c** and **e** induce the same orientation on the sphere. Summing over all the subregions proves (15.11) and justifies the notation. We have sketched the proof of the justification of our notation for the case of a sphere. Of course it works in far greater generality. For any oriented surface in three-space (or more generally for any oriented submanifold M in $\mathbb{R}^n$) the integral $\int_M \omega$ makes sense and does not depend on how we cut M up into pieces so as to write it as the union of images of cells.

(We have defined the integral of a two-form over the oriented two-sphere by 'cutting it up into pieces' and then showing that the result is independent of how we cut up the sphere. There is an alternative (and equivalent) definition which involves 'covering the sphere with patches' rather than cutting it up into pieces. We shall sketch this definition at the end of the next section.)

15.4. Stokes' theorem

The general form of Stokes' theorem that we will prove in this section says the following: Let K be a complex. Let ϕ be a smooth map of K into $\mathbb{R}^n$. Let ω be a $(k-1)$-form on $\mathbb{R}^n$, and let **c** be a k-chain. Then

$$\int_{\mathbf{c}} \phi^* \mathrm{d}\omega = \int_{\partial \mathbf{c}} \phi^*\omega. \tag{15.12}$$

If we think of ϕ and ω as fixed, both sides of (15.12) are linear functions of **c**. Since every chain is a linear combination of cells, it is enough to prove (15.12) for the case where **c** is a cell. We will now follow the proof given in Chapter 8 for the two-dimensional case, but with a slight twist at the end.

The first step is to prove the theorem for the case that the cell is a cube. We might as well assume that the cell C is the unit cube in $\mathbb{R}^k$. Let $u^1, \ldots, u^k$ be the coordinates on $\mathbb{R}^k$. Then we can write

$$\phi^*\omega = a_1\,du^2 \wedge \cdots \wedge du^k + a_2\,du^1 \wedge du^3 \wedge \cdots \wedge du^k + \cdots + a_k\,du^1 \wedge \cdots \wedge du^{k-1}, \tag{15.13}$$

so that the left-hand side of (15.12) becomes

$$\int_C \phi^*\,d\omega = \int_C d\phi^*\omega = \int_C (\partial a_1/\partial u^1 - \partial a_2/\partial u^2 + \cdots \pm \partial a_k/\partial u^k)\,du^1 \wedge \cdots \wedge du^k. \tag{15.14}$$

We now recall the signs associated to the ∂ operator applied to any cell. Our procedure was as follows:

Recall that the orientation of a cell C determines whether a k-tuple of basis vectors is 'correctly' or 'incorrectly' ordered. An 'incorrect' set of basis vectors is converted to a 'correct' set either by interchanging two vectors with consecutive numbers or by changing the sign of one vector. Correct and incorrect sets for $k = 1$, 2, and 3 are illustrated in figure 15.10.

The boundary of an oriented k-cell consists of a sum of oriented $(k-1)$-cells, and we gave a general prescription for determining the sign of each cell in the boundary: At a point on the boundary of C, construct a set of vectors in which the first vector points *out of* the cell C and the remaining $k-1$ vectors are a correctly oriented set for the boundary cell. If the resulting set of k vectors is a correctly oriented set for the cell C then the boundary cell appears with a plus sign in ∂C; otherwise it appears with a minus sign.

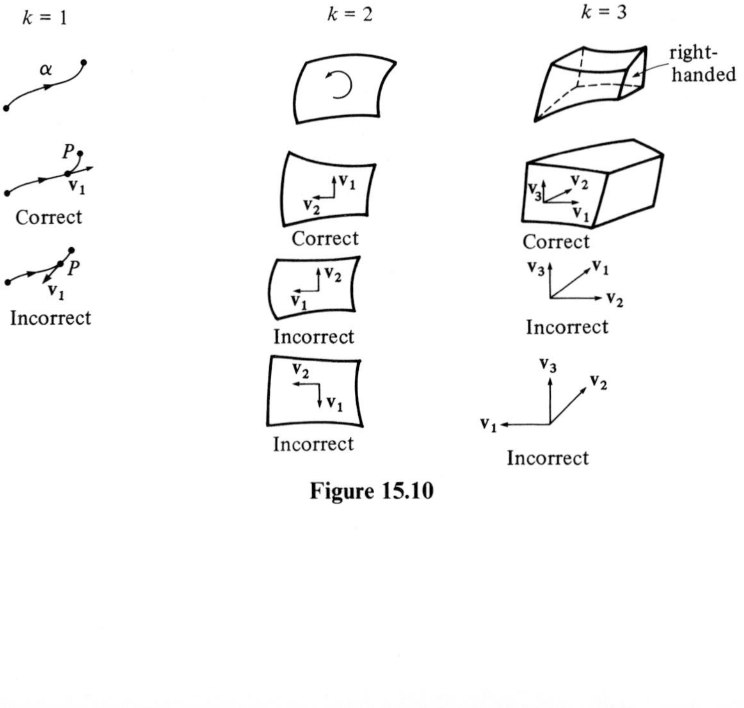

Figure 15.10

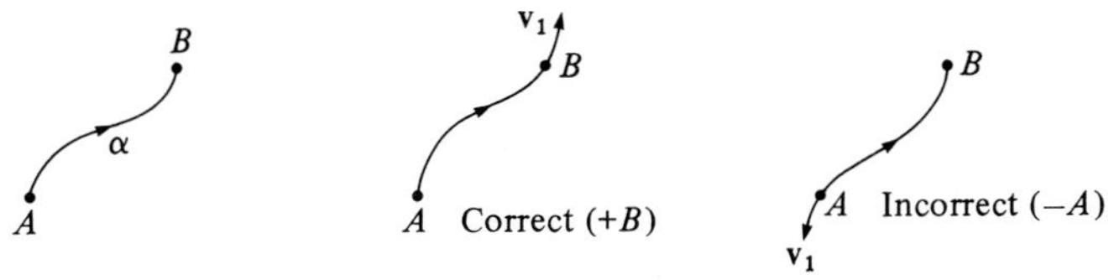

Figure 15.11

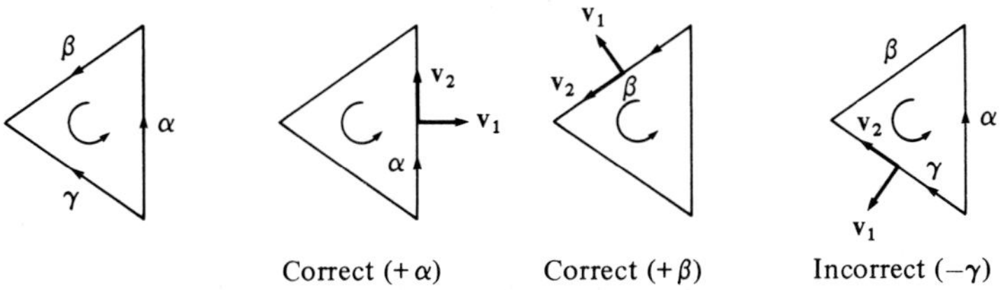

Figure 15.12

Figure 15.11 illustrates the application of this rule in the case $k = 1$. The vector $\mathbf{v}_1$ points out, and there are no other vectors. At B, the vector $\mathbf{v}_1$ has the correct orientation for the branch α, but at A it has the wrong orientation. Hence $\partial\alpha = B - A$.

The case $k = 2$ is illustrated in figure 15.12 for a triangular cell C with counterclockwise orientation.

For each branch, the vector $\mathbf{v}_2$ is chosen to lie tangent to the branch, its direction determined by the orientation of the branch. The vector $\mathbf{v}_1$ always points out of the cell C. For branches α and β, the resulting pair of vectors is correctly ordered (according to the counterclockwise orientation of C) but for branch γ it is incorrectly ordered. Hence $\partial C = \alpha + \beta - \gamma$.

We return to the case where $C = I^k$ is a cube. Let us consider the various terms obtained by substituting (15.13) into $\int_{\partial C}\phi^*\omega$. We consider the various faces of ∂C. We begin with the two faces for which $u_1 =$ constant. On the face B_1 where $u_1 = 1$, the vector $\mathbf{v}_1$ points *out* and the remaining vectors $\mathbf{v}_2, \ldots, \mathbf{v}_k$ are correctly oriented. On the face A_1 where $u_1 = 0$, the vector $\mathbf{v}_1$ points *in*, and so $\mathbf{v}_2, \ldots, \mathbf{v}_k$ are incorrectly oriented. This situation is illustrated for the cases $k = 2$ and $k = 3$ in figure 15.13.

When we come to evaluate $\phi^*\omega$ on these two faces with $u^1 =$ constant, all the terms in τ which include a factor $\mathrm{d}u^1$ give zero, and all that remains is $a_1(u^1, u^2, \ldots, u^k)\,\mathrm{d}u^2 \wedge \cdots \wedge \mathrm{d}u^k$. Evaluation of this term on face B_1, where $\mathbf{v}_2, \ldots, \mathbf{v}_k$ are correctly ordered, gives $\int_{I^{k-1}}\alpha_1(1, u^2, \ldots, u^k)\,\mathrm{d}u^2 \ldots \mathrm{d}u^k$. Evaluation on face A_1, where the vectors $\mathbf{v}_2, \ldots, \mathbf{v}_k$ are incorrectly ordered, gives $-\int_{I^{k-1}}(0, u^2, \ldots, u^k)\,\mathrm{d}u^2 \ldots \mathrm{d}u^k$. The two terms together yield the integral

$$\int_{I^{k-1}} [a_1(1, u^2, \ldots, u^k) - a_1(0, u^2, \ldots, u^k)]\,\mathrm{d}u^2 \ldots \mathrm{d}u^k.$$

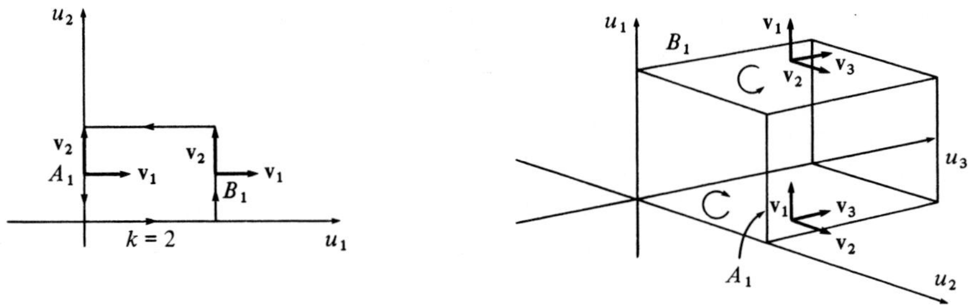

Figure 15.13

But by the fundamental theorem of calculus

$$a_1(1,u^2,\dots,u^k) - a_1(0,u^2,\dots,u^k) = \int_0^1 \frac{\partial a_1}{\partial u_1}(u^1,u^2,\dots,u^k)\,du^1$$

and so the combined contribution of the two faces with $u_1 = \text{constant}$ is $\int_{I^k}(\partial a_1/\partial u^1)\,du^1\dots du^k$, where the integral is now over the unit cube in $\mathbb{R}^k$.

In carrying through the same procedure for the faces where $u_2 = \text{constant}$, we note that the set of vectors $\mathbf{v}^2, \mathbf{v}^1, \mathbf{v}^3, \dots, \mathbf{v}^k$ is *incorrectly* ordered. On the face where $u^2 = 1$, $\mathbf{v}_2$ points out, and so remaining vectors $\mathbf{v}^1, \mathbf{v}^3, \dots, \mathbf{v}^k$ are incorrectly ordered. On the face where $u^2 = 0$, $\mathbf{v}_2$ points in, and $\mathbf{v}^1, \mathbf{v}^3, \dots, \mathbf{v}^k$ are correctly ordered. In the evaluation of τ, only the term $a_2(u^1, u^2, \dots, u^k)\,du^1 \wedge du^3 \wedge \cdots \wedge du^k$ contributes, and we obtain

$$\int_{I^{k-1}} [-a_2(u^1,1,\dots,u^k) + a_2(u^1,0,\dots,u^k)]\,du^1\,du^3\dots du^k = -\int_{I^k} \frac{\partial a}{\partial u^2}\,du^1\,du^2\dots du^k.$$

A similar argument shows that the faces where u^j is constant contribute $\pm\int_{I^k}(\partial a_j/\partial u^j)\,du^1\dots du^k$, with a $+$ sign for odd j, a $-$ sign for even j. On summing the contributions from all the $2k$ faces in ∂I^k, we obtain (15.12).

This completes our proof of Stokes' theorem for a cube. We must now consider more general cells. In the plane, as we pointed out in Chapter 8, every polygon can be decomposed into triangles. Indeed, we can decompose any polygon into convex polygons and any convex polygon can be decomposed into triangles by choosing a point in the interior and joining it to all the vertices.

Similarly, in three dimensions, every polyhedron can be decomposed into tetrahedra. Indeed, we may, after a preliminary decomposition, assume that C is convex. We may also assume that the faces of C have been decomposed into triangles. Again, just pick a point p in the interior. C will be decomposed into tetrahedra whose bases are the facial triangles and with apex p. Thus for two-forms, it is sufficient to prove Stokes' theorem for tetrahedra.

By induction we can do the same in k dimensions: Let $S \in \mathbb{R}^k$ be the set defined by

$$S = \{x \mid x_i \geqslant 0 \quad \text{all } i \text{ and } \textstyle\sum x_i \leqslant 1\}.$$

A k-simplex is the subset of $\mathbb{R}^k$ which is the image of S under an invertible affine map. Thus a 1-simplex is an interval, a 2-simplex is a triangle and a 3-simplex is a tetrahedron.

Now suppose that we have C decomposed into simplices$[\Delta_i]$. We claim that if we know Stokes' theorem for all simplices then we know it for C. Indeed, consider the complex whose cells are all the simplices going into the decomposition of C, together with all their lower dimensional faces. We can then identify C with a chain in this complex, i.e. we can write

$$C = \sum \Delta_i$$

and

$$\partial C = \sum \partial \Delta_i.$$

Therefore

$$\begin{aligned} \int_\partial \omega &= \sum \int_{\partial \Delta_i} \omega \\ &= \sum \int_{\Delta_i} d\omega \quad \text{since we know Stokes' theoremfor the } \Delta_i. \\ &= \int_C d\omega. \end{aligned}$$

By the preceding argument, it is enough to prove Stokes' theorem for the case of a simplex.

We will make use of the invariance of the integral under pullback to reduce the proof for a simplex to that for a cube. Here is the idea for the case of a triangle.

To prove Stokes' theorem for a triangle, it is enough to prove it for an equilateral triangle Δ since we can find an affine transformation which carried any triangle into an equilateral one. We may assume that the edge length of Δ is 1. Suppose the linear differential form τ were identically zero outside a disk whose center is at one of the vertices and which does not tough the opposite side.

Then $d\tau$ is also identically zero outside this disk. Now as far as Stokes' theorem is

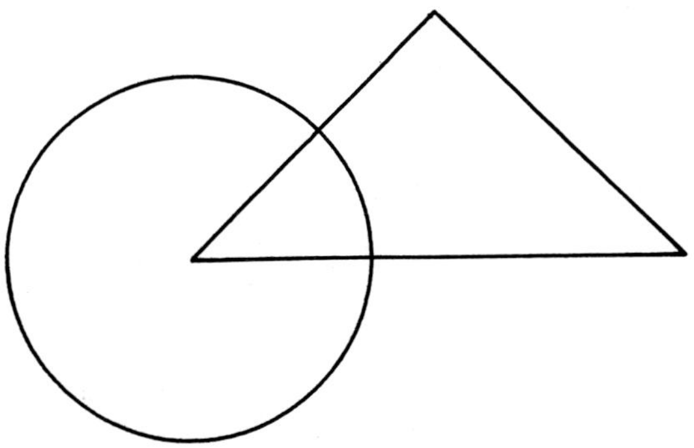

Figure 15.14

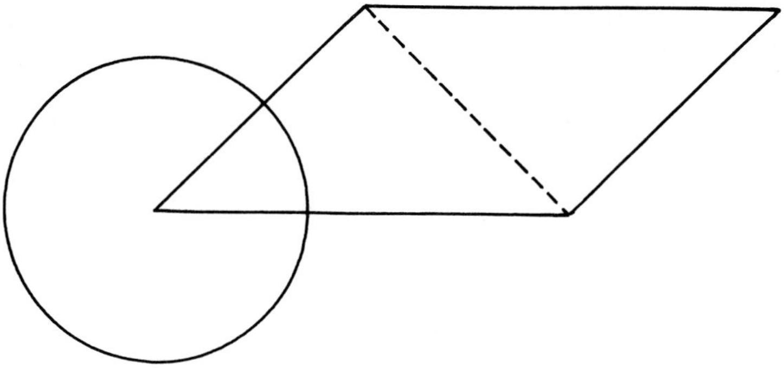

Figure 15.15

concerned, the only contributions to the integral come from within the circle and we can 'close up' this portion of the triangle by making it a parallelogram:

That is, if $\square$ denotes the parallelogram and Δ the triangle, then because τ and $d\tau$ vanish outside the circle

$$\int_{\partial\square} \tau = \int_{\partial\Delta} \tau$$

and

$$\int_{\square} d\tau = \int_{\Delta} d\tau.$$

But, up to a change of variables, the parallelogram is just a square. Hence $\int_{\partial\square} \tau = \int_{\square} d\tau$ and therefore $\int_{\partial\Delta} \tau = \int_{\Delta} d\tau$.

We will have proved Stokes' theorem for Δ if we can establish the following:

Every smooth linear differential form (defined in a neighborhood of Δ) can be written as a sum of three terms

$$\tau = \tau_1 + \tau_2 + \tau_3$$

where each τ_i is identically zero outside the disk D_i, where D_i is the disk centered at the Cth vertex and of radious R, where $\frac{2}{3} < \mathrm{R} < \frac{3}{4}$ (so that the three disks cover the triangle, but each disk does not meet the opposite edge).

To prove this we first establish a fact about a function of one variable.

Lemma. *The function*

$$f(u) = \begin{cases} e^{-1/u} & u \geqslant 0 \\ 0 & u \leqslant 0 \end{cases}$$

is infinitely differentiable at all points.

Proof. For $u \neq 0$ it is clear that f has derivatives of all orders and that $f^{(k)}(u) \equiv 0$ for $u < 0$. So we must prove that $f^{(k)}(u)/u \to 0$ as $u \to 0$ for any k. But

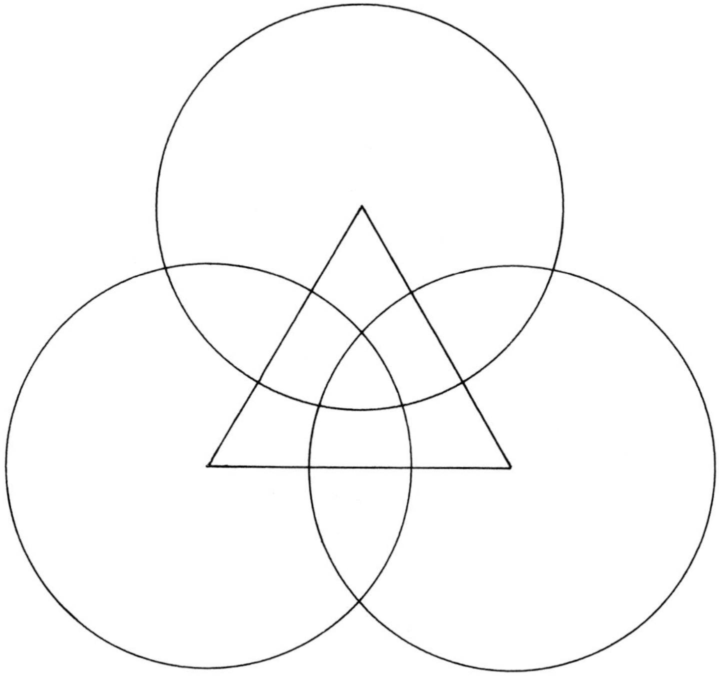

Figure 15.16

$f^{(k)}(u) = P_k(1/u)e^{-1/u}$ where P_k is some polynomial and hence

$$\lim_{u \to 0} (1/u)^{(k)}(u) = \lim_{s \to \infty} sP_k(s)\,e^{-s} = 0$$

since e^s goes to infinity faster than any polynomial.

Notice that the function f is $\equiv 0$ for $u \leqslant 0$ and is strictly positive for $u > 0$.

Let r_i denote the distance from the ith vertex and define the function g_i by

$$g_i(x) = \begin{cases} f(R - r_i(x)) & r_i(x) \leqslant R \\ 0 & r_i(x) > R. \end{cases}$$

Then g_i is infinitely differentiable and

$$g_i(\mathbf{x}) \begin{cases} > 0 & r_i(x) < R \\ = 0 & r_i(x) \geqslant 0. \end{cases}$$

Let

$$g = g_1 + g_2 + g_3.$$

Since each point in Δ is interior to at least one of the disks D_i – that is – since $r_i(x) < R$ for at least one of $1 = 1, 2, 3$ we know that

$$g(x) > 0$$

for every $\mathbf{x}\in\Delta$. So we can divide by g and set

$$\phi_L = \frac{g_i}{g}, \quad L = 1, 2, 3.$$

Then

$$\phi_i(x)\begin{cases} >0 & r_i(x) < R \\ =0 & r_i(x) \geqslant 0 \end{cases}$$

and

$$\phi_1(\mathbf{x}) + \phi_2(\mathbf{x}) + \phi_3(\mathbf{x}) \equiv 1 \quad \text{for all} \quad \mathbf{x}\in\Delta.$$

Multiplying this last equation by τ gives

$$\tau = \phi_1\tau + \phi_2\tau + \phi_3\tau.$$

If we set $\tau_i = \phi_i\tau$ we get the desired decomposition

$$\tau = \tau_1 + \tau_2 + \tau_3.$$

It is also clear that the above proof works in n dimensions. Just replace 3 by $n+1$ and $\frac{3}{4}$ by $\frac{1}{2}+\frac{1}{2}n$. This completes the proof of Stokes' theorem.

Indeed, we don't have to be so precise. Just cover the simplex by a finite number of balls D_i such that the intersection of D_i with the simplex looks like a bit of a cube – break up into pieces and reduce to the cube case.

Example

As an explicit example of Stokes' theorem and the evaluation of forms on cells, we consider a two-form in $\mathbb{R}^3$, $\tau = (x^2 + y^2 + z^2)\,\mathrm{d}x \wedge \mathrm{d}y$, and a cell C which is the solid hemisphere of radius R shown in figure 15.17, with a right-handed orientation. We will evaluate $\int_{\partial C}\tau$ and $\int_C \mathrm{d}\tau$ explicitly.

The solid hemisphere is bounded by two cells: the hemispherical surface A and the disk B in the equatorial plane. If we assign both of these cells a counterclockwise orientation as seen from outside, then $\partial C = A + B$.

For the hemispherical surface, we use the spherical coordinates θ and ϕ as parameters. The two-dimensional parameter space is shown in figure 15.16. In principle we could define parameters $u = 2\theta/\pi$ and $v = \phi/2\pi$ so that the region of integration is the unit square, but in practice a rectangle is just as convenient.

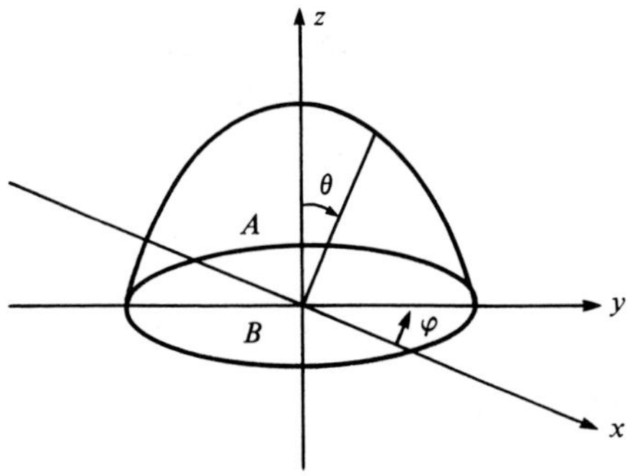

Figure 15.17

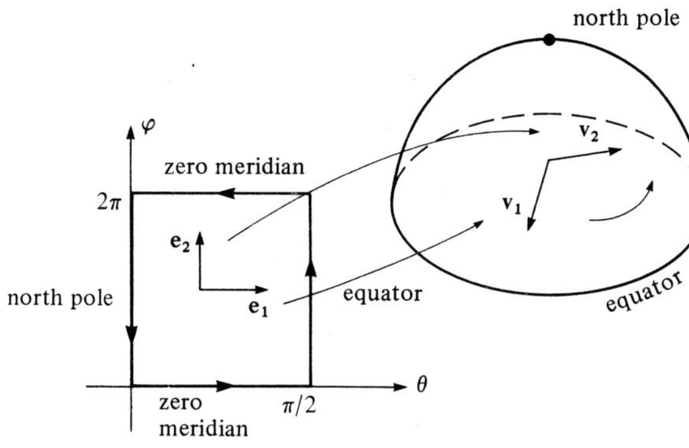

Figure 15.18

The pullback of x, y and z under the parameterization α is $\alpha^*x = R\sin\theta\cos\phi$, $\alpha^*y = R\sin\theta\sin\phi$, $\alpha^*z = R\cos\theta$. Thus

$$\alpha^*\mathrm{d}x = R\cos\theta\cos\phi\,\mathrm{d}\theta - R\sin\theta\sin\phi\,\mathrm{d}\phi,$$
$$\alpha^*\mathrm{d}y = R\cos\theta\sin\phi\,\mathrm{d}\theta + R\sin\theta\cos\phi\,\mathrm{d}\phi,$$

so that

$$\begin{aligned}\alpha^*\tau &= \alpha^*(x^2+y^2+z^2)(\alpha^*\mathrm{d}x)\wedge(\alpha^*\mathrm{d}y)\\ &= R^2(R^2\sin\theta\cos\theta\cos^2\phi\,\mathrm{d}\theta\wedge\mathrm{d}\phi - R^2\sin\theta\cos\theta\sin^2\phi\,\mathrm{d}\phi\wedge\mathrm{d}\theta)\\ &= R^4\sin\theta\cos\theta\,\mathrm{d}\theta\wedge\mathrm{d}\phi.\end{aligned}$$

To check that α preserves orientation, we look at the images of the ordered set of vectors $\mathbf{e}_1$ and $\mathbf{e}_2$. The images $\mathbf{v}_1$ and $\mathbf{v}_2$ agree with the orientation of the hemisphere A, as shown in figure 15.18.

Now, to evaluate τ on the hemisphere A, we just calculate the double integral $\int \alpha^*\tau = \int_{\theta=0}^{\pi/2}\int_{\phi=0}^{2\pi} R^4\sin\theta\cos\theta\,\mathrm{d}\theta\,\mathrm{d}\phi = \pi R^4$.

For the disk B, a convenient choice of parameters consists of $r = (x^2 + y^2)$ and the angle ϕ. In order to obtain the disk B with the correct orientation we choose the ordering ϕ, r, so that the parameter space is as shown in figure 15.19. Then the images of $\mathbf{e}_1$ and $\mathbf{e}_2$ have the correct ordering for the orientation of B (counterclockwise as seen from below – clockwise as seen from above).

This parameterization β is specified by the pullback $\beta^*x = r\cos\phi$, $\beta^*y = r\sin\phi$, $\beta^*z = 0$, so that $\beta^*\tau = r^2(\cos\phi\,\mathrm{d}r - r\sin\phi\,\mathrm{d}\phi)\wedge(\sin\phi\,\mathrm{d}r + r\cos\phi\,\mathrm{d}\phi)$ or $\beta^*\tau = r^3\,\mathrm{d}r\wedge\mathrm{d}\phi = -r^3\,\mathrm{d}\phi\wedge\mathrm{d}r$. Thus the value of τ on the oriented disk B is

$$\int_{r=0}^{R}\int_{\phi=0}^{2\pi} -r^3\,\mathrm{d}\phi\,\mathrm{d}r = -\tfrac{1}{2}\pi R^4.$$

Combining the results for the two cells in the boundary, we find

$$\int_{\partial C} = \int_A \tau + \int_B \tau = \pi R^4 - \tfrac{1}{2}\pi R^4 = \tfrac{1}{2}\pi R^4.$$

Now we evaluate $\int_C \mathrm{d}\tau$. Since $\tau = (x^2 + y^2 + z^2)\,\mathrm{d}x\wedge\mathrm{d}y$, $\mathrm{d}\tau = 2z\,\mathrm{d}z\wedge\mathrm{d}x\wedge\mathrm{d}y =$

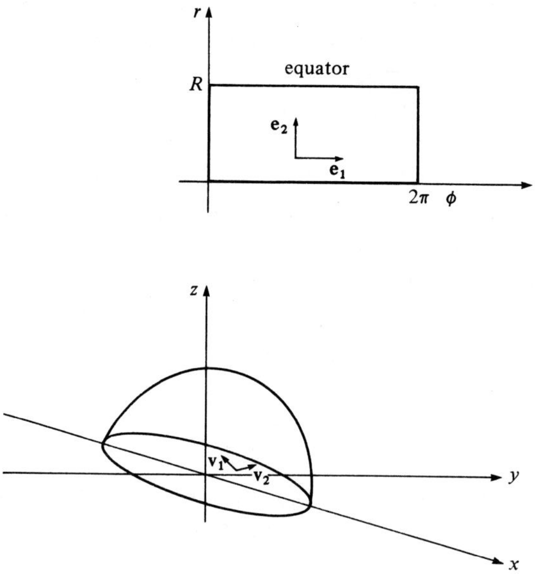

Figure 15.19

$2z\,dx \wedge dy \wedge dz$. We parameterize the cell C by using spherical coordinates, r, θ, ϕ. The pullback under this parameterization ψ is $\psi^*x = r\sin\theta\cos\phi$, $\psi^*y = r\sin\theta\sin\phi$, $\psi^*z = r\cos\theta$ and, after some calculation, we find $\psi^*(d\tau) = 2r^3\sin\theta \times \cos\theta\,dr \wedge d\theta \wedge d\phi$. To determine the value of $d\tau$ on C we integrate over a rectangular solid in parameter space:

$$\int_C d\tau = \int_{r=0}^{R}\int_{\theta=0}^{\pi/2}\int_{\phi=0}^{2\pi} 2r^3\sin\theta\cos\theta\,dr\,d\theta\,d\phi = \tfrac{1}{2}\pi R^4.$$

Let us now briefly indicate how we can use the *proof* that we gave of Stokes' theorem to suggest an alternative *definition* of integration of a k-form over a k-dimensional submanifold. This alternative definition will coincide with the definition of integration over a k-chain that we gave above in the case that the submanifold is given as the image of a k-chain. This will then sketch out in more detail how to prove the fact that the integral over a k-chain is independent of the cellular decomposition, as we illustrated at the end of the preceding section for the case of a sphere. The material presented from here to the end of this section can be omitted on first reading.

Patching

In our proof of Stocke's theorem we made use of a collection of functions $\{\phi_1, \ldots, \phi_n\}$ with the property that each of the ϕs was continuously differentiable, non-negative, and

$$\phi_1 + \cdots + \phi_n \equiv 1 \quad \text{in the region of interest (the simplex).}$$

Furthermore each ϕ_i vanishes outside a ball D_i where we choose the ball D_i to have a desirable property. (In our proof of Stokes' theorem this desirable property was that the intersection of D_i with the simplex could be transformed into an open subset of the cube by a differentiable map.)

We now recall the definition of a submanifold, M, of $\mathbb{R}^n$. A subset M of $\mathbb{R}^n$ was called a submanifold of a dimension k if about each point, p, of M we can find a ball D and a diffeomorphism f of D with an open subset, U, of $\mathbb{R}^n$ such that $f(D \cap M) = U \cap \mathbb{R}^k$ where $\mathbb{R}^k$ is identified with the k-dimensional subspace of $\mathbb{R}^n$ given by setting the last n-k coordinates equal to zero. Suppose that we can cover M by a finite number of such balls, $D_1, \ldots, D_r$. (This will always be the case if M is 'compact', i.e. a closed and bounded subset of $\mathbb{R}^n$.) Each D_i comes with its own f_i and

$$f_i(D_i \cap M) = U_i \cap \mathbb{R}^k.$$

Let us call this subset W_i so

$$W_i = U_i \cap \mathbb{R}^k$$

is an open subset of $\mathbb{R}^k$. On any overlap $D_i \cap D_j$ we can consider the restriction of either f_i or f_j. So $f_i(D_i \cap D_j)$ is an open subset of U_i and $f_j(D_i \cap D_j)$ is an open subset of U_j and

$$f_i \circ f_j^{-1} \quad \text{maps} \quad f_j(D_i \cap D_j) \quad \text{onto} \quad f_i(D_i \cap D_j)$$

and carries

$$W_j \cap f_j(D_i \cap D_j) \quad \text{onto} \quad W_i \cap f_i(D_i \cap D_j).$$

So our situation is as follows: We have covered M with sets $O_1 = D_1 \cap M$. Let g_1 denote the restriction of f_i to O_i. Then g_i is a one-to-one map of O_i onto the open

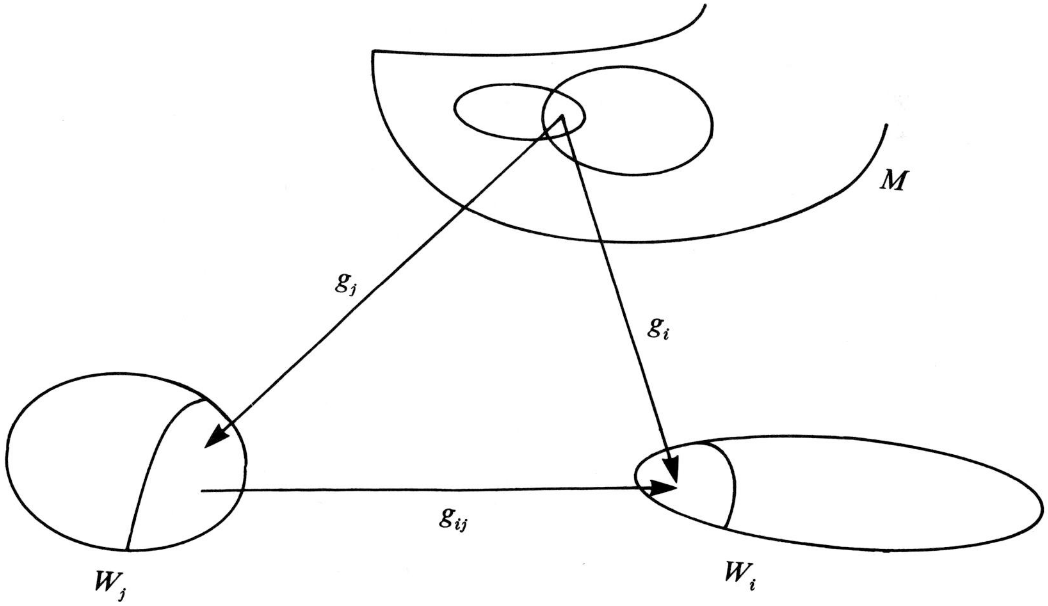

Figure 15.20

subset W_i of $\mathbb{R}^k$. Furthermore, the map

$$g_{ij} = g_i \circ g_j^{-1} \quad \text{of} \quad W_j \cap g_j(O_i \cap O_j) \quad \text{onto} \quad W_i \cap g_i(O_i \cap O_j)$$

is differentiable with differentiable inverse given by $g_{ji} = g_j \circ g_i^{-1}$.

We can think of the manifold M as being covered with 'patches': The maps g_i^{-1} tell how the W_i cover the manifold, and the maps g_{ij} tell how the W_i and W_j patch together. Now each map g_{ij} can be either orientation preserving or reversing. Suppose that we are in a situation where all the g_{ij} are orientation preserving. (This then determines an 'orientation' on M. It is intuitively clear that if M is connected there are then only two orientations.) Now let $\phi_1, \ldots, \phi_r$ be a collection of functions as above so that

$$0 \leqslant \phi_i \leqslant 1, \quad \phi_i = 0 \text{ outside of } D_i \quad \text{and} \quad \phi_1 + \cdots + \phi_r \equiv 1 \text{ on } M.$$

Now for any k-form ω we can write

$$\omega = \phi_1 \omega + \cdots + \phi_r \omega.$$

Each summand on the right vanishes outside D_i. We would then define

$$\int_M \omega = \sum \int_{W_i} (g_i^{-1})^* \phi_i \omega.$$

In other words, instead of cutting M up into pieces, we write any form as a sum of small pieces each of which lives only on one coordinate patch on M which we can then integrate by pulling it back to a subset of $\mathbb{R}^k$. A repeated use of the change of variables formula easily shows that this definition of integral does not depend on the choice of the ϕ_i or of the patching (i.e. of the choice of O_i and g_i) but only on the choice of orientation of M. For further details on this definition we refer once again to the book *Advanced Calculus* by Loomis and Sternberg, Chapter 9–12.

15.5. Differential forms and cohomology*

The k-forms which we have defined will function as incipient k-cochains for any complex situated in $\mathbb{R}^n$, and the differential operator d is consistent with the coboundary operator d. Corresponding to the subspaces which were defined in terms of the image and kernel of the coboundary operator, we can now construct subspaces of the (infinite-dimensional) spaces $\Omega^{(k)}(\mathbb{R}^n)$ by using the differential operator d.

If a k-form ω satisfies $\mathrm{d}\omega = 0$, it is called *closed.* Such a k-form becomes, by integration over cells, a k-cochain W which satisfies $\mathrm{d}W = 0$ and therefore belongs to the space Z^k of k-cocycles. In other words, a *closed* k-form is an incipient k-cocycle.

If a k-form ω can be expressed as $\omega = \mathrm{d}\tau$, it is called *exact.* In this case, by integration over the cells of a complex, τ gives rise to a $(k-1)$-cochain T, ω to a k-cochain W, such that $W = \mathrm{d}T$. The cochain W therefore lies in the subspace B^k

* Can be omitted on first reading.

of k-coboundaries. In other words, any *exact* k-form is an incipient coboundary.

For any complex, we know that B^k is a subspace of Z^k. The corresponding statement about k-forms is that any exact k-form (incipient element of B^k) is also closed (incipient element of Z^k). The proof is simple: if ω is exact, $\omega = \mathrm{d}\tau$ and so $\mathrm{d}\omega = \mathrm{d}(\mathrm{d}\tau) = 0$ and ω is closed.

For any complex, the quotient spaces (cohomology spaces) $H^k = Z^k/B^k$ depend just upon the underlying space and not upon how it is cut up to form a complex. We might reasonably expect, therefore, to obtain the spaces H^k by considering differential forms: we take the quotient of the infinite-dimensional space of closed forms ($\mathrm{d}\omega = 0$) by the space of exact forms ($\omega = \mathrm{d}\tau$). The resulting spaces define the so-called *de Rham cohomology* of the underlying space on which the differential forms are defined.

As a clue in constructing a basis for the spaces H^k, we recall that, for any complex, the space H^k is dual to the homology space H_k. It is therefore reasonable to expect that a basis element for H^k will be determined by its values on the k-chains which form a basis for H_k in any convenient complex in the space.

Let us consider some extremely simple examples which illustrate these rather abstract considerations.

Example 1A. The underlying space is a single line segment. On the space we construct zero-forms, which are differentiable functions $f(t)$, and one-forms, which are all of the form $\omega = g(t)\,\mathrm{d}t$.

In this case, the closed zero-forms are the constant functions, which define a one-dimensional subspace. There can be no exact zero-forms, so the quotient space H^0 of closed zero-forms by exact zero-forms is one-dimensional. This reflects the fact that no matter how we cut up the line segment to obtain a complex, the complex is always a *connected* one, with $\dim H_0 = 1$.

On considering one-forms, we discover that every one-form is exact: given $\omega = g(t)\,\mathrm{d}t$ we form an antiderivative $G(t) = \int g(t)\,\mathrm{d}t$, so that $\omega = \mathrm{d}G$. Hence H^1 is zero-dimensional in this case.

Example 1B. The underlying space now consists of two disjoint line segments. In this case the space $H^0 = Z^0$ is two-dimensional: it consists of functions which have one constant value on the interval $[a, b]$ and another, possibly different, constant value on the interval $[c, d]$. This reflects the fact that a complex constructed on this space will have two connected components.

Figure 15.21

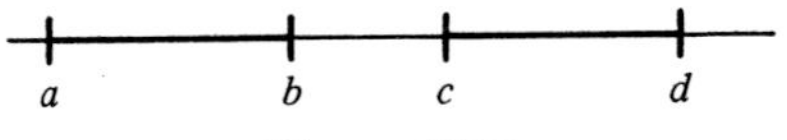

Figure 15.22

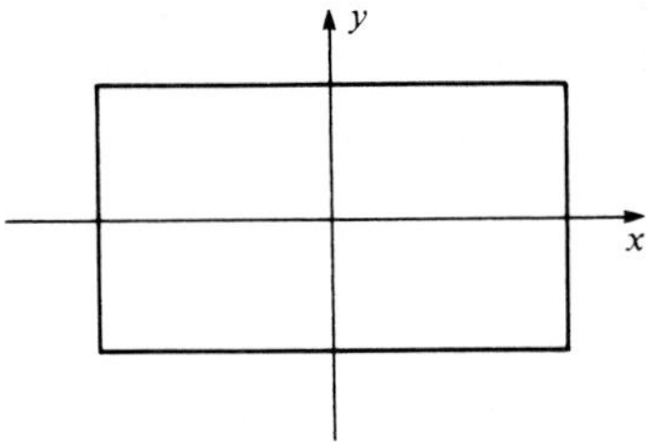

Figure 15.23

More generally, whatever the dimension of the underlying space, the closed zero-forms will be functions which are constant on each connected component, and these will form a space $H^0 = Z^0$ whose dimension equals the number of connected components in the underlying space. From now on, we shall consider only connected spaces, and we need say nothing more about H^0.

Example 2A. The underlying space is a rectangle in the plane. Because the space is connected, H^0 is one-dimensional. On considering one-forms, we note that any closed one-form ω is also exact, a result which we proved when considering line integrals. The space H^1 is therefore zero-dimensional.

Looking now at two-forms, we note that every two-form is exact. The most general two-form can be expressed as $\tau = f(x, y)\,\mathrm{d}x \wedge \mathrm{d}y$. We form a function

$$F(x, y) = \int_0^x f(t, y)\,\mathrm{d}t$$

so that

$$\frac{\partial F}{\partial x} = f(x, y)$$

and then

$$\mathrm{d}(F\mathrm{d}y) = f(x, y)\,\mathrm{d}x \wedge \mathrm{d}y = \tau.$$

We conclude that H^2 is zero-dimensional.

Example 2B. The underlying space is again a rectangle in the plane, but with one point, which we take to be the origin 0, deleted. Again, H^0 is one-dimensional because the space is connected, and H^2 is zero-dimensional because every two-form

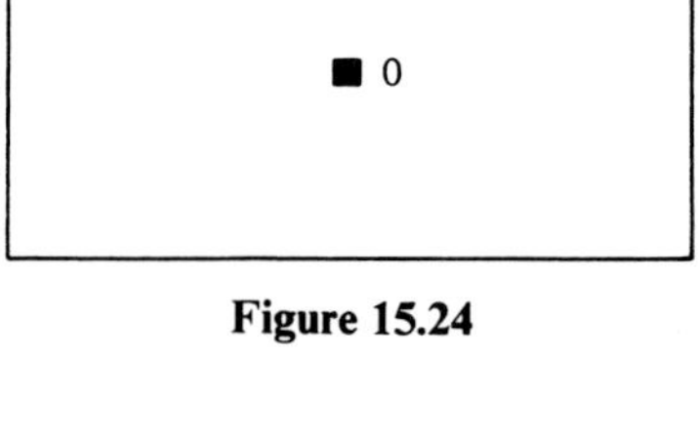

Figure 15.24

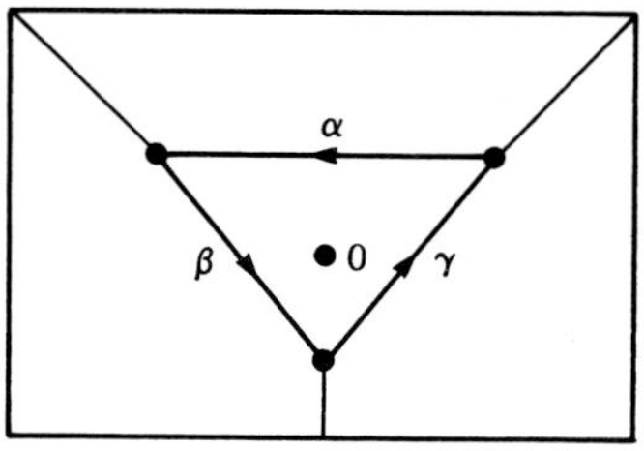

Figure 15.25

is exact. When we turn to one-forms, however, a new phenomenon arises: there exist one-forms which are closed but not exact, so that H^1 is not empty in this case.

We can guess a basis for H^1 by first considering the homology space H_1 for a two-complex situated in this space, as shown in figure 15.25. Because the origin is not part of the space, there can be no cell in the complex which includes the origin. As a result, a cycle like $\alpha + \beta + \gamma$ that encircles the origin cannot be a boundary. The equivalence class of such a cycle (modulo the space B_1 of boundaries) forms a basis for the one-dimensional space H_1. For the same complex, the cohomology space H^1 is dual to H_1. Its basis element will be the equivalence class of a one-cocycle $\mathbf{W}$ which is *not* a coboundary. Because H^1 is dual to H_1, we expect $\mathbf{W}$ to have a fixed non-zero value on any cycle such as $\alpha + \beta + \gamma$ which encircles the origin once in a counterclockwise sense. To find a basis for the de Rham cohomology space H^1, we must discover an incipient cocycle for $\mathbf{W}$: a closed one-form ω_0 with a fixed non-zero value on any curve which encircles the origin once. We can obtain such a one-form by considering the function $\theta = \arctan(y/x)$, which is defined everywhere except at the origin and which can be made continuous except on one line proceeding outward from the origin (usually the negative x-axis, so that $-\pi < \theta \leqslant \pi$). We then form the differential

$$\omega_0 = \mathrm{d}\theta = \frac{x\,\mathrm{d}y - y\,\mathrm{d}x}{x^2 + y^2}.$$

This one-form is defined and continuous everywhere except at the origin. It is closed: $\mathrm{d}\omega_0 = 0$, as you can verify by direct computation. On the other hand, it is not exact: there is no continuous function $f(x, y)$ such that $\omega_0 = \mathrm{d}f$. For any curve α which encircles the origin n times in a counterclockwise sense,

$$\int_\alpha \omega_0 = 2\pi n.$$

A basis for H^1 in this case is therefore the equivalence class

$$\overline{\omega_0} = \frac{x\,\mathrm{d}y - y\,\mathrm{d}x}{x^2 + y^2} + \mathrm{d}g$$

where $g(x, y)$ is any differentiable function.

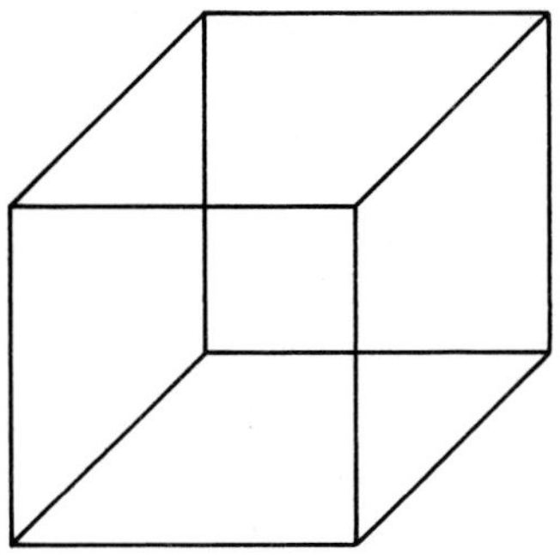

Figure 15.26

Example 3A. The underlying space is a rectangular parallelepiped region in $\mathbb{R}^3$. We state the results without proof:

H^0 is one-dimensional (the region is connected);

H^1 is $\{0\}$ (every closed one-form is exact);

H^2 is $\{0\}$ (every closed two-form is exact);

H^3 is $\{0\}$ (every three-form is exact).

The results for H^1, H^2, H^3 follow as consequences of a general theorem, known as *Poincaré's lemma*, which states, loosely speaking, that, on any space that can be continuously contracted to a point, every closed differential form is also exact. The interesting cases are those in which the space has 'holes' and so cannot be contracted to a point. Of course, these are also the cases in which there exist cycles which cannot be boundaries, so that the dual spaces $H_1, H_2, \ldots$ are non-empty.

Example 3B. The underlying space is a rectangular region in $\mathbb{R}^3$ with the origin excluded. The results for H^0, H^1, and H^3 are unchanged, but now H^2 is one-dimensional. The differential form

$$\tau_0 = \frac{x\,\mathrm{d}y \wedge \mathrm{d}z + y\,\mathrm{d}z \wedge \mathrm{d}x + z\,\mathrm{d}x \wedge \mathrm{d}y}{(x^2 + y^2 + z^2)^{3/2}}$$

is defined everywhere except at the origin. It is closed, as you can verify by a rather tedious computation. However, $\int_S \tau_0 = 4\pi$ for any closed surface S which encloses the origin. Such a surface, of course, is a cycle which is not a boundary, and so it is not surprising to find it belongs to the equivalence class which is dual to H^2.

Example 3C. The underlying space is a rectangular region in $\mathbb{R}^3$ with the z-axis excluded. In this case H^1 is one-dimensional, with basis

$$\bar{\omega}_0 = \frac{x\,\mathrm{d}y - y\,\mathrm{d}x}{x^2 + y^2} + \mathrm{d}g$$

but H^2 is zero-dimensional.

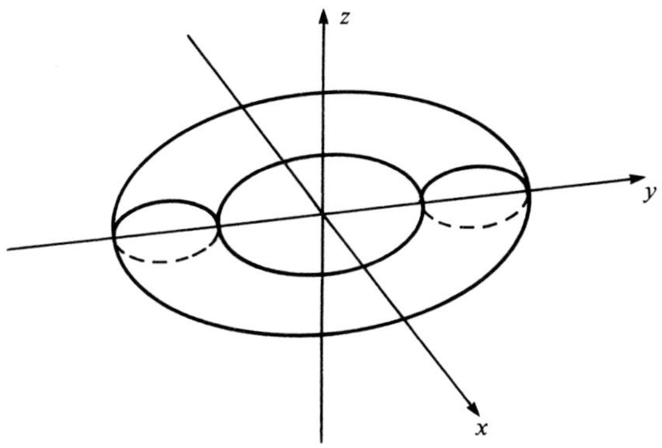

Figure 15.27

Example 3D. The underlying space is the surface of a torus situated in $\mathbb{R}^3$ as suggested by figure 15.27. In this case H^1 is two-dimensional. One basis element is again

$$\bar{\omega}_0 = \frac{x\,\mathrm{d}y - y\,\mathrm{d}x}{x^2 + y^2} + \mathrm{d}g$$

which measures how many times a closed curve encircles the z-axis. The other basis element is

$$\omega_1 = \frac{(R-1)\,\mathrm{d}z - z\,\mathrm{d}R}{z^2 + (R-1)^2} + \mathrm{d}g \quad R = \sqrt{(x^2 + y^2)}.$$

The value of this form on any closed curve measures how many times the curve winds around the unit circle in the xy-plane. Thus the basis elements ω_0 and ω_1 are dual to the two 'interesting' types of cycles for any complex on a torus.

In support of some of the statements made in the above examples, we shall now prove Poincaré's lemma by giving an explicit prescription for constructing a $(k-1)$-form ϕ with the property that $\mathrm{d}\phi = \omega$, where ω is a specified closed k-form. The theorem will be proved for the case where ω is defined on an open set in $\mathbb{R}^n$ which has the property that each point in the set can be joined to the origin by a straight line which lies entirely within the set. Such a set is called *star-shaped.* Later we can easily extend the proof to the case of a set which can be put into one-to-one correspondence with a star-shaped set by a smooth mapping. Examples are shown in figure 15.28.

Given a star-shaped region Q, with coordinates $x^1, \ldots, x^n$, we can construct a line segment joining the origin to an arbitrary point p. Because Q is star-shaped, this line segment is guaranteed to lie within Q. We define a function β which, for each point p, maps the interval $[0, 1]$ into this line segment. Thus, if the coordinates

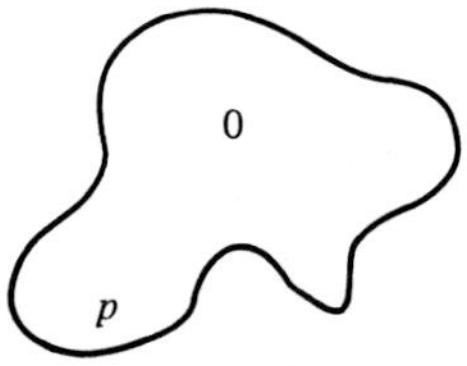

Star-shaped relative to 0 (though not with respect to p).

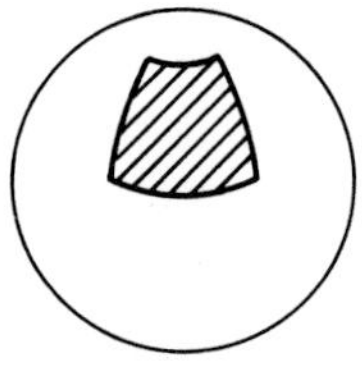

Region on surface of sphere– not star-shaped, but the image of a star-shaped region.

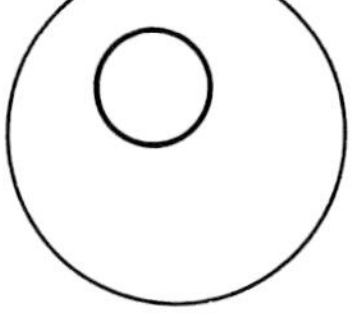

Not star-shaped with respect to any origin, because of the hole.

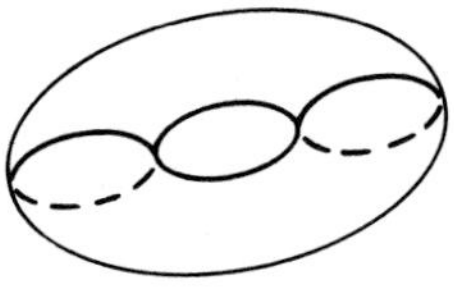

The surface of the entire torus is not the image of any star-shaped region.

Figure 15.28

of p are $\begin{pmatrix} x^1 \\ \vdots \\ x^n \end{pmatrix}$, the line segment is specified parametrically by

$$\begin{pmatrix} u^1 \\ \vdots \\ u^n \end{pmatrix} = \begin{pmatrix} tx^1 \\ \vdots \\ tx^n \end{pmatrix} \qquad 0 \leqslant t \leqslant 1.$$

In terms of pullback, we have $\beta^*(u^1) = tx^1$, etc.

Any function g defined on the region Q can be pulled back to the region $Q \times [0, 1]$. For example, the function $g(u^1, \ldots, u^n)$ pulls back to the function $\beta^* g = g(tx^1, \ldots, tx^n)$. This means, in effect, that $\beta^* g$ is a function of $p = (x^1, \ldots, x^n)$ and t which assigns, as t ranges from 0 to 1, the values assumed by g along the line joining the origin to p. Of course $\beta^* g$ is also a function of the coordinates of p; if p is changed, we look at values of g on a different line segment. Study figure 15.29 until you visualize the significance of $\beta^* g$.

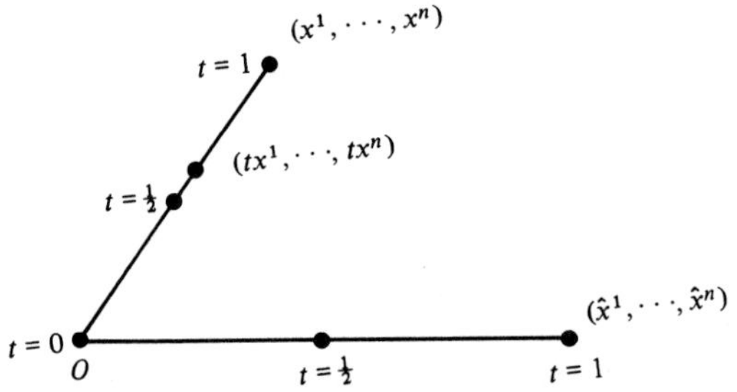

Figure 15.29

By the usual rules, we can pull back any basis one-form,

$$\beta^* \mathrm{d}u^i = \mathrm{d}(tx^i) = x^i \mathrm{d}t + t\,\mathrm{d}x^i,$$

of any k-form. Given a k-form ω defined on Q, we can write it in terms of the coordinates $u^1, \ldots, u^n$:

$$\omega = \sum_{i_1 < \cdots < i_k} a_{i_1 \ldots i_k}(u^1 \ldots u^n)\, \mathrm{d}u^{i_1} \wedge \cdots \wedge \mathrm{d}u^{i_k}.$$

On pulling back such a form we obtain

$$\beta^*\omega = \sum_{i_1 < \cdots < i_k} a_{i_1 \ldots i_k}(tx^1 \ldots tx^n)(t\,\mathrm{d}x^{i_1} + x^{i_1}\,\mathrm{d}t) \wedge \cdots \wedge (t\,\mathrm{d}x^{i_k} + x^{i_k}\,\mathrm{d}t).$$

This k-form $\beta^*\omega$ is a sum of terms of the form $\tau_1(t) = A(t, x^1, \ldots, x^n)\,\mathrm{d}x^{i_1} \wedge \cdots \wedge \mathrm{d}x^{i_k}$ or $\tau_2(t) = B(t, x^1, \ldots, x^n)\,\mathrm{d}t \wedge \mathrm{d}x^{i_1} \wedge \cdots \wedge \mathrm{d}x^{i_{k-1}}$. We now define a linear operator L which acts on these terms as follows:

$$L\tau_1 = 0$$

$$L\tau_2 = \left(\int_0^1 B(t, x^1, \ldots, x^n)\,\mathrm{d}t \right) \mathrm{d}x^{i_1} \wedge \cdots \wedge \mathrm{d}x^{i_{k-1}}.$$

This operator has the remarkable property that

$$\mathrm{d}L\tau_1 + L\mathrm{d}\tau_1 = \tau_1(1) - \tau_1(0),$$
$$\mathrm{d}L\tau_2 + L\mathrm{d}\tau_2 = 0.$$

This proof is a matter of straightforward computation.

We start with τ_1. Of course $\mathrm{d}L\tau_1 = 0$. Also

$$\mathrm{d}\tau_1 = \frac{\partial A}{\partial t}\mathrm{d}t \wedge \mathrm{d}x^{i_1} \wedge \cdots \wedge \mathrm{d}x^{i_k} + \text{terms with no } \mathrm{d}t$$

so that

$$L\mathrm{d}\tau_1 = \int_0^1 \frac{\partial A}{\partial t}\,\mathrm{d}x^{i_1} \wedge \cdots \wedge \mathrm{d}x^{i_k}$$
$$= \int A(1, x^1, \ldots, x^n) - A(0, x^1 \ldots x^n)\,\mathrm{d}x^{i_1} \wedge \cdots \wedge \mathrm{d}x^{i_k}.$$

For τ_2 we have

$$L\tau_2 = \left(\int_0^1 B(t, x^1, \ldots, x^n)\,\mathrm{d}t \right) \mathrm{d}x^{i_1} \wedge \cdots \wedge \mathrm{d}x^{i_{k-1}},$$

which is not a function of t. In forming $\mathrm{d}L\tau_2$, we may differentiate under the integral sign to obtain

$$\mathrm{d}L\tau_2 = \sum_{j=1}^{n} \left(\int_0^1 \frac{\partial B}{\partial x^j}\mathrm{d}t \right) \mathrm{d}x^j \wedge \mathrm{d}x^{i_1} \wedge \cdots \wedge \mathrm{d}x^{i_{k-1}}.$$

On the other hand

$$\mathrm{d}\tau_2 = \sum_{j=1}^{n} \frac{\partial B}{\partial x^j}\mathrm{d}x^j \wedge \mathrm{d}t \wedge \mathrm{d}x^{i_1} \wedge \cdots \wedge \mathrm{d}x^{i_{k-1}}$$

or

$$d\tau_2 = -\sum_{j=1}^{n} \frac{\partial B}{\partial x^j} dt \wedge dx^j \wedge dx^{i_1} \wedge \cdots \wedge dx^{i_{k-1}}$$

so that

$$Ld\tau_2 = -\sum_{j=1}^{n} \left(\int_0^1 \frac{\partial B}{\partial x^j} dt \right) dx^j \wedge dx^{i_1} \wedge \cdots \wedge dx^{i_{k-1}}$$

and we conclude that $dL\tau_2 + Ld\tau_2 = 0$.

It follows that if we consider $dL(\beta^*\omega) + Ld(\beta^*\omega)$, we can ignore all the terms of type τ_2, which involve a factor of dt. As long as $k > 0$, we have $\tau_1(0) = 0$ because of the factor of t which accompanies each dx^i. Thus, setting $t = 1$, we obtain

$$dL(\beta^*\omega) + Ld\beta^*\omega = \sum_{i_1 < \cdots < i_k} a_{i_1 \ldots i_k}(x^1, \ldots, x^n)\, dx^{i_1} \wedge \cdots \wedge dx^{i_k} = \omega.$$

But we know that $d\beta^*\omega = \beta^* d\omega$, so we have

$$dL\beta^*\omega + L\beta^* d\omega = \omega.$$

Thus the linear operator $S = L\beta^*$ satisfies the identity

$$dS\omega + Sd\omega = \omega.$$

If ω is closed, so that $d\omega = 0$, we have

$$dS\omega = \omega$$

and we have proved that ω is exact.

So far we have assumed that ω is defined on a star-shaped region Q. We now extend the above result to the case where ω is defined on a region $\Psi(Q)$ which is the image of a star-shaped region under a smooth one-to-one mapping Ψ. If ω is closed, so is $\Psi^*\omega$, since $d\Psi^*\omega = \Psi^* d\omega = 0$. We therefore can write $\Psi^*\omega = d(S\Psi^*\omega)$. Now, since Ψ is invertible, we can apply the inverse pullback $(\Psi^*)^{-1}$ to obtain

$$\omega = (\Psi^*) d(S\Psi^*\omega)$$

or

$$\omega = (\Psi^*)^{-1} S\Psi^*\omega.$$

Thus, for a form ω defined on a region D, we have a whole family of antiderivative operators, $(\Psi^*)^{-1}S\Psi^*\omega$, corresponding to various ways of expressing D as the image of a star-shaped region: $D = \Psi(Q)$. Another way of looking at this state of affairs is to note that we may introduce various coordinate systems on D in such a way the region in *coordinate space* which gets mapped into D is star-shaped. The operator $(\Psi^*)^{-1}S\Psi^*$ then corresponds to integrating along straight lines in coordinate space. If D is the surface of the Gulf of Mexico, for example, then we form S by integrating along straight lines joining points of D to the center of the Earth (or other origin). By introducing spherical coordinates, we construct a different antiderivative $(\Psi^*)^{-1}S\Psi^*$ which corresponds to joining each point to the origin by a path which appears straight when drawn on a Mercator projection map.

Here is a summary of the procedure for forming an antiderivative of a differential form

$$\omega = \sum_{i_1 < \cdots < i_k} a_{i_1 \ldots i_k}(x^1 \ldots x^n)\,\mathrm{d}x^{i_1} \wedge \cdots \wedge \mathrm{d}x^{i_k}.$$

Step 1. Form the pullback $\beta^*\omega$ by making the replacement $x^i \to tx^i$ in the arguments of all the coefficient functions, so that

$$a(x^1, \ldots, x^n) \to a(tx^1, \ldots, tx^n)$$

and make the replacement

$$\mathrm{d}x^i \to x^i\mathrm{d}t + t\mathrm{d}x^i.$$

Step 2. Throw out all terms which do not involve $\mathrm{d}t$. Move $\mathrm{d}t$ to the left in all other terms, keeping track of signs carefully.

Step 3. Treat the $\mathrm{d}t$ as in an ordinary integral, and integrate over t from 0 to 1. The following examples show the procedure in action:

Example 1. $\omega = y^2\mathrm{d}x + 2xy\mathrm{d}y$ (a closed one-form).

Step 1. $\beta^*\omega = t^2y^2(x\mathrm{d}t + t\mathrm{d}x) + 2t^2xy(y\mathrm{d}t + t\mathrm{d}y)$.

Step 2. $\beta^*\omega = (t^2xy^2 + 2t^2xy^2)\,\mathrm{d}t$ + other terms.

Step 3. $S\omega = xy^2\int_0^1 3t^2\mathrm{d}t = xy^2$.

Check. $\mathrm{d}(xy^2) = y^2\mathrm{d}x + 2xy\mathrm{d}y$.

Example 2. $\omega = \sin x\,\mathrm{d}x \wedge \mathrm{d}y$.

Step 1. $\beta^*\omega = \sin(tx)(x\mathrm{d}t + t\mathrm{d}x) \wedge (y\mathrm{d}t + t\mathrm{d}y)$.

Step 2. $\beta^*\omega = xt\sin tx\,\mathrm{d}t \wedge \mathrm{d}y - yt\sin tx\,\mathrm{d}t \wedge \mathrm{d}x$ + term without $\mathrm{d}t$.

Step 3. $S\omega = (x\mathrm{d}y - y\mathrm{d}x)\int_0^1 t\sin xt\,\mathrm{d}t$

$$= (x\mathrm{d}y - y\mathrm{d}x)\left(\frac{\sin x - x\cos x}{x^2}\right)$$

$$= \left(\frac{\sin x - x\cos x}{x}\right)\mathrm{d}y - y\left(\frac{\sin x - x\cos x}{x^2}\right)\mathrm{d}x$$

or

$$S\omega = -\cos x\,\mathrm{d}y + y\left(\frac{x\cos x - \sin x}{x^2}\right)\mathrm{d}x + \frac{\sin x}{x}\mathrm{d}y.$$

The answer differs from the 'obvious' antiderivative $-\cos x\,\mathrm{d}y$ by the exact one-form

$$\mathrm{d}\left(\frac{y}{x}\sin x\right) = \frac{x\cos x - \sin x}{x^2}\,y\,\mathrm{d}x + \frac{\sin x}{x}\,\mathrm{d}y.$$

Example 3. $\beta^*\omega = (y^2 - x^2)z\mathrm{d}x \wedge \mathrm{d}y + (x^2 - z^2)y\mathrm{d}z \wedge \mathrm{d}x + (z^2 - y^2)x\,\mathrm{d}y \wedge \mathrm{d}z$.

Step 1. $\beta^*\omega = t^3(y^2 - x^2)z(x\mathrm{d}t + t\mathrm{d}x) \wedge (y\mathrm{d}t + t\mathrm{d}y)$

$$+ t^3(x^2 - z^2)y(z\mathrm{d}t + t\mathrm{d}z) \wedge (x\mathrm{d}t + t\mathrm{d}x)$$
$$+ t^3(z^2 - y^2)x(y\mathrm{d}t + t\mathrm{d}y) \wedge (z\mathrm{d}t + t\mathrm{d}z).$$

Step 2. $\beta^*\omega = t^4(-(y^2 - x^2)zy + (x^2 - z^2)yz)\,\mathrm{d}t \wedge \mathrm{d}x$
$$+ ((y^2 - x^2)zx - (z^2 - y^2)xz)\,\mathrm{d}t \wedge \mathrm{d}y$$
$$+ (-(x^2 - z^2)yx + (z^2 - y^2)xy)\,\mathrm{d}t \wedge \mathrm{d}z$$
$$+ \text{terms without } \mathrm{d}t.$$

Step 3. $S\omega = \int_0^1 t^4\mathrm{d}t(2x^2yz - y^3z - yz^3)\,\mathrm{d}x$
$$+ (2y^2zx - x^3z - xz^3)\,\mathrm{d}y + (2z^2xy - xy^3 - x^3y)\,\mathrm{d}z$$

or

$$S\omega = \tfrac{2}{5}(x^2yz\,\mathrm{d}x + y^2zx\mathrm{d}y + z^2xy\,\mathrm{d}z).$$

We have just shown that the kernel of $S: \Omega^k \to \Omega^{k-1}$ is a subspace of the image of $S: \Omega^{k+1} \to \Omega^k$. Indeed, it is the entire image. Suppose, for example, that a k-form satisfies $S\omega = 0$. Then, since $\omega = S\mathrm{d}\omega + \mathrm{d}S\omega$ we know that $\omega = S\,\mathrm{d}\omega$ so that ω is in the image of S. To summarize: $\mathbf{S}\boldsymbol{\omega} = \mathbf{0}$ *if and only if* $\boldsymbol{\omega} = \mathbf{S}\boldsymbol{\phi}$ *for some* $\boldsymbol{\phi}$.

Summary

A Exterior algebra and calculus

You should be able to define the spaces $\Lambda^k(V^*)$ for an n-dimensional vector space V, and to write down a basis for each of these spaces.

You should be able to state and apply the properties of the d operator for differential forms of arbitrary degree, including its relationship to pullback.

B Integration of differential forms

You should be able to evaluate the integral of a k-form over a k-cell which is expressed as the image of the unit k-cube.

You should be able to state and apply Stokes' theorem and outline a proof of it.

C Differential forms and cohomology

You should know how to construct cochains by interaction of differential forms and be able to identify forms that define basis elements of H^k for 2-complexes or 3-complexes.

Given a differential k-form ϕ with $\mathrm{d}\phi = 0$, defined on a star-shaped region, you should be able to construct the $(k-1)$-form $S\phi$ with the property $\mathrm{d}S\phi = \phi$.

Exercises

15.1(a) Let $\tau(\mathbf{v}_1, \mathbf{v}_2, \mathbf{v}_3)$ be an alternating trilinear function (i.e., a three-form). *Without* invoking properties of determinants, prove that, if $\mathbf{v}_1, \mathbf{v}_2$, and $\mathbf{v}_3$ are linearly dependent, then $\tau(\mathbf{v}_1, \mathbf{v}_2, \mathbf{v}_3) = 0$.

(b) Let $\omega^1, \omega^2, \omega^3$ be three elements of V^*. Without invoking properties of determinants, prove that, if $\omega^1, \omega^2, \omega^3$ are linearly dependent, then $\omega^1 \wedge \omega^2 \wedge \omega^3 = 0$.

15.2. In four-dimensional spacetime let E be the two-form $E = \mathrm{d}t \wedge (E_x \mathrm{d}x + E_y \mathrm{d}y + E_z \mathrm{d}z)$. Let $B = B_z \mathrm{d}x \wedge \mathrm{d}y - B_y \mathrm{d}x \wedge \mathrm{d}z + B_x \mathrm{d}y \wedge \mathrm{d}z$. Calculate $E \wedge B$.

15.3. Suppose we define the determinant of a linear transformation $A: \mathbb{R}^n \to \mathbb{R}^n$ by

$$\mathrm{Det}\, A = \mathrm{d}x^1 \wedge \mathrm{d}x^2 \wedge \cdots \wedge \mathrm{d}x^n[A\mathbf{e}_1, A\mathbf{e}_2, \ldots, A\mathbf{e}_n]$$

where $\{\mathrm{d}x^1, \mathrm{d}x^2, \ldots, \mathrm{d}x^n\}$ are dual to $\{\mathbf{e}_1, \mathbf{e}_2, \ldots, \mathbf{e}_n\}$.

(a) Prove that if A is the identity matrix, $\mathrm{Det}\, A = 1$.

(b) Prove that if A^* denotes the adjoint of A, then

$$\mathrm{Det}\, A = A^*\mathrm{d}x^1 \wedge A^*\mathrm{d}x^2 \wedge \cdots \wedge A^*\mathrm{d}x^n(\mathbf{e}_1, \mathbf{e}_2, \ldots, \mathbf{e}_n).$$

15.4. Evaluate the following determinants by using the results of Exercises 15.3(c), (b):

(a) $$\mathrm{Det}\begin{pmatrix} 2 & -1 & 3 \\ -1 & 3 & 2 \\ 3 & 2 & -1 \end{pmatrix} = \mathrm{d}x \wedge \mathrm{d}y \wedge \mathrm{d}z \left[\begin{pmatrix} 2 \\ -1 \\ 3 \end{pmatrix}, \begin{pmatrix} -1 \\ 3 \\ 2 \end{pmatrix}, \begin{pmatrix} 3 \\ 2 \\ -1 \end{pmatrix}\right].$$

(b) $$\mathrm{Det}\begin{pmatrix} a & -b & 0 & 0 \\ b & a & -b & 0 \\ 0 & b & a & -b \\ 0 & 0 & b & a \end{pmatrix}.$$

15.5. Let $\Lambda^3(V)$ denote the space of all *alternating multilinear* functions from $V \times V \times V$ to $\mathbb{R}$. Suppose that V is four-dimensional, and let $\varepsilon^1, \varepsilon^2, \varepsilon^3, \varepsilon^4$ be a basis for the dual space V^*. Using the wedge notation, write down a basis for $\Lambda^3(V)$ and write a formula for the action of one basis element on a triplet of vectors $(\mathbf{v}_1, \mathbf{v}_2, \mathbf{v}_3)$.

15.6. Let $\omega = f\mathrm{d}x + \mathrm{d}y$, $\tau = g\mathrm{d}x + \mathrm{d}z$ where f and g are differentiable functions of x, y, and z. Calculate $\mathrm{d}(\omega \wedge \tau)$, expressing your answer as a multiple of $\mathrm{d}x \wedge \mathrm{d}y \wedge \mathrm{d}z$.

15.7. Let β be a differentiable mapping from an affine space A (dimension m) to B (dimension n). Let $\tau(q; \mathbf{w})$ be a one-form on B, where q is a point of B, $\mathbf{w}$ a vector at that point. Then $\mathrm{d}\tau$ may be *defined* as

$\mathrm{d}\tau(q; \mathbf{w}_1, \mathbf{w}_2) =$ best linear approximation to

$$\tau(q + \mathbf{w}_1, \mathbf{w}_2) - \tau(q, \mathbf{w}_2) - \tau(q + \mathbf{w}_2, \mathbf{w}_1) + \tau(q, \mathbf{w}_1).$$

The pullback of a one-form is *defined* by

$$(\beta^*\tau)(q; \mathbf{v}) = \tau(\beta q; \mathrm{d}\beta(\mathbf{v}))$$

while for a two-form σ,

$$(\beta^*\sigma)(q; \mathbf{v}_1, \mathbf{v}_2) = \sigma(\beta q; \mathrm{d}\beta(\mathbf{v}_1), \mathrm{d}\beta(\mathbf{v}_2)).$$

(a) Prove directly from the above definitions (and the chain rule, if you need it)

$$\mathrm{d}\beta^*\tau = \beta^*(\mathrm{d}\tau).$$

(b) Introduce affine coordinates $u^1,\ldots,u^m$ on A and $x^1,\ldots,x^n$ on B. Then β may be specified by the differentiable pullback functions

$$\beta^*x^1 = F^1(u^1,\ldots,u^m),$$
$$\beta^*x^2 = F^2(u^1,\ldots,u^m),$$
$$\vdots$$
$$\beta^*x^n = F^n(u^1,\ldots,u^m)$$

while

$$\tau = \sum_{j=1}^{n} G^j(x^1,\ldots,x^n)\mathrm{d}x^j$$

where the functions G are differentiable. Calculate $\mathrm{d}\beta^*\tau$ and $\beta^*(\mathrm{d}\tau)$ explicitly and show that they are equal.

(c) Let τ and λ both be one-forms. Show that if $\Omega = \tau \wedge \lambda$ then $\beta^*\Omega = (\beta^*\tau) \wedge (\beta^*\lambda)$.

15.8. Spherical coordinates r, θ, ϕ and Cartesian coordinates x, y, z are related by

$$\alpha^*x = r\sin\theta\cos\phi,$$
$$\alpha^*y = r\sin\theta\sin\phi,$$
$$\alpha^*z = r\cos\theta$$

where α is the mapping which carries the point $\begin{pmatrix} r \\ \theta \\ \phi \end{pmatrix}$ in spherical coordinate space into the point $\begin{pmatrix} x \\ y \\ z \end{pmatrix}$ whose coordinates are r, θ, and ϕ.

Calculate the pullback of the following differential forms (i.e., express them in spherical coordinates).

(a) $\alpha^*(x\mathrm{d}y - y\mathrm{d}x)$;

(b) $\alpha^*\dfrac{x\mathrm{d}x + y\mathrm{d}y + z\mathrm{d}z}{(x^2+y^2+z^2)^3}$;

(c) $\alpha^*\dfrac{x\mathrm{d}y\wedge\mathrm{d}z + y\mathrm{d}z\wedge\mathrm{d}x + z\mathrm{d}x\wedge\mathrm{d}y}{(x^2+y^2+z^2)^{3/2}}$;

(d) $\alpha^*\mathrm{d}x\wedge\mathrm{d}y\wedge\mathrm{d}z$.

15.9(a) Let f be a differentiable function on $\mathbb{R}^3$. Let A be the two-form

$$A = \frac{\partial f}{\partial x}\mathrm{d}y\wedge\mathrm{d}z + \frac{\partial f}{\partial y}\mathrm{d}z\wedge\mathrm{d}x + \frac{\partial f}{\partial z}\mathrm{d}x\wedge\mathrm{d}y.$$

Show that $\mathrm{d}A = \Delta f\mathrm{d}x\wedge\mathrm{d}y\wedge\mathrm{d}z$, where Δ denotes the Laplacian $\partial^2/\partial x^2 + \partial^2/\partial y^2 + \partial^2/\partial z^2$.

(b) Let F_x, F_y, F_z be differentiable functions on $\mathbb{R}^3$ which satisfy

$\dfrac{\partial F_x}{\partial x} + \dfrac{\partial F_y}{\partial y} + \dfrac{\partial F_z}{\partial z} = 0$. Define

$$B = \left(\frac{\partial F_z}{\partial y} - \frac{\partial F_y}{\partial z}\right)dx + \left(\frac{\partial F_x}{\partial z} - \frac{\partial F_z}{\partial x}\right)dy + \left(\frac{\partial F_y}{\partial x} + \frac{\partial F_x}{\partial y}\right)dz.$$

Show that $dB = \Delta F_x dy \wedge dz + \Delta F_y dz \wedge dx + \Delta F_z dx \wedge dy$.

15.10. A solid of revolution about the z-axis is bounded below by the disk $x^2 + y^2 \leqslant 1, z = 0$, on the side by the cylinder $x^2 + y^2 = 1, 0 < z < 2$, and on top by the paraboloid $z = 1 + x^2 + y^2$ described parametrically by

$$\beta^*x = r\cos\theta,$$
$$\beta^*y = r\sin\theta,$$
$$\beta^*z = 1 + r^2.$$

All these surfaces are oriented counterclockwise as viewed from outside, as shown in Fig. 15.30.

(a) Evaluate the integral of the two-form $\tau = x dy \wedge dz - y dx \wedge dz$ over the paraboloidal top surface.

(b) Evaluate the integral of τ over the cylindrical surface.

(c) Use Stokes' theorem to express the *volume* of the solid in terms of the integrals which you have just calculated.

15.11. Let $\Omega = x dy \wedge dz$.

(a) Evaluate $\int\Omega$ over the disk $x^2 + y^2 \leqslant R^2, z = 0$.

(b) Consider the hemisphere $x^2 + y^2 + z^2 = R^2, z \geqslant 0$, which can be parametrized by spherical coordinates as follows (see Fig. 15.31):

$$z = R\cos\theta,$$
$$x = R\sin\theta\cos\phi,$$
$$y = R\sin\theta\sin\phi.$$

Express Ω in terms of the constant R and the coordinates θ and ϕ, and thereby evaluate $\int\Omega$ over the hemisphere described above.

(c) Use Stokes' theorem to write down an integral of a three-form which must equal the difference of the two surface integrals in (b) and (a). State the geometrical significance of this integral, and thereby evaluate it by inspection.

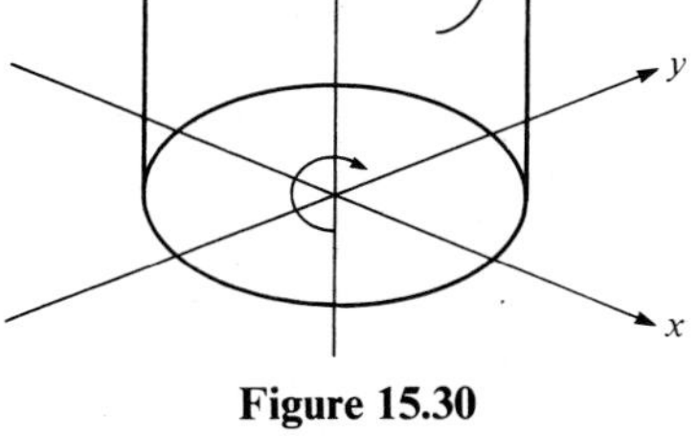

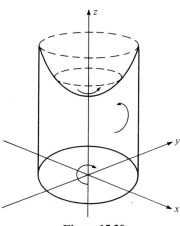

Figure 15.30

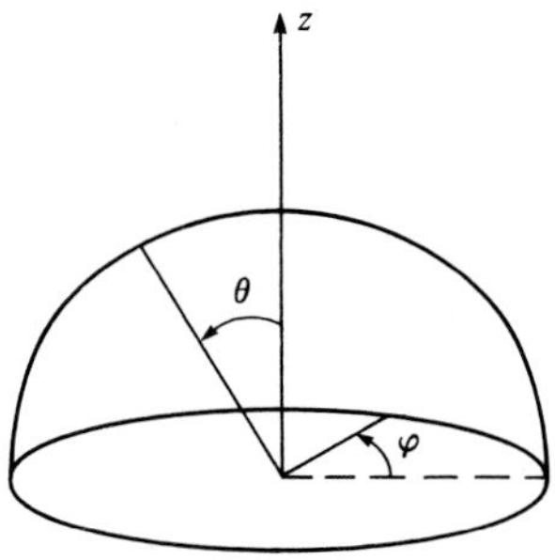

Figure 15.31

15.12(a) Consider a solid of uniform density ρ which occupies a region Q in $\mathbb{R}^3$. Show that the integral of the two-form

$$\tau = \tfrac{1}{3}(x^3 dy \wedge dz - y^3 dx \wedge dz)$$

over the *boundary* of Q gives the moment of inertia I_{zz} for the solid.

(b) Invent a two-form σ whose integral over ∂Q gives the product of inertia I_{xy}. Why is your answer not unique?

(c) Use the result of part (a) to calculate the moment of inertia of a sphere of radius a about a diameter.
Hint: Use spherical coordinates.

15.13(a) Let W be the region in $\mathscr{R}^3$ occupied by a solid body of uniform density ρ. Show that the z-component of the center of mass of the body, $\bar{z}$, can be determined as a quotient of two surface integrals evaluated over the boundary of the body, ∂W, as follows:

$$\bar{z} = \frac{\displaystyle\int_{\partial W} \rho \tfrac{1}{2} r^2 \, dx \wedge dy}{\displaystyle\int_{\partial W} z dx \wedge dy}$$

where $r^2 = x^2 + y^2 + z^2$.

(b) Spherical coordinates are defined by the pullback $\alpha^* x = r \sin\theta \cos\phi$, $\alpha^* y = r \sin\theta \sin\phi$, $\alpha^* z = r\cos\theta$. Calculate $\alpha^*(dx \wedge dy)$.

(c) Using the result of (b), evaluate

$$\int_{\partial W} \tfrac{1}{2} r^2 dx \wedge dy$$

for the case where W is the interior of a hemisphere of radius a, with a right-handed orientation.

15.14. Let C be the half-cylinder bounded by the planes $z = 0$ (Bottom) and $z = 1$ (Top), the plane $x = 0$ (Flat) and the surface $x^2 + y^2 = a^2$ (Curve). C has a right-handed orientation, so that $\mathbf{e}_x, \mathbf{e}_y, \mathbf{e}_z$, as shown in figure 15.32, are a correctly ordered basis. Each face in the boundary of C has been given an ordered basis, as shown, which defines its orientation.

(a) Write down an expression, with appropriate signs, for ∂C.

(b) Let τ be the two-form $\tau = x^2 dy \wedge dz - 2xz dx \wedge dy$. Calculate the pullback of τ under the mapping defined by $\beta^* x = a \cos u$,

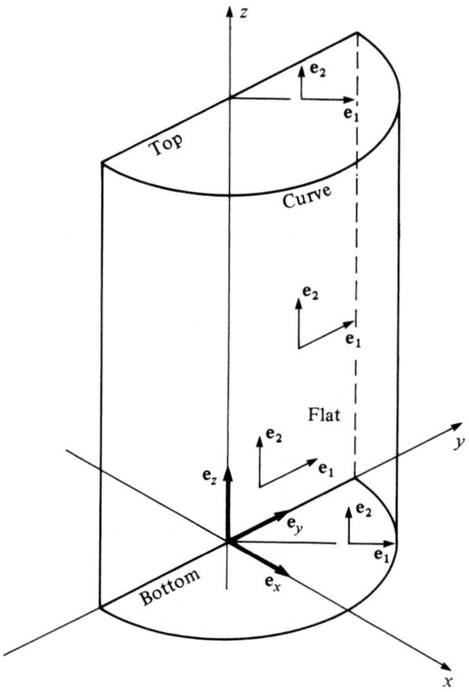

Figure 15.32

$\beta^*y = a \sin u$, $\beta^*z = v$, and thereby calculate the integral of τ over the face Curve.

(c) Using the mapping defined by $\alpha^*x = r\cos\theta, \alpha^*y = r\sin\theta, \alpha^*z = 1$, evaluate the integral of τ over the face Top.

(d) What is the integral of τ over the other two faces? Explain.

(e) Calculate $d\tau$. Using Stokes' theorem, explain how this result provides a check on your answers to parts (b), (c) and (d).

15.15. Consider the two-complex shown in figure 15.33.

(a) Let W be a one-cochain. Express the two-cochain dW in terms of W; i.e., express $\int_{S_1} dW$ and $\int_{S_2} dW$ in terms of $W_\alpha, \ldots$.

(b) Consider the one-form $\omega = y\mathrm{d}x + x\mathrm{d}y = \mathrm{d}(xy)$. Calculate the one-cochain W determined by ω by integrating ω over the branches. Verify that $dW = 0$.

(c) Do the same for the one-form $\omega = (x\mathrm{d}y - y\mathrm{d}x)/r^2$. Show that it determines a cochain W which is a cocycle but *not* a coboundary. Would W still be a cocycle if S_3, the interior of the circle bounded by α and ε, were included in the complex?

15.16. The unit cube shown in figure 15.34 determines a three-complex with six faces:

$$S_1 : x = 0; \quad 0 \leqslant y, z \leqslant 1$$
$$S_2 : x = 1; \quad 0 \leqslant y, z \leqslant 1$$
$$S_3 : y = 0; \quad 0 \leqslant x, z \leqslant 1$$

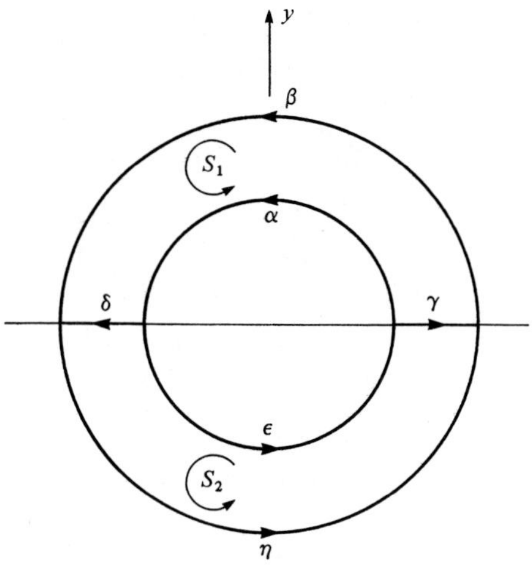

Figure 15.33

$$S_4: y = 1; \quad 0 \leqslant y, z \leqslant 1$$
$$S_5: z = 0; \quad 0 \leqslant x, y \leqslant 1$$
$$S_6: z = 1; \quad 0 \leqslant x, y \leqslant 1.$$

Each face is given a counterclockwise orientation as seen from the outside, so that the interior of the cube, R, has as its boundary $\partial R = S_1 + S_2 + S_3 + S_4 + S_5 + S_6$.

(a) Let T be any two-cochain. Find an expression for $\mathrm{d}T$.

(b) Let τ be the two-form $\tau = x^2 \mathrm{d}y \wedge \mathrm{d}z + yx\mathrm{d}z \wedge \mathrm{d}x + xz\mathrm{d}x \wedge \mathrm{d}y$. Construct the corresponding two-cochain T by integrating τ over each face. Evaluate $\mathrm{d}\tau$ and construct the corresponding three-cochain: check that it equals $\mathrm{d}T$.

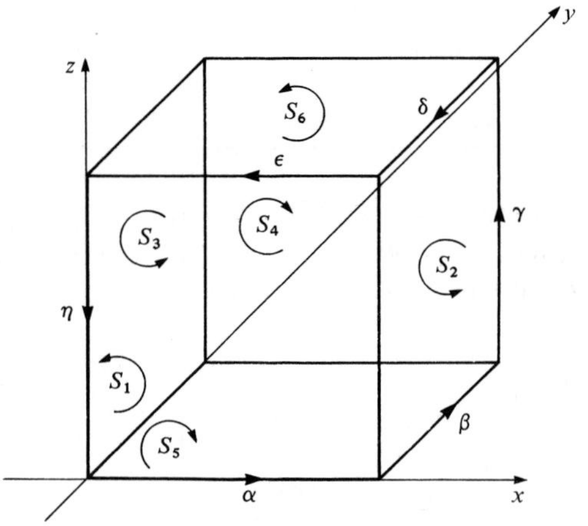

Figure 15.34

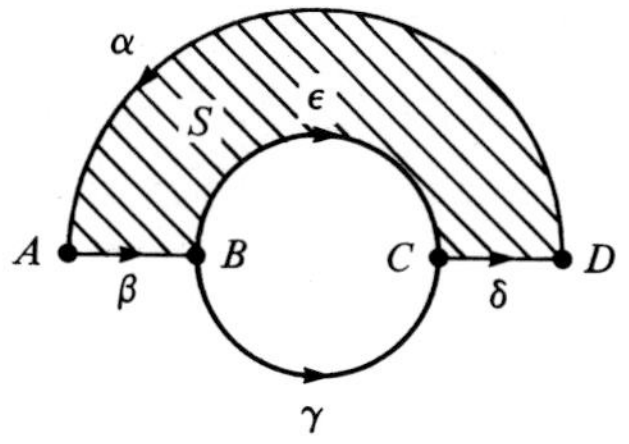

Figure 15.35

(c) Let ω be the one-form $\omega = x^2y\mathrm{d}z$. Compute $\mathrm{d}\omega$. Evaluate the one-cochain W corresponding to ω on branches $\alpha, \beta, \gamma, \delta, \varepsilon, \eta$. Evaluate the two-cochain T corresponding to $\mathrm{d}\omega$ on faces S_2 and S_3. Check that T, evaluated on $S_2 + S_3$, equals W evaluated on the boundary of $S_2 + S_3$.

15.17(a) Let τ be a two-form on $\mathbb{R}^3$: $\tau = a\mathrm{d}x \wedge \mathrm{d}y + b\mathrm{d}x \wedge \mathrm{d}z + c\mathrm{d}y \wedge \mathrm{d}z$, where a, b, c are differentiable functions of x, y, z. Let ϕ be a mapping from $\mathbb{R}^2$ (coordinates u, v) to $\mathbb{R}^3$ (coordinates x, y, z) defined by

$$\phi^*x = F(u, v), \quad \phi^*y = G(u, v), \quad \phi^*z = H(u, v).$$

Verify by explicit computation that $\phi^*(\mathrm{d}\tau) = \mathrm{d}(\phi^*\tau)$. You will of course have to use the chain rule.

(b) Let $\mathbf{v}_1, \mathbf{v}_2$ be vectors in $\mathbb{R}^2$. Then the pullback $\phi^*\tau$ can be defined by

$$\phi^*\tau(\mathbf{v}_1, \mathbf{v}_2) = \tau[\mathrm{d}\phi(\mathbf{v}_1), \mathrm{d}\phi(\mathbf{v}_2)]$$

(remember, $\mathrm{d}\phi$ is represented at each point by a 3×2 Jacobian matrix). Use this definition and the chain rule to prove that $\phi^*\mathrm{d}\tau = \mathrm{d}(\phi^*\tau)$.

15.18. Using the S-operator described in section 15.4:

(a) Find a function $f(x, y, z)$ such that

$$\mathrm{d}f = ye^{yz}\mathrm{d}x + (x + xyz)e^{yz}\mathrm{d}y + xy^2e^{yz}\mathrm{d}z.$$

(b) Find a one-form ω such that

$$\mathrm{d}\omega = x\mathrm{d}y \wedge \mathrm{d}z + y\mathrm{d}x \wedge \mathrm{d}z.$$

(c) Find a two-form τ such that

$$\mathrm{d}\tau = xy^2z^3\mathrm{d}x \wedge \mathrm{d}y \wedge \mathrm{d}z.$$

15.19. Let $B = z^2\mathrm{d}x \wedge \mathrm{d}y + yz\mathrm{d}x \wedge \mathrm{d}z - xz\mathrm{d}y \wedge \mathrm{d}z$.

(a) Let W be any region in $\mathbb{R}^3$. Show that $\int_{\partial W} B = 0$.

(b) Find a one-form A such that $\mathrm{d}A = B$.

15.20. Consider the two-complex shown in figure 15.35, with four nodes, five branches, and one two-cell S, which is shaded in the diagram. Let

$$\omega = \mathrm{d}\theta = \frac{x\mathrm{d}y - y\mathrm{d}x}{x^2 + y^2}.$$

Determine the one-chain W which corresponds to ω. (Calculate W^α, W^β, W^γ, W^δ, W^ε. Show that W is a cocycle but not a coboundary.

15.21. Let $\mathbb{R}^{1,3}$ denote four-dimensional spacetime, with affine coordinates ordered t, x, y, z.

(a) Write down a basis for $\Lambda^2(\mathbb{R}^{1,3})$.

(b) Consider the two form

$$W = A(x, y, t)\mathrm{d}t \wedge \mathrm{d}x + B(x, y, t)\mathrm{d}t \wedge \mathrm{d}y + C(x, y, t)\mathrm{d}x \wedge \mathrm{d}y.$$

If $\mathrm{d}W = 0$, what relation among the partial derivatives of A, B, and C must hold?

16

Electrostatics in $\mathbb{R}^3$

Chapter 16 is devoted to electrostatics. We suggest that the dielectric properties of the vacuum give the continuous analog of the capacitance of a network, and that these dielectric properties are what determine Euclidean geometry in three-dimensional space. The basic facts of potential theory are presented.

16.1. From the discrete to the continuous

We let $A^0 = \Omega^0(\mathbb{R}^3)$ denote the space of smooth functions on $\mathbb{R}^3$ (thought of as forms of 'degree zero'); we let $A^1 = \Omega^1(\mathbb{R}^3)$ denote the space of smooth linear differential forms, A^2 denote the space of smooth forms of degree two and A^3 the space of smooth forms of degree three. We have been thinking of A^i as *incipient cochains* of degree i; that is, if we are given a complex in $\mathbb{R}^3$, each form of degree i defines a cochain of degree i by a process of integration. We thus have been thinking of i-forms as rules which assign numbers to chains. We can turn this picture around. If we fix a definite chain c of degree i, then we can think of c as defining a linear function on the space A^i: to each form ω of degree i we assign the number $\int_c \omega$. If ω_1 and ω_2 are two forms of degree i and if a and b are real numbers, then it follows from the properties of integration that

$$\int_c (a\omega_1 + b\omega_2) = a\int_c \omega_1 + b\int_c \omega_2;$$

in other words, the rule which assigns to each ω the number $\int_c \omega$ is a linear function.

The simplest illustration of this is when we take the chain c to be a zero-chain, say the zero-chain which consists of the single point $P \in \mathbb{R}^3$ with the orientation $+$. In this case, the integration reduces simply to evaluation: the chain c gives rise to the linear function on A^0 which assigns to each $f \in A^0$ the value $f(P)$. If we think of c as a unit charge placed at the point P and f as a potential, then $f(P)$ is the energy corresponding to the charge distribution which places a unit charge at P when the potential is given by f. This reversed viewpoint, where we consider P as a linear function of f, has the following advantage. Suppose that we want to replace the discrete charge distribution concentrated at P by a 'smeared-out' or 'continuous' charge distribution, say with density ρ. Here ρ is assumed to be a smooth function of compact support. (Compact support means that ρ vanishes outside some bounded set.) We can do so as follows. Consider the three-form (of compact support) $\rho \mathrm{d}x \wedge \mathrm{d}y \wedge \mathrm{d}z$. For each $f \in A^0$ we can multiply f with $\rho \mathrm{d}x \wedge \mathrm{d}y \wedge \mathrm{d}z$ to obtain the three-form $f\rho \mathrm{d}x \wedge \mathrm{d}y \wedge \mathrm{d}z$ which is a three-form of compact support on $\mathbb{R}^3$. Having fixed the orientation on $\mathbb{R}^3$, we can integrate this three-form to obtain a number. Thus the form $\rho \mathrm{d}x \wedge \mathrm{d}y \wedge \mathrm{d}z$ defines a linear function on A^0, the linear function which assigns to each $f \in A^0$ the number $\int_{\mathbb{R}^3} f\rho \mathrm{d}x \wedge \mathrm{d}y \wedge \mathrm{d}z$. (Notice that if the total integral of the function ρ equals 1 and if ρ gets more and more concentrated about the point P – in other words, if we let the continuous charge distribution approach the point charge concentrated at P – then the value of this integral will approach the number $f(P)$.) Let A_3 denote the space of three-forms of compact support on $\mathbb{R}^3$. Then we have shown that every $\rho \mathrm{d}x \wedge \mathrm{d}y \wedge \mathrm{d}z \in A_3$ defines a linear function on A^0. We can think of an element of A_3 as a kind of smeared-out zero-chain.

In reality, as we know, the electric charges are discrete. But we can now get a grip on the notion of *approximating the discrete by the continuous*. Suppose we had some densely packed distribution of small charges, $c(P_i)$. Then we would get a linear function

$$f \rightsquigarrow \sum c_i f(P_i) \quad \rightsquigarrow$$

Figure 16.1

on functions. This linear function might be well-approximated (for a broad class of functions f) by the *smeared-out charge distribution* $\rho \mathrm{d}x \wedge \mathrm{d}y \wedge \mathrm{d}z$ in the sense that

$$\int f\rho \mathrm{d}x \wedge \mathrm{d}y \wedge \mathrm{d}z$$

Figure 16.2

is close to

$$\sum c_i f(P_i).$$

We shall now introduce *continuous approximations* for one-chains, two-chains, etc. Indeed, let A_2 denote the space of smooth two-forms of compact support. We claim that we can regard an element Ω of A_2 as a kind of smeared-out one-chain, i.e. that Ω defines a linear function on A^1. To see this let ω be any element of A^1 and form the product $\omega \wedge \Omega$. This is a three-form of compact support which we can again integrate over $\mathbb{R}^3$ to obtain a number. In other words, $\Omega \in A_2$ defines a linear function on A^1 which assigns to each ω the number $\int_{\mathbb{R}^3} \omega \wedge \Omega$. Similarly, if we let A_1 denote the space of smooth one-forms of compact support, then each element of A_1 defines a linear function on A^2, so can be thought of as a kind of smeared-out two-chain, and, if A_0 denotes the space of zero-forms (i.e., smooth functions) of compact support, then each element of A_0 defines a linear function on A^3 and so an element of A_0 can be thought of as a kind of smeared-out three-chain. To summarize:

A^0 is paired with A_3,
A^1 is paired with A_2,
A^2 is paired with A_1,
A^3 is paired with A_0,

in the sense that, if $\omega \in A^i$ and $\Omega \in A_{3-i}$, we get the number $\int_{\mathbb{R}^3} \omega \wedge \Omega$. For fixed Ω this is a linear function of ω; also, for fixed ω it is a linear function of Ω. Also this pairing is invariant under any orientation-preserving one-to-one map ϕ of $\mathbb{R}^3$ onto itself with differentiable inverse. If ϕ is such a map, then, for any three-form, τ, with compact support, the change of variables formula for integration says that $\int_{\mathbb{R}^3} \phi^* \tau = \int_{\mathbb{R}^3} \tau$. On the other hand, $\phi^*(\omega \wedge \Omega) = \phi^* \omega \wedge \phi^* \Omega$. Therefore

$$\int \phi^* \omega \wedge \phi^* \Omega = \int \omega \wedge \Omega.$$

In other words, the pairing is invariant under orientation-preserving changes of coordinates.

16.2. The boundary operator

Since we have the map $\mathrm{d}: A^i \to A^{i+1}$ we have the adjoint map, ∂, which assigns, to each linear function, c, on A^{i+1} the linear function ∂c on A^i defined by $(\partial c)(\omega) = c(\mathrm{d}\omega)$. This suggests that we should be able to find a map ∂ from A_{i+1} to A_i such that for any $\sigma \in A_{i+1}$ and any $\omega \in A^i$ we have

$$\int_{\mathbb{R}^3} \omega \wedge \partial \sigma = \int_{\mathbb{R}^3} \mathrm{d}\omega \wedge \sigma.$$

To find out what the operator ∂ actually is we observe that

$$\mathrm{d}(\omega \wedge \sigma) = \mathrm{d}\omega \wedge \sigma + (-1)^i \omega \wedge \mathrm{d}\sigma,$$

and, by Stokes' theorem (and the fact that $\omega \wedge \sigma$ has compact support, so that we can replace integration over $\mathbb{R}^3$ by integration over some finite region on whose boundary the form $\omega \wedge \sigma$ vanishes), the integral $\int_{\mathbb{R}^3} \mathrm{d}(\omega \wedge \sigma)$ vanishes. Thus

$$(-1)^{i+1} \int_{\mathbb{R}^3} \omega \wedge \mathrm{d}\sigma = \int_{\mathbb{R}^3} \mathrm{d}\omega \wedge \sigma.$$

From this we see that

$$\partial = (-1)^{i+1} \mathrm{d}.$$

In other words,

$$\begin{aligned}
\partial &= -\mathrm{d} && \text{on} \quad A_0,\\
\partial &= \mathrm{d} && \text{on} \quad A_1,\\
\partial &= -\mathrm{d} && \text{on} \quad A_2,\\
\partial &= \mathrm{d} = 0 && \text{on} \quad A_3.
\end{aligned}$$

We thus have the series of maps

$$A^0 \xrightarrow{\mathrm{d}} A^1 \xrightarrow{\mathrm{d}} A^2 \xrightarrow{\mathrm{d}} A^3,$$

$$A_3 \xleftarrow{\partial} A_2 \xleftarrow{\partial} A_1 \xleftarrow{\partial} A_0.$$

where $\partial = (-1)^{i+1}\mathrm{d}$ on A_{i+1}.

The operator ∂ is sometimes called the *formal transpose* of d. The reason for the word formal is the fact that we can express ∂ purely in terms of d and that the formula

$$\int \omega \wedge \partial\sigma = \int \mathrm{d}\omega \wedge \sigma$$

is only valid when the integration is over a region for which there are no boundary contributions. We have arranged this by having the integration carried out over all space and by insisting that σ have compact support. We could equally well arrange that the same formula hold by insisting that ω have compact support. That is, we could think of d as a map from A_i to A_{i+1} and ∂ as a map from A^i to A^{i+1}. That is, we might want to consider the forms of compact support as the fundamental objects, and consider that the map ∂ is defined on the space of linear functions on A_i.

16.3. Solid angle

As an illustration of this reversed point of view, we shall do a basic computation. Let P be some point in $\mathbb{R}^3$ with coordinates x_P, y_P, z_P. Let τ_P be the *solid angle form subtended from* P which is defined to be the two-form

$$\tau_P = \frac{(x-x_P)\mathrm{d}y \wedge \mathrm{d}z - (y-y_P)\mathrm{d}x \wedge \mathrm{d}z + (z-z_P)\mathrm{d}x \wedge \mathrm{d}y}{r_P^3}.$$

Here r_P denotes the distance to the point P. The form τ_P is not defined on all of $\mathbb{R}^3$; it is only defined on the space $\mathbb{R}^3 - \{P\}$, i.e., on three-space with the point P removed. Nevertheless, the form τ_P does define a linear function on the space A_1. Indeed, let $\omega = a\mathrm{d}x + b\mathrm{d}y + c\mathrm{d}z$ be a linear differential form with smooth coefficients and compact support. Then

$$\omega \wedge \tau_P = r_P^{-3}[a(x-x_P) + b(y-y_P) + c(z-z_P)]\mathrm{d}x \wedge \mathrm{d}y \wedge \mathrm{d}z.$$

The coefficient of $\mathrm{d}x \wedge \mathrm{d}y \wedge \mathrm{d}z$ is not defined at P but the singularity at P is only of order r_P^{-2} (since the expression inside the bracket vanishes to first order at P).

Hence, the point P represents no problem in the computation of the integral. Since ω is of compact support, the integral over all space is well defined. In order to evaluate the integral, it is convenient to pass to spherical coordinates centered at P. If r_P, θ, and ϕ are such coordinates, then

$$\tau_P = \sin\theta\, \mathrm{d}\theta \wedge \mathrm{d}\phi.$$

Figure 16.3

(Thus τ_P gives the solid angle subtended by a surface over which it is integrated.) Suppose that $\omega = F\mathrm{d}r_P + G\mathrm{d}\theta + H\mathrm{d}\phi$ in terms of these spherical coordinates. Then

$$\omega \wedge \tau_P = F \sin\theta\, \mathrm{d}r_P \wedge \mathrm{d}\theta \wedge \mathrm{d}\phi.$$

Thus

$$\int_{\mathbb{R}^3} \omega \wedge \tau_P = \int_0^{2\pi}\int_0^{\pi} \sin\theta \left\{ \int_0^{\infty} F(r_P, \theta, \phi)\mathrm{d}r_P \right\} \mathrm{d}\theta\, \mathrm{d}\phi.$$

Let us now consider the case where $\omega = \mathrm{d}u$ for some smooth function u of compact support. Thus in the preceding formula $F = \partial u/\partial r_P$ in terms of the spherical coordinates and the innermost integral on the right reduces to the constant value $-u(P)$ so that integration with respect to θ and ϕ just multiplies by 4π, the area of the sphere, and we have proved the basic formula

$$\int \mathrm{d}u \wedge \tau_P = -4\pi u(P). \tag{16.1}$$

Let δ_P denote the zero-chain which assigns to each function the value $u(P)$, i.e., δ_P represents a unit charge concentrated at P. Then the right-hand side of equation (16.1) is the value of the zero-chain $-4\pi\delta_P$ when evaluated on u while the left-hand side is the value of τ_P when evaluated on du. We can thus write (16.1) as

$$\partial\tau_P = -4\pi\delta_P. \tag{16.1a}$$

(Notice that in the preceding equation we may *not* equate ∂ with d. The form τ_P is not defined at P and hence $\mathrm{d}\tau_P$ is not defined there. At all points where τ_P is defined, we have $\mathrm{d}\tau_P = 0$. Nevertheless (16.1a) is fine as an equality of zero-chains – that is, as linear functions on functions.)

Suppose we have a finite number of points, P_1, P_2, P_3, etc., and we place the charges c_1 at P_1, c_2 at P_2, etc. That is, we consider the zero-chain

$$Q = c_1\delta_{P_1} + c_2\delta_{P_2} + \cdots.$$

Let us set

$$D = -(c_1\tau_{P_1} + c_2\tau_{P_2} + \cdots).$$

Then D gives a well-defined linear function on A_1 and we have the formula

$$\partial D = -4\pi Q.$$

If, instead of a finite discrete charge distribution, we are given a charge distribution with smooth charge density ρ of compact support, then we can replace the preceding definition of D, which involves a sum, by a corresponding integral. That is, we can write $D = \int \rho(P)\tau_P \mathrm{d}P$ and $Q = \rho \mathrm{d}x \wedge \mathrm{d}y \wedge \mathrm{d}z$ and again the preceding equation holds. In this case of continuous superposition, it is not hard to show, and we shall do so shortly, that D is a smooth two-form that is defined throughout all space.

16.4. Electric field strength and dielectric displacement

We are now in a position to define the fundamental objects in the theory of electrostatics. The *electric field strength* E is a linear differential form, which, when integrated along any path, gives the voltage drop across the path. Thus the units of E will be voltage/length. (Since voltage has units energy/charge and force has units energy/length we can also write the units of E as force/charge.) The basic equation satisfied by E is

$$\mathrm{d}E = 0.$$

Locally, this is equivalent to the existence of a potential u, i.e., to the equation

$$E = -\mathrm{d}u. \tag{16.2}$$

For most of the regions that we shall consider, this is the form of the equation that we shall use.

We want to think of E as an incipient one-cochain and of u as an incipient zero-cochain. We now also need an object which should give a smeared-out one-chain. This will be a two-form, D. It is called the *dielectric displacement*. It is to represent a smeared-out version of the one-chain giving the branch charges. We also want a smeared-out version of the zero-chain representing the node charges. It will be some three-form $\rho \mathrm{d}x \wedge \mathrm{d}y \wedge \mathrm{d}z$. In our network model we had

$$\partial(\text{branch charges}) = -\text{node charges}.$$

So we expect that

$$\partial D = -\rho \mathrm{d}x \wedge \mathrm{d}y \wedge \mathrm{d}z.$$

In fact, the standard definition of the units of D and ρ (in the cgs system) are such that this is true up to a factor of 4π:

$$\boxed{\partial D = -4\pi\rho \mathrm{d}x \wedge \mathrm{d}y \wedge \mathrm{d}z.} \tag{16.3}$$

This is known as *Gauss's Law*. By Stokes' theorem, and the fact that $\partial = -\mathrm{d}$ on two-forms, we can rewrite this as

$$\int_{\partial W} D = 4\pi \int_W \rho \mathrm{d}x \wedge \mathrm{d}y \wedge \mathrm{d}z$$

relating the surface integral of D over the boundary of a region W to the total charge inside.

Of course, D is not determined by the equation $\partial D = -4\pi\rho\mathrm{d}x \wedge \mathrm{d}y \wedge \mathrm{d}z$. Adding any closed form to D will not change this equation. The two-form D is part of the data of an electrostatic system. In principle, it can be measured by the following procedure. Write

$$D = D_x\mathrm{d}y \wedge \mathrm{d}z + D_y\mathrm{d}z \wedge \mathrm{d}x + D_z\mathrm{d}x \wedge \mathrm{d}y.$$

To measure D_z, insert small parallel metal plates, lying in the xy-plane, into a cavity in the surrounding medium. Touch the plates together, then separate them. They acquire charges $\pm Q$. Then

$$D_z = 4\pi \lim_{\text{area}\to 0} \frac{\text{charge on top plane (toward } +z)}{\text{area of plates}}.$$

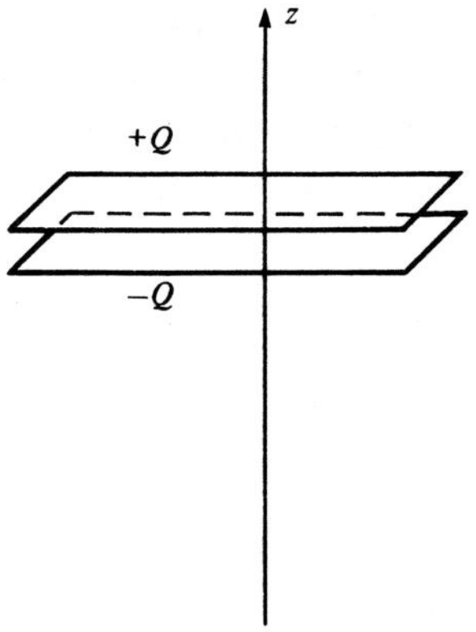

Figure 16.4

(The 4π is the fault of the cgs units.) This definition works in any dielectric, even if we do not know ε, and it *makes no mention of an electric field.*

In fact, the preceding definition extends to a coordinate-free definition of D: given any pair of vectors, $\mathbf{v}_1$ and $\mathbf{v}_2$, form a parallel-plate capacitor whose plates are parallelograms determined by $h\mathbf{v}_1$ and $h\mathbf{v}_2$, then define

$$D(\mathbf{v}_1, \mathbf{v}_2) = 4\pi \lim_{h\to 0} \frac{\text{charge on top plate}}{h^2}.$$

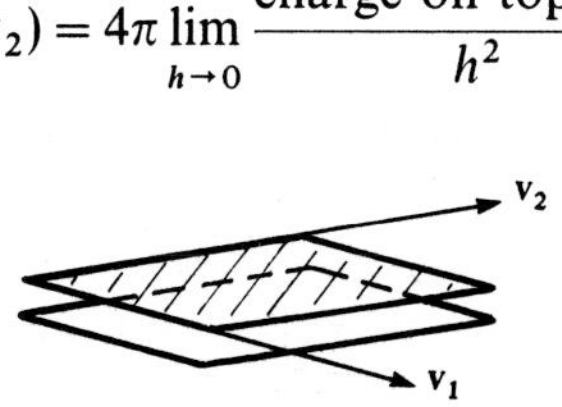

Figure 16.5

Which is the top plane depends on orientation, since orientation reverses if $\mathbf{v}_1$ and $\mathbf{v}_2$ are interchanged.

$$D(\mathbf{v}_2, \mathbf{v}_1) = -D(\mathbf{v}_1, \mathbf{v}_2).$$

Since charge is additive, D is bilinear in $\mathbf{v}_1$ and $\mathbf{v}_2$, so by its definition it is a two-form.

We have now generalized the topological equations for capacitive networks to electrostatics in general.

$$\mathbf{V} = -\,\mathrm{d}\Phi \text{ becomes } E = -\,\mathrm{d}u;$$
$$\partial\mathbf{Q} = -\,\rho \text{ becomes } \partial D = -\,4\pi\rho\mathrm{d}x \wedge \mathrm{d}y \wedge \mathrm{d}z.$$

If D is a *smooth* two-form, $\mathrm{d}D = +\,4\pi\rho\mathrm{d}x \wedge \mathrm{d}y \wedge \mathrm{d}z$.

We should recall that (having picked an orientation – say, the standard one) we can regard E as a linear function of D that assigns to D the number

$$\int_{R^3} E \wedge D.$$

In this formula, we have, so far, regarded *either* D or E as having compact support. In fact, this integral may be defined, for a particular pair E and D, so long as the product $E \wedge D$ vanishes sufficiently rapidly at ∞. For example, if the product goes to zero as r^{-4} or faster, the integral will converge. We shall need to keep this degree of flexibility in mind.

We now have the smeared-out versions of the topological part of our theory of capacitive networks. We still need a version of the matrix C giving the relationship $\mathbf{Q} = C\mathbf{V}$. So we want a map

$$D = C(E)$$

sending electric fields into dielectrics. We will study this map C in some detail in the next section. Here we will draw some consequences from the following two assumptions about C.

Recall that in network theory the map C was diagonal in terms of the branch voltages. In particular, if a voltage across a particular branch vanishes, then the corresponding branch charge vanishes. As a very mild analogue to this condition, we will assume that

(a) C is local in the sense that, if E vanishes identically in the interior of some region, $D = C(E)$ also vanishes identically there.

In analogy to the network case, we will assume that

(b) The map C is symmetric. That is, if E_1 and E_2 are two electric fields, then

$$\int E_1 \wedge C(E_2) = \int E_2 \wedge C(E_1).$$

In the next section we shall describe the map C as it actually occurs in nature. We shall see that there is an intimate connection between the map C for *vacuum* and Euclidean geometry. In *all* media, conditions (a) and (b) hold. For now, we shall derive some consequences of (a) and (b).

Conducting material, by definition, offers no resistance to the passage of electric charge. In other words, in a conductor the form E, which measures the work done in moving an electric charge, must vanish. Thus, using (a), we see that

In the interior of any conductor, the forms E and D must both vanish. Hence the function u must be constant on each connected conducting body.

As $D=0$ in the interior of a conductor, $\rho=0$ there as well. So there can be no charges (in electrostatics) in the interior of a conductor. Or, in the words of Faraday,

All the charges of a conductor lie on its surface.

The density of this surface charge is, of course, given by D. The total charge, ρ_i, on a particular conductor, is given by integrating D over its surface.

Let us suppose that we have introduced charges *only* on the conductors. So we are assuming that $\mathrm{d}D=0$ outside the conductors. Then for two such potentials, u and $\hat{u}$, with corresponding D and $\hat{D}$, we have

$$\begin{aligned}\mathrm{d}(u\hat{D}) &= \mathrm{d}u \wedge \hat{D} + u\mathrm{d}\hat{D} \\ &= -E \wedge \hat{D}\end{aligned}$$

over the exterior of the conductors, as $\mathrm{d}\hat{D}=0$ there. Thus, by Stokes' theorem

$$(E,\hat{E}) \stackrel{\text{def}}{=} \int_{\substack{\text{region}\\ \text{exterior to all}\\ \text{conductors}}} E \wedge \hat{D} = \int_{\substack{\text{all}\\ \text{conductor}\\ \text{surfaces}}} u\hat{D}.$$

On the ith conductor surface, u takes on a constant value – say Φ_i. We can thus pull u out of the surface integral and so

$$(E,\hat{E}) = \sum_{\text{conductors}} \Phi\hat{\rho}.$$

This gives Green's reciprocity theorem as used in section 8 of Chapter 13.

A third property of the map C is that the scalar product (,) is positive-definite. Thus if we take $E=\hat{E}$ in the preceding equation, we get

$$(E,E) = \sum_{\text{conductors}} \Phi\rho.$$

If all the conductor charges are zero, the right-hand side is zero. By the positive-definiteness property, this would imply that $E=0$ and hence that all the Φs are equal. So if we consider one conductor (say an imaginary conductor at infinity) as grounded, the charges ρ uniquely determine the potentials Φ. Similarly, the potentials uniquely determine the charges. We would then have a map from the space of all conductor potentials $\mathbf{\Phi}$ to all conductor charges ρ and we would be able to use the results of Chapter 13.

All of this is under the assumption that, *given* an assignment, Φ, of potentials to the various conductors, we can find the (unique) corresponding E and D. At this point the mathematician and the physicist part company. From the point of view of mathematics, an interesting and non-trivial problem has been posed: *the Dirichlet problem.* It is a question of an existence theorem in the theory of partial differential equations. It occupied the efforts of many gifted mathematicians in the last third of the nineteenth century. It was finally resolved positively, by several different techniques, each of which has given rise to interesting mathematical developments – some with important consequences to physics. The corresponding problem for more general partial differential equations is still an active area of research. We shall say something about this mathematical problem in the next sections, but its complete discussion is outside the scope of this book. For the physicist, the question of existence is a peripheral issue, relating to the logical consistency of a mathematical idealization of the physical model. The physicist knows that the hypothesis of static charge distributions is only an approximation to the underlying physical reality. If the model is appropriate – if the idealizations of static charges on the surfaces of conductors at constant potentials and the corresponding electric fields give a reasonable description of the situation – then 'of course' the existence theorem must hold – because we can 'put' whatever charges we like on each conductor. If the existence theorem is false, it is either because the mathematicians allow weirdly-shaped conductors which cannot be constructed in practice (or some other 'physically unreasonable' conditions) or because there is something fundamentally wrong with the theory. Better to deduce logical consequences of the assumed existence of solutions and observe if their physical predictions hold. If they do, then a failure of the existence theorem probably represents an esoteric fine point in the theory of no great physical moment.

Let us adopt the physicist's point of view for the remainder of this section. We assume the existence of solutions for given charge or potential distributions on conductors. We also assume that the solution depends continuously on the given boundary conditions.

We can now draw several important conclusions.

The principle of electrical screening. Suppose that several of our conductors – say conductors 1 through 7 – are completely surrounded by another conductor – say conductor number 8. Suppose that no charges are placed on the surrounded conductors. Consider any distribution of charges on the remaining conductors. There will be some electric field $E = -\mathrm{d}u$, with corresponding D and induced

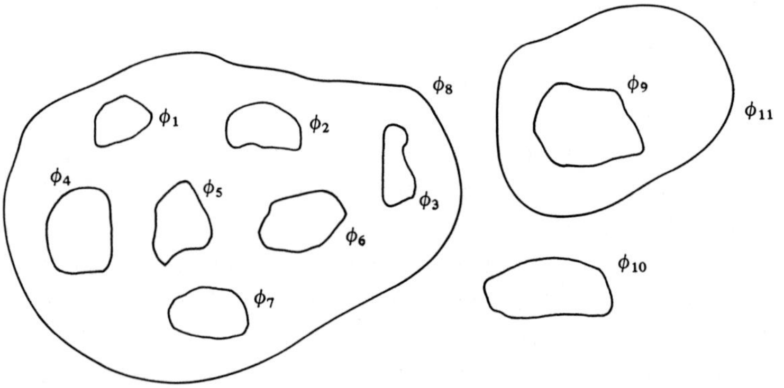

Figure 16.6

potentials on all conductors. Conductor number 8 will be at some constant potential, Φ_8, as a result of these charges. *A priori*, we may not know what the induced charges and potentials are on the interior conductors. But suppose we consider a new function $u' = u$ outside and $u' \equiv \Phi_8$ inside the cavity of conductor number 8. This *is* a solution to our problem, with $E \equiv D \equiv 0$ inside the cavity – and zero charges on all interior conductors. By *uniqueness* this is our original solution – as no charges have been introduced to any interior conductor; similarly, if we adjusted the potentials of any of the exterior conductors. In short (up to a meaningless, overall constant in u), the interior cavity of a conductor is electrically screened from all electrostatic systems outside the conductor. This principle is used in the construction of electrostatic measuring instruments where we do not want any interference from external electrical actions other than the effect we wish to observe. (As we want to be able to see inside the cavity, we may leave a little window open – or replace a portion of the conductor by a wire mesh which is almost as effective a screen as a solid conductor.) In addition to its practical importance, the principle of electrical screening is a helpful computational device when used in conjunction with the following principle.

Replacement of an equipotential by a conducting sheet. Suppose that we have a solution $E = -\,du$, $D = CE$ of some system of charges or potentials. Let S be some surface on which u is a constant, Φ. (S is called an equipotential surface.) Now suppose we replace S by a thin conducting sheet inserted at potential Φ. The nature of the map C is such that it is essentially unaffected by the insertion of such a conductor. Thus the insertion of the conductor has practically no effect. In the interior of the thin conducting sheet, E and D have become zero, and charge has accumulated along its two surfaces, but elsewhere everything is as before the sheet was inserted.

For example, suppose our medium is rotationally invariant. Then the potential of a charge Q placed at the origin must be of the form

$$u = Af(r).$$

(In the next section, we shall see that, if the system is invariant under all Euclidean motions, we must have

$$f(r) = \frac{c}{r}$$

for some constant c.) Now insert a thin spherical conducting sheet of radius a. This does not affect the fields either inside or outside the sphere, but D vanishes in the interior of the sheet. If we draw a spherical surface inside the sheet, the total charge enclosed must be zero. Thus a total charge $-\,Q$ has accumulated over the inside surface of the sphere (and a total charge Q on the outside). We can now discharge the interior (by conducting a wire between the charge at the origin and the interior surface of the conductor). This has no effect outside the sphere by the principle of electrical screening, while the interior potential becomes constant. Thus the field for the exterior of a charged spherical conductor is given by

$$\begin{aligned} &Qf(r) \quad r \geqslant a, \\ &Qf(a) \quad r \leqslant a, \end{aligned}$$

where $f(r)$ denotes the potential of unit charge at the origin. This method allows us to compute the field and the capacitance of a spherical capacitor whose plates are a pair of concentric spherical conductors:

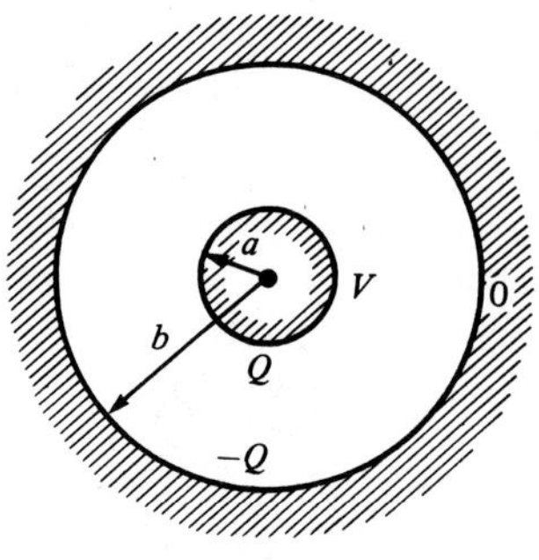

Figure 16.7

Again start with a charge Q at the origin, with the conductors absent. Insert the spherical conductors. Charges accumulate along the inner and outer surfaces. Discharge the interior surface of the inner conductor by connecting a wire to the origin and discharge the outer surface of the outer conductor by grounding it to infinity. The fields inside the inner sphere and outside the outer sphere vanish. The function u between the two spheres remains the same as before with a charge Q on the outer surface of the inner sphere and $-Q$ on the inner surface of the outer surface.

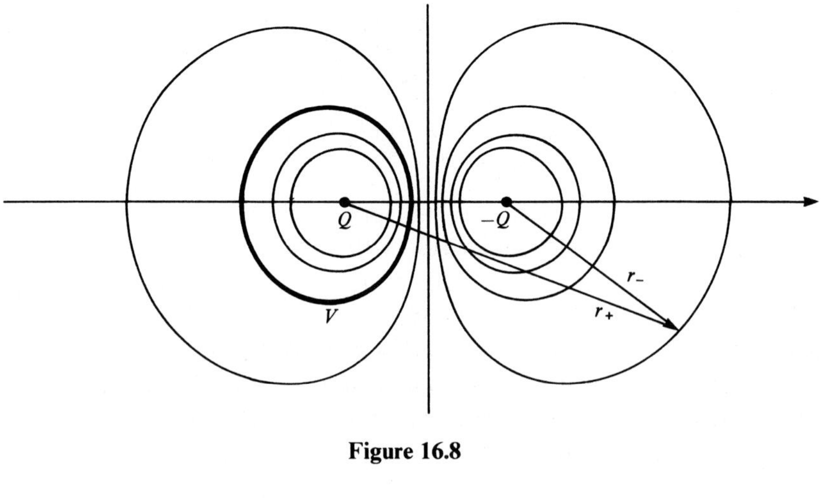

Figure 16.8

Here is another illustration. Suppose we place a charge Q at the point $\begin{pmatrix} -1 \\ 0 \\ 0 \end{pmatrix}$ of the x-axis and $-Q$ at the point $\begin{pmatrix} 1 \\ 0 \\ 0 \end{pmatrix}$. Then the potential at all points will be given by

$$Q(f(r_-) - f(r_+))$$

where r_+ and r_- denote the distances to these two points. Under the assumption of Euclidean invariance, $f(r) = c/r$. For this situation, the equipotential surfaces look like figure 16.8.

After we insert a conducting sheet at one of these equipotentials, we can abolish the field on one side of the surface by discharging the charge $-Q$. We then get the field of a point charge Q and one of these surfaces. For example, if we take

Figure 16.9

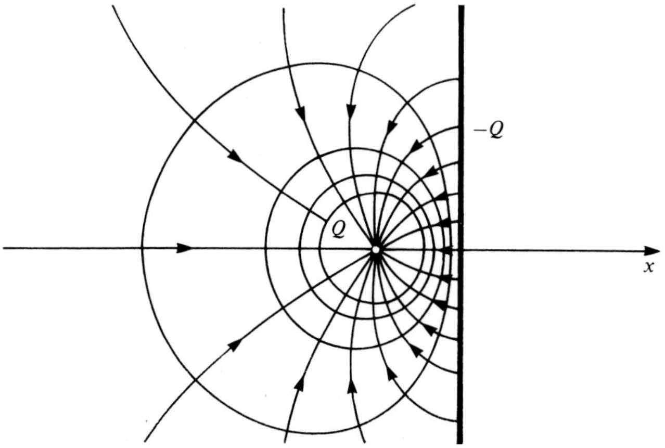

Figure 16.10

the surface to be the plane $x = 0$, at zero potential, we obtain the solution for the problem of a point charge in the presence of grounded plane conductor. A charge $-Q$ is induced on the inner surface. The field on the other side of the conductor vanishes, while the field on the side of the charge is *as if* a charge $-Q$ was placed on the other side of the conductor. This is an example of the method of images.

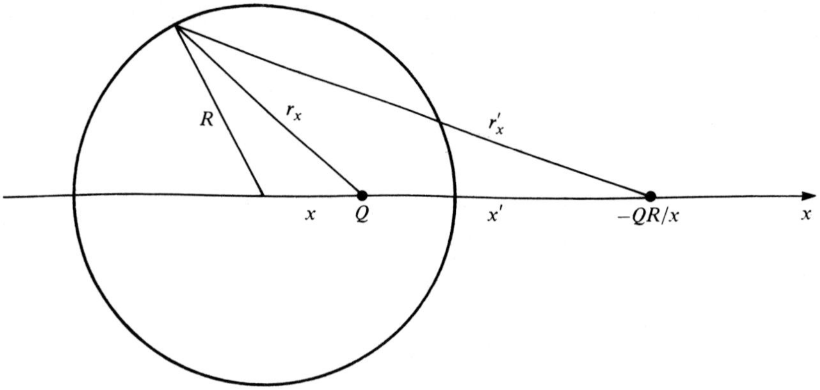

Figure 16.11

Here is one final example that we shall use later. Consider a sphere of radius R. Place a charge Q at a point $\begin{pmatrix} x \\ 0 \\ 0 \end{pmatrix}$. Let x' be such that $xx' = R^2$. By similar triangles

$$\frac{x'}{r_{x'}} = \frac{R}{r_x}$$

or

$$\frac{1}{r_{x'}} = \frac{R}{x' r_x}$$

for any point on the sphere of radius R.

So placing a charge $-Q(R/x)$ at the point $\begin{pmatrix} x' \\ 0 \\ 0 \end{pmatrix}$ gives potential zero on the sphere. This allows us to compute the potential for a point charge placed anywhere in the interior of a grounded conducting sphere.

16.5. The dielectric coefficient

We now must turn to the problem of finding the analog, in electrostatics, of the mapping from C^1 to C_1 given by the capacitance matrix, of the map $\mathbf{Q} = C\mathbf{V}$.

The first property we recall of the matrix C is that it was *diagonal* in terms of the basis given by the branches. In other words, the branch charge contribution of any branch voltage is localized at that branch. We shall make a similar hypothesis in electrostatics: we want a linear map $E \rightsquigarrow D$ such that the value of D at any point depends only on the value of E at that same point. In other words, if

$$E = E_x \mathrm{d}x + E_y \mathrm{d}y + E_z \mathrm{d}z$$

and

$$D = D_x \mathrm{d}y \wedge \mathrm{d}z + D_y \mathrm{d}z \wedge \mathrm{d}x + D_z \mathrm{d}x \wedge \mathrm{d}y$$

(where the coefficients $E_x, \ldots, D_z$ are functions), then the relationship between the E and D should be given by a matrix of functions

$$(\varepsilon_{ij})$$

so that

$$D_x = \varepsilon_{11}E_x + \varepsilon_{12}E_y + \varepsilon_{13}E_z,$$
$$D_y = \varepsilon_{21}E_x + \varepsilon_{22}E_y + \varepsilon_{23}E_z,$$
$$D_z = \varepsilon_{31}E_x + \varepsilon_{32}E_y + \varepsilon_{33}E_z.$$

Each of the entries ε_{ij} will be a function, and the matrix function $\boldsymbol{\varepsilon} = (\varepsilon_{ij})$ gives the *dielectric* properties of the medium. We can write the relationship between D and E as

$$D = \boldsymbol{\varepsilon} E.$$

So now the equations of electrostatics become

$$E = -\mathrm{d}u, \quad \partial D = -4\pi\rho\, \mathrm{d}x \wedge \mathrm{d}y \wedge \mathrm{d}z$$

and

$$D = \boldsymbol{\varepsilon} E.$$

We may combine these equations by introducing the Laplace operator defined by

$$(\Delta u)\mathrm{d}x \wedge \mathrm{d}y \wedge \mathrm{d}z = -\partial \boldsymbol{\varepsilon}\, \mathrm{d}u$$

so we get Poisson's equation

$$\Delta u = -4\pi\rho.$$

To make any further progress, we must get some further information about the matrix $\boldsymbol{\varepsilon}$.

For general media, all we can say about $\boldsymbol{\varepsilon}$ is that it is a symmetric matrix, for much the same reason that C was. Beyond that, we can say nothing. There do exist situations – crystals under stress, for example – where $\boldsymbol{\varepsilon}$ is a variable symmetric matrix.

The medium is called *homogeneous* if $\boldsymbol{\varepsilon}$ is invariant under translations, and so is a *constant*. The medium is called *isotropic* if the relationship between E and D is invariant under rotations. We now examine what are the possible forms of $\boldsymbol{\varepsilon}$ if the medium is homogeneous and isotropic.

16.6. The star operator in Euclidean three-dimensional space

We begin by exhibiting a linear map

$$\star : \Lambda^1(\mathbb{R}^{3*}) \to \Lambda^2(\mathbb{R}^{3*})$$

which *is* rotation invariant. Define

$$\star \mathrm{d}x = \mathrm{d}y \wedge \mathrm{d}z,$$
$$\star \mathrm{d}y = \mathrm{d}z \wedge \mathrm{d}x,$$
$$\star \mathrm{d}z = \mathrm{d}x \wedge \mathrm{d}y$$

and extend by linearity; that is, define $\star:\Lambda^1(\mathbb{R}^{3*})\to\Lambda^2(\mathbb{R}^{3*})$ by

$$\star(a\mathrm{d}x + b\mathrm{d}y + c\mathrm{d}z) = a\,\mathrm{d}y \wedge \mathrm{d}z + b\,\mathrm{d}z \wedge \mathrm{d}x + c\,\mathrm{d}x \wedge \mathrm{d}y.$$

Let

$$\omega = a\mathrm{d}x + b\mathrm{d}y + c\mathrm{d}z$$

and

$$\sigma = A\mathrm{d}x + B\mathrm{d}y + C\mathrm{d}z$$

and notice that

$$\star\omega \wedge \sigma = (aA + bB + cC)\mathrm{d}x \wedge \mathrm{d}y \wedge \mathrm{d}z.$$

We can write this last equation as follows. The Euclidean scalar product on $\mathbb{R}^3$ gives rise to a scalar product on $\mathbb{R}^{3*}$ – on one-forms – given by

$$(\omega, \sigma) = aA + bB + cC.$$

The scalar product, together with the *orientation*, picks out a preferred volume form $\mathrm{d}x \wedge \mathrm{d}y \wedge \mathrm{d}z$. The last equation reads

$$(\star\omega \wedge \sigma) = (\omega, \sigma)\,\mathrm{d}x \wedge \mathrm{d}y \wedge \mathrm{d}z. \tag{16.5}$$

A moment's reflection shows that this equation *determines* the map $\star$ uniquely. But the right-hand side of this equation involves (as a function of ω and σ) only the scalar product and the orientation. Any rotation preserves these. Hence

The $\star$ operator is rotation invariant.

We claim that, up to a scalar factor, $\star$ is the *only* map of $\Lambda^1(\mathbb{R}^{3*})\to\Lambda^2(\mathbb{R}^{3*})$ which is rotationally invariant. That is, we claim that, if $r:\Lambda^1(\mathbb{R}^{3*})\to\Lambda^2(\mathbb{R}^{3*})$ is some other map which is a rotation invariant, then $r = a\star$ for some scalar a. Indeed, we claim, first, that either $r = 0$ or r has zero kernel. Indeed, $\ker r$ is a subspace of $\mathbb{R}^3$. If r is invariant under rotations, this subspace would have to be invariant under rotations. But there are no rotation-invariant subspaces of $\mathbb{R}^3$ other than the trivial spaces $\{0\}$ and $\mathbb{R}^3$. Thus either $r = 0$ or r is an isomorphism. If $r = 0$, there is nothing to prove. If r is an isomorphism, consider the map

$$l = r^{-1}\star, \quad l:\mathbb{R}^{3*}\to\mathbb{R}^{3*}.$$

By hypothesis, the map l is rotation invariant, i.e.,

$$l(R\omega) = Rl(\omega)$$

for any rotation R. Let us write

$$l\omega = a\omega + L\omega$$

where $L\omega$ is perpendicular to ω. In other words, decompose $l\omega$ into components along and perpendicular to ω. By rotational invariance, the coefficient a given by

$$(l\omega, \omega) = a\,\|\,\omega\,\|^2$$

is independent of ω. We claim that $L\omega = 0$. Indeed, we must have $LR\omega = RL\omega$ for any rotation R. Choose some rotation R which fixes ω, but rotates non-trivially in the $\omega^{\perp}$ plane. Then $R\omega = \omega$ but $RL\omega \neq L\omega$ if $L\omega \neq 0$. Thus $L\omega = 0$. Thus

$$l\omega = a\omega$$

or

$$r\omega = a\star\omega$$

which is what was to be proved.

Thus, up to scalar multiples, $\star$ is the only rotation-invariant map from $\Lambda^1(\mathbb{R}^{3*})$ to $\Lambda^2(\mathbb{R}^{3*})$.

(The converse is also worth noting. The $\star$ operator determines the scalar product (,) occurring on the right-hand side of (16.5) – hence giving the $\star$ operator in $\mathbb{R}^{3*}$ *determines* the scalar product and orientation.)

We can now let

$$\omega = a\mathrm{d}x + b\mathrm{d}y + c\mathrm{d}z$$

be a differential form. That is, we can let a, b and c be functions. Define the $\star$ operator as before

$$\star\omega = a\mathrm{d}y \wedge \mathrm{d}z + b\mathrm{d}z \wedge \mathrm{d}x + c\mathrm{d}x \wedge \mathrm{d}y.$$

Once again

$$\star\omega \wedge \tau = (\omega, \tau)\mathrm{d}x \wedge \mathrm{d}y \wedge \mathrm{d}z$$

where ω and τ are now differential forms.

We can now use the assumption that our medium is homogeneous and isotropic. This implies that

$$\boldsymbol{\varepsilon} = \varepsilon\star$$

where ε is a *constant* called the dielectric constant of the medium. For instance, it is a property of the vacuum that $\boldsymbol{\varepsilon} = \varepsilon_0\star$ where ε_0 is a constant. Conversely, since the $\star$ operator from $\Lambda^1(\mathbb{R}^3)$ to $\Lambda^2(\mathbb{R}^3)$ determines the scalar product, we may say that the *dielectric property of the vacuum determines the Euclidean geometry of space.*

In what follows, we will assume that the units of E, D and ε have been chosen so that the ε is absorbed into the $\star$ operator. Thus the equations of electrostatics have become

$$\boxed{E = -\mathrm{d}u, \quad D = \star E, \quad \mathrm{d}D = 4\pi\rho\mathrm{d}x \wedge \mathrm{d}y \wedge \mathrm{d}z}$$

so

$$\boxed{\Delta u = -4\pi\rho}$$

where a direct computation shows that

$$\Delta u = \frac{\partial^2 u}{\partial x^2} + \frac{\partial^2 u}{\partial y^2} + \frac{\partial^2 u}{\partial z^2}.$$

It is important to observe that the $\star$ operator is *not* invariant under differentiable maps. We do *not* have

$$\phi^*(\star\omega) = \star\phi^*(\omega)$$

unless ϕ is an orientation-preserving Euclidean transformation. Nevertheless, we can compute the $\star$ operator in more general coordinate systems by using its definition. For example, suppose we wish to compute the $\star$ operator in polar coordinates given by

$$r = \sqrt{(x^2 + y^2 + z^2)}, \quad \theta = \arctan(\sqrt{(x^2 + y^2)}/z), \quad \phi = \arctan(y/x).$$

If we calculate the differentials of these functions and express the results in terms of the basis $\mathrm{d}x$, $\mathrm{d}y$, $\mathrm{d}z$ with coordinates expressed for convenience in terms of r, θ and ϕ, we find

$$\mathrm{d}r = \sin\theta\cos\phi\,\mathrm{d}x + \sin\theta\sin\phi\,\mathrm{d}y + \cos\theta\,\mathrm{d}z,$$

$$\mathrm{d}\theta = \frac{1}{r}(\cos\theta\cos\phi\,\mathrm{d}x + \cos\theta\sin\phi\,\mathrm{d}y - \sin\theta\,\mathrm{d}z),$$

$$\mathrm{d}\phi = \frac{1}{r\sin\theta}(-\sin\phi\,\mathrm{d}x + \cos\phi\,\mathrm{d}y).$$

Direct calculation, using the orthonormality of $\mathrm{d}x, \mathrm{d}y$ and $\mathrm{d}z$, shows that $\mathrm{d}r, \mathrm{d}\theta$, and $\mathrm{d}\phi$ are orthogonal elements of $\mathbb{R}^{3*}$ at each point (r, θ, ϕ), and that

$$(\mathrm{d}r, \mathrm{d}r) = 1, \quad (\mathrm{d}\theta, \mathrm{d}\theta) = \frac{1}{r^2}, \quad (\mathrm{d}\phi, \mathrm{d}\phi) = \frac{1}{r^2\sin^2\theta}.$$

The best way to summarize all of these results is to notice that $\{\mathrm{d}r, r\mathrm{d}\theta, r\sin\theta\,\mathrm{d}\phi\}$ forms an orthonormal basis for $\Lambda^1(\mathbb{R}^3)$ so that we can calculate with these three differentials just as we do with $\mathrm{d}x, \mathrm{d}y$ and $\mathrm{d}z$. Thus

$$\star\mathrm{d}r = r^2\sin\theta\,\mathrm{d}\theta \wedge \mathrm{d}\phi,$$
$$\star(r\mathrm{d}\theta) = -r\sin\theta\,\mathrm{d}r \wedge \mathrm{d}\phi,$$
$$\star(r\sin\theta\,\mathrm{d}\phi) = r\mathrm{d}r \wedge \mathrm{d}\theta.$$

Of course, using linearity, we can then compute $\star\mathrm{d}\theta$ or $\star\mathrm{d}\phi$.

16.7. Green's formulas

In this section we give the continuous version of the Green's formulas of Chapter 13. Let U be a bounded region in $\mathbb{R}^3$ with boundary ∂U. Let

$$E = E_x\mathrm{d}x + E_y\mathrm{d}y + E_z\mathrm{d}z \quad \text{and} \quad F = F_x\mathrm{d}x + F_y\mathrm{d}y + F_z\mathrm{d}z$$

be linear differential forms defined on U. We define their *scalar product* $(E, F)_U$ by

$$(E, F)_U = \int_U E \wedge \star F = \int_U (E_xF_x + E_yF_y + E_zF_z)\mathrm{d}x\,\mathrm{d}y\,\mathrm{d}z = (F, E)_U.$$

We define the corresponding *electrostatic energy* to be $\frac{1}{2}\| E \|_U^2$ where

$$\| E \|_U^2 = (E, E)_U = \int_U (E_x^2 + E_y^2 + E_z^2)\mathrm{d}x\,\mathrm{d}y\,\mathrm{d}z.$$

Now suppose that

$$E = -\mathrm{d}u \quad \text{and} \quad F = -\mathrm{d}v$$

where u and v are smooth functions. Then

$$\begin{aligned} d(u\star dv) &= du \wedge \star dv + u d\star dv \\ &= du \wedge \star dv + u\Delta v\, dx \wedge dy \wedge dz \end{aligned}$$

or

$$du \wedge \star dv = d(u\star dv) - (u\Delta v) dx \wedge dy \wedge dz.$$

Then

$$(du, dv)_U = \int_U du \wedge \star dv = \int_U d(u\star dv) - \int_U u\Delta v\, dx \wedge dy \wedge dz$$

so

$$(du, dv)_U = \int_{\partial U} u\star dv - \int_U u\Delta v\, dx \wedge dy \wedge dz. \tag{16.6}$$

This is known as *Green's first formula*. Since $(E, F)_U = (F, E)_U$, we can interchange u and v in Green's first formula and subtract. We get

$$\int_{\partial U} (u\star dv - v\star du) = \int_U (u\Delta v - v\Delta u) dx \wedge dy \wedge dz \tag{16.7}$$

which is called *Green's second formula*. Let P be a point of U and let us draw a small ball, B_ε, of radius ε centered at P and completely contained in U. Let us

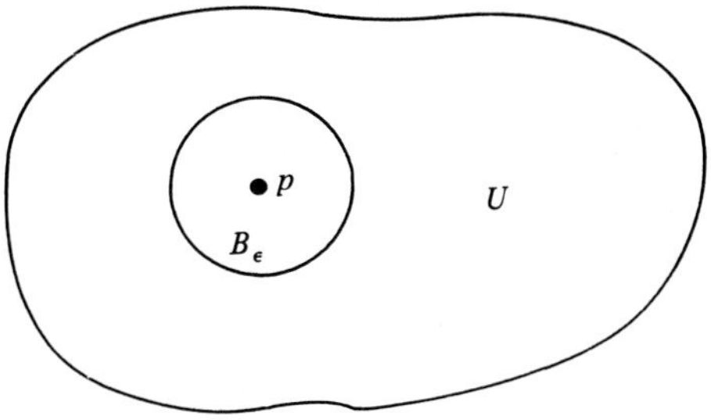

Figure 16.12

apply Green's second formula to the region U with B_ε removed. The boundary of this region will consist of ∂U and of ∂B_ε with the reversed orientation. We will take

$$v = r_p^{-1},$$

so

$$dv = -\frac{dr}{r_P^2},$$

$$\star dv = -\tau_P,$$

and

$$\Delta v = 0 \quad \text{in} \quad U - B_\varepsilon.$$

On the surface ∂B_ε the function v takes on the constant value ε^{-1}. We thus obtain

$$-\int_{\partial U}\left(u\tau_P + \frac{\star du}{r_P}\right) + \int_{\partial B_\varepsilon} u\tau_P + \varepsilon^{-1}\int_{\partial B_\varepsilon} \star du = -\int_{U - B_\varepsilon} \frac{\Delta u}{r_P} dx \wedge dy \wedge dz.$$

Now by Stokes' theorem

$$\int_{\partial B_\varepsilon} \star du = \int_{\partial B_\varepsilon} d\star du$$

so that integral is of order ε^3. Hence the third term on the left goes to zero, in fact is $O(\varepsilon^2)$. The function u is assumed differentiable and so $u(x) = u(P) + O(\varepsilon)$ on ∂B_ε so we can replace

$$\int_{\partial B_\varepsilon} u\tau_P \quad \text{by} \quad u(P)\int_{\partial B_\varepsilon} \tau_P$$

with an error which is $O(\varepsilon)$. But

$$\int_{\partial B_\varepsilon} \tau_P = 4\pi,$$

the total area of the unit sphere. Thus, letting $\varepsilon \to 0$, we get

$$u(P) = \frac{1}{4\pi}\int_{\partial U}\left(u\tau_P + \frac{\star du}{r_P}\right) - \frac{1}{4\pi}\int_U \frac{\Delta u}{r_P} dx\, dy\, dz. \qquad (16.8)$$

16.8 Harmonic functions

A function u is called *harmonic* if $\Delta u = 0$. For harmonic functions the last term in (16.8) vanishes, and we have

$$u(P) = \frac{1}{4\pi}\int_{\partial U}\left(u\tau_P + \frac{\star du}{r_P}\right). \qquad (16.9)$$

This shows that a harmonic function u in the interior of a region is determined by its value, and the value of its derivatives, at the boundary. In fact, as we shall prove, u is determined by its values on ∂U alone. To begin, let us apply the preceding formula to a ball W centered at P (and lying inside U):

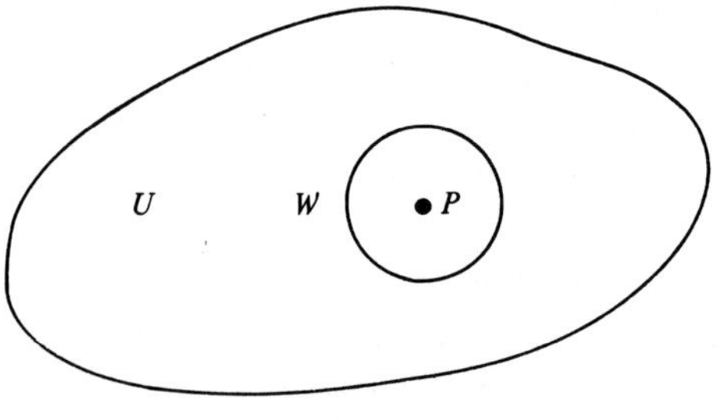

Figure 16.13

$$u(P) = \frac{1}{4\pi}\int_{\partial W} u\tau_P + \int_{\partial W} \frac{\star du}{r_P}.$$

Now r_P is constant on ∂W and, by Stokes' theorem,

$$\int_{\partial W} \star du = \int_W d\star du = 0 \quad \text{since} \quad \Delta u = 0.$$

Thus

$$u(P) = \frac{1}{4\pi}\int_{\partial W} u\tau_P. \tag{16.10}$$

But the expression on the right is just the average of u over the sphere ∂W. We have thus proved:

> If u is a function harmonic in some domain, then the value of u at any point is equal to its average value on any sphere centered at that point, whose interior is completely contained in the domain.

We can draw a number of startling consequences. Suppose there is a point $\mathbf{x}_0$ and a neighborhood, Z, of $\mathbf{x}_0$ such that

$$u(\mathbf{x}) \leqslant u(\mathbf{x}_0) \quad \text{at all points in } Z.$$

Let S_a be a sphere of radius a lying in Z and centered at $\mathbf{x}_0$. Then $u(\mathbf{x}) \leqslant u(\mathbf{x}_0)$ at all parts of S_a. Thus the average of u over S_a is $\leqslant u(\mathbf{x}_0)$. Now suppose that there were some point y on S_a at which $u(\mathbf{y}) < u(\mathbf{x}_0)$. Then $u(\mathbf{x}) < u(\mathbf{x}_0)$ at all points $\mathbf{x}$ near $\mathbf{y}$, and therefore the average of u over S_a would be strictly less than $u(\mathbf{x}_0)$. But this is impossible. Thus

$$u(\mathbf{y}) = u(\mathbf{x}_0) \quad \text{at all points of } S_a.$$

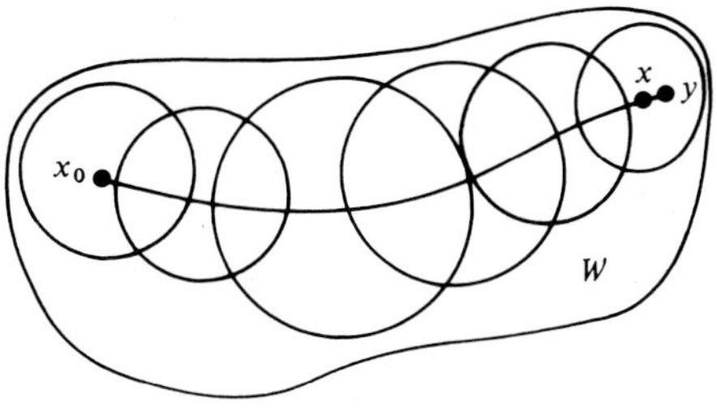

Figure 16.14

Now suppose that W is an open set that is *connected*; that is, suppose that any two points of W can be joined by a continuous curve lying entirely in W. Suppose u achieves its maximum at some $\mathbf{x}_0 \in W$. Let $\mathbf{y}$ be any point of W, and let C be a curve joining $\mathbf{x}_0$ to $\mathbf{y}$. About each $\mathbf{x}$ on the curve we can find a sufficiently small ball with center $\mathbf{x}$ lying entirely in W. By the compactness of C we can choose a finite number of these balls which cover C. We can therefore formulate the following. There are a finite number of spheres $S_{a_1}, \ldots, S_{a_k}$ such that each sphere and its interior lie entirely in W, S_{a_1} has center $\mathbf{x}_0$, the center $\mathbf{x}_i$ of S_{a_i+1} lies on

S_{a_i}, and $\mathbf{y} \in S_{a_k}$. (See figure 16.14). But this implies that $u(\mathbf{x}_0) = u(\mathbf{x}_1) = \cdots = u(\mathbf{x}_k) = u(\mathbf{y})$. In other words, we have established:

Let u be harmonic in a connected open set W, and suppose that u achieves its maximum value at some $\mathbf{x}_0 \in W$. Then u is constant on W.

An immediate corollary of this result is:

Let U be a connected open set and let $\bar{U}$ denote its closure, so $\bar{U} = U \cup \partial U$, where ∂U denotes the boundary of U. Suppose that $\bar{U}$ bounded. Then if u is a function that is continuous $\bar{U}$ and harmonic in U,

$$u(\mathbf{x}) < \max_{\mathbf{y} \in \partial U} u(\mathbf{y})$$

unless u is a constant.

Proof. In fact, since $\bar{U}$ is closed and bounded it is 'compact' and u is continuous. It is a standard theorem from real analysis that u must achieve its maximum at *some* point $\mathbf{x}_0$ of $\bar{U}$. If we could actually choose $\mathbf{x}_0 \in U$, then u would have to be a constant by our preceding results. If u is not a constant, then $\mathbf{x}_0 \in \partial U$, and we have proved the proposition.

From this, we deduce the following:

Let U be a connected open set with $\bar{U}$ compact. Let u and v be functions that are continuous in $\bar{U}$ and harmonic in U. Suppose that

$$u(\mathbf{y}) = v(\mathbf{y}) \quad \text{for all} \quad \mathbf{y} \in \partial U.$$

Then

$$u(\mathbf{x}) = v(\mathbf{x}) \quad \text{for all} \quad \mathbf{x} \in U.$$

Proof. In fact, $u - v$ is harmonic and vanishes on ∂U. Thus $u(\mathbf{x}) - v(\mathbf{x}) \leqslant 0$ for $x \in U$. Similarly $v(\mathbf{x}) - u(\mathbf{x}) \leqslant 0$, which implies the proposition.

An alternative way of formulating the last proposition is to say that on a domain U a harmonic function is completely determined by its boundary values. This is a *uniqueness theorem*: there is at most one harmonic function with given boundary values. It suggests the problem of deciding whether the corresponding existence theorem is true. This problem is known as *Dirichlet's problem.*

Dirichlet's problem. Given a continuous function f defined on ∂U, does there exist a function u that is continuous in $\bar{U}$ and harmonic in U and such that $u(y) = f(y)$ for all $y \in \partial U$?

16.9. The method of orthogonal projection

Let us give an alternative proof of the *uniqueness* of the solution of Dirichlet's problem for a bounded domain, U. We will use Green's first formula (16.6)

$$(\mathrm{d}u, \mathrm{d}v)_U = \int_{\partial U} u \star \mathrm{d}v - \int_U u \Delta v \mathrm{d}x \wedge \mathrm{d}y \wedge \mathrm{d}z.$$

Let C^0_{int} denote the space of differentiable functions which vanish on ∂U. Let H denote the space of harmonic functions on U – functions v that are continuous on $\bar{U}$ and satisfy $\Delta v = 0$ in U. Then, if $u \in C^0_{\text{int}}$, then, since $u = 0$ on ∂U,

$$\int_{\partial U} u\star \mathrm{d}v = 0 \quad \text{for any smooth function } v.$$

If $v \in H$, then $u\Delta v \equiv 0$ for any function u. Thus

If $u \in C^0_{\text{int}}$ and $v \in H$, then

$$(\mathrm{d}u, \mathrm{d}v)_U = 0.$$

In other words,

The spaces $\mathrm{d}C^0_{\text{int}}$ and $\mathrm{d}H$ are orthogonal under $(\ ,\ )_U$.

In particular, suppose u is harmonic and vanishes at the boundary. Then

$$(\mathrm{d}u, \mathrm{d}u)_U = 0.$$

But

$$(\mathrm{d}u, \mathrm{d}u)_U = \int_U \left(\left(\frac{\partial u}{\partial x}\right)^2 + \left(\frac{\partial u}{\partial y}\right)^2 + \left(\frac{\partial u}{\partial z}\right)^2\right)\mathrm{d}x \wedge \mathrm{d}y \wedge \mathrm{d}z.$$

The only way this can happen is if all the partial derivatives of u vanish identically in U. But then u must be a constant, and since $u = 0$ on ∂U, $u \equiv 0$. Thus a harmonic function which vanishes on ∂U must vanish identically. As before, this proves the *uniqueness* part of Dirichlet's problem.

Now the spaces $\mathrm{d}C^0$, $\mathrm{d}H$, etc., are infinite-dimensional vector spaces. We have not, in this book, developed a theory of such spaces, or of projection operators π, on such spaces. Such a theory can be developed, and it can be proved that the following principle, due to Hermann Weyl, holds:

The space $\mathrm{d}H$ is the entire orthogonal complement of $\mathrm{d}C^0_{\text{int}}$ inside $\mathrm{d}C^0$.

In other words,

If $E = \mathrm{d}v$ and $(E, \mathrm{d}u) = 0$ for all $u \in C^0_{\text{int}}$, then $\Delta v = 0$ in U.

Therefore, given any function ψ, we can break $\mathrm{d}\psi$ up into its components

$$\mathrm{d}\psi = \mathrm{d}u + \mathrm{d}v \quad \text{where} \quad u \in C^0_{\text{int}} \quad \text{and} \quad v \in H.$$

In other words, $\mathrm{d}u = \pi \mathrm{d}\psi$ *where* π *denotes the projection onto the* $\mathrm{d}C^0_{\text{int}}$ *component*, and $\mathrm{d}v = (1 - \pi)\mathrm{d}\psi$. We can now solve the *Dirichlet problem*: given a (smooth) function ϕ defined on ∂U, find a function v such that $\Delta v = 0$ and $v = \phi$ on ∂U. *Solution*: choose any ψ which agrees with ϕ on ∂U and is defined throughout U. Decompose $\mathrm{d}\psi$ into its components as above. Then, since u vanishes on ∂U, we know that v agrees with ψ on ∂U and hence is the desired solution.

Notice that since $(\mathrm{d}u, \mathrm{d}v)_U = 0$, we have

$$\|\mathrm{d}\psi\|^2_U = \|\mathrm{d}u\|^2_U + \|\mathrm{d}v\|^2_U$$

so

$$\|\mathrm{d}\psi\|^2_U \geqslant \|\mathrm{d}v\|^2_U.$$

Thus, among all functions which take on the values ψ on the boundary, the solution, v, of the Dirichlet problem can be characterized as the function with the smallest $\|\mathrm{d}v\|^2_U$. This is known as *Dirichlet's principle*. If we think of $\frac{1}{2}\|E\|^2_U$ as the *energy* of the electric field, then Dirichlet's principle can be thought of as saying that the solution, v, of the Dirichlet problem is the function v whose dv has minimum energy compatible with the constraint that $v = \phi$ on ∂U.

We shall not prove that the solution of Dirichlet's problem exists for general domains. (For a proof, see Loomis & Sternberg, Chapter 12.) However, we shall show how to solve this problem for certain kinds of domains. We shall also derive some general principles of electrostatics.

16.10. Green's functions

Suppose that U is a domain for which the Dirichlet problem can be solved. We shall now develop, for U, the analog of the Green's function that we have discussed in Chapter 13 for the discrete case.

For each $\mathbf{x}$ in $\mathbb{R}^3$, let us set

$$K(\mathbf{x},\mathbf{y}) = \frac{1}{4\pi}\frac{1}{\|\mathbf{x}-\mathbf{y}\|}.$$

As a function of $\mathbf{y}$ (for fixed $\mathbf{x}$) we have

$$\star \mathrm{d}K(\mathbf{x},\cdot) = \frac{1}{4\pi}\tau_{\mathbf{x}}.$$

We can rewrite equation (16.8) of section 16.6 in terms of K as

$$u(\mathbf{x}) = \int_{\partial U}(u\star \mathrm{d}K(\mathbf{x},\cdot) + K(\mathbf{x},\cdot)\star \mathrm{d}u) - \int_U K(\mathbf{x},\cdot)\Delta u\,\mathrm{d}V.$$

Now for fixed $\mathbf{x}\in U$ the function $K(\mathbf{x},\mathbf{y})$ is a differentiable function of $\mathbf{y}$ as $\mathbf{y}$ varies on ∂U. By assumption, we can therefore find a function $h(\mathbf{x},\cdot)$ that is harmonic in U and continuous in $\bar{U}$ and such that

$$h(\mathbf{x},\mathbf{y}) = -K(\mathbf{x},\mathbf{y}) \quad \text{for all } \mathbf{y}\in\partial U.$$

Furthermore, for fixed $\mathbf{y}, h(\mathbf{x},\mathbf{y})$ is a continuous function of $\mathbf{x}$. In fact, by the maximum principle, the harmonic function $h(\mathbf{x}_1,\mathbf{y}) - h(\mathbf{x}_2,\mathbf{y})$ achieves its maximum on ∂U, and hence

$$|h(\mathbf{x}_1,\mathbf{y}) - h(\mathbf{x}_2,\mathbf{y})| \leqslant \max_{\mathbf{z}\in\partial U}|K(\mathbf{x}_1,\mathbf{z}) - K(\mathbf{x}_2,\mathbf{z})|$$

and $K(\mathbf{x},\mathbf{z})$ is clearly uniformly continuous in $\mathbf{x}$ for all $\mathbf{z}\in\partial U$ so long as $\mathbf{x}$ stays a fixed distance away from ∂U. We have thus constructed a function h_U such that

(i) $h_U(\mathbf{x},\mathbf{y})$ is a continuous function of $\mathbf{x}$ and $\mathbf{y}$ for $\mathbf{x},\mathbf{y}\in\bar{U}$, and is differentiable in $\mathbf{y}$ for $\mathbf{y}\in U$;

(ii) For each fixed $\mathbf{x}, \Delta_{\mathbf{y}} h_U = 0$; i.e., $\Delta h_U(\mathbf{x},\cdot) = 0$;

(iii) $G_U(\mathbf{x},\mathbf{y}) = K(\mathbf{x},\mathbf{y}) + h_U(\mathbf{x},\mathbf{y}) = 0$ for $\mathbf{y}\in\partial U$.

The function G_U is called *the Green's function* of the domain U. Let us suppose for a moment that G_U exists, and let us derive some of its properties. We write $G = G_U$ when there is no possibility of confusion. We first show that for $\mathbf{x}\in U$ and $\mathbf{y}\in U$, we have

$$G(\mathbf{x},\mathbf{y}) = G(\mathbf{y},\mathbf{x}).$$

Let $B_{\mathbf{x},\varepsilon}$ and $B_{\mathbf{y},\varepsilon}$ be small balls about $\mathbf{x}$ and $\mathbf{y}$. Let $u = G(\mathbf{x},\cdot)$ and $v = G(\mathbf{y},\cdot)$ in Green's second formula where U is replaced by $U - B_{\mathbf{x},\varepsilon} - B_{\mathbf{y},\varepsilon}$. Since both functions are harmonic in the domain and vanish on ∂U, we obtain

$$\int_{\partial B_{\mathbf{x},\varepsilon}} G(\mathbf{x},\cdot)\star dG(\mathbf{y},\cdot) + \int_{\partial B_{\mathbf{y},\varepsilon}} G(\mathbf{x},\cdot)\star dG(\mathbf{y},\cdot)$$
$$= \int_{\partial B_{\mathbf{x},\varepsilon}} G(\mathbf{y},\cdot)\star dG(\mathbf{x},\cdot) + \int_{\partial B_{\mathbf{y},\varepsilon}} G(\mathbf{y},\cdot)\star dG(\mathbf{x},\cdot).$$

We will show that the left-hand side approaches $G(\mathbf{x},\mathbf{y})$ and the right-hand side approaches $G(\mathbf{y},\mathbf{x})$ as $\varepsilon \to 0$. By symmetry, it suffices to look at the left-hand side.

Now

$$\int_{\partial B_{\mathbf{y},\varepsilon}} G(\mathbf{x},\cdot)\star dG(\mathbf{y},\cdot) = \int_{\partial B_{\mathbf{y},\varepsilon}} G(\mathbf{x},\cdot)\star dK(\mathbf{y},\cdot) + \int_{\partial B_{\mathbf{y},\varepsilon}} G(\mathbf{x},\cdot)\star dh.$$

The first term on the right-hand side approaches $G(\mathbf{x},\mathbf{y})$ as in section 16.6, since $4\pi\star dK$ is the solid angle about $\mathbf{y}$. The second term tends to zero, since $G(\mathbf{x},\cdot)$ and h are smooth functions on $B_{\mathbf{y},\varepsilon}$. On the other hand,

$$\int_{\partial B_{\mathbf{x},\varepsilon}} G(\mathbf{x},\cdot)\star dG(\mathbf{y},\cdot) = \int_{\partial B_{\mathbf{x},\varepsilon}} K(\mathbf{x},\cdot)\star dG(\mathbf{y},\cdot) + \int_{\partial B_{\mathbf{x},\varepsilon}} h(\mathbf{x},\cdot)\star dG(\mathbf{y},\cdot).$$

The second term tends to zero, as above. The first term can be written

$$\frac{1}{4\pi}\int_{\partial B_{\mathbf{z},\varepsilon}} \star dG(\mathbf{y},\cdot) = \frac{1}{4\pi}\int_{B_{\mathbf{x},\varepsilon}} d\star dG(\mathbf{y},\cdot) = 0,$$

since $G(\mathbf{y},\cdot)$ is harmonic in $B_{\mathbf{x},\varepsilon}$. This proves that $G(\mathbf{x},\mathbf{y}) = G(\mathbf{y},\mathbf{x})$.

Let u be any smooth function on U. Apply Green's second formula to u and $v = G(\mathbf{x},\cdot)$ on $U - B_{\mathbf{x},\varepsilon}$. We get (since $G(\mathbf{x},\cdot) = 0$ on ∂U),

$$\int_{\partial U} u\star dG(\mathbf{x},\cdot) - \int_{\partial B_{\mathbf{x},\varepsilon}} u\star dG(\mathbf{x},\cdot) + \int_{\partial B_{\mathbf{x},\varepsilon}} G(\mathbf{x},\cdot)\star du = \int_U G(\mathbf{x},\cdot)\Delta u\, d\mathbf{x}^1 \wedge \cdots \wedge d\mathbf{x}^n.$$

The third integral on the left-hand side can be written as

$$\int_{\partial B_{\mathbf{x},\varepsilon}} K(\mathbf{x},\cdot)\star du + \int_{\partial B_{\mathbf{x},\varepsilon}} h(\mathbf{x},\cdot)\star du = \frac{1}{4\pi\varepsilon}\int_{\partial B_{\mathbf{x},\varepsilon}} \star du + \int_{\partial B_{\mathbf{x},\varepsilon}} h(\mathbf{x},\cdot)\star du$$
$$= \frac{1}{4\pi\varepsilon} O(\varepsilon) + O(\varepsilon^2)$$

and so tends to zero. We get

$$u(\mathbf{x}) = \int_U G(\mathbf{x},\cdot)\Delta u\, d\mathbf{x}^1 \wedge \cdots \wedge d\mathbf{x}^n + \int_{\partial U} u\star d\,G(\mathbf{x},\cdot).$$

We observe that this equation shows that *if we know that there exists* a solution $\Delta F = f$ with the boundary conditions $F = 0$, then it is given by

$$F(\mathbf{x}) = \int G(\mathbf{x},\mathbf{y}) f(\mathbf{y})\, d\mathbf{y} \quad \text{solution of the Poisson problem.} \qquad (16.11)$$

Similarly, if we know that there exists a smooth solution to the problem

$$\Delta u = 0, u(\mathbf{x}) = f(\mathbf{x}) \quad \text{for} \quad \mathbf{x} \in \partial U,$$

then it is given by

$$u(\mathbf{x}) = \int_{\partial U} u \star \mathrm{d}G(\mathbf{x}, \cdot) \quad \text{solution of the Dirichlet problem} \qquad (16.12)$$

16.11. The Poisson integral formula

It is important to observe that these formulas are consequences of the existence of Green's function for U. Thus they are valid whenever we can find the function h such that properties (ii) and (iii) hold. For example, we can explicitly construct the Green's function for a ball, B_R, of radius R. For simplicity, let us assume that the ball is centered at the origin.

For $\mathbf{x} \neq 0$, let $\mathbf{x}'$ be its image under *inversion* with respect to the sphere of radius R:

$$\mathbf{x}' = \frac{R^2}{\|\mathbf{x}\|^2}\mathbf{x}. \quad \text{(see figure 16.11.)}$$

Define the function G_R by

$$4\pi G_R(\mathbf{x}, \mathbf{y}) = \begin{cases} \dfrac{1}{\|\mathbf{y} - \mathbf{x}\|} - \dfrac{R}{\|\mathbf{x}\| \, \|\mathbf{y} - \mathbf{x}'\|} & \text{if } \mathbf{x} \neq \mathbf{0}, \\[2ex] \dfrac{1}{\|\mathbf{y}\|} - \dfrac{1}{R} & \text{if } \mathbf{x} = \mathbf{0}. \end{cases}$$

If $\mathbf{x} \in B_R$, then $\mathbf{x}' \in B_R$, and so the second terms on the right-hand side of the equation are continuous and harmonic on $\overline{B_R}$. We must merely check that property (iii) holds. Now for $\|\mathbf{y}\| = R$ we have, by similar triangles (or direct computation),

$$\frac{R}{\|\mathbf{x}\|} = \frac{\|\mathbf{y} - \mathbf{x}'\|}{\|\mathbf{y} - \mathbf{x}\|},$$

so that

$$G_R(\mathbf{x}, \mathbf{y}) = 0 \quad \text{for} \quad \|\mathbf{y}\| = R.$$

This is (iii), so we have verified that G_R is the Green's function for the ball of radius R.

To apply our formula for the solution of Dirichlet's problem using the Green's function, we must compute $\star \mathrm{d}G_R$ on the sphere of radius R. Now

$$\begin{aligned} 4\pi \star \mathrm{d}G_R(\mathbf{x}, \cdot) &= \tau_{\mathbf{x}} - \frac{R}{\|\mathbf{x}\|}\tau_{\mathbf{x}'} \\ &= \sum \frac{\mathbf{y}^i - \mathbf{x}^i}{\|\mathbf{y} - \mathbf{x}\|^3} - \frac{R}{\|\mathbf{x}\|} \frac{\mathbf{y}^i - \mathbf{x}'^i}{\|\mathbf{y} - \mathbf{x}'\|^3} \star \mathrm{d}\mathbf{y}^i. \end{aligned}$$

But

$$\frac{R}{\|\mathbf{x}\|\,\|\mathbf{y}-\mathbf{x}'\|^3}=\frac{\|\mathbf{x}\|^2}{R^2\|\mathbf{y}-\mathbf{x}\|^3}\quad \text{if}\quad \|\mathbf{y}\|=R.$$

We thus see that for $\|\mathbf{y}\|=R$,

$$\begin{aligned}(4\pi\star\mathrm{d}G_R)&=\frac{1}{\|\mathbf{y}-\mathbf{x}\|^3}\sum\left(\mathbf{y}^i-\mathbf{x}^i-\frac{\|\mathbf{x}\|^2}{R^2}\mathbf{y}^i-\frac{R^2}{\|\mathbf{x}\|^2}\mathbf{x}^i\right)\star\mathrm{d}\mathbf{y}^i\\&=\frac{R^2-\|\mathbf{x}\|^2}{\|\mathbf{y}-\mathbf{x}\|^3R^2}\sum\mathbf{y}^i\star\mathrm{d}\mathbf{y}^i\\&=\frac{R^2-\|\mathbf{x}\|^2}{R^2\|\mathbf{y}-\mathbf{x}\|^3}\star r\,\mathrm{d}r.\end{aligned}$$

But on the sphere of radius R, the form $\star r\mathrm{d}r$ is just $R\mathrm{d}S_R$, where $\mathrm{d}S_R$ is the area element on the sphere of radius R. If we now substitute into our solution of Dirichlet's problem:

$$u(\mathbf{x})=\int_{\partial U}u(\mathbf{y})\star\mathrm{d}G(\mathbf{x},\mathbf{y})$$

we get

$$4\pi u(\mathbf{x})=\int_{S_R}\frac{R^2-\|\mathbf{x}\|^2}{R^2|\mathbf{y}-\mathbf{x}\|^3}u(\mathbf{y})\star r\,\mathrm{d}r.$$

But $(\star r\mathrm{d}r)=R\mathrm{d}S_R$, where $\mathrm{d}S_R$ is the volume element of the sphere of radius R. If we substitute into the last formula, we obtain

$$u(\mathbf{x})=\frac{R^2-\|\mathbf{x}\|^2}{4\pi R}\int_{S_R}\frac{u(\mathbf{y})}{\|\mathbf{y}-\mathbf{x}\|^3}\mathrm{d}S_R\ \text{(the Poisson integral formula).}\quad(16.13)$$

In the proof of (16.13) we used the assumption that the function u is differentiable in some neighborhood of the ball B_R and is harmonic for $\|\mathbf{x}\|<R$. Actually, all that we need to assume is that u is differentiable and harmonic for $\|\mathbf{x}\|<R$ and continuous on the closed ball $\|\mathbf{x}\|\leqslant R$. In fact, for any $\|\mathbf{x}\|<R$, equation (16.13) will be valid with R replaced by R_a, where $\|\mathbf{x}\|<R_a<R$. If we then let R_a approach R, we recover (16.13) by virtue of the assumed continuity of u.

Equation (16.13) gives the solution of Dirichlet's problem for a ball *provided we know that the solution exists.* That is, if u is any function that is harmonic on the open ball and continuous on the closed ball, it satisfies (16.13). Now let us show that (16.13) *is* actually a solution of Dirichlet's problem for prescribed boundary values. Thus suppose we are given a continuous function u defined on the sphere S_R. Then we are given $u(\mathbf{y})$ for all $\mathbf{y}\in S_R$. Define $u(\mathbf{x})$ for $\|\mathbf{x}\|<R$ by (16.13). We must show that

(a) u is harmonic for $\|\mathbf{x}\|<R$, and
(b) $u(\mathbf{x})\to u(\mathbf{y}_0)$ if $\mathbf{x}\to\mathbf{y}_0$ and $\|\mathbf{y}_0\|=R$.

To prove (a) we observe that $G_R(\mathbf{x},\mathbf{y})$ is a differentiable function of $\mathbf{x}$ and $\mathbf{y}$ in the range $\|\mathbf{x}\|<R_1<R, R_1<\|\mathbf{y}\|<R^2/R_1$, and is, by construction, a harmonic

function of $\mathbf{y}$. For $\|\mathbf{x}\| < R$ and $\|\mathbf{y}\| < R$, we know that

$$G_R(\mathbf{x}, \mathbf{y}) = G_R(\mathbf{y}, \mathbf{x}).$$

Thus for fixed $\mathbf{y}$ with $\|\mathbf{y}\| < R$, $G_R(\cdot, \mathbf{y})$ is a harmonic function on $B_R - \{\mathbf{y}\}$. Letting $\|\mathbf{y}\| \to R$, we see that $G_R(\mathbf{x}, \mathbf{y})$ is a harmonic function of $\mathbf{x}$ for $\|\mathbf{x}\| < R_1 < R$ for each fixed $\mathbf{y} \in S_R$. Thus $\partial G_R(\mathbf{x}, \mathbf{y})/\partial \mathbf{y}^i$ is a harmonic function of $\mathbf{x}$ for each $\mathbf{y} \in S_R$. In other words, all the coefficients of $\star \mathrm{d} G_R(\cdot, \mathbf{y})$ are harmonic functions of $\mathbf{x}$ for each $\mathbf{y} \in S_R$, and therefore so is each coefficient of $u(\mathbf{y}) \star \mathrm{d} G_R(\cdot, \mathbf{y})$. It follows that the function $u(\mathbf{x}) = \int_{S_R} u \star \mathrm{d} G_R(\mathbf{x}, \cdot)$ is a harmonic function of $\mathbf{x}$, since the integral converges uniformly (as do the integrals of the various derivatives with respect to $\mathbf{x}$) for $\|\mathbf{x}\| < R_1 < R$. This proves (a).

To prove (b), we first remark that the constant one is a harmonic function everywhere, so equation (16.13) is applicable to it. We thus have

$$\frac{R^2 - \|\mathbf{x}\|^2}{4\pi R} \int_{S_R} \frac{\mathrm{d}S_R}{\|\mathbf{y} - \mathbf{x}\|^3} = 1 \quad \text{for any } \|\mathbf{x}\| < R. \tag{16.14}$$

Now let $\mathbf{y}_0$ be some point of S_R, and let u be a continuous function on S_R. For any $\varepsilon > 0$ we can find a $\delta > 0$ such that

$$|u(\mathbf{y}) - u(\mathbf{y}_0)| < \varepsilon \quad \text{for} \quad \|\mathbf{y} - \mathbf{y}_0\| \leqslant 2\delta \quad \mathbf{y} \in S_R.$$

Let $Z_1 = \{\mathbf{y} \in S_R : \|\mathbf{y} - \mathbf{y}_0\| > 2\delta\}$ and $Z_2 = \{\mathbf{y} \in S_R : \|\mathbf{y} - \mathbf{y}_0\| \leqslant 2\delta\}$. Then by (16.13) and (16.14) we have, for $\|\mathbf{x}\| < R$,

$$u(\mathbf{x}) - u(\mathbf{y}_0) = \frac{R^2 - \|\mathbf{x}\|^2}{4\pi R} \int_{S_R} \frac{u(\mathbf{y}) - u(\mathbf{y}_0)}{\|\mathbf{y} - \mathbf{x}\|^3} \mathrm{d}S_R,$$

so

$$|u(\mathbf{x}) - u(\mathbf{y}_0)| \leqslant I_1 + I_2,$$

where

$$I_1 = \frac{R^2 - \|\mathbf{x}\|^2}{4\pi R} \int_{Z_1} \frac{|u(\mathbf{y}) - u(\mathbf{y}_0)|}{\|\mathbf{y} - \mathbf{x}\|^3} \mathrm{d}S_R$$

and

$$I_2 = \frac{R^2 - \|\mathbf{x}\|^2}{4\pi R} \int_{Z_2} \frac{|u(\mathbf{y}) - u(\mathbf{y}_0)|}{\|\mathbf{y} - \mathbf{x}\|^3} \mathrm{d}S_R.$$

Now if $\|\mathbf{y}_0 - \mathbf{x}\| < \delta$, then for all $\mathbf{y} \in Z_1$, we have $\|\mathbf{y} - \mathbf{x}\| > \|\mathbf{y} - \mathbf{y}_0\| - \|\mathbf{x} - \mathbf{y}_0\|$, so that $\|\mathbf{y} - \mathbf{x}\| > \delta$. Thus for all $\mathbf{x}$ such that $\|\mathbf{x} - \mathbf{y}_0\| < \delta$ the integral occurring in I_1 is uniformly bounded. Since $\|\mathbf{x}\| \to R$ as $\mathbf{x} \to \mathbf{y}_0$, we conclude that $I_1 \to 0$ as $\mathbf{x} \to \mathbf{y}_0$.

With respect to I_2, we know that $|u(\mathbf{y}) - u(\mathbf{y}_0)| < \varepsilon$ for all $\mathbf{y} \in Z_2$, so that

$$I_2 = \frac{R^2 - \|\mathbf{x}\|^2}{4\pi R} \int_{Z_2} \frac{\varepsilon \mathrm{d}S_R}{\|\mathbf{y} - \mathbf{x}\|^3} < \varepsilon \frac{R^2 - \|\mathbf{x}\|^2}{4\pi R} \int_{S_R} \frac{\mathrm{d}S_R}{\|\mathbf{y} - \mathbf{x}\|^3} = \varepsilon$$

by (16.14). This proves (b).

We have thus proved:

Theorem: Let u be a continuous function defined on the sphere S_R. There is a unique continuous function defined for $\|\mathbf{x}\| \leqslant R$ which coincides with the given function on the sphere S_R and is harmonic for $\|\mathbf{x}\| < R$. This function is given by (16.13) *for all $\|\mathbf{x}\| < R$.*

We have exerted a substantial amount of effort in giving a detailed proof of the existence of a solution of the Dirichlet problem for the simple case of the sphere. We will not give the details of any of the various proofs which work for more general regions. But here is an outline of a method of proof which can be made to work with some effort.

Suppose we approximate Euclidean space by a network. Let us place a node at every vector $\mathbf{v}$ whose coordinates are dyadic fractions with denominator 2^N – that is, at all $\mathbf{v}$ of the form

$$\mathbf{v} = \frac{1}{2^N}\begin{pmatrix} k \\ l \\ m \end{pmatrix} \quad \text{where } k, l, m \text{ are integers.}$$

Here N is some large, fixed integer. Let us join every node to each of its six nearest neighbor nodes. So our network looks like a cubic scaffolding over all space. For this network, let us assign capacitance 2^N to each branch. Let Δ_N denote the Laplace operator for this network. For any twice-differentiable function u, it is easy to see that

$$\Delta_N u \to \Delta u \quad \text{as} \quad N \to \infty.$$

If we are now given a bounded region U with boundary ∂U, we can consider the network of all nodes in U and all branches joining these nodes. We can declare the nodes closest to the boundary to be boundary nodes. By our results for finite networks, we can solve the Dirichlet problem for this finite approximation. Given any continuous function ϕ on ∂U we can assign values, ϕ_N, on the boundary of our finite approximation by taking $\phi_N(\mathbf{n}) = \phi(\mathbf{x})$ where $\mathbf{x}$ is the nearest point to $\mathbf{n}$ on the boundary. (If there are several points equally near, choose $\mathbf{x}$ to be one of them.) In this way, we have approximated the continuous Dirichlet problem by a

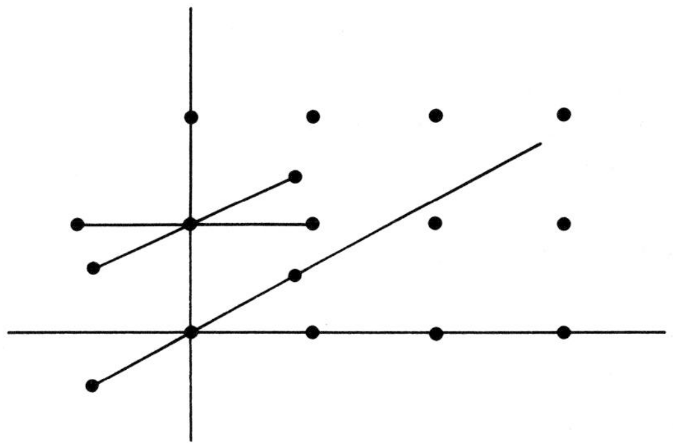

Figure 16.15

discrete, finite problem we know how to solve. We expect, as $N \to \infty$, that the sequence of solutions U_N will converge to a function u (defined on all the dyadic vectors and extended by continuity to all) which solves the honest Dirichlet problem. In fact, this method works, at least if the shape of the boundary is not too weird. The proof requires careful estimates which are beyond the scope of this book.

Summary

A Electrostatics and differential forms

You should be able to state the laws of electrostatics in terms of the differential forms D and E.

You should know how to state, prove, and apply Green's formulas for electrostatics in $\mathbb{R}^3$.

B The star operator in $\mathbb{R}^3$

Given a differential form ω on $\mathbb{R}^3$, expressed in terms of coordinates whose differentials are mutually orthogonal you should be able to construct an orthonormal basis and use it to construct $*\omega$.

You should be able to describe the relation between electrostatics in $\mathbb{R}^3$ and capacitive networks, explaining how the star operator plays the same role as the capacitance matrix.

Exercises

16.1. Let $D = \tau_0 = (x\mathrm{d}y \wedge \mathrm{d}z + y\mathrm{d}z \wedge \mathrm{d}x + z\mathrm{d}x \wedge \mathrm{d}y)/r^3$ where $r^2 = x^2 + y^2 + z^2$. This represents the electric displacement for a unit positive charge at the origin.

(a) Evaluate $\int_{R_1} D$ where R_1 is the disk $z = z_0$, $x^2 + y^2 \leqslant a^2$.

(b) Evaluate $\int_{R_2} D$ where R_2 is the curved surface of a cylinder:

$$-z_0 < z < z_0,\ x^2 + y^2 = a^2.$$

(c) Using the results of (a) and (b), check that $\int_{R_3} D = 4\pi$, where R_3 is the cylinder with curved surface R_2 plus disks like R_1 at its top and bottom.

16.2. Let A be the one-form $A_x\mathrm{d}x + A_y\mathrm{d}y + A_z\mathrm{d}z$. The quantities A_x, A_y, and A_z are all functions of x, y, and z. Using the star operator $\star$ defined to be linear and to satisfy

$$\begin{aligned}
&\star\mathrm{d}x = \mathrm{d}y \wedge \mathrm{d}z, && \star\mathrm{d}y = \mathrm{d}z \wedge \mathrm{d}x, && \star\mathrm{d}z = \mathrm{d}x \wedge \mathrm{d}y,\\
&\star\mathrm{d}y \wedge \mathrm{d}z = \mathrm{d}x, && \star\mathrm{d}z \wedge \mathrm{d}x = \mathrm{d}y, && \star\mathrm{d}x \wedge \mathrm{d}y = \mathrm{d}z,\\
&\star 1 = \mathrm{d}x \wedge \mathrm{d}y \wedge \mathrm{d}z, && \star\mathrm{d}x \wedge \mathrm{d}y \wedge \mathrm{d}z = 1,
\end{aligned}$$

compute the following:

(a) $\star\mathrm{d}A$;

(b) $\star\mathrm{d}\star A$;

(c) $\star d \star dA$;
(d) $\star d \star df$ where $f(x, y, z)$ is a scalar field;
(e) $\star(A \wedge B)$ where A and B are one-forms;
(f) $\star(A \wedge \star B)$ where A and B are one-forms.

16.3. Consider spherical coordinates r, θ, ϕ defined by $x = r\sin\theta\cos\phi, y = r\sin\theta\sin\phi, z = r\cos\theta$.
(a) Express dx, dy, dz, $dy \wedge dz$, $dz \wedge dx$, $dx \wedge dy$, and $dx \wedge dy \wedge dz$ in spherical coordinates (i.e., use the pullback equations for x, y, and z to pull back these differential forms).
(b) The star operator defined in Exercise 16.2 is a linear transformation specified by its action with respect to basis vectors $dx, dy, dz, dy \wedge dz, \ldots$. Express this operator in spherical coordinates; i.e., calculate $\star dr$, $\star d\theta$, and $\star d\phi$ in terms of $d\theta \wedge d\phi$, $d\phi \wedge dr$, and $dr \wedge d\theta$, and calculate $\star(d\theta \wedge d\phi)$, $\star(d\phi \wedge dr)$, and $\star(dr \wedge d\theta)$ in terms of dr, $d\theta$, and $d\phi$.
(c) Express the two-form τ_0 of exercise 16.1 in spherical coordinates, and show that it equals $\star d(1/r)$.

16.4. If u represents an electric potential function on $\mathbb{R}^3$, its Laplacian Δu may be defined by

$$\Delta u \, dx \wedge dy \wedge dz = d \star du.$$

(a) By regarding u as a function of x, y, and z, confirm that

$$\Delta u = \frac{\partial^2 u}{\partial x^2} + \frac{\partial^2 u}{\partial y^2} + \frac{\partial^2 u}{\partial z^2}.$$

(b) By regarding u as a function of r, θ, and ϕ, develop a formula for Δu in terms of partial derivatives of u with respect to r, θ, and ϕ.

16.5. Consider a two-form D expressed in spherical coordinates by $D = r^3(1-r)\sin\theta \, d\theta \wedge d\phi$ $(r \leqslant 1)$, $D = 0$ $(r > 1)$.
(a) Verify that if $E = r(1-r)dr$ for $r \leqslant 1$, then $D = \star E$. (Note that $dr \wedge \star dr = dx \wedge dy \wedge dz$.)
(b) Find a potential u such that $E = -du$. Also evaluate ∂D (which equals $-dD$). Now check that $\int_D E = -\int_{\partial D} u$, as required by Stokes' theorem. (Recall that $\int_D E \equiv \int_{\mathbb{R}^3} E \wedge D$.)

16.6. The differential form that represents D for a unit positive charge may be expressed in spherical coordinates as $\tau = \sin\theta \, d\theta \wedge d\phi$.
(a) Evaluate $\int_S \tau$ where S is the sphere $x^2 + y^2 + z^2 = 1$.
(b) Show that $d\tau = 0$.
(c) Using Stokes' theorem, explain why there *cannot* exist a one-form ω, defined everywhere in $\mathbb{R}^3$ except possibly at the origin, for which $d\omega = \tau$.
(d) Find a one-form ω, defined on the octant of $\mathbb{R}^3$ where $x > 0$, $y > 0$, $z > 0$, for which $\tau = d\omega$. You will probably want to use spherical coordinates to invent a suitable ω, but then express it in terms of x, y, and z.

16.7. Let $u(r, \theta)$ be a smooth function on $\mathbb{R}^2$ with compact support. Evaluate

$$\int_{\mathbb{R}^2} du \wedge \omega \quad \text{where} \quad \omega = d\theta = \frac{x\,dy - y\,dx}{x^2 + y^2}.$$

(Because ω is undefined at the origin, you should first integrate over the region outside a disk of radius ε centered at the origin, then take the limit as $\varepsilon \to 0$.)

16.8. Let ϕ be a rotation in $\mathbb{R}^3$, so that ϕ is represented by a matrix A for which $AA^{\mathrm{T}} = I$ and $\operatorname{Det} A = +1$. Let ω be a one-form $\omega = F_x \mathrm{d}x + F_y \mathrm{d}y + F_z \mathrm{d}z$. Prove that in this case

$$\star(\phi^* \omega) = \phi^*(\star\omega).$$

16.9. Consider the five functions

$$v_1 = r^2, \quad v_2 = r^2 \cos^2 \theta (= z^2), \quad v_3 = r^4, \quad v_4 = r^4 \cos^2 \theta (= r^2 z^2),$$
$$v_5 = r^4 \cos^4 \theta (= z^4).$$

These span a five-dimensional vector space which we shall call C^0.

(a) Calculate $\mathrm{d}v$, $\star\mathrm{d}v$, and Δv for each of these five functions.

(b) Calculate all the scalar products of the form $(\mathrm{d}v_1, \mathrm{d}v_2)$, letting U be the interior of the unit sphere, $r \leqslant 1$.

(c) Find a basis for the two-dimensional subspace H of C^0, which consists of functions with $\Delta v = 0$.

(d) Construct the matrix $(I - \pi)$, which is the orthogonal projection of $\mathrm{d}C^0$ onto $\mathrm{d}H$. Thereby you can also write down π.

(e) Use $I - \pi$ to find a function u which satisfies $\Delta u = 0$ and which equals $\cos^4\theta + \cos^2\theta$ on the sphere $r = 1$. (This is a Dirichlet problem.)

(f) Use π to find a function ϕ which vanishes on the unit sphere, for which $\Delta\phi = 5r^2 - 4r^2 \cos^2\theta + 1$. (Now you are solving the Poisson equation.)

17

Currents, flows and magnetostatics

Chapter 17 continues the study of the exterior differential calculus. The main topics are vector fields and flows, interior products and Lie derivatives. These are applied to magnetostatics.

17.1. Currents

At first glance, it would seem to be a straightforward matter to give the continuous version of resistive networks. All we need to do is make some minor changes in the discussion of the preceding chapter. We keep the form E and the equation $E = -\,\mathrm{d}u$. We replace energy by power. We replace the two-form D of electrostatics by a two-form, J, which should represent the smeared-out version of the branch currents. We have a pairing between the current J and an electric field E given by

$$\int E \wedge J$$

measured in units of power. Kirchhoff's current law said that $\partial I = 0$. So the corresponding smeared-out version says

$$\partial J = 0$$

or, since $\partial = -\,\mathrm{d}$, that

$$\mathrm{d}J = 0.$$

Instead of $D = \varepsilon \star E$ for an isotropic electrostatic medium, we will have

$$J = \sigma \star E$$

where σ is called the *conductivity*. If we define $r = 1/\sigma$ and call it the *resistivity*, then we can write the preceding equation

$$E = r\star^{-1} J$$

and this is the smeared-out version of Ohm's law

$$V = RI.$$

We should now pause for a moment to see what the form J represents. Suppose we consider some surface S. What is the meaning of $\int_S J$? Suppose we had a finite network and the surface crossed certain branches, say α, β and γ, oriented as indicated in figure 17.1. Then, taking orientation into account, we can think of S as defining the one-cochain which assigns to each current I the number $I_\alpha - I_\beta + I_\gamma$. More generally, S defines the cochain which assigns to each current I the *total current across* S, taking orientation into account. Thus, in our smeared-out version, we should think of $\int_S J$ as the current across S.

If we think of current as charge in motion, then we can think of $\int_S J$ as the rate of charge transport across S. If S is a closed surface, $S = \partial U$, then

$$\int_S J = \int_U \mathrm{d}J = 0$$

so the total rate of transport across S is zero, no charge can be created in the interior. This fits with our original discussion of Kirchhoff's current law, so all works out fine.

There *is* a point of geometric interpretation which *does* require further explanation. We know that in reality the electric current consists of electrons in motion. Each electron, thought of as a point particle, moves along some curve. Thus the form J, integrated over a surface S, is really counting the number of electrons crossing S (in either direction) per unit time. We should have some geometric object, more directly related to the family of curves of the individual electrons, which describes their motion, and from which we can reconstruct the two-form J. As we shall see, the correct notion here is that of a *vector field*.

17.2. Flows and vector fields

In this section, to avoid an accumulation of indices, we shall work for the most part in $\mathbb{R}^3$. But we will not make any special use of three dimensions, so our formulations are valid in any dimension.

Let U be an open region in $\mathbb{R}^k$. Let I be some interval in $\mathbb{R}$ containing the

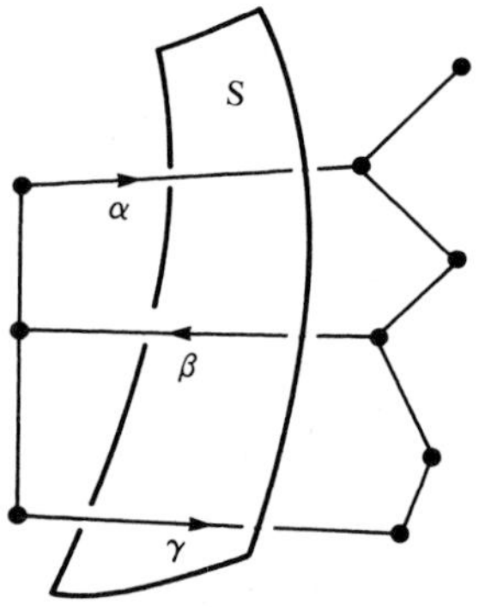

Figure 17.1

origin. Let $\phi: I \times U \to \mathbb{R}^k$ be a differentiable map. We think of $t \in I$ as representing an instant of time. We can then regard ϕ in two different ways: Suppose we fix some $\mathbf{p} \in U$. Then the map

$$\phi(\cdot, \mathbf{p}): t \rightsquigarrow \phi(t, \mathbf{p})$$

is a curve in $\mathbb{R}^k$. Of course, for each different $\mathbf{p}$, we get a different curve. We can consider the tangent vector, $\phi'(t, \mathbf{p}) = (\mathrm{d}/\mathrm{d}t)\phi(t, \mathbf{p})$. On the other hand, for each fixed t, we can consider the map $\phi_t: U \to \mathbb{R}^k$ given by

$$\phi_t(\mathbf{p}) = \phi(t, \mathbf{p}).$$

Let us assume that

$$\phi_0 = \mathrm{id},$$

i.e., that

$$\phi(0,\mathbf{p}) = \mathbf{p}$$

for all $\mathbf{p}$. Then for each $\mathbf{p}$, the tangent $\phi'(0, \mathbf{p})$ to the curve $\phi(\cdot, \mathbf{p})$ is a tangent vector at $\mathbf{p}$. We thus have a rule which assigns to each point $\mathbf{p}$ of U a tangent vector at $\mathbf{p}$. Such a rule is called a *vector field* on U. We will denote a vector field by a symbol such as ξ. Thus $\xi(\mathbf{p})$ is a tangent vector at the point $\mathbf{p}$.

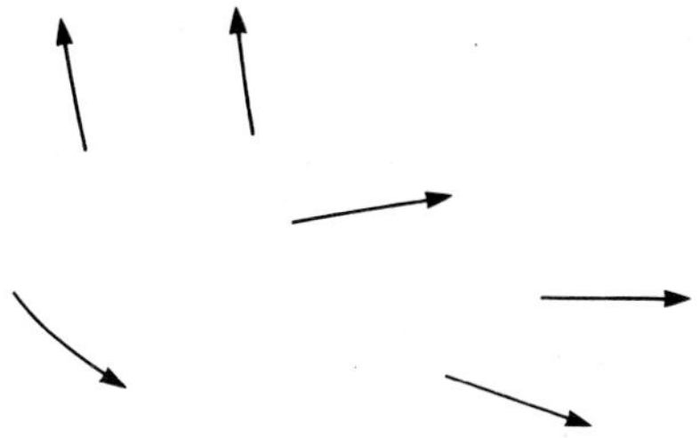

Figure 17.2

For any $\mathbf{p}$, the assumption $\phi_0 = \mathrm{id}$ implies that $\mathrm{d}\phi_0 = \mathrm{id}$ so $\mathrm{d}(\phi_t)_{\mathbf{p}}$ is non-singular for small enough t. Thus by the implicit function theorem, ϕ_t is one-to-one with differentiable inverse in some neighborhood of $\mathbf{p}$. In order to avoid accumulation of notation, let us assume that $\phi_t: U \to R^k$ has a differentiable inverse on all of U. Fix some time $s \in I$. Then define

$$\psi_t = \phi_t \circ \phi_s^{-1}.$$

Then ψ_t has all the properties of ϕ_t, but now

$$\psi_s = \mathrm{id}.$$

Thus for each $s \in I$, we get a vector field, call it ξ_s. In short, we have shown that, given $\phi: I \times U \to \mathbb{R}^k$, we get a one-parameter family of vector fields ξ_s. For each $\mathbf{p} \in U$ there will be some neighborhood W of $\mathbf{p}$ and some neighborhood k of 0 such that ξ_s is a vector field defined on W.

Example:

(a) Suppose $\mathbf{w}$ is some fixed vector. Let

$$\phi(t, \mathbf{p}) = \mathbf{p} + t\mathbf{w}.$$

Then $\phi'(0,\mathbf{p}) = \mathbf{w}$ for all $\mathbf{p}$. Also

$$\begin{aligned}\phi_t \circ \phi_s^{-1}(\mathbf{p}) &= \phi_t(\mathbf{p} - s\mathbf{w})\\ &= \mathbf{p} + (t-s)\mathbf{w}\\ &= \phi_{t-s}(\mathbf{w})\end{aligned}$$

so

$$\xi_s(\mathbf{p}) \equiv \mathbf{w} \quad \text{for all } \mathbf{v} \text{ and } s.$$

(b) Let A be a matrix. Define

$$\phi(t,\mathbf{p}) = \mathrm{e}^{tA}\mathbf{p}.$$

Then

$$\phi'(0,\mathbf{p}) = A\mathbf{p}.$$

Also

$$\begin{aligned}\phi_t \circ \phi_s^{-1}(\mathbf{p}) &= \mathrm{e}^{tA}(\mathrm{e}^{-sA}\mathbf{p})\\ &= \mathrm{e}^{(t-s)A}\mathbf{p}\\ &= \phi_{t-s}(\mathbf{p})\end{aligned}$$

so

$$\xi_s(\mathbf{p}) = A\mathbf{p} \quad \text{for all } s.$$

The vector field in this case assigns to each point $\mathbf{p}$ the vector $A\mathbf{p}$.

Notice that, in both examples, the vector field ξ_s did not depend on s. This was a consequence of the identity

$$\phi_t \circ \phi_s^{-1} = \phi_{t-s},$$

or, replacing s by $-s$,

$$\phi_{s+t} = \phi_s \circ \phi_t.$$

A map ϕ satisfying this identity whenever both sides are defined is called a (stationary) *flow*. In case $\phi_t(U) = U$ and $I = \mathbb{R}$, we also speak of a *one-parameter group of transformations of* U.

Thus, for flows, we get a fixed vector field, $\xi = \xi_s$ all s.

Suppose we *start* with a *linear* vector field

$$\xi(\mathbf{p}) = A\mathbf{p}$$

where A is some given matrix. Then, by the results of Chapter 3, we know that we can form

$$\mathrm{e}^{tA}$$

and then we have ϕ defined by

$$\phi(t,\mathbf{p}) = \mathrm{e}^{tA}\mathbf{p}.$$

Thus we have an existence theorem in ordinary differential equations. Starting with ξ, we have found the ϕ. The corresponding theorem is true in general:

Let ξ_s be a vector field depending smoothly on s and $\mathbf{p}$ defined for $s \in I$ and $\mathbf{p} \in U$. Then each $\mathbf{p}$ has a neighborhood W of $\mathbf{p}$ in $\mathbb{R}^k$ and a neighborhood K of 0 in $\mathbb{R}$ such *that* $\phi: K \times W \to \mathbb{R}^k$, $\phi_0 = \mathrm{id}$ *and is smooth and the s-dependent vector fields defined by* ϕ *are precisely* ξ_s. *If the* $\xi_s \equiv \xi$ *(i.e., do not depend on s), then* ϕ *satisfies the identity*

$$\phi_{s+k} = \phi_s \circ \phi_t$$

whenever both sides are defined.

The proof of this theorem is not difficult. One proceeds by a method of successive approximations. We will not present the proof here. It is standard and can be found in any text on ordinary differential equations. For example, a proof is given in Loomis & Sternberg Chapter 6.

Suppose that ξ is a vector field and f is a differentiable function. At each point $\mathbf{p}$, we can construct the directional derivative

$$\mathrm{D}_{\xi(\mathbf{p})} f.$$

This gives a new *function*: the function which assigns to each point $\mathbf{p}$ the value $\mathrm{D}_{\xi(\mathbf{p})} f$. For example, suppose we are in $\mathbb{R}^3$ and the vector field is given by

$$\xi(\mathbf{p}) = \begin{pmatrix} z(\mathbf{p}) \\ b(\mathbf{p}) \\ c(\mathbf{p}) \end{pmatrix}$$

where a, b and c are functions. Then

$$\mathrm{D}_{\xi(\mathbf{p})} f = a(\mathbf{p})\frac{\partial f}{\partial x}(\mathbf{p}) + b(\mathbf{p})\frac{\partial f}{\partial y}(\mathbf{p}) + c(\mathbf{p})\frac{\partial f}{\partial z}(\mathbf{p}).$$

Thus

$$\mathrm{D}_\xi f = a\frac{\partial f}{\partial x} + b\frac{\partial f}{\partial y} + c\frac{\partial f}{\partial z}.$$

For this reason we shall use the following notation: we shall write

$$\xi = a\frac{\partial}{\partial x} + b\frac{\partial}{\partial y} + c\frac{\partial}{\partial z}.$$

In this notation the symbol $\partial/\partial x$, for example, is thought of as the constant vector field

$$\mathbf{p} \rightsquigarrow \begin{pmatrix} 1 \\ 0 \\ 0 \end{pmatrix}.$$

But

$$\mathrm{D}_{\partial/\partial x} f = \frac{\partial f}{\partial x}$$

so this constant vector field acts like the partial derivative. So, similarly, the vector field

$$\xi = \begin{pmatrix} a \\ b \\ c \end{pmatrix}$$

is written as

$$\xi = a\frac{\partial}{\partial x} + b\frac{\partial}{\partial y} + c\frac{\partial}{\partial z}$$

and it acts on functions by

$$D_\xi f = a\frac{\partial f}{\partial x} + b\frac{\partial f}{\partial y} + c\frac{\partial f}{\partial z}.$$

Another way of putting this is as follows: the differential

$$df = \frac{\partial f}{\partial x}dx + \frac{\partial f}{\partial y}dy + \frac{\partial f}{\partial z}dz$$

can be thought of as the function which assigns to each point the row vector

$$\left(\frac{\partial f}{\partial x}(\mathbf{p}), \frac{\partial f}{\partial y}(\mathbf{p}), \frac{\partial f}{\partial z}(\mathbf{p})\right).$$

We can (at each point) evaluate this row vector on the column vector $\begin{pmatrix} a \\ b \\ c \end{pmatrix}$ to get the value

$$\langle \xi, df \rangle = \left(\frac{\partial f}{\partial x}, \frac{\partial f}{\partial y}, \frac{\partial f}{\partial z}\right)\begin{pmatrix} a \\ b \\ c \end{pmatrix} = a\frac{\partial f}{\partial x} + b\frac{\partial f}{\partial y} + c\frac{\partial f}{\partial z}.$$

So

$$\boxed{D_\xi f = \langle \xi, df \rangle.}$$

There is still another way of thinking about $D_\xi f$ which is very instructive. Let $\phi: I \times U \to \mathbb{R}^k$ be as above and ξ_s the corresponding vector fields. For each t we can consider the function $\phi_t^* f$. Recall that

$$\phi_t^* f(\mathbf{p}) = f(\phi_t(\mathbf{p})) = f(\phi(t, \mathbf{p})).$$

Now

$$\frac{1}{t}(\phi_t^* f - f)$$

is again a perfectly good function. Evaluating this function at some point $\mathbf{v}$ and letting $t \to 0$, we see that

$$\lim_{t\to 0}\frac{1}{t}(\phi_t^* f - f)(\mathbf{p}) = \lim \frac{1}{t}[f(\phi(t, \mathbf{p})) - f(\mathbf{p})]$$
$$= D_{\xi(\mathbf{p})} f.$$

So, as functions,

$$\boxed{D_\xi f = \lim_{t\to 0}\frac{1}{t}(\phi_t^* f - f).}$$

We have now seen that a natural geometric object to attach to a flow is a vector field. But we have seen in the preceding section that the electric current (which is, after all, a flow of electrons) is given by a two-form. So we need a mathematical operation which allows us to pass from the vector field describing the flow of the electrons to the two-form describing the current. The two-form will depend also on the density of the electrons. Obviously a higher density of electrons moving along the same flow lines will produce a larger current. So we really need a mathematical operation which will pass from vector fields, ξ, and three-forms $\rho \mathrm{d}x \wedge \mathrm{d}y \wedge \mathrm{d}z$ to two-forms. We describe this operation in the next section.

17.3. The interior product

Let V be a vector space and ω an element of $\Lambda^k(V^*)$. Recall that ω is a function (multilinear and antisymmetric) of k vectors in V. For every $\mathbf{v}_1, \ldots, \mathbf{v}_k$, we get a number

$$\omega(\mathbf{v}_1, \mathbf{v}_2, \ldots, \mathbf{v}_k).$$

Now let $\mathbf{v}$ be a vector in V. We will define a function of $k-1$ vectors of V by simply substituting $\mathbf{v}$ into the first position of ω. That is, we consider the function of $\mathbf{w}_1, \ldots, \mathbf{w}_{k-1}$ given by

$$\omega(\mathbf{v}, \mathbf{w}_1, \ldots, \mathbf{w}_{k-1}).$$

This function (of the $\mathbf{w}$'s) is again multilinear and antisymmetric. Hence it is an element of $\Lambda^{k-1}(V^*)$. We call this function the *interior product* of $\mathbf{v}$ and ω and denote it by

$$i(\mathbf{v})\omega.$$

Thus

$$i(\mathbf{v})\omega(w_1, \ldots, w_{k-1}) = \omega(\mathbf{v}, w_1, \ldots, w_{k-1}).$$

Let us see what the interior product looks like in various cases. Suppose we are in $\mathbb{R}^3$ and take a basis:

$$\left(\frac{\partial}{\partial x} = \begin{pmatrix}1\\0\\0\end{pmatrix}\right) \text{ and dual basis element } \mathrm{d}x = (1,0,0),$$

$$\left(\frac{\partial}{\partial y} = \begin{pmatrix}0\\1\\0\end{pmatrix}\right) \text{ and dual basis element } \mathrm{d}y = (0,1,0),$$

$$\left(\frac{\partial}{\partial z} = \begin{pmatrix}0\\0\\1\end{pmatrix}\right) \text{ and dual basis element } \mathrm{d}z = (0,0,1).$$

Then if $k = 1$, so ω is a one-form, $i(\mathbf{v})\omega$ is just $\omega(\mathbf{v})$, a number. Thus

$$i\left(\frac{\partial}{\partial x}\right)\mathrm{d}x = 1, \quad i\left(\frac{\partial}{\partial x}\right)\mathrm{d}y = 0, \quad i\left(\frac{\partial}{\partial x}\right)\mathrm{d}z = 0$$

etc.

If $k=2$, then

$$i\left(\frac{\partial}{\partial x}\right)(\mathrm{d}x\wedge\mathrm{d}y)(\mathbf{w})=\mathrm{d}x\wedge\mathrm{d}y\left(\frac{\partial}{\partial x},\mathbf{w}\right)$$

$$=\mathrm{d}x\left(\frac{\partial}{\partial x}\right)\mathrm{d}y(\mathbf{w})-\mathrm{d}y\left(\frac{\partial}{\partial x}\right)\mathrm{d}x(\mathbf{w})$$

$$=\mathrm{d}y(\mathbf{w})$$

since $\mathrm{d}y(\partial/\partial x)=0$. So

$$i\left(\frac{\partial}{\partial x}\right)\mathrm{d}x\wedge\mathrm{d}y=\mathrm{d}y.$$

Similarly

$$i\left(\frac{\partial}{\partial x}\right)\mathrm{d}x\wedge\mathrm{d}z=\mathrm{d}z$$

and

$$i\left(\frac{\partial}{\partial x}\right)\mathrm{d}y\wedge\mathrm{d}z=0.$$

Also

$$i\left(\frac{\partial}{\partial y}\right)(\mathrm{d}x\wedge\mathrm{d}y)(\mathbf{w})=\mathrm{d}x\left(\frac{\partial}{\partial y}\right)\mathrm{d}y(\mathbf{w})-\mathrm{d}x(\mathbf{w})\mathrm{d}y\left(\frac{\partial}{\partial y}\right)$$

$$=-\mathrm{d}x(\mathbf{w})$$

so

$$i\left(\frac{\partial}{\partial y}\right)\mathrm{d}x\wedge\mathrm{d}y=-\mathrm{d}x,\quad i\left(\frac{\partial}{\partial y}\right)\mathrm{d}x\wedge\mathrm{d}z=0,\quad i\left(\frac{\partial}{\partial y}\right)\mathrm{d}y\wedge\mathrm{d}z=\mathrm{d}z.$$

And similarly

$$i\left(\frac{\partial}{\partial z}\right)\mathrm{d}x\wedge\mathrm{d}y=0,\quad i\left(\frac{\partial}{\partial z}\right)\mathrm{d}x\wedge\mathrm{d}z=-\mathrm{d}x,\quad i\left(\frac{\partial}{\partial z}\right)\mathrm{d}y\wedge\mathrm{d}z=-\mathrm{d}y.$$

Also

$$i\left(\frac{\partial}{\partial x}\right)(\mathrm{d}x\wedge\mathrm{d}y\wedge\mathrm{d}z)(\mathbf{w}_1,\mathbf{w}_2)=(\mathrm{d}x\wedge\mathrm{d}y\wedge\mathrm{d}z)\left(\frac{\partial}{\partial x},\mathbf{w}_1,\mathbf{w}_2\right)$$

$$=\mathrm{Det}\begin{pmatrix}\mathrm{d}x\left(\frac{\partial}{\partial x}\right) & \mathrm{d}y\left(\frac{\partial}{\partial x}\right) & \mathrm{d}z\left(\frac{\partial}{\partial x}\right)\\ \mathrm{d}x(\mathbf{w}_1) & \mathrm{d}y(\mathbf{w}_1) & \mathrm{d}z(\mathbf{w}_1)\\ \mathrm{d}x(\mathbf{w}_2) & \mathrm{d}y(\mathbf{w}_2) & \mathrm{d}z(\mathbf{w}_2)\end{pmatrix}$$

$$=\mathrm{Det}\begin{pmatrix}1 & 0 & 0\\ \mathrm{d}x(\mathbf{w}_1) & \mathrm{d}y(\mathbf{w}_1) & \mathrm{d}z(\mathbf{w}_1)\\ \mathrm{d}x(\mathbf{w}_2) & \mathrm{d}y(\mathbf{w}_2) & \mathrm{d}z(\mathbf{w}_2)\end{pmatrix}$$

$$=\mathrm{Det}\begin{pmatrix}\mathrm{d}y(\mathbf{w}_1) & \mathrm{d}z(\mathbf{w}_1)\\ \mathrm{d}y(\mathbf{w}_2) & \mathrm{d}z(\mathbf{w}_2)\end{pmatrix}.$$

So

$$i\left(\frac{\partial}{\partial x}\right)(\mathrm{d}x \wedge \mathrm{d}y \wedge \mathrm{d}z) = \mathrm{d}y \wedge \mathrm{d}z.$$

Similarly

$$i\left(\frac{\partial}{\partial y}\right)(\mathrm{d}x \wedge \mathrm{d}y \wedge \mathrm{d}z) = -\,\mathrm{d}x \wedge \mathrm{d}z$$

and

$$i\left(\frac{\partial}{\partial z}\right)(\mathrm{d}x \wedge \mathrm{d}y \wedge \mathrm{d}z) = \mathrm{d}x \wedge \mathrm{d}y.$$

Now $i(\mathbf{v})\omega$ is clearly bilinear in $\mathbf{v}$ and ω. So, if

$$\xi = a\frac{\partial}{\partial x} + b\frac{\partial}{\partial y} + c\frac{\partial}{\partial z}$$

and

$$\omega = \rho \mathrm{d}x \wedge \mathrm{d}y \wedge \mathrm{d}z,$$

then

$$\mathrm{i}(\xi)\omega = \rho a\,\mathrm{d}y \wedge \mathrm{d}z - \rho b\,\mathrm{d}x \wedge \mathrm{d}z + \rho c\,\mathrm{d}x \wedge \mathrm{d}y.$$

We have defined the interior product $i(\xi)\omega$ for vectors, ξ, and for ω in $\Lambda^k(V^*)$. But, using this definition at each point, we can equally well define the interior product $i(\xi)\omega$ where ξ is a vector field, ω is a differential form of degree k. In $\mathbb{R}^3$, all the preceding formulas hold.

In particular, if we start with a vector field, ξ, representing the stationary flow of the electrons, and a three-form $\omega = \rho \mathrm{d}x \wedge \mathrm{d}y \wedge \mathrm{d}z$ giving (a smeared-out approximation to) the electron density, we get a two-form

$$J = i(\xi)\omega$$

giving the current. The explicit formula for J in coordinates was given by the preceding formula. In order to understand why J represents the current, we can use the basic definition of interior product. Let us look at a point p and a small parallelogram spanned by vectors $\mathbf{w}_1$ and $\mathbf{w}_2$ at p. The vector $\mathbf{v} = h\xi(p)$ represents

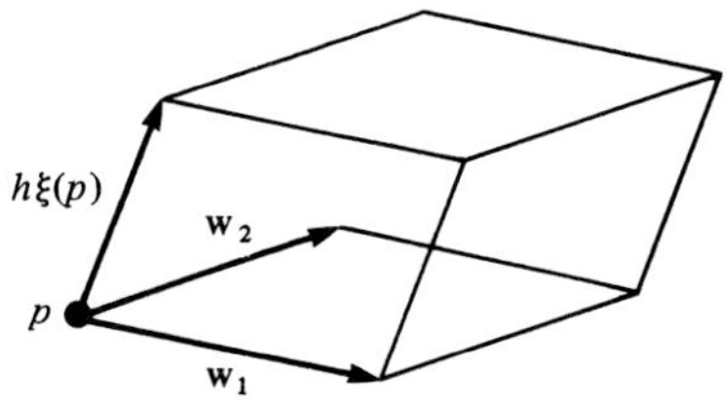

Figure 17.3

a (linear) approximation to the motion of a particle situated near p in a small time interval, h. In other words, the parallelepiped spanned by $\mathbf{v}$, $\mathbf{w}_1$ and $\mathbf{w}_2$ gives (approximately) the region swept out by the parallelogram in time h. The total charge in this parallelogram is given by

$$\begin{aligned}\omega(\mathbf{v},\mathbf{w}_1,\mathbf{w}_2) &= [i(\mathbf{v})\omega](\mathbf{w}_1,\mathbf{w}_2)\\ &= h[i(\xi(p))\omega](\mathbf{w}_1,\mathbf{w}_2)\\ &= hJ(p)(\mathbf{w}_1,\mathbf{w}_2).\end{aligned}$$

Thus $J(p)(\mathbf{w}_1,\mathbf{w}_2)$ does indeed represent the rate of flow of charge across the parallelogram spanned by $\mathbf{w}_1$ and $\mathbf{w}_2$. So J, when integrated over a surface, gives the rate of flow of charge across that surface.

Let us conclude this section by proving the following important formula concerning the interior product:

$$i(\xi)[\omega_1 \wedge \omega_2] = (i(\xi)\omega_1) \wedge \omega_2 + (-1)^{\deg \omega_1}\omega_1 \wedge i(\xi)\omega_2. \tag{17.1}$$

(A check shows that this formula is true in all the cases in $\mathbb{R}^3$ that we have computed.) In (17.1), ξ is a vector field and ω_1 and ω_2 are differential forms. But since (17.1) is purely algebraic, it is enough to prove it at each point, i.e., to prove it when $\xi = \mathbf{v}$ is a vector in V and ω_1 is an element of $\Lambda^k(V^*)$ and ω_2 an element of $\Lambda^l(V^*)$. Both sides of the expression are linear in ξ, ω_1 or ω_2 when the other two are held fixed. We may write $\xi = a_1\mathbf{v}_1 + \cdots + a_n\mathbf{v}_n$ where $\{\mathbf{v}_1,\ldots,\mathbf{v}_n\}$ is a basis. By linearity in ξ, it is enough to verify it for $\xi = \mathbf{v}_i$, a basis element. With no loss of generality, we may assume

$$\xi = \mathbf{v}_1.$$

Let us first prove the formula when ω_1 is a one-form. By linearity, we may assume that

$$\omega_1 = \mathbf{v}_i^* \quad i = 1,\ldots,n,$$

an element of the dual basis. We may also assume that ω_2 is a basis element of the form

$$\omega_2 = \mathbf{v}_{j_1}^* \wedge \cdots \wedge \mathbf{v}_{j_l}^*.$$

We can now consider various possibilities.

(a) $i = 1$, $j_1 > 1$. Then $\mathbf{v}_j^*(\mathbf{v}_1) = 0$ for all $j = j_1,\ldots,j_l$ and in the determinant expression for

$$(\omega_1 \wedge \omega_2)(\mathbf{v}_1,\mathbf{w}_1,\ldots,\mathbf{w}_l)$$

all elements in the top row vanish except the upper left-hand corner. For example, if $\omega_2 = \mathbf{v}_2^* \wedge \mathbf{v}_3^* \wedge \mathbf{v}_4^*$, then

$$(\omega_1 \wedge \omega_2)(\mathbf{v},\mathbf{w}_1,\mathbf{w}_2,\mathbf{w}_3) = \operatorname{Det}\begin{pmatrix} \mathbf{v}_1^*(\mathbf{v}_1) & \mathbf{v}_2^*(\mathbf{v}_1) & \mathbf{v}_3^*(\mathbf{v}_1) & \mathbf{v}_4^*(\mathbf{v}_1)\\ \mathbf{v}_1^*(\mathbf{w}_1) & \mathbf{v}_2^*(\mathbf{w}_1) & \mathbf{v}_3^*(\mathbf{w}_1) & \mathbf{v}_4^*(\mathbf{w}_1)\\ \mathbf{v}_1^*(\mathbf{w}_2) & \mathbf{v}_2^*(\mathbf{w}_2) & \mathbf{v}_3^*(\mathbf{w}_2) & \mathbf{v}_4^*(\mathbf{w}_2)\\ \mathbf{v}_1^*(\mathbf{w}_3) & \mathbf{v}_2^*(\mathbf{w}_3) & \mathbf{v}_3^*(\mathbf{w}_3) & \mathbf{v}_4^*(\mathbf{w}_3)\end{pmatrix}$$

$$= \operatorname{Det}\begin{pmatrix} 1 & 0 & 0 & 0\\ \mathbf{v}_1^*(\mathbf{w}_1) & \mathbf{v}_2^*(\mathbf{w}_1) & \mathbf{v}_3^*(\mathbf{w}_1) & \mathbf{v}_4^*(\mathbf{w}_1)\\ \mathbf{v}_1^*(\mathbf{w}_2) & \mathbf{v}_2^*(\mathbf{w}_2) & \mathbf{v}_3^*(\mathbf{w}_2) & \mathbf{v}_4^*(\mathbf{w}_2)\\ \mathbf{v}_1^*(\mathbf{w}_3) & \mathbf{v}_2^*(\mathbf{w}_3) & \mathbf{v}_3^*(\mathbf{w}_3) & \mathbf{v}_4^*(\mathbf{w}_3)\end{pmatrix}$$

$$= \mathrm{Det}\begin{pmatrix} \mathbf{v}_2^*(\mathbf{w}_1) & \mathbf{v}_3^*(\mathbf{w}_1) & \mathbf{v}_4^*(\mathbf{w}_1) \\ \mathbf{v}_2^*(\mathbf{w}_2) & \mathbf{v}_3^*(\mathbf{w}_2) & \mathbf{v}_4^*(\mathbf{w}_2) \\ \mathbf{v}_2^*(\mathbf{w}_3) & \mathbf{v}_3^*(\mathbf{w}_3) & \mathbf{v}_4^*(\mathbf{w}_3) \end{pmatrix}$$

$$= (\mathbf{v}_2^* \wedge \mathbf{v}_3^* \wedge \mathbf{v}_4^*)(\mathbf{w}_1, \mathbf{w}_2, \mathbf{w}_3).$$

Thus the determinant expression for $(\omega_1 \wedge \omega_2)(\mathbf{v}, \mathbf{w}_1, \ldots, \mathbf{w}_l)$ is the same as the expression for $\omega_2(\mathbf{w}_1, \ldots, \mathbf{w}_l)$. Thus

$$i(\mathbf{v})(\omega_1 \wedge \omega_2) = \omega_2 = [i(\mathbf{v})\omega_1]\omega_2 - \omega_1 \wedge i(\mathbf{v})\omega_2$$

since $i(\mathbf{v})\omega_1 = 1$ and $i(\mathbf{v})\omega_2 = 0$ in this case.

We now turn to case (b): $i = 1, j_1 = 1$. Then $\omega_1 \wedge \omega_2 = 0$. Now $i(\mathbf{v}_1)\omega_1 = 1$ so

$$(i(\mathbf{v}_1)\omega_1) \wedge \omega_2 = \omega_2,$$

while

$$i(\mathbf{v}_1)\omega_2 = \mathbf{v}_{j_2}^* \wedge \cdots \wedge \mathbf{v}_{j_l}^*$$

so

$$\begin{aligned} \omega_1 \wedge i(\mathbf{v}_1)\omega_2 &= \mathbf{v}_1^* \wedge \mathbf{v}_{j_2}^* \wedge \cdots \wedge \mathbf{v}_{j_l}^* \\ &= \omega_2. \end{aligned}$$

Thus

$$i(\mathbf{v}_1)\omega_1 \wedge \omega_2 - \omega_1 \wedge i(\mathbf{v}_1)\omega_2 = \omega_2 - \omega_2 = 0.$$

Now consider case (c) where $i > 1$ and $j_1 = 1$. Then we can write $\omega_2 = \mathbf{v}_1^* \wedge \omega_3$ and the left hand side of (17.1) can be written as

$$\begin{aligned} i(\mathbf{v})[\omega_1 \wedge \mathbf{v}_1^* \wedge \omega_3] &= -i(\mathbf{v})[\mathbf{v}_1^* \wedge (\omega_1 \wedge \omega_3)] \quad \text{(by interchanging } \mathbf{v}_1 \text{ with } \omega_1\text{)} \\ &= -\omega_1 \wedge \omega_3 \end{aligned}$$

by case (a). The first term on the right hand side of (17.1) vanishes and the second term equals $-\omega_1 \wedge \omega_3$ by another application of case (a).

Finally, if (d) $i > 1$ and $j_1 > 1$, then

$$i(\mathbf{v}_1)(\omega_1 \wedge \omega_2) = 0$$

$$i(\mathbf{v}_1)\omega_1 = 0$$

and

$$i(\mathbf{v}_1)\omega_2 = 0$$

so both sides of our equation vanish. We have now proved formula (17.1) for the case that $\deg \omega_1 = 1$. But then the associative law for exterior multiplication will allow us to prove it in general. For example, suppose that $\omega_1 = \sigma_1 \wedge \sigma_2$ where $\deg \sigma_1 = \deg \sigma_2 = 1$. Then

$$\begin{aligned} i(\mathbf{v})(\sigma_1 \wedge \sigma_2 \wedge \omega_2) &= i(\mathbf{v})\sigma_1 \wedge (\sigma_2 \wedge \omega_2) - \sigma_1 \wedge i(\mathbf{v})(\sigma_2 \wedge \omega_2) \\ &= i(\mathbf{v})\sigma_1 \wedge (\sigma_2 \wedge \omega_2) - \sigma_1 \wedge i(\mathbf{v})\sigma_2 \wedge \omega_2 + \sigma_1 \wedge \sigma_2 \wedge i(\mathbf{v})\omega_2 \\ &= [i(\mathbf{v})\sigma_1 \wedge \sigma_2 - \sigma_1 \wedge i(\mathbf{v})\sigma_2] \wedge \omega_2 + \sigma_1 \wedge \sigma_2 \wedge i(\mathbf{v})\omega_2 \\ &= i(\mathbf{v})(\sigma_1 \wedge \sigma_2) \wedge \omega_2 + (\sigma_1 \wedge \sigma_2) \wedge i(\mathbf{v})\omega_2. \end{aligned}$$

This proves the formula when $\omega_1 = \sigma_1 \wedge \sigma_2$ and hence, by linearity, for all cases with $\deg \omega_1 = 2$. We can continue to prove the general formula by induction. For example, if $\deg \sigma_1 = 1$ and $\deg \sigma_2 = 2$,

$$\begin{aligned}i(\mathbf{v})(\sigma_1 \wedge \sigma_2 \wedge \omega_2) &= i(\mathbf{v})\sigma_1 \wedge (\sigma_2 \wedge \omega_2) - \sigma_1 \wedge i(\mathbf{v})(\sigma_2 \wedge \omega_2)\\ &= i(\mathbf{v})\sigma_1 \wedge \sigma_2 \wedge \omega_2 - \sigma_1 \wedge i(\mathbf{v})\sigma_2 \wedge \omega_2 - \sigma_1 \wedge \sigma_2 \wedge i(\mathbf{v})\omega_2\\ &= i(\mathbf{v})(\sigma_1 \wedge \sigma_2) \wedge \omega_2 - \sigma_1 \wedge \sigma_2 \wedge i(\mathbf{v})\omega_2\end{aligned}$$

so

$$i(\mathbf{v})\omega_1 \wedge \omega_2 = i(\mathbf{v})\omega_1 \wedge \omega_2 - \omega_1 \wedge i(\mathbf{v})\omega_2$$

if $\deg \omega_1 = 3$ and so on.

17.4. Lie derivatives

Let $\phi: I \times U \to \mathbb{R}^k$ be as in section 17.2. Thus we have, for each $t \in I$, the map

$$\phi_t: U \to \mathbb{R}^k, \quad \phi_t(\mathbf{p}) = \phi(t, \mathbf{p})$$

and we assume that

$$\phi_0 = \mathrm{id}.$$

Let $\xi = \xi_0$ be the vector field associated to ϕ_t at $t = 0$.

Let ω be a differential form of degree k. We can consider the form $\phi_t^*\omega$ and hence the form

$$\frac{1}{t}(\phi_t^*\omega - \omega)$$

for $t \neq 0$. We claim that the limit of this expression exists as $t = 0$. Indeed, if $\omega = f$ is a function, this limit exists and is just the expression

$$\mathrm{D}_\xi f = \lim_{t \to 0} \frac{1}{t}(\phi^* f - f)$$

as we have seen. If $\omega = \mathrm{d}f$,

$$\phi_t^* \mathrm{d}f = \mathrm{d}\phi_t^* f$$

so

$$\frac{1}{t}(\phi_t^* \mathrm{d}f - \mathrm{d}f) = \mathrm{d}\left[\frac{1}{t}(\phi_t^* f - f)\right].$$

Interchanging the limits involved in the partial derivatives expressing d and D_ξ is legitimate and so we see that

$$\lim_{t \to 0} \frac{1}{t}(\phi_t^* \mathrm{d}f - \mathrm{d}f) = \mathrm{d}D_\xi f.$$

Now the most general linear differential form is a sum of expressions of the form $f\,\mathrm{d}g$ and

$$\begin{aligned}\phi_t^*(f\,\mathrm{d}g) - f\,\mathrm{d}g &= (\phi_t^* f)\phi_t^* \mathrm{d}g - f\,\mathrm{d}g\\ &= \phi_t^* f\, \phi_t^* \mathrm{d}g - f\,\phi_t^* \mathrm{d}g + f(\phi_t^* \mathrm{d}g)\end{aligned}$$

so

$$\lim_{t \to 0} \frac{1}{t}(\phi_t^* f\,\mathrm{d}g - f\,\mathrm{d}g) = (\mathrm{D}_\xi f)\mathrm{d}g + f\,\mathrm{d}(\mathrm{D}_\xi g).$$

Thus $\lim_{t=0}(1/t)(\phi^*\omega - \omega)$ exists for any one-form. Furthermore, this limit can be expressed in terms of ξ and ω. We shall denote this limit by $D_\xi\omega$. Now if

$$\omega = \omega_1 \wedge \omega_2$$

where ω_1 and ω_2 are one-forms we have

$$\begin{aligned}\phi_t^*(\omega_1 \wedge \omega_2) - \omega_1 \wedge \omega_2 &= \phi_t^*\omega_1 \wedge \phi_t^*\omega_2 - \omega_1 \wedge \omega_2\\ &= (\phi_t^*\omega_1 - \omega_1) \wedge \phi_t^*\omega_2 + \omega_1 \wedge (\phi_t^*\omega_2 - \omega_2).\end{aligned}$$

Dividing by t and passing to the limit, we see that

$$\lim_{t\to 0}\frac{1}{t}(\phi_t^*(\omega_1 \wedge \omega_2) - \omega_1 \wedge \omega_2)$$

exists and equals

$$D_\xi\omega_1 \wedge \omega_2 + \omega_1 \wedge D_\xi\omega_2.$$

As every two-form can be written as a sum of terms like $\omega_1 \wedge \omega_2$, we see that

$$\lim_{t\to 0}\frac{1}{t}(\phi^*\omega - \omega)$$

exists for any two-form ω. We may call this limit $D_\xi\omega$, as the limit depends only on ξ and ω. Proceeding in this way, we see that this limit exists for all ω and depends only on ξ and ω. We denote it by $D_\xi\omega$. It is called the *Lie derivative* of ω with respect to ξ. The proof shows that, for any forms ω_1 and ω_2, we have

$$\boxed{D_\xi(\omega_1 \wedge \omega_2) = D_\xi\omega_1 \wedge \omega_2 + \omega_1 \wedge D_\xi\omega_2.}$$

The definition, and the fact that $d\phi^* = \phi^* d$, shows that

$$\boxed{dD_\xi\omega = D_\xi d\omega.}$$

There is a formula which relates the interior product, $i(\xi)$, the exterior derivative, d, and the Lie derivative, D. Before stating the formula, we make an observation. Recall that d maps l-forms into $(l+1)$-forms and satisfies

$$d(\omega_1 \wedge \omega_2) = d\omega_1 \wedge \omega_2 + (-1)^k\omega_1 \wedge d\omega_2,$$

if degree $\omega_1 = k$. also $i(\xi)$ maps l-forms into $(l-1)$-forms and satisfies

$$i(\xi)(\omega_1 \wedge \omega_2) = i(\xi)\omega_1 \wedge \omega_2 + (-1)^k\omega_1 \wedge i(\xi)\omega_2.$$

Therefore, $d \circ i(\xi)$ maps l-forms into l-forms and satisfies

$$\begin{aligned}[d\circ i(\xi)](\omega_1 \wedge \omega_2) &= d(i(\xi)\omega_1 \wedge \omega_2 + (-1)^k\omega_1 \wedge i(\xi)\omega_2)\\ &= di(\xi)\omega_1 \wedge \omega_2 + (-1)^{k-1}i(\xi)\omega_1 \wedge d\omega_2\\ &\quad + (-1)^k d\omega_1 \wedge i(\xi)\omega_2 + (-1)^{2k}\omega_1 \wedge di(\xi)\omega_2.\end{aligned}$$

Now $(-1)^{2k} = 1$. Also we have the corresponding formula

$$\begin{aligned}[i(\xi)\circ d](\omega_1 \wedge \omega_2) &= i(\xi)d\omega_1 \wedge \omega_2 + (-1)^{k+1}d\omega_1 \wedge i(\xi)\omega_2\\ &\quad + (-1)^k i(\xi)\omega_1 \wedge d\omega_2 + \omega_1 \wedge i(\xi)d\omega_2.\end{aligned}$$

If we add these two expressions, the middle terms cancel, and we have the *derivation* or *Leibnitz rule* property

$$[i(\xi)\circ d + d\circ i(\xi)](\omega_1 \wedge \omega_2) = [i(\xi)d + di(\xi)]\omega_1 \wedge \omega_2 + \omega_1 \wedge [i(\xi)d + di(\xi)]\omega_2.$$

We can now prove the following fundamental formula of differential calculus:

$$D_\xi = i(\xi)d + di(\xi). \tag{17.2}$$

Proof. Both sides of this equation are operators which satisfy the Leibnitz identity when acting on products. Every form is a sum of products of functions f and differentials dg. So we need only to verify the identity

$$D_\xi\omega = i(\xi)d\omega + di(\xi)\omega$$

for $\omega = f$ and for $\omega = df$. For the case $\omega = f$, $i(\xi)f = 0$ by convention. (There are no (-1)-forms.) The formula reduces to

$$D_\xi f = i(\xi)df = df(\xi),$$

which we know from section 17.2. For the case $\omega = dg$, we have

$$\begin{aligned} D_\xi dg &= dD_\xi g \\ &= d(i(\xi)dg) \\ &= (i(\xi)d + di(\xi))dg \end{aligned}$$

since $d^2 = 0$. So the formula is true in general. We will give an alternative proof of a generalization of this formula in the appendix to this chapter.

For example, suppose

$$\omega = \rho\, dx \wedge dy \wedge dz$$

in $\mathbb{R}^3$. Then $d\omega = 0$, since there are no non-zero four-forms in three dimensions. Thus

$$D_\xi\omega = di(\xi)\omega.$$

If we set

$$J = i(\xi)\omega,$$

then the condition

$$dJ = 0$$

is equivalent to

$$D_\xi\omega = 0.$$

This is the infinitesimal way of asserting that $\phi_t^*\omega = \omega$ for all t. Thus *Kirchhoff's current law* can be formulated as, say, that the flow ϕ_t preserves the charge density ω in the sense that $D_\xi\omega = 0$.

17.5. Magnetism

In a very primitive form, certain manifestations of the phenomena of electricity and magnetism were known to the ancients. It is one of the wonders of history that the investigation of these two obscure effects, by scientists of the seventeenth and eighteenth centuries, led in the nineteenth century to the understanding of the fundamental role that electromagnetic forces have in nature and to the revolutionary change in society brought about by electrical technology.

Amber, a fossilized tree resin, was known to the ancients to have the property that, when rubbed by cloth, it could make small bits of light material jump up and stick to it. The Greek word for amber is 'elektron'. Gilbert, in his book *On the Magnet*, published in London in 1600, introduced the term 'electricity' to mean the property of attracting like amber. He devoted a whole long chapter to amber 'to show the nature of the attachment of bodies to it and to point out the vast difference between this and the magnetic actions' – to distinguish between the 'pure bond of sympathy uniting iron and lodestones from the promiscuous behavior of amber'. The ancients had regarded the attractions of amber and of the lodestone as similar phenomena giving a similar explanation for both. Cardano (1550) had recognized a number of distinctions between electric and magnetic attractions. But Gilbert discovered that the electrostatic attractive property of amber could be reproduced in a number of other hard substances. Of course, the electrostatic attraction of a charged body for a light neutral body involving induced polarization and stronger attraction for the nearer opposite charge – is a difficult phenomenon both for precise theoretical computation and for experiment. The light bits of paper fly to the amber in a jerky irregular manner, stick to it, then fall off after a while, certainly 'promiscuous' behavior when compared to the steady, regular attraction between magnets and iron or magnets for one another. In fact, it took nearly two more centuries to realize that the fundamental attraction and repulsion to be studied in electrostatics should be between charged bodies and to discover the correct law of force. This was done in careful experiments with the torsion balance by Coulomb in the 1780s.

Ancient Greek writings referred to certain stones that had the property of attracting iron. A large such stone near the city of Magnesia in Asia Minor was reported to pull at the iron tips of shepherd's staffs and the nails in their shoes. From this city's name, we get the term *magnet*. It was discovered that this stone could impart a similar attractive quality to an iron needle. The needle would acquire a directional quality or 'load' from which the term lodestone was derived. The attractive and repulsive forces between magnetic poles were also measured by Coulomb with his torsion balance. (He once again found the inverse square law.) However, our true understanding of the nature of magnetic forces had to await the work of Ampère. It turned out that the correct object of study was not the attraction of magnets for iron or of magnets for one another. Rather it was the force that a magnet exerts on a moving charge (or, better yet, the force that one moving charge exerts on another).

In the spring of 1820 while delivering a lecture on electric currents, Oersted noticed that a compass needle, accidently placed nearby, was deflected by the current. The announcement of this result – that an electric current has magnetic effects – astounded the scientific world. Electricity and magnetism, separated for over two centuries, once again became related. Ampère realized that, if a current deflects a magnet, then a magnet should exert a force on a current, and two currents should exert a force on one another. In a few breathless weeks, from September 18 to October 9, 1820, Ampère conducted a series of brilliant experiments and geometric deductions. In some sense, the notions of differential forms and line integrals arose in Ampère's mathematical analysis. In any event, Ampère discovered the basic laws of magnetostatics.

A magnetic field exerts a force on a current. We can think of a current as moving charges. Thus a magnetic field produces a force on charges. But this force depends on the velocity of the charge, in the following way. The *magnetic induction* is a two-form, B, in $\mathbb{R}^3$. Suppose that the charge e moves past the point $\mathbf{p}$ with velocity $\mathbf{v}$. Then the force is given by

$$ei(\mathbf{v})B_{\mathbf{p}}.$$

In this expression B is a two-form, so $B_{\mathbf{p}}$, the value of B at $\mathbf{p}$, is an element of $\Lambda^2(\mathbb{R}^3)$. Thus $i(\mathbf{v})B_{\mathbf{p}}$ is a linear function on tangent vectors at $\mathbf{p}$ – and that is what we have been calling a force field, cf. page 248 of volume 1.

Thus, if a particle with charge e moves along a curve γ, then at time t it will be subject to the force

$$ei(\gamma'(t))B_{\gamma(t)}.$$

It follows immediately from the definition of the interior product that

$$i(\mathbf{v})i(\mathbf{w})\omega = -i(\mathbf{w})i(\mathbf{v})\omega$$

for any $\mathbf{v}$, $\mathbf{w}$ and ω. In particular,

$$i(\mathbf{v})i(\mathbf{v})\omega = 0,$$

or, if $\omega \in \Lambda^2(V^*)$, then

$$[i(\mathbf{v})\omega](\mathbf{v}) = 0.$$

This shows that there is no component of force along the direction of motion. Thus the force on a conducting wire carrying a current will be perpendicular to the wire. (This was verified by Ampère in a clever experiment involving a movable wire connecting two basins of mercury through which a current flowed.)

At each point, the form B, if $\neq 0$, will determine a direction in space – the line given by the equation

$$i(\mathbf{w})B = 0.$$

(You should convince yourself that this really is a pair of linearly independent equations for $\mathbf{w}$.) Iron filings placed in the magnetic field (free to rotate but not to move) will align themselves in these directions, producing the *magnetic lines of force*. (These are precisely the directions in which a current will feel no force.)

Ampère not only discovered the force that a magnetic field produces on a current; he also deduced the quantitative formulation of the fact (first discovered by Oersted) that a current produces a magnetic field. For this it is useful to introduce a *one-form*

H which, following Sommerfeld, we will call the *magnetic excitation*. The second of Ampère's laws then says that, if S is any surface bounded by a curve $\gamma = \partial S$, then

$$\int_\gamma H = 4\pi \int_S J.$$

In other words, the integral of the magnetic excitation around any closed curve bounding a surface equals the flow of current through that surface.

By Stokes' theorem, this is the same as

$$\mathrm{d}H = 4\pi J.$$

In contrast to electrostatics, we can formulate the next law of magnetostatics as

$$\int_S B = 0 \text{ for any closed surface, } S.$$

('There are no magnetic poles' to quote Hertz.) Finally, we need a law relating B to H. In non-ferromagnetic materials, it is given by

$$B = \mu \star H$$

where μ is called the *permeability*. To summarize the laws of magnetostatics,

(i) *the force on a current element* $\mathbf{I} = e\mathbf{v}$ *at a point* $\mathbf{p}$ *is given by* $i(\mathbf{I})B_{\mathbf{p}}$;
(ii) $\int_S B = 0$ *for a closed surface* S; *so*

$$\mathrm{d}B = 0;$$

(iii) $\int_{\partial S} H = 4\pi \int_S J$ *or*

$$\mathrm{d}H = 4\pi J \quad \text{Ampère's law}$$

and
(iv) $B = \mu \star H.$

Suppose that we are in a region U where $H^2(U) = 0$. Then the condition $\mathrm{d}B = 0$ implies that

$$B = \mathrm{d}A$$

for some one-form, A. Of course

$$A = A_x \mathrm{d}x + A_y \mathrm{d}y + A_z \mathrm{d}z$$

is not determined by the equation $B = \mathrm{d}A$. For any function, ψ, $A + \mathrm{d}\psi$ will satisfy the same equation. Let us choose ψ so that

$$\frac{\partial^2 \psi}{\partial x^2} + \frac{\partial^2 \psi}{\partial y^2} + \frac{\partial^2 \psi}{\partial z^2} = -\left(\frac{\partial A_x}{\partial x} + \frac{\partial A_y}{\partial y} + \frac{\partial A_z}{\partial z}\right).$$

Once we have chosen some A, this amounts to solving the Poisson equation*. Then,

* The following computations take on a cleaner form once we introduce the general form of the $\star$ operator. See the next chapter.

replacing A by $A + d\psi$, we have arranged that

$$\frac{\partial A_x}{\partial x} + \frac{\partial A_y}{\partial y} + \frac{\partial A_z}{\partial z} = 0.$$

Now

$$H = \frac{1}{\mu}\star^{-1}B = \frac{1}{\mu}\star^{-1}\mathrm{d}A$$

and

$$\mathrm{d}H = J.$$

Suppose we are in a region where μ is constant. Then

$$\boxed{\mathrm{d}\star^{-1}\mathrm{d}A = \mu J.}$$

Now

$$\mathrm{d}(A_x\mathrm{d}x) = -\frac{\partial A_x}{\partial y}\mathrm{d}x \wedge \mathrm{d}y + \frac{\partial A_x}{\partial z}\mathrm{d}z \wedge \mathrm{d}x$$

so

$$\star^{-1}\mathrm{d}(A_x\mathrm{d}x) = -\frac{\partial A_x}{\partial y}\mathrm{d}z + \frac{\partial A_x}{\partial z}\mathrm{d}y$$

and

$$(\mathrm{d}\star^{-1}\mathrm{d})(A_x\mathrm{d}x) = -\frac{\partial^2 A_x}{\partial x\,\partial y}\mathrm{d}x \wedge \mathrm{d}z - \frac{\partial^2 A_x}{\partial y^2}\mathrm{d}y \wedge \mathrm{d}z + \frac{\partial^2 A_x}{\partial z\,\partial x}\mathrm{d}x \wedge \mathrm{d}y - \frac{\partial^2 A_x}{\partial z^2}\mathrm{d}y \wedge \mathrm{d}z.$$

Doing the same computation for $A_y\mathrm{d}y$ and $A_z\mathrm{d}z$ and adding gives

$$(\mathrm{d}\star^{-1}\mathrm{d})A = -(\Delta A_x\mathrm{d}y \wedge \mathrm{d}z + \Delta A_y\mathrm{d}z \wedge \mathrm{d}x + \Delta A_z\mathrm{d}x \wedge \mathrm{d}y)$$

where

$$\Delta = \frac{\partial^2}{\partial x^2} + \frac{\partial^2}{\partial y^2} + \frac{\partial^2}{\partial z^2}$$

is the usual Laplace operator. Then, if we write

$$J = J_x\mathrm{d}y \wedge \mathrm{d}z + J_y\mathrm{d}z \wedge \mathrm{d}x + J_z\mathrm{d}x \wedge \mathrm{d}y$$

the equations

$$\boxed{\begin{aligned} \mathrm{d}A &= B, \\ B &= \mu\star H, \\ \mathrm{d}H &= 4\pi J \end{aligned}}$$

reduce to solving three Poisson equations

$$\begin{aligned} \Delta A_x &= -4\pi J_x, \\ \Delta A_y &= -4\pi J_y, \\ \Delta A_z &= -4\pi J_z. \end{aligned}$$

For example, suppose we wish to determine B and H in the exterior of a long, thin, straight wire carrying a steady current I. We may assume that the wire is of radius a and is along the z-axis. So $J_x = J_y = 0$ and

$$J_z = \begin{cases} 0 & \text{outside the wire,} \\ \dfrac{I}{\pi a^2} & \text{inside the wire.} \end{cases}$$

Then $A_x = A_y = 0$ and we solve to get

$$A_z = -2\mu I \log \sqrt{(x^2 + y^2)}$$

for points outside the wire. Thus

$$\begin{aligned} B = \mathrm{d}A &= -\frac{\partial A_z}{\partial x}\mathrm{d}x \wedge \mathrm{d}z - \frac{\partial A_z}{\partial y}\mathrm{d}y \wedge \mathrm{d}z \\ &= 2\mu I\left(\frac{x}{r^2}\mathrm{d}x \wedge \mathrm{d}z + \frac{y}{r^2}\mathrm{d}y \wedge \mathrm{d}z\right) \\ &= \frac{2\mu I}{r}\mathrm{d}r \wedge \mathrm{d}z \end{aligned}$$

in terms of cylindrical coordinates

$$x = r\cos\theta,\ r = \sqrt{(x^2 + y^2)},$$
$$y = r\sin\theta.$$
$$z.$$

The lines of force are circles centered at the wire, and $H = 1/\mu\star^{-1}B$ is proportional to $\mathrm{d}\theta$. If we knew this fact in advance, i.e., that

$$H = f\,\mathrm{d}\theta,$$

then we could conclude from symmetry that $f = f(r)$ and then from

$$\int_\gamma H = 4\pi I$$

we could conclude that

$$f(r) = \frac{2I}{r}.$$

Since $\mathrm{d}\theta/r = \mathrm{d}r \wedge \mathrm{d}z$ we would then know that

$$B = \mu\star H = 2\mu T\,\mathrm{d}r \wedge \mathrm{d}z.$$

More generally, the solution of the Poisson equation shows that A is given by the volume integral

$$A_x(\mathbf{v}) = \int \frac{1}{\|\mathbf{v} - \mathbf{w}\|} j_x(\mathbf{w})\mathrm{d}^3\mathbf{w}$$

with similar formulas for A_y and A_z.

Appendix: an alternative proof of the fundamental formula of differential calculus

We give an alternative proof of the basic formula (17.2). In fact, we will prove it in somewhat greater generality. Let $W \subset \mathbb{R}^m$ and $Z \subset \mathbb{R}^n$ be open sets. Suppose that we are given a differentiable map $\phi: W \times I \to Z$, where I is some interval in $\mathbb{R}$ containing the origin. For each fixed $\mathbf{w} \in W$, we get a curve, $t \rightsquigarrow \phi(\mathbf{w}, t)$. We let $\xi_t(\mathbf{w})$ denote the tangent vector to this curve at $\phi(\mathbf{w}, t)$. It is a tangent vector in the image space, $\mathbb{R}^n$, but depending on t and $\mathbf{w} \in W$. We also let $\phi_t: W \to Z$ be the map given by

$$\phi_t(\mathbf{w}) = \phi(\mathbf{w}, t).$$

We think if ϕ_t as a one-parameter family of maps of W into Z.

Let σ be a differential $(k+1)$-form on Z. For each t let us define the k-form $\phi_t^* i(\xi_t)\sigma$ on W by the formula

$$[\phi_t^* i(\xi_t)\sigma](\eta_1, \ldots, \eta_k) = \sigma(\xi_t(\mathbf{w}), \mathrm{d}\phi_t \eta_1, \ldots, \mathrm{d}\phi_t \eta_k),$$

where $\eta_1, \ldots, \eta_k$ are tangent vectors at $\mathbf{w}$. All $k+1$ vectors occurring inside the σ on the right of this equation are tangent vectors at $\phi_t(\mathbf{w})$. So the right-hand side makes sense. We take it as the definition of the left-hand side. In generalization of (17.2) we wish to prove the following:

> Let σ_t be a smooth one-parameter family of forms on Z. Then $\phi_t^* \sigma_t$ is a smooth family of forms on W and the basic formula of the differential calculus of forms asserts that

$$\frac{\mathrm{d}}{\mathrm{d}t} \phi_t^* \sigma_t = \phi_t^* \frac{\mathrm{d}\sigma_t}{\mathrm{d}t} + \phi_t^*(i(\xi_t)\mathrm{d}\sigma_t) + \mathrm{d}\phi_t^* i(\xi_t)\sigma_t. \tag{17.3}$$

We first prove the formula in the special case where $W = Z = M \times I$ where M is an open subset of $\mathbb{R}^m$ and ϕ_t is the map $\psi_t: M \times I \to M \times I$ given by

$$\psi_t(\mathbf{x}, s) = (\mathbf{x}, s + t).$$

The most general differential form on $M \times I$ can be written as

$$\mathrm{d}s \wedge a \quad + \quad b,$$

where a and b are forms on M that may depend on t and s. (In terms of local coordinates, $s, x^1, \ldots, x^n$, these forms are sums of terms that look like

$$c\,\mathrm{d}x^{i_1} \wedge \cdots \wedge \mathrm{d}x^{i_k},$$

where c is a function of t, s, and $\mathbf{x}$.) To show the dependence on $\mathbf{x}$ and s we shall rewrite the above expression as

$$\sigma_t = \mathrm{d}s \wedge a(\mathbf{x}, s, t)\,\mathrm{d}x + b(\mathbf{x}, s, t)\,\mathrm{d}x.$$

With this notation it is clear that

$$\psi_t^* \sigma_t = \mathrm{d}s \wedge a(\mathbf{x}, s+t, t)\,\mathrm{d}x + b(\mathbf{x}, s+t, t)\,\mathrm{d}x$$

and therefore

$$\frac{\mathrm{d}\psi_t^* \sigma_t}{\mathrm{d}t} = \mathrm{d}s \wedge \frac{\partial a}{\partial s}(\mathbf{x}, s+t, t)\,\mathrm{d}x + \frac{\partial b}{\partial s}(\mathbf{x}, s+t, t)\,\mathrm{d}x$$

$$+ \mathrm{d}s \wedge \frac{\partial a}{\partial t}(\mathbf{x}, s+t, t)\,\mathrm{d}x + \frac{\partial b}{\partial t}(\mathbf{x}, s+t, t)\,\mathrm{d}x,$$

so that

$$\frac{\mathrm{d}\psi_t^*\sigma_t}{\mathrm{d}t} - \psi_t^*\left(\frac{\mathrm{d}\sigma_t}{\mathrm{d}t}\right) = \mathrm{d}s \wedge \frac{\partial a}{\partial s}(\mathbf{x}, s+t, t)\,\mathrm{d}x + \frac{\partial b}{\partial s}(\mathbf{x}, s+t, t)\mathrm{d}x. \tag{17.4}$$

It is also clear that in this case the tangent to $\psi_t(\mathbf{x}, s)$ is $\partial/\partial s$ evaluated at $(\mathbf{x}, s+t)$.

In this case $\partial/\partial s$ is a vector field and

$$i\left(\frac{\partial}{\partial s}\right)\sigma_t = a\,\mathrm{d}x$$

so

$$\psi_t^*\left(i\left(\frac{\partial}{\partial s}\right)\sigma_t\right) = a(\mathbf{x}, s+t, t)\,\mathrm{d}x$$

and therefore

$$\mathrm{d}\psi_t^*\left(i\left(\frac{\partial}{\partial s}\right)\sigma_t\right) = \frac{\partial a}{\partial s}(\mathbf{x}, s+t, t)\,\mathrm{d}s \wedge \mathrm{d}x + \mathrm{d}_x a(\mathbf{x}, s+t, t)\,\mathrm{d}x \tag{17.5}$$

(where d_x denotes the exterior derivative of the form $a(\mathbf{x}, s+t, t)\mathrm{d}x$ on M, holding s fixed). Similarly,

$$\mathrm{d}\sigma_t = -\,\mathrm{d}s \wedge \mathrm{d}_x a\,\mathrm{d}x + \frac{\partial b}{\partial s}\mathrm{d}s \wedge \mathrm{d}x + \mathrm{d}_x b\,\mathrm{d}x$$

so

$$i\left(\frac{\partial}{\partial s}\right)\mathrm{d}\sigma_t = -\,\mathrm{d}_x a\,\mathrm{d}x + \frac{\partial b}{\partial s}\mathrm{d}x$$

and

$$\psi_t^* i\left(\frac{\partial}{\partial s}\right)\mathrm{d}\sigma_t = -\,\mathrm{d}_x a(\mathbf{x}, s+t, t)\,\mathrm{d}x + \frac{\partial b}{\partial s}(\mathbf{x}, s+t, t)\,\mathrm{d}x. \tag{17.6}$$

Adding equations (17.4)–(17.6) proves (17.3) for ψ_t.

Now let ϕ: $W \times I \to Z$ be given by $\phi(\mathbf{w}, s) = \phi_s(\mathbf{w})$. Then the image under ϕ of the lines parallel to I through $\mathbf{w}$ in $W \times I$ are just the curves $\phi_s(\mathbf{w})$ in Z. In other words

$$\mathrm{d}\phi\left(\frac{\partial}{\partial s}\right)_{(\mathbf{w},t)} = \zeta_t(\mathbf{w}).$$

If we let $\iota: W \to W \times I$ be given by $\iota(\mathbf{w}) = (\mathbf{w}, 0)$, then we can write the map ϕ_t as $\phi \circ \psi_t \circ \iota$. Thus $\phi_t^*\sigma_t = \iota^*\psi_t^*\phi^*\sigma_t$ and, since ι and ϕ do not vary with t,

$$\frac{\mathrm{d}}{\mathrm{d}t}\phi_t^*\sigma_t = \iota^*\frac{\mathrm{d}}{\mathrm{d}t}\psi_t^*(\phi^*\sigma_t).$$

At the point w, t of $W \times I$, we have

$$i\left(\frac{\partial}{\partial s}\right)\phi^*\sigma_t = (\mathrm{d}\phi)^*\left\{\left(i\left(\mathrm{d}\phi\frac{\partial}{\partial s}\right)\sigma_t\right)\right\} = (\mathrm{d}\phi)^* i(\xi_t)\sigma_t$$

and thus

$$\iota^*\psi_t^*\left(i\left(\frac{\partial}{\partial s}\right)\phi^*\sigma_t\right) = \iota^*\psi_t^*\phi^*(i(\xi_t)\sigma_t) = \phi_t^*(i(\xi_t)\sigma_t).$$

Substituting into the formula for $(\mathrm{d}/\mathrm{d}t)[\psi_t^*\phi^*\sigma_t]$ yields (17.3).

Summary

A Differential forms, vector fields and magnetism

Given a vector field ξ and a differential form ω, you should know how to define and compute the Lie derivative $D_\xi \omega$.

You should be able to define and compute the interior product of a vector and a differential form and use it to formulate the magnetic force law.

You should know the formulation of the laws of magnetostatics, and the technique for calculating B for a given current J, in terms of differential forms.

Exercises

17.1. Let $\{\mathbf{e}_1, \mathbf{e}_2, \mathbf{e}_3\}$ be a basis for $\mathbb{R}^3$ so that

$$\mathrm{d}x[\mathbf{e}_1] = \mathrm{d}y[\mathbf{e}_2] = \mathrm{d}z[\mathbf{e}_3] = 1.$$

Let $\xi = x\mathbf{e}_2 - y\mathbf{e}_1$. This vector field ξ describes the velocity associated with uniform rotation about the z-axis. We can also denote it by $\xi = x(\partial/\partial y) - y(\partial/\partial x)$.

(a) For uniform charge density, $\omega = \rho_0 \, \mathrm{d}x \wedge \mathrm{d}y \wedge \mathrm{d}z$. Using the relation $J = i(\xi)\omega$, determine the current associated with a rotating uniform distribution of charge.

(b) Let $\omega = z \, \mathrm{d}x \wedge \mathrm{d}y$. Confirm explicitly the identities $D_\xi \omega = i(\xi)\mathrm{d}\omega + \mathrm{d}i(\xi)\omega$ and $\mathrm{d}D_\xi \omega = D_\xi \mathrm{d}\omega$, where $\xi = x(\partial/\partial y) - y(\partial/\partial x)$.

17.2. Suppose current I per unit length flows axially outward from the z-axis in such a way that the current passing through the wall of a cylinder of radius r centered on the z-axis is independent of r. In this case the velocity of charge carriers is described by

$$\xi = \frac{x\mathbf{e}_1 + y\mathbf{e}_2}{x^2 + y^2} \quad \text{or} \quad \xi = \frac{1}{x^2 + y^2}\left(x\frac{\partial}{\partial x} + y\frac{\partial}{\partial y}\right).$$

Calculate the two-form J using $J = i(\xi)\,\mathrm{d}x \wedge \mathrm{d}y \wedge \mathrm{d}z$ and confirm that $\mathrm{d}J = 0$.

17.3. Suppose that $B = B_0 \mathrm{d}z$, where B_0 is a constant. Find A satisfying the conditions

$$B = \mathrm{d}A \quad \text{and} \quad \frac{\partial A_x}{\partial x} + \frac{\partial A_y}{\partial y} + \frac{\partial A_z}{\partial z} = 0.$$

17.4. Suppose that within a wire of radius a there is an axially symmetric current J whose magnitude is proportional to r^2. Explicitly, $j_x = j_y = 0$ and

$$j_z = \begin{cases} 0 & \text{outside the wire,} \\ \dfrac{2I(x^2 + y^2)}{\pi a^4} & \text{inside the wire.} \end{cases}$$

(a) Find a function A_z, depending only on $r = \sqrt{(x^2 + y^2)}$, such that

$$\Delta A_z = -4\pi j_z.$$

(b) Calculate B and H, in both cylindrical and Cartesian coordinates. Confirm that $dH = 4\pi J$.

17.5. A very long cylinder of radius a, centered on the z-axis, has a uniform charge density ρ_0. Outside the cylinder there is no charge. The cylinder is set into rotation about its axis with angular velocity ω, so that the magnitude of J is $\omega r \rho_0$ for $r < a$.

(a) Express J in cylindrical coordinates and in Cartesian coordinates. (See Exercise 17.1.)

(b) By symmetry, B is of the form $B = f(r)\,dx \wedge dy$, and $H = 1/\mu f(r)\,dz$. Determine H so that $dH = 4\pi J$.

(c) Determine A both inside and outside the rotating cylinder. In cylindrical coordinates, A will be of the form $g(r)\,d\theta$.

18

The star operator

In this chapter, we discuss the star operator in general. This will allow us to formulate Maxwell's equation in the next chapter and to understand the relation between Maxwell's equations and the geometry of spacetime. It will also explain the various vector calculus operators and identities.

18.1. Scalar products and exterior algebra

Let V be a finite-dimensional vector space equipped with a non-degenerate scalar product. We do not assume that this scalar product, (,), is positive-definite. However, it might be good to keep the positive-definite case in mind to have an intuitive idea of what is going on. What we wish to explain in this section is that (,) induces a scalar product on each of the spaces $\Lambda^k(V^*)$. The idea is that the notion of length of a line segment implies an area for parallelograms, a volume for parallelepipeds, etc.

We begin by pointing out that a scalar product on V induces a scalar product on V^*. Indeed, any scalar product on V induces a map $L: V \to V^*$ given by

$$L(\mathbf{v})(\mathbf{w}) = (\mathbf{v}, \mathbf{w}) \quad \mathbf{v}, \mathbf{w} \in V.$$

In this equation $L(\mathbf{v})$ is an element of V^*, i.e., a function on V, and so may be evaluated on any $\mathbf{w} \in V$. The map L is defined so that this evaluation yields $(\mathbf{v}, \mathbf{w})$. Now if (,) is non-degenerate, then $(\mathbf{v}, \mathbf{w}) = 0$ for all $\mathbf{w}$ implies $\mathbf{v} = 0$. Hence $L(\mathbf{v})(\mathbf{w}) = 0$ for all $\mathbf{w}$ implies $\mathbf{v} = 0$ so $L(\mathbf{v}) = 0$ implies $\mathbf{v} = 0$. Since $\dim V^* = \dim V$, we conclude that (,) is nonsingular if and only if L is an isomorphism. Since we are assuming that (,) is non-degenerate, we conclude that L is an isomorphism. We can now define a scalar product on V^* by

$$(\alpha, \beta)V^* = (L^{-1}\alpha, L^{-1}\beta)V^* \quad \alpha, \beta \in V^*.$$

For example, suppose that $V = \mathbb{R}^n$ consists of column vectors, so $V^* = \mathbb{R}^{n*}$ consists of row vectors. The general scalar product on $\mathbb{R}^n$ is given by a symmetric matrix, S. Suppose S is diagonal, so

$$(\mathbf{v}, \mathbf{w})V^* = \sum s_i x_i y_i$$

if

$$\mathbf{v} = \begin{pmatrix} x_1 \\ \vdots \\ x_n \end{pmatrix} \quad \text{and} \quad \mathbf{w} = \begin{pmatrix} y_1 \\ \vdots \\ y_n \end{pmatrix}.$$

Then

$$L\mathbf{v} = (s_1 x_1, \ldots, s_n x_n).$$

Thus, if

$$\alpha = (\alpha_1, \ldots, \alpha_n),$$
$$\beta = (\beta_1, \ldots, \beta_n),$$

then

$$L^{-1}\alpha = \begin{pmatrix} \alpha_1/s_1 \\ \vdots \\ \alpha_n/s_n \end{pmatrix}, \quad L^{-1}\beta = \begin{pmatrix} \beta_1/s_1 \\ \vdots \\ \beta_n/s_n \end{pmatrix}$$

so

$$(\alpha, \beta)V^* = \sum \frac{1}{s_i} \alpha_i \beta_i.$$

Notice that the s_i now occur in the denominator. We must assume that no $s_i = 0$. This is the non-degeneracy assumption on (,).

If (,) is positive-definite and $\{\mathbf{e}_1, \ldots, \mathbf{e}_n\}$ is an orthonormal basis of V, let $\{\mathbf{e}_1^*, \ldots, \mathbf{e}_n^*\}$ be the dual basis. An examination of the definition of L will show that

$$L(\mathbf{e}_i) = \mathbf{e}_i^*$$

and hence that $\mathbf{e}_1^*, \ldots, \mathbf{e}_n^*$ form an orthonormal basis of V^*. More generally, if we do not assume that (,) is positive-definite (but do assume that it is non-degenerate), then we choose $\{\mathbf{e}_1, \ldots, \mathbf{e}_n\}$ to be an orthogonal basis with $(\mathbf{e}_i, \mathbf{e}_i) = \pm 1$. Then $\{\mathbf{e}_1^*, \ldots, \mathbf{e}_n^*\}$ is an orthogonal basis of V^* and

$$(\mathbf{e}_i^*, \mathbf{e}_i^*)_{V^*} = (\mathbf{e}_i, \mathbf{e}_i)_V.$$

Of course, we may now define (,) for any two linear differential forms. For example, in the case of Euclidean space $\mathbb{R}^n$, the basis one-forms $\mathrm{d}x^1, \ldots, \mathrm{d}x^n$ are orthonormal with

$$(\mathrm{d}x^i, \mathrm{d}x^i) = +1.$$

For example, in $\mathbb{R}^3$, if $\omega = A_x\mathrm{d}x + A_y\mathrm{d}y + A_z\mathrm{d}z$ and $\tau = B_x\mathrm{d}x + B_y\mathrm{d}y + B_z\mathrm{d}z$, then $(\omega, \tau) = A_xB_x + A_yB_y + A_zB_z$. Notice that (ω, τ) is a *function*.

Let us now consider the space $\Lambda^k(V^*)$. Every element of $\Lambda^k(V^*)$ is a sum of the form

$$\alpha = \alpha^1 \wedge \cdots \wedge \alpha^k + \beta^1 \wedge \cdots \wedge \beta^k + \cdots.$$

We claim that there is a unique scalar product defined on $\Lambda^k(V^*)$ such that

$$(\alpha^1 \wedge \cdots \wedge \alpha^k, \gamma^1 \wedge \cdots \wedge \gamma^k) = \mathrm{Det}((\alpha^i, \gamma^j)).$$

It is clear that if such a scalar product exists, it is uniquely determined by the preceding equation, since one extends (,) by bilinearity to all αs and γs which are sums of decomposable elements, i.e., expressions like $\alpha^1 \wedge \cdots \wedge \alpha^k$ and $\gamma^1 \wedge \cdots \wedge \gamma^k$. The problem is to show that the definition is consistent. That is, to show that if α and/or γ are written as sums of decomposable elements in two different ways, we shall get the same value for (α, γ). This proof becomes quite straightforward once we use a slightly more abstract definition of the space $\Lambda^k(V^*)$ than we have been using so far. We give this definition and proof in the appendix to this chapter, so as not to interrupt the flow of discussion here.

We now get a (function-valued) scalar product on the space of k-forms. For example, for two-forms,

$$(\omega^1 \wedge \omega^2, \tau^1 \wedge \tau^2) = (\omega^1, \tau^1)(\omega^2, \tau^2) - (\omega^1, \tau^2)(\omega^2, \tau^1).$$

Thus, for example, in $\mathbb{R}^3$ with its Euclidean scalar product,

$$(\mathrm{d}x \wedge \mathrm{d}y, \mathrm{d}x \wedge \mathrm{d}y) = +1.$$

For four-dimensional spacetime, with the Lorentz scalar product, we have

$$(\mathrm{d}t \wedge \mathrm{d}x, \mathrm{d}t \wedge \mathrm{d}x) = -1$$

but

$$(\mathrm{d}x \wedge \mathrm{d}y, \mathrm{d}x \wedge \mathrm{d}y) = +1.$$

As an illustration of how to do these calculations in more general coordinate systems, consider spherical coordinates in $\mathbb{R}^3$, which may be defined by

$$r = \sqrt{(x^2 + y^2 + z^2)}, \quad \theta = \arctan((\sqrt{(x^2 + y^2)})/z), \quad \phi = \arctan(y/x).$$

If we calculate the differentials of these functions and express the results in terms of the basis $\{\mathrm{d}x, \mathrm{d}y, \mathrm{d}z\}$ with coefficients expressed for convenience in terms of r, θ, and ϕ, we find

$$\mathrm{d}r = \sin\theta\cos\phi\,\mathrm{d}x + \sin\theta\sin\phi\,\mathrm{d}y + \cos\theta\,\mathrm{d}z,$$

$$\mathrm{d}\theta = \frac{1}{r}(\cos\theta\cos\phi\,\mathrm{d}x + \cos\theta\sin\phi\,\mathrm{d}y - \sin\theta\,\mathrm{d}z),$$

$$\mathrm{d}\phi = \frac{1}{r\sin\theta}(-\sin\phi\,\mathrm{d}x + \cos\phi\,\mathrm{d}y).$$

Direct calculation, using the orthonormality of $\mathrm{d}x$, $\mathrm{d}y$, and $\mathrm{d}z$, shows that $\mathrm{d}r$, $\mathrm{d}\theta$, and $\mathrm{d}\phi$ are orthogonal elements of V^* at each point (r, θ, ϕ), and that

$$(\mathrm{d}r, \mathrm{d}r) = 1,\ (\mathrm{d}\theta, \mathrm{d}\theta) = 1/r^2,\ (\mathrm{d}\phi, \mathrm{d}\phi) = 1/r^2\sin^2\theta.$$

It follows immediately that $\mathrm{d}r \wedge \mathrm{d}\theta$, $\mathrm{d}r \wedge \mathrm{d}\phi$, and $\mathrm{d}\theta \wedge \mathrm{d}\phi$ constitute an orthogonal basis for $\Lambda^2(\mathbb{R}^3)$, with scalar products such as

$$(\mathrm{d}r \wedge \mathrm{d}\theta, \mathrm{d}r \wedge \mathrm{d}\theta) = 1/r^2.$$

Finally, we can calculate the scalar product of the basis element $dr \wedge d\theta \wedge d\phi$ with itself:

$$(dr \wedge d\theta \wedge d\phi, dr \wedge d\theta \wedge d\phi) = 1/r^4 \sin^2\theta.$$

The best way to summarize all of these results is to notice that

$$dr, r d\theta, \quad \text{and} \quad r \sin\theta d\phi$$

form an *orthonormal* basis for $\Lambda^1(\mathbb{R}^{3*})$ at each point so that we can calculate with these three differentials just as we do with dx, dy, and dz.

18.2. The star operator

Let V be a finite-dimensional vector space. In Chapter 8 we discussed the notion of an *orientation* of V. We recall that our definition there was as follows. Any two bases $\{\mathbf{e}_1, \ldots, \mathbf{e}_n\}$ and $\{\mathbf{f}_1, \ldots, \mathbf{f}_n\}$ differ from one another by a change of basis matrix, B. The two bases are called *equivalent* or *determine the same orientation* if $\operatorname{Det} B > 0$. Otherwise, they determine opposite orientations. The set of all bases breaks up into two equivalence classes, and each equivalence class is called an *orientation.* Using the space $\Lambda^n(V)$ we can put this somewhat differently. The basis $\{\mathbf{e}_1, \ldots, \mathbf{e}_n\}$ determines a basis – simply a non-zero vector – in the one-dimensional vector space $\Lambda^n(V)$, namely the element

$$\mathbf{e}_1 \wedge \cdots \wedge \mathbf{e}_n.$$

A second basis, $\mathbf{f}_1, \ldots, \mathbf{f}_n$, determines the element $\mathbf{f}_1 \wedge \cdots \wedge \mathbf{f}_n$. These two elements differ by the scalar multiple $\operatorname{Det} B$. Thus a choice of orientation on V is the same as the choice of one of the two equivalence classes of non-zero elements of $\Lambda^n(V)$, when we regard two such elements as equivalent if they differ from one another by a *positive* multiple.

Now suppose that V has a non-degenerate scalar product. Let $\{\mathbf{e}_1, \ldots, \mathbf{e}_n\}$ be an orthogonal basis of V with $(\mathbf{e}_i, \mathbf{e}_i) = \pm 1$. Any two such bases differ from one another by a change of basis matrix B with $\operatorname{Det} B = \pm 1$. Thus a choice of orientation on V, together with the scalar product, determines a unique element, $\mathbf{e}_1 \wedge \cdots \wedge \mathbf{e}_n$ of $\Lambda^n(V)$. We therefore also get a unique dual element of $\Lambda^n(V^*)$. We shall denote this element by σ.

Suppose we are considering differential forms. Then at each point of V we get a unique element of $\Lambda^n(V^*)$ at each point, in other words an n-form. To avoid cumbersome notation (and at the risk of some confusion), we shall also denote this n-form by σ. Thus, if $V = \mathbb{R}^n$ is Euclidean n-space, then

$$\sigma = dx^1 \wedge dx^2 \wedge \cdots \wedge dx^n.$$

If we are considering spacetime, we adopt the convention that time is the *first* coordinate, so that, in four dimensions,

$$\sigma = dt \wedge dx \wedge dy \wedge dz.$$

(Notice that in this case $(\sigma, \sigma) = -1$.)

We are now in a position to define a linear mapping from $\Lambda^k(V^*)$ to $\Lambda^{n-k}(V^*)$ called the *star operator*. To achieve this, we shall identify both $\Lambda^k(V^*)$ and $\Lambda^{n-k}(V^*)$ with the space of linear functions on $\Lambda^{n-k}(V^*)$, then identify elements of $\Lambda^k(V^*)$ and $\Lambda^{n-k}(V^*)$ which correspond to the same linear function. This identification of $\Lambda^k(V^*)$ with $\Lambda^{n-k}(V^*)$ will be called the star operator.

We first show how the *wedge* product, together with our choice of $\sigma\in\Lambda^n(V^*)$, assigns to each $\lambda\in\Lambda^k(V^*)$ a linear function on $\Lambda^{n-k}(V^*)$. Indeed, if $\omega\in\Lambda^{n-k}(V^*)$ then $\lambda\wedge\omega$ is an element of $\Lambda^n(V^*)$. Hence it must be some multiple of σ. In other words, we can write.

$$\lambda\wedge\omega = f(\omega)\sigma.$$

Thus each λ defines a linear function

$$\omega\mapsto f(\omega).$$

Now there is a unique element of $\Lambda^{n-k}(V^*)$, which we shall denote $\star\lambda$, which determines the same function $f(\omega)$ from $\Lambda^{n-k}(V^*)$ to $\mathbb{R}$ via the *scalar* product:

$$(\star\lambda,\omega) = f(\omega).$$

Thus, given any $\lambda\in\Lambda^k(V^*)$, we may define $\star\lambda\in\Lambda^{n-k}(V^*)$ by the condition that

$$\lambda\wedge\omega = (\star\lambda,\omega)\sigma$$

for all $\omega\in\Lambda^{n-k}(V^*)$. Notice that this definition of the star operator depends on the choice of orientation; reversing orientation changes the sign of σ and hence the sign of $\star\lambda$.

To calculate $\star\lambda$, we must in general apply the above definition using each basis element of $\Lambda^{n-k}(V^*)$ in turn as ω. If, however, λ is a basis element of $\Lambda^k(V^*)$, of the form $dx^{i_1}\wedge\cdots\wedge dx^{i_k}$, we need to consider only one ω, the basis element of $\Lambda^{n-k}(V^*)$ which involves the $n-k$ factors which do *not* occur in λ. In this case $\lambda\wedge\omega$ is the wedge product of dx^1 through dx^n in some order; so it is $\pm\sigma$. Then $\star\lambda=\pm\omega$. The only difficulty lies in determining the sign correctly.

The general definition

$$\lambda\wedge\omega = (\star\lambda,\omega)\sigma$$

needs to be supplemented slightly when $k=n$ or $k=0$. We denote the basis element for $\Lambda^0(\mathbb{R}^n)$ by 1, with the scalar product $(1,1)=1$ and the trivial wedge product $1\wedge\omega=\omega\wedge 1=\omega$. Then, if $\lambda=\sigma$, we need consider only the basis element $\omega=1$, and

$$\sigma\wedge 1 = (\star\lambda,1)\sigma$$

so that $\star\sigma=1$. On the other hand, if $\lambda=1$, we consider $\omega=\sigma$, so that

$$1\wedge\sigma = (\star\lambda,\sigma)\sigma$$

and $\star 1=(\sigma,\sigma)\sigma$. This means that whenever we work with a Euclidean scalar product, for which $(\sigma,\sigma)=1$, we have $\star\sigma=1, \star 1=\sigma$. However, in the important case of two-dimensional or four-dimensional spacetime, with the Lorentz scalar product, $(\sigma,\sigma)=-1$, so that

$$\star\sigma=+1 \quad\text{but}\quad \star 1=-\sigma.$$

The computation of the star operator is best illustrated by a few examples.

Example 1. $\mathbb{R}^2$ with the Euclidean scalar product: basis for $\Lambda^1(V^*)$ is dx, dy, with $(dx, dx) = 1$, $(dy, dy) = 1$ and $\sigma = dx \wedge dy$.

$$\star(dx \wedge dy) = 1 \text{ (always true)};$$
$$dx \wedge dy = (\star dx, dy) dx \wedge dy \quad \text{so} \quad \star dx = dy;$$
$$dy \wedge dx = (\star dy, dx) dx \wedge dy \quad \text{so} \quad \star dy = -dx;$$
$$\star 1 = dx \wedge dy \text{ because } (\sigma, \sigma) = +1.$$

Example 2. Two-dimensional spacetime, with the Lorentz scalar product: basis for $\Lambda^1(V)$ is $\{dt, dx\}$, with $(dt, dt) = +1$, $(dx, dx) = -1$ and $\sigma = dt \wedge dx$.

$$\star dt \wedge dx = 1;$$
$$dt \wedge dx = (\star dt, dx) dt \wedge dx \quad \text{so} \quad \star dt = -dx;$$
$$dx \wedge dt = (\star dx, dt) dt \wedge dx \quad \text{so} \quad \star dx = -dt;$$
$$\star 1 = -dt \wedge dx \text{ because } (\sigma, \sigma) = -1.$$

Example 3. $\mathbb{R}^3$ with the Euclidean scalar product: basis for $\Lambda^1(V)$ is $\{dx, dy, dz\}$, with $(dx, dx) = (dy, dy) = (dz, dz) = 1$ and $\sigma = dx \wedge dy \wedge dz$.

$$\star(dx \wedge dy \wedge dz) = 1,$$
$$(dx \wedge dy) \wedge dz = (\star(dx \wedge dy), dz)\sigma \quad \text{so} \quad \boxed{\star(dx \wedge dy) = dz;}$$
$$(dx \wedge dz) \wedge dy = (\star(dx \wedge dz), dy)\sigma \quad \text{so} \quad \boxed{\star(dx \wedge dz) = -dy;}$$
$$(dy \wedge dz) \wedge dx = (\star(dy \wedge dz), dx)\sigma \quad \text{so} \quad \boxed{\star(dy \wedge dz) = dx;}$$
$$dx \wedge (dy \wedge dz) = (\star dx, dy \wedge dz)\sigma \quad \text{so} \quad \boxed{\star dx = dy \wedge dz;}$$
$$dy \wedge (dx \wedge dz) = (\star dy, dx \wedge dz)\sigma \quad \text{so} \quad \boxed{\star dy = -dx \wedge dz;}$$
$$dz \wedge (dx \wedge dy) = (\star dz, dx \wedge dy)\sigma \quad \text{so} \quad \boxed{\star dz = dx \wedge dy;}$$
$$\star 1 = dx \wedge dy \wedge dz.$$

Example 4. Four-dimensional spacetime with the Lorentz scalar product: basis for $\Lambda^1(V)$ is $\{dt, dx, dy, dz\}$ with $(dt, dt) = 1$, $(dx, dx) = (dy, dy) = (dz, dz) = -1$, and $\sigma = dt \wedge dx \wedge dy \wedge dz$.

$$\star(dt \wedge dx \wedge dy \wedge dz) = 1,$$

Now

$$(dt \wedge dx \wedge dy) \wedge dz = (\star(dt \wedge dx \wedge dy), dz)\sigma,$$

so

$$\star(dt \wedge dx \wedge dy) = -dz.$$

Similarly

$$\star(dt \wedge dx \wedge dz) = +dy;$$
$$\star(dt \wedge dy \wedge dz) = -dx;$$

Also

$$(dx \wedge dy \wedge dz) \wedge dt = (\star(dx \wedge dy \wedge dz), dt)\sigma;$$

so

$$\star(dx \wedge dy \wedge dz) = -dt;$$

Now consider two-forms:

$$(\mathrm{d}t \wedge \mathrm{d}x) \wedge (\mathrm{d}y \wedge \mathrm{d}z) = (\star(\mathrm{d}t \wedge \mathrm{d}x), (\mathrm{d}y \wedge \mathrm{d}z))\sigma;$$

so

$$\star(\mathrm{d}t \wedge \mathrm{d}x) = \mathrm{d}y \wedge \mathrm{d}z.$$

Similarly

$$\star(\mathrm{d}t \wedge \mathrm{d}y) = -\mathrm{d}x \wedge \mathrm{d}z,$$
$$\star(\mathrm{d}t \wedge \mathrm{d}z) = \mathrm{d}x \wedge \mathrm{d}y;$$

$$(\mathrm{d}x \wedge \mathrm{d}y) \wedge (\mathrm{d}t \wedge \mathrm{d}z) = (\star(\mathrm{d}x \wedge \mathrm{d}y), \mathrm{d}t \wedge \mathrm{d}z)\sigma;$$

so

$$\star(\mathrm{d}x \wedge \mathrm{d}y) = -\mathrm{d}t \wedge \mathrm{d}z.$$

Similarly

$$\star(\mathrm{d}x \wedge \mathrm{d}z) = \mathrm{d}t \wedge \mathrm{d}y,$$
$$\star(\mathrm{d}y \wedge \mathrm{d}z) = -\mathrm{d}t \wedge \mathrm{d}x;$$

Next consider one-forms

$$\mathrm{d}t \wedge (\mathrm{d}x \wedge \mathrm{d}y \wedge \mathrm{d}z) = (\star\mathrm{d}t, \mathrm{d}x \wedge \mathrm{d}y \wedge \mathrm{d}z)\sigma;$$

so

$$\star\mathrm{d}t = -\mathrm{d}x \wedge \mathrm{d}y \wedge \mathrm{d}z;$$

Also

$$\mathrm{d}x \wedge (\mathrm{d}t \wedge \mathrm{d}y \wedge \mathrm{d}z) = (\star\mathrm{d}x, \mathrm{d}t \wedge \mathrm{d}y \wedge \mathrm{d}z)\sigma;$$

so

$$\star\mathrm{d}x = -\mathrm{d}t \wedge \mathrm{d}y \wedge \mathrm{d}z.$$

Similarly

$$\star\mathrm{d}y = \mathrm{d}t \wedge \mathrm{d}x \wedge \mathrm{d}z,$$
$$\star\mathrm{d}z = -\mathrm{d}t \wedge \mathrm{d}x \wedge \mathrm{d}y.$$

Finally

$$\star 1 = -\mathrm{d}t \wedge \mathrm{d}x \wedge \mathrm{d}y \wedge \mathrm{d}z.$$

It is apparent that, in the above examples, $\star(\star\lambda) = \pm\lambda$. To discover the general rule, we choose an orthogonal basis $\mathbf{e}_1, \ldots, \mathbf{e}_n$ of V with $(\mathbf{e}_i, \mathbf{e}_i) = \pm 1$. We consider the case where $\lambda \in \Lambda^k$ is the product of k of the $\mathbf{e}_i^*$'s and ω is $\pm$ the product of the remaining $(n-k)$ $\mathbf{e}_\nu^*$'s. Suppose we have chosen λ and ω so that $\lambda \wedge \omega = \sigma$ (e.g., $\lambda = \mathrm{d}x, \omega = \mathrm{d}y \wedge \mathrm{d}z$ in $\mathbb{R}^3$). Then $\star\lambda = \pm\omega$ and $\star\omega = \pm\lambda$. To determine the signs, set $\star\lambda = c_1\omega$ and $\star\omega = c_2\lambda$. Now

$$\sigma = \lambda \wedge \omega = (\star\lambda, \omega)\sigma$$

so

$$(\star\lambda, \omega) = c_1(\omega, \omega) = 1.$$

Also, by the rule for interchanging factors in a wedge product,

$$(-1)^{k(n-k)}\sigma = \omega \wedge \lambda = (\star\omega, \lambda)\sigma$$

so

$$(\star\omega, \lambda) = c_2(\lambda, \lambda) = (-1)^{k(n-k)}.$$

But because λ and ω are basis elements,

$$(\lambda, \lambda)(\omega, \omega) = (\lambda \wedge \omega, \lambda \wedge \omega) = (\sigma, \sigma)$$

so

$$c_1 c_2(\sigma, \sigma) = (-1)^{k(n-k)}.$$

Since $(\sigma, \sigma) = \pm 1$ we can move it to the other side of the equation. Thus

$$\star(\star\lambda) = c_1 \star\omega = c_1 c_2 \lambda$$

where

$$c_1 c_2 = (-1)^{k(n-k)}(\sigma, \sigma).$$

But by linearity of the $\star$ operator, if the above equations hold for each of the basis elements of $\Lambda^k(V^*)$ they must hold for all of $\Lambda^k(V^*)$. Thus we have proved

$$\boxed{\star\star = (-1)^{k(n-k)}(\sigma, \sigma) \text{ on } \Lambda^k(V^*)}$$

If n is odd and $(\sigma, \sigma) = +1$, as for $\mathbb{R}^3$ with the Euclidean scalar product, then $\star(\star\lambda) = \lambda$, but in general this is not the case.

Using the result just derived, that

$$\star\star\omega = (-1)^{k(n-k)}(\sigma, \sigma)\omega,$$

we can derive a useful explicit expression for the scalar product of any two k-forms. Notice first that, because $\star\star\omega = \pm\omega$, *four* successive applications of the star operator yield the identity: i.e.,

$$\star\star\star\star\omega = \omega.$$

By definition of the star operator,

$$(\star\star\star\lambda, \omega)\sigma = \star\star(\star\lambda \wedge \omega) = (-1)^{k(n-k)}(\sigma, \sigma)\star\lambda \wedge \omega$$

for any two k-forms λ and ω, so

$$\begin{aligned}(\lambda, \omega)\sigma &= (-1)^{k(n-k)}(\sigma, \sigma)\star\lambda \wedge \omega \\ &= \omega \wedge \star\lambda(\sigma, \sigma).\end{aligned}$$

But

$$\star\star\star\sigma = \star\star(1) = (\sigma, \sigma)1.$$

So by applying $\star\star\star$ to both sides we have

$$(\lambda, \omega)(\sigma, \sigma)1 = \star\star\star(\omega \wedge \star\lambda)(\sigma, \sigma)1$$

or

$$\boxed{(\lambda, \omega) = \star\star\star(\omega \wedge \star\lambda) = \star\star\star(\lambda \wedge \star\omega).}$$

A similar useful result follows from

$$(\star\lambda, \star\omega)\sigma = \lambda \wedge \star\omega.$$

Simply applying $\star$ to both sides, we have

$$(\star\lambda, \star\omega) = \star(\lambda \wedge \star\omega) = \star(\omega \wedge \star\lambda).$$

Using this result and the preceding one we can express scalar products in terms of the star operator.

18.3. The Dirichlet integral and the Laplacian

Let ω be a k-form and λ be a $(k-1)$-form. Then

$$d(\lambda \wedge \star\omega) = d\lambda \wedge \star\omega + (-1)^{k-1}\lambda \wedge d(\star\omega).$$

So, for any (bounded) domain U we have

$$\int_U d\lambda \wedge \star\omega + (-1)^{(k-1)} \int_U \lambda \wedge d\star\omega = \int_U d(\lambda \wedge \star\omega).$$

If U is unbounded and if λ or ω have compact support (or if they vanish sufficiently rapidly at infinity), then by Stokes' theorem

$$\int_U d\lambda \wedge \star\omega + \int_U (-1)^{k-1}\lambda \wedge d(\star\omega) = 0.$$

Now applying $*$ to the rule $(\rho, \tau) = \star\star\star(\rho \wedge \star\tau)$, we see that

$$d\lambda \wedge \star\omega = (d\lambda, \omega)\star 1.$$

On the other hand, ω is a k-form so $\star\omega$ is an $(n-k)$-form so $d\star\omega$ is an $(n-k+1)$-form. Therefore

$$\begin{aligned}\lambda \wedge d\star\omega &= (-1)^{(k-1)(n-k+1)} d\star\omega \wedge \lambda \\ &= (-1)^{(k-1)(n-k+1)}(\star^{-1} d\star\omega, \lambda)\sigma \\ &= (-1)^{(k-1)(n-k+1)}(\lambda, \star^{-1} d\star\omega)\sigma.\end{aligned}$$

Now $\star d\star\omega$ is a form of degree $n-(n-k+1) = k-1$. So, since $\star\star\star\star = \text{id}$,

$$\star^{-1} d\star\omega = \star\star(\star d\star\omega) = (-1)^{(k-1)(n-k+1)}(\sigma, \sigma)\star d\star\omega,$$

and

$$\sigma = (\sigma, \sigma)\star 1.$$

Thus

$$\lambda \wedge d\star\omega = (\lambda, \star\star\star d\star\omega)\star 1.$$

So we can write

$$d\lambda \wedge \star\omega + (-1)^{k-1}\lambda \wedge d\star\omega = [(d\lambda, \omega) + (-1)^{(k-1)}(\lambda, \star^{-1} d\star\omega)]\star 1.$$

Thus, if λ or ω vanish at ∂U (if U is bounded) or have compact support (or vanish sufficiently rapidly at infinity),

$$\int_U (d\lambda, \omega)\sigma = \int_U (\lambda, (-1)^k \star\star\star d\star\omega)\sigma.$$

Let us define the operator d* by

$$d^*\omega = (-1)^k \star\star\star d\star\omega = (-1)^k \star^{-1} d\star\omega.$$

Thus d* maps k-forms into $(k-1)$-forms. Let us also write, for any k-forms λ and τ,

$$(\lambda, \tau)_U = D_U(\lambda, \tau) = \int_U (\lambda, \tau)\sigma.$$

Then we have

$$D_U(d\lambda, \omega) = D_U(\lambda, d^*\omega)$$

for λ or ω vanishing at ∂U (or sufficiently rapidly at infinity).

Notice that

$$d^*d^* = 0$$

since

$$dd = 0.$$

Let us define the *Laplace operator* $\square$ mapping k-forms into k-forms by[†]

$$\square = dd^* + d^*d.$$

We claim that

$$\square d = d\square,$$
$$\square d^* = d^*\square,$$
$$\square\star = \star\square.$$

To see that $\square d = d\square$, observe that

$$\square d = (dd^* + d^*d)d = dd^*d$$

and

$$d\square = d(dd^* + d^*d) = dd^*d$$

since $dd = 0$. A similar computation shows that

$$\square d^* = d^*\square.$$

The proof that $\square\star = \star\square$ is trickier, requiring careful attention to signs. We start with

$$d^*\omega = (-1)^k \star\star\star d\star\omega \quad \text{for a } k\text{-form } \omega.$$

On replacing the k-form ω by the $(n-k)$-form $\star\omega$, we have

$$d^*\star\omega = (-1)^{n-k} \star\star\star d\star\star\omega.$$

Since $d(\star\star\omega)$ is a $(k+1)$-form,

$$\star\star d(\star\star\omega) = (-1)^{(k+1)(n-k-1)}(\sigma, \sigma) d(\star\star\omega).$$

[†] This operator $\square$ differs by a minus sign from what we have called the Laplacian in previous chapters.

Since ω is a k-form,

$$\star\star\omega = (-1)^{k(n-k)}(\sigma,\sigma)\omega.$$

Combining these results, and using $(\sigma,\sigma)^2 = 1$, we find

$$\mathrm{d}^*\star\omega = (-1)^{n-k}(-1)^{(k+1)(n-k-1)}(-1)^{k(n-k)}\star\mathrm{d}\omega$$

or

$$\mathrm{d}^*\star\omega = (-1)^{(k+1)(n-k-1)}(-1)^{(k+1)(n-k)}\star\mathrm{d}\omega$$

or

$$\mathrm{d}^*\star\omega = (-1)^{k-1}\star\mathrm{d}\omega.$$

With this rule and the rule $\star\mathrm{d}^*\omega = (-1)^k\mathrm{d}\star\omega$, the proof becomes easy:

$$\star\Box\omega = \star\mathrm{d}(\mathrm{d}^*\omega) + \star\mathrm{d}^*(\mathrm{d}\omega)$$
$$\star\Box\omega = (-1)^k\mathrm{d}^*\star\mathrm{d}\omega + (-1)^{k+1}\mathrm{d}\star\mathrm{d}\omega$$

while

$$\Box\star\omega = \mathrm{d}\mathrm{d}^*\star\omega + \mathrm{d}^*\mathrm{d}\star\omega$$
$$\Box\star\omega = (-1)^{k+1}\mathrm{d}\star\mathrm{d}\omega + (-1)^k\mathrm{d}\star\mathrm{d}^*\omega$$

so that

$$\star\Box = \Box\star.$$

The $\Box$ operator on $\mathbb{R}^3$

We compute the Laplacian explicitly in $\mathbb{R}^3$. For a function f,

$$\Box f = \mathrm{d}\mathrm{d}^*f + \mathrm{d}^*\mathrm{d}f.$$

But $\star f$ is a three-form and $\mathrm{d}\Omega = 0$ for any three-form on $\mathbb{R}^3$. Thus $\Box f = \mathrm{d}^*\mathrm{d}f$. Now

$$\mathrm{d}f = \frac{\partial f}{\partial x}\mathrm{d}x + \frac{\partial f}{\partial y}\mathrm{d}y + \frac{\partial f}{\partial z}\mathrm{d}z$$

and

$$\star\frac{\partial f}{\partial x}\mathrm{d}x = \frac{\partial f}{\partial x}\mathrm{d}y \wedge \mathrm{d}z$$

and so

$$\mathrm{d}\star\frac{\partial f}{\partial x}\mathrm{d}x = \frac{\partial^2 f}{\partial x^2}\mathrm{d}x \wedge \mathrm{d}y \wedge \mathrm{d}z$$

and

$$\mathrm{d}^*\left(\frac{\partial f}{\partial x}\mathrm{d}x\right) = -\star^{-1}\mathrm{d}\star\frac{\partial f}{\partial x}\mathrm{d}x = -\frac{\partial^2 f}{\partial x^2}.$$

Similar computations for the remaining two terms show that

$$\Box f = -\left(\frac{\partial^2 f}{\partial x^2} + \frac{\partial^2 f}{\partial y^2} + \frac{\partial^2 f}{\partial z^2}\right).$$

Now let us compute the Laplacian for one-forms: it is enough to compute $\Box\omega$ for $\omega = a\mathrm{d}x$. We can rotate to interchange any of the axes and $\Box$, being defined purely by the scalar product and orientation, will commute with all rotations. So, if we have a formula for $\square(a\mathrm{d}x)$, a similar formula will hold for $\square(b\mathrm{d}y)$ and $\square(c\mathrm{d}z)$. By linearity, we will then get a formula for $a\mathrm{d}x + b\mathrm{d}y + c\mathrm{d}z$.

Now

$$\mathrm{d}(a\mathrm{d}x) = \mathrm{d}a \wedge \mathrm{d}x = \frac{\partial a}{\partial y}\mathrm{d}y \wedge \mathrm{d}x + \frac{\partial a}{\partial z}\mathrm{d}z \wedge \mathrm{d}x.$$

So

$$\star(\mathrm{d}(a\mathrm{d}x)) = -\frac{\partial a}{\partial y}\mathrm{d}z + \frac{\partial a}{\partial z}\mathrm{d}y.$$

Thus

$$\begin{aligned}\mathrm{d}\star\mathrm{d}(a\mathrm{d}x) = &-\frac{\partial^2 a}{\partial x \partial y}\mathrm{d}x \wedge \mathrm{d}z - \frac{\partial^2 a}{\partial y^2}\mathrm{d}y \wedge \mathrm{d}z \\ &+ \frac{\partial^2 a}{\partial x\,\partial z}\mathrm{d}x \wedge \mathrm{d}y + \frac{\partial^2 a}{\partial z^2}\mathrm{d}z \wedge \mathrm{d}y.\end{aligned}$$

Hence, using the fact that $\star = \star^{-1}$ for two-forms on $\mathbb{R}^3$,

$$\mathrm{d}^*\mathrm{d}(a\mathrm{d}x) = \star\mathrm{d}\star\mathrm{d}(a\mathrm{d}x) = \frac{\partial^2 a}{\partial x\,\partial y}\mathrm{d}y - \frac{\partial^2 a}{\partial y^2}\mathrm{d}x + \frac{\partial^2 a}{\partial x\,\partial z}\mathrm{d}z - \frac{\partial^2 a}{\partial z^2}\mathrm{d}x.$$

Now

$$\begin{aligned}\mathrm{d}^*(a\mathrm{d}x) &= -\star\mathrm{d}\star(a\mathrm{d}x) = -\star\mathrm{d}(a\mathrm{d}y \wedge \mathrm{d}z) \\ &= -\star\left(\frac{\partial a}{\partial x}\mathrm{d}x \wedge \mathrm{d}y \wedge \mathrm{d}z\right) \\ &= -\frac{\partial a}{\partial x}.\end{aligned}$$

So

$$\mathrm{d}\mathrm{d}^*(a\mathrm{d}x) = -\frac{\partial^2 a}{\partial x^2}\mathrm{d}x - \frac{\partial^2 a}{\partial x\,\partial y}\mathrm{d}y - \frac{\partial^2 a}{\partial x\,\partial z}\mathrm{d}z.$$

Thus

$$\begin{aligned}\square(a\mathrm{d}x) &= (\square a)\mathrm{d}x \\ &= -\left(\frac{\partial^2 a}{\partial x^2} + \frac{\partial^2 a}{\partial y^2} + \frac{\partial^2 a}{\partial z^2}\right)\mathrm{d}x.\end{aligned}$$

If $\omega = a\mathrm{d}x + b\mathrm{d}y + c\mathrm{d}z$, we may compute $\square\omega$ by applying $\square$ to each of the coefficients.

Any two-form on $\mathbb{R}^3$ can be written as $\star\omega$ where ω is a one-form. Since $\star\square = \square\star$ we see that, for

$$\begin{aligned}\Omega &= A\mathrm{d}x \wedge \mathrm{d}y + B\mathrm{d}z \wedge \mathrm{d}x + C\mathrm{d}y \wedge \mathrm{d}z, \\ \square\Omega &= (\square A)\mathrm{d}x \wedge \mathrm{d}y + (\square B)\mathrm{d}z \wedge \mathrm{d}x + (\square C)\mathrm{d}y \wedge \mathrm{d}z.\end{aligned}$$

Similarly

$$\square(\rho\mathrm{d}x \wedge \mathrm{d}y \wedge \mathrm{d}z) = (\square\rho)\mathrm{d}x \wedge \mathrm{d}y \wedge \mathrm{d}z.$$

In short, for the *standard rectangular coordinates* in $\mathbb{R}^3$, we can compute $\square$ of any form by applying $\square$ to each of the coefficients.

We can now redo the argument in the last section of Chapter 17 – reducing the equations of magnetism to Poisson's equation. We assume that units have been chosen so that $\mu = 1$ and we set

$$j = -\star J.$$

Then the equations can be written as

$$\mathrm{d}B = 0,$$
$$\mathrm{d}^* B = 4\pi j.$$

We choose some A' with

$$\mathrm{d}A' = B$$

(possible as we are assuming $H^1 = 0$). Solve the Poisson equation

$$\square\psi = \mathrm{d}^* A'.$$

Replace A' by

$$A = A' - \mathrm{d}\psi.$$

We shall have

$$\mathrm{d}A = B.$$

Then, since ψ is a function,

$$\square\psi = \mathrm{d}^*\mathrm{d}\psi = \mathrm{d}^* A'$$

so

$$\mathrm{d}^* A = \mathrm{d}^* A' - \mathrm{d}^*\mathrm{d}\psi = 0.$$

Thus our equations become

$$\mathrm{d}^*\mathrm{d}A = 4\pi j,$$

or, since $\mathrm{d}^* A = 0$,

$$\square A = 4\pi j.$$

In the above argument, we seemed to have used the fact that B was a *two-form* so A was a one-form and ψ a function. Actually, the argument is more general. Suppose B is a k-form in n-dimensional space and we wish to solve

$$\mathrm{d}B = 0,$$
$$\mathrm{d}^* B = 4\pi j.$$

We assume the region in question is such that $B = \mathrm{d}A'$ for some A'. Then solve the *Poisson equation*

$$\square\psi = \mathrm{d}^* A'.$$

Write

$$A = A' - \mathrm{d}\psi.$$

Then

$$d^*A = dd^*\psi$$

so

$$dd^*A = 0$$

so

$$\Box A = (dd^* + d^*d)A = 4\pi j$$

and again we are reduced to solving the Poisson equation.

18.4. The □ operator in spacetime

Let us consider four-dimensional spacetime with the coordinates t, x, y, z and scalar product with $(dt, dt) = 1$, $(dx, dx) = (dy, dy) = (dz, dz) = -1$. For a function,

$$\begin{aligned}\Box f &= d^*df \\ &= -\star^{-1}d\star\left(\frac{\partial f}{\partial t}dt + \frac{\partial f}{\partial x}dx + \frac{\partial f}{\partial y}dy + \frac{\partial f}{\partial z}dz\right) \\ &= -\star^{-1}d\left(-\frac{\partial f}{\partial t}dx \wedge dy \wedge dz - \frac{\partial f}{\partial x}dt \wedge dy \wedge dz\right. \\ &\quad \left. + \frac{\partial f}{\partial y}dt \wedge dx \wedge dz - \frac{\partial f}{\partial z}dt \wedge dx \wedge dy\right) \\ &= -\star^{-1}\left(-\frac{\partial^2 f}{\partial t^2} + \frac{\partial^2 f}{\partial x^2} + \frac{\partial^2 f}{\partial y^2} + \frac{\partial^2 f}{\partial z^2}\right)dt \wedge dx \wedge dy \wedge dz.\end{aligned}$$

So

$$\Box f = -\frac{\partial^2 f}{\partial t^2} + \frac{\partial^2 f}{\partial x^2} + \frac{\partial^2 f}{\partial y^2} + \frac{\partial^2 f}{\partial z^2}.$$

For one-forms, let us first consider $a dt$. Then

$$d(a dt) = \frac{\partial a}{\partial x}dx \wedge dt + \frac{\partial a}{\partial y}dy \wedge dt + \frac{\partial a}{\partial z}dz \wedge dt$$

so

$$\star d(a dt) = -\frac{\partial a}{\partial x}dy \wedge dz + \frac{\partial a}{\partial y}dx \wedge dz - \frac{\partial a}{\partial z}dx \wedge dy$$

and

$$\begin{aligned}d\star d(a dt) = &-\frac{\partial^2 a}{\partial t\,\partial x}dt \wedge dy \wedge dz - \frac{\partial^2 a}{\partial x^2}dx \wedge dy \wedge dz \\ &+ \frac{\partial^2 a}{\partial y\,\partial t}dt \wedge dx \wedge dz - \frac{\partial^2 a}{\partial y^2}dx \wedge dy \wedge dz \\ &- \frac{\partial^2 a}{\partial z\,\partial t}dt \wedge dx \wedge dy - \frac{\partial^2 a}{\partial z^2}dx \wedge dy \wedge dz.\end{aligned}$$

So

$$\begin{aligned}
d^*d(a\mathrm{d}t) &= \star^{-1}\mathrm{d}\star\mathrm{d}(a\mathrm{d}t)\\
&= \left(\frac{\partial^2 a}{\partial x^2}+\frac{\partial^2 a}{\partial y^2}+\frac{\partial^2 a}{\partial z^2}\right)\mathrm{d}t+\frac{\partial^2 a}{\partial x\,\partial t}\mathrm{d}x+\frac{\partial^2 a}{\partial y\,\partial t}\mathrm{d}y+\frac{\partial^2 a}{\partial z\,\partial t}\mathrm{d}z.
\end{aligned}$$

On the other hand,

$$\begin{aligned}
\mathrm{d}^*(a\mathrm{d}t) &= -\star^{-1}\mathrm{d}\star(a\mathrm{d}t)\\
&= \star^{-1}\mathrm{d}(a\mathrm{d}x\wedge\mathrm{d}y\wedge\mathrm{d}z)\\
&= \star^{-1}\frac{\partial a}{\partial t}\mathrm{d}t\wedge\mathrm{d}x\wedge\mathrm{d}y\wedge\mathrm{d}z\\
&= -\frac{\partial a}{\partial t}.
\end{aligned}$$

So

$$\mathrm{d}\mathrm{d}^*(a\mathrm{d}t) = -\frac{\partial^2 a}{\partial t^2}\mathrm{d}t-\frac{\partial^2 a}{\partial t\,\partial x}\mathrm{d}x-\frac{\partial^2 a}{\partial t\,\partial y}\mathrm{d}y-\frac{\partial^2 a}{\partial t\,\partial z}\mathrm{d}z,$$

and thus

$$\Box(a\mathrm{d}t) = (\Box a)\,\mathrm{d}t = \left(-\frac{\partial^2 a}{\partial t^2}+\frac{\partial^2 a}{\partial x^2}+\frac{\partial^2 a}{\partial y^2}+\frac{\partial^2 a}{\partial z^2}\right)\mathrm{d}t.$$

A similar calculation (or rotational invariance) shows that

$$\begin{aligned}
\Box(b\mathrm{d}x) &= (\Box b)\,\mathrm{d}x,\\
\Box(c\mathrm{d}y) &= (\Box c)\,\mathrm{d}y,
\end{aligned}$$

etc.

The same is true for two-forms. Consider, for example, $a(\mathrm{d}t\wedge\mathrm{d}x)$. Then

$$\begin{aligned}
\mathrm{d}^*(a\mathrm{d}t\wedge\mathrm{d}x) &= \star^{-1}\mathrm{d}\star(a\mathrm{d}t\wedge\mathrm{d}x)\\
&= \star^{-1}\mathrm{d}(a\mathrm{d}y\wedge\mathrm{d}z)\\
&= \star^{-1}\left(\frac{\partial a}{\partial t}\mathrm{d}t\wedge\mathrm{d}y\wedge\mathrm{d}z\right)+\star^{-1}\left(\frac{\partial a}{\partial x}\mathrm{d}x\wedge\mathrm{d}y\wedge\mathrm{d}z\right)\\
&= -\frac{\partial a}{\partial t}\mathrm{d}x-\frac{\partial a}{\partial x}\mathrm{d}t.
\end{aligned}$$

So

$$\begin{aligned}
\mathrm{d}\mathrm{d}^*(a\mathrm{d}t) = \left(-\frac{\partial^2 a}{\partial t^2}+\frac{\partial^2 a}{\partial x^2}\right)&\mathrm{d}t\wedge\mathrm{d}x\\
&-\frac{\partial^2 a}{\partial t\partial y}\mathrm{d}y\wedge\mathrm{d}x-\frac{\partial^2 a}{\partial t\partial z}\mathrm{d}z\wedge\mathrm{d}x\\
&-\frac{\partial^2 a}{\partial x\partial y}\mathrm{d}t\wedge\mathrm{d}y-\frac{\partial^2 a}{\partial x\partial z}\mathrm{d}t\wedge\mathrm{d}z.
\end{aligned}$$

On the other hand,

$$\mathrm{d}(a\mathrm{d}t\wedge\mathrm{d}x) = \frac{\partial a}{\partial y}\mathrm{d}y\wedge\mathrm{d}t\wedge\mathrm{d}x+\frac{\partial a}{\partial z}\mathrm{d}z\wedge\mathrm{d}t\wedge\mathrm{d}x$$

so

$$\star \mathrm{d}(a\mathrm{d}t \wedge \mathrm{d}x) = -\frac{\partial a}{\partial y}\mathrm{d}z + \frac{\partial a}{\partial z}\mathrm{d}y,$$

$$\mathrm{d}\star\mathrm{d}(a\mathrm{d}t \wedge \mathrm{d}x) = -\left(\frac{\partial^2 a}{\partial y^2} + \frac{\partial^2 a}{\partial z^2}\right)\mathrm{d}y \wedge \mathrm{d}z - \frac{\partial^2 a}{\partial y \partial x}\mathrm{d}x \wedge \mathrm{d}z$$
$$- \frac{\partial^2 a}{\partial y \partial t}\mathrm{d}t \wedge \mathrm{d}z + \frac{\partial^2 a}{\partial z \partial t}\mathrm{d}t \wedge \mathrm{d}y + \frac{\partial^2 a}{\partial x \partial z}\mathrm{d}x \wedge \mathrm{d}y.$$

Then

$$\mathrm{d}^*\mathrm{d}(a\mathrm{d}t \wedge \mathrm{d}x) = -\star^{-1}\mathrm{d}\star\mathrm{d}(a\mathrm{d}t \wedge \mathrm{d}x)$$

and

$$\Box(a\mathrm{d}t \wedge \mathrm{d}x) = (\Box a)\mathrm{d}t \wedge \mathrm{d}x = \left(-\frac{\partial^2 a}{\partial t^2} + \frac{\partial^2 a}{\partial x^2} + \frac{\partial^2 a}{\partial y^2} + \frac{\partial^2 a}{\partial z^2}\right)\mathrm{d}t \wedge \mathrm{d}x.$$

We will leave to you the similar verification that

$$\Box(b\mathrm{d}x \wedge \mathrm{d}z) = (\Box b)\mathrm{d}x \wedge \mathrm{d}z.$$

It then follows that for any two-form, expressed in terms of the $\mathrm{d}t, \mathrm{d}x, \mathrm{d}y, \mathrm{d}z$, applying $\Box$ is the same as applying $\Box$ to the coefficients. Having verified this for zero-, one-, and two-forms, it now follows for three- and four-forms from $\star\Box = \Box\star$.

18.5. The Clifford algebra

By now you should be reasonably convinced that, in *Euclidean coordinates* on $\mathbb{R}^n$, we can compute the operator $\Box$ applied to any k-form by applying $\Box$ to each of its coefficients. You might also be somewhat apprehensive that, if we try to prove this fact by the methods we have been using in the past two sections, we will end up in a mess. The purpose of this section is to introduce an algebraic formalism that will allow a simple direct proof. We will not need the results of this section again in this book – we have already proved the results for the important cases of three and four dimensions. On the other hand, the ideas we will introduce here have proved to be important in modern physics.

Our proof is going to make use directly of the property that

$$D_U(\mathrm{d}\omega, \tau) = D_U(\omega, \mathrm{d}^*\tau) + \text{boundary terms},$$

which we can phrase by saying that d^* is the *formal adjoint* of d. The word formal refers to the fact that there are boundary terms: we only have

$$D_U(\mathrm{d}\omega, \tau) = D_U(\omega, \mathrm{d}^*\tau)$$

when various hypotheses are made that guarantee that there are no contributions from the boundary. Let us study this situation in slightly more generality.

Let E and F be vector spaces, each equipped with its own scalar product. We may then identify E with E^* and F with F^*. In particular, if $A: E \to F$ is a linear map, then we may regard $A^*: F^* \to E^*$ as a linear map $A^*: F \to E$ defined by

$$(e, A^*f)_E = (Ae, f)_F$$

where $(\, , \,)_E$ and $(\, , \,)_F$ denote the scalar products on E and F respectively. Now suppose that we have chosen a coordinate system, $x_1, \ldots, x_n$, on a vector space V, and so the corresponding *volume form* is;

$$\sigma = \mathrm{d}x_1 \wedge \cdots \wedge \mathrm{d}x_n.$$

We can then consider the space $\mathscr{F}(V, E)$ of smooth functions from V to E. If u_1 and u_2 are such functions, and U is some bounded domain, we can define

$$D_U(u_1, u_2) = \int_U (u_1, u_2)_E \sigma.$$

(If U is unbounded, we can still define D_U on those pairs of functions which vanish sufficiently rapidly at infinity.) We can consider operators, L, from $\mathscr{F}(V, E)$ to $\mathscr{F}(V, F)$. Thus, for each $u \in \mathscr{F}(V, E)$, Lu is an element of $\mathscr{F}(V, F)$. The operator L is called a *first-order linear differential operator* if there are linear maps $A_1 : E \to F$, $A_2 : E \to F, \ldots, A_n : E \to F$ and $B : E \to F$ such that

$$Lu = A_1 \frac{\partial u}{\partial x_1} + \cdots + A_n \frac{\partial u}{\partial x_n} + Bu.$$

In general, the maps $A_1, \ldots, A_n$ and B can depend on the point of V. That is, the A_i and the B are functions on V whose values are linear maps. Thus the previous equation might be written as

$$(Lu)(x) = A_1(x) \frac{\partial u}{\partial x_1}(x) + \cdots + A_n(x) \frac{\partial u}{\partial x_n}(x) + B(x)u(x).$$

If the As and B are constant, we say that L is a first-order linear differential operator with constant coefficients.

Now let $u \in \mathscr{F}(V, E)$ and $v \in \mathscr{F}(V, F)$. Then

$$\begin{aligned} \frac{\partial}{\partial x_i}(A_i u, v)_F &= \left(\frac{\partial A_i}{\partial x_i} u, v \right)_F + \left(A_i \frac{\partial u}{\partial x_i}, v \right)_F + \left(A_i u, \frac{\partial v}{\partial x_i} \right)_F \\ &= \left(A_i \frac{\partial u}{\partial x_i}, v \right)_F + \left(u, \left(\frac{\partial A_i}{\partial x_i} \right)^* v \right)_E + \left(u, A_i^* \frac{\partial v}{\partial x_i} \right)_E. \end{aligned}$$

So

$$\left(A_i \frac{\partial U}{\partial X_C}, V \right)_F = \frac{\partial}{\partial x_i}(A_i u, v)_F - \left(u, A_i^* \frac{\partial v}{\partial x_i} \right)_E - \left(u, \left(\frac{\partial A_i}{\partial x_i} \right)^* v \right)_E.$$

Thus, if we define

$$L^* v = -A_1^* \frac{\partial v}{\partial x_i} - A_2^* \frac{\partial v}{\partial x_2} - \cdots - A_n^* \frac{\partial v}{\partial x_n} + \left(B^* - \sum \left(\frac{\partial A_i}{\partial x_i} \right)^* \right) v,$$

we see that

$$(Lu, v)_F = (u, L^* v)_E + \sum \frac{\partial}{\partial x_i}(A_i u, v)_F.$$

Now the last term on the right integrates out to a boundary term multiplication by σ and an integration. Indeed, if we write

$$\tau = \frac{\partial}{\partial x_1}(A_1 u, v)_F \mathrm{d}x_2 \wedge \cdots \wedge \mathrm{d}x_n - \frac{\partial}{\partial x_2}(A_2 u, v)_F \mathrm{d}x_1 \wedge \mathrm{d}x_3 \wedge \cdots + \cdots$$

then

$$\mathrm{d}\tau = \sum \frac{\partial}{\partial x_i}(A_i u, v)_F \mathrm{d}x_1 \wedge \cdots \wedge \mathrm{d}x_n.$$

So

$$\int_U (Lu, v)_F \sigma = \int_U (u, L^* v)_E \sigma + \int_{\partial U} \tau.$$

Thus L^* is the formal adjoint of L.

Now, strictly speaking, a k-form ω is not a function from V to a fixed vector space. (This is because $\omega(x)$ is an element of the space of k-antisymmetric linear functions on the tangent space at x, and the tangent space varies from point to point.) However, once we have chosen a coordinate system, we can regard a k-form ω as a function from V to $\Lambda^k(V^*)$. Thus, we may take $E = \Lambda^k(V^*)$. If we want to take our differential operator L to be d, then we set $F = \Lambda^{k+1}(V^*)$. Then we can write.

$$\mathrm{d} = \varepsilon_1 \frac{\partial}{\partial x_1} + \cdots + \varepsilon_n \frac{\partial}{\partial x_n}$$

where ε_k denotes the linear map of $\Lambda^k(V^*) \to \Lambda^{k+1}(V^*)$ consisting of exterior multiplication by $\mathrm{d}x_k$, so

$$\varepsilon_k \omega = \mathrm{d}x_k \wedge \omega.$$

Now to compute d^* all we need is a formula for ε_k^*. We claim that

$$\varepsilon_k^* = \pm \mathrm{i}\left(\frac{\partial}{\partial x_k}\right) \quad \text{with} + \text{if } (\mathrm{d}x_k, \mathrm{d}x_k) = 1 \quad \text{and} - \text{if } (\mathrm{d}x_k, \mathrm{d}x_k) = -1.$$

To prove this, assume that $k = 1$ (this is only a matter of notation) and examine what happens on basis elements

$$\omega = \mathrm{d}x_{i_1} \wedge \cdots \wedge \mathrm{d}x_{j_k}$$

of $\Lambda^k(V)$ and those of $\Lambda^{k+1}(V^*)$. There are tow possibilities.

(a) $i_1 = 1$. Then $\varepsilon_1 \omega = 0$. On the other hand, for

$$\tau = \mathrm{d}x_{j_1} \wedge \cdots \wedge \mathrm{d}x_{j_n}$$

we have

$$\mathrm{i}\left(\frac{\partial}{\partial x_1}\right)\tau = 0 \quad (\text{if } j_1 > 1)$$

and

$$\mathrm{i}\left(\frac{\partial}{\partial x_1}\right)\tau = \mathrm{d}x_{j_2} \wedge \cdots \wedge \mathrm{d}x_{j_n} \quad \text{if } j_1 = 1.$$

In either event, we have

$$(\varepsilon_1 \omega, \tau) = \left(\omega, \mathrm{i}\left(\frac{\partial}{\partial x_1}\right)\tau\right) = 0.$$

(b) $i_1 > 1$. Then

$$\varepsilon_1 \omega = \mathrm{d}x_1 \wedge \mathrm{d}x_{i_1} \wedge \cdots \wedge \mathrm{d}x_{i_k}$$

and

$$\mathrm{i}\left(\frac{\partial}{\partial x_1}\right)\tau = \begin{cases} 0 & \text{if } j_1 > 1, \\ \mathrm{d}x_{j_2} \wedge \cdots \wedge \mathrm{d}x_{j_{k+1}} & \text{if } j_1 > 1. \end{cases}$$

Then

$$(\varepsilon_1 \omega, \tau) = 0 = \left(\omega, \mathrm{i}\left(\frac{\partial}{\partial x_1}\right)\tau\right) \quad \text{if } j_1 > 1.$$

If $j_1 = 1$, we still have

$$(\varepsilon_1 \omega, \tau) = 0 = \left(\omega, \mathrm{i}\left(\frac{\partial}{\partial x_1}\right)\tau\right)$$

unless $j_2 = i_1, j_3 = i_2$, etc. If $j_2 = i_1, j_3 = i_2, \ldots$, etc., then

$$(\varepsilon_1 \omega, \tau) = (\mathrm{d}x_1, \mathrm{d}x_1)(\omega, \omega)$$

while

$$\left(\omega, \mathrm{i}\left(\frac{\partial}{\partial x_1}\right)\tau\right) = (\omega, \omega).$$

This proves that

$$\varepsilon_1^* = (\mathrm{d}x_1, \mathrm{d}x_1)\mathrm{i}\left(\frac{\partial}{\partial x_1}\right).$$

(More generally, the proof shows the following. Let $L: V \to V^*$ denote the linear map determined by the scalar product on V. Let θ be any one form and $\varepsilon(\theta)$ denote exterior multiplication by θ. Then

$$(\varepsilon(\theta))^* = \mathrm{i}(L^{-1}\theta).)$$

Let us define

$$\iota_p = \pm \mathrm{i}\left(\frac{\partial}{\partial x_p}\right) = \varepsilon_p^*.$$

We have thus shown that

$$\boxed{\mathrm{d}^* = -\sum_{p=1}^{n} \iota_p \frac{\partial}{\partial x_p}.}$$

Now it follows from the anticommutativity of exterior multiplication that

$$\varepsilon_p \varepsilon_q + \varepsilon_q \varepsilon_p = 0 \quad \text{for all } p \text{ and } q$$

and hence (or directly from properties of interior multiplication) that

$$\iota_p \iota_q + \iota_q \iota_p = 0 \quad \text{all } p \text{ and } q.$$

If $p \neq q$, then exterior multiplication by $\mathrm{d}x_p$ and interior multiplication by $i(\partial/\partial x_q)$ satisfy

$$\mathrm{i}\left(\frac{\partial}{\partial x_q}\right)\varepsilon(\mathrm{d}x_p) + \varepsilon(\mathrm{d}x_p)\mathrm{i}\left(\frac{\partial}{\partial x_q}\right) = 0 \quad p \neq q.$$

Indeed, it is enough to check this for $p=1$, $q=2$, and we leave this routine verification to you. On the other hand, for basis elements ω,

$$\mathrm{i}\left(\frac{\partial}{\partial x_1}\right)\varepsilon(\mathrm{d}x_1)\omega=\begin{cases}0 & \text{if } \omega=\mathrm{d}x_1\wedge\cdots,\\ \omega & \text{if } \omega=\mathrm{d}x_2\wedge\cdots,\end{cases}$$

while

$$\varepsilon(\mathrm{d}x_1)\mathrm{i}\left(\frac{\partial}{\partial x_1}\right)\omega=\begin{cases}\omega & \text{if } \omega=\mathrm{d}x_1\wedge\cdots,\\ 0 & \text{if } \omega=\mathrm{d}x_2\wedge\cdots.\end{cases}$$

Thus

$$\mathrm{i}\left(\frac{\partial}{\partial x_1}\right)\varepsilon(\mathrm{d}x_1)+\varepsilon(\mathrm{d}x_1)\mathrm{i}\left(\frac{\partial}{\partial x_1}\right)=\mathrm{id}.$$

As there is nothing general about 1, we conclude that

$$\iota_p\varepsilon_p+\varepsilon_p\iota_p=\pm\,\mathrm{id},\quad \text{where}\pm=\operatorname{sign}(\mathrm{d}x_p,\mathrm{d}x_p).$$

Finally, since the operators ι_p and ε_q are all constant (in our coordinate description), we have

$$\begin{aligned}\mathrm{dd}^*+\mathrm{d}^*\mathrm{d}&=-\left(\sum\varepsilon_p\frac{\partial}{\partial x_p}\right)\left(\sum\iota_q\frac{\partial}{\partial x_q}\right)-\left(\sum\iota_q\frac{\partial}{\partial x_q}\right)\left(\sum\varepsilon_p\frac{\partial}{\partial x_p}\right)\\ &=-\sum_{p,q}(\varepsilon_p\iota_q+\iota_q\varepsilon_p)\frac{\partial^2}{\partial x_p\partial x_q}\\ &=-\sum\pm\frac{\partial^2}{\partial x_p^2}.\end{aligned}$$

This completes our proof of the formula for $\square$ in Euclidean coordinates.

For the case of positive-definite scalar product, the operators ε_p and ι_q have an important significance in quantum physics where they are known as the *creation* and *annihilation* operators for fermions. The relations

$$\varepsilon_p\varepsilon_q+\varepsilon_q\varepsilon_p=0,\quad \iota_p\iota_q+\iota_q\iota_p=0,$$

$$\varepsilon_p\iota_q+\iota_q\varepsilon_p=\begin{cases}0 & \text{if } p\neq q,\\ 1 & \text{if } p=q,\end{cases}$$

are known as the *anticommutation relations* for these operators.

In mathematics, the algebra generated by these operations – that is, the set of all sums and products of the εs and ιs – is a special case of a family of algebras constructed by Clifford in the last century. These algebras – i.e., sets of objects in which addition and multiplication are defined – were created by Clifford as a generalization of the complex number system and of the quaternions of Hamilton. As these algebras play an important role in modern physics, we take this space here to describe them.

We begin by reformulating the preceding example of the anticommutation relations of the creation and annihilation operators for fermions. Let V be a vector space, and V^* its dual space. Let us consider the direct sum space $W=V\oplus V^*$. Thus a vector of W is a pair $\mathbf{w}=\begin{pmatrix}\mathbf{v}\\ \alpha\end{pmatrix}$ where $\mathbf{v}\in V$ and $\alpha\in V^*$. Let us define a scalar product on W by

$$(\mathbf{w}, \mathbf{w}') = \tfrac{1}{2}(\alpha(\mathbf{v}') + \alpha'(\mathbf{v}))$$

if

$$\mathbf{w} = \begin{pmatrix} \mathbf{v} \\ \alpha \end{pmatrix} \quad \text{and} \quad \mathbf{w}' = \begin{pmatrix} \mathbf{v}' \\ \alpha' \end{pmatrix}.$$

Let $\mathbf{v}_1, \ldots, \mathbf{v}_n$ be a basis of V and let $\alpha^1, \ldots, \alpha^n$ be the dual basis of V^*. Then the vectors

$$\begin{pmatrix} \mathbf{v}_1 \\ 0 \end{pmatrix}, \ldots, \begin{pmatrix} \mathbf{v}_n \\ 0 \end{pmatrix}, \quad \begin{pmatrix} 0 \\ \alpha^1 \end{pmatrix}, \ldots, \begin{pmatrix} 0 \\ \alpha^n \end{pmatrix}$$

form a basis of W. Let us call these vectors by the names $\mathbf{e}_p$ and $\mathbf{i}_q$. That is, we define:

$$\mathbf{e}_p = \begin{pmatrix} \mathbf{v}_p \\ 0 \end{pmatrix}, \quad \mathbf{i}_q = \begin{pmatrix} 0 \\ \alpha^q \end{pmatrix}.$$

Thus

$$(\mathbf{e}_p, \mathbf{e}_q) = 0, \quad (\mathbf{i}_p, \mathbf{i}_q) = 0,$$

$$(\mathbf{e}_p, \mathbf{i}_q) = \begin{cases} \tfrac{1}{2} & \text{if } p = q, \\ 0 & \text{if } p \neq q. \end{cases}$$

We can now write the anticommutation relations for the εs and ιs in a succinct way. Let us think of the ε_p as being associated to the $\mathbf{e}_p$ and the ι_q as being associated to the $\mathbf{i}_q$. That is, we consider the linear map γ which is defined by

$$\gamma(\mathbf{e}_p) = \varepsilon_p,$$
$$\gamma(\mathbf{i}_q) = \iota_q.$$

Then, for

$$\mathbf{w} = v_1\mathbf{e}_1 + \cdots + v_n\mathbf{e}_n + \alpha^1\mathbf{i}_1 + \cdots + \alpha^n\mathbf{i}_n,$$
$$\gamma(\mathbf{w}) = v_1\varepsilon_1 + \cdots + v_n\varepsilon_n + \alpha^1\iota_1 + \cdots + \alpha^n\iota_n$$

is an element of our algebra and

$$\gamma(\mathbf{w})\gamma(\mathbf{w}') + \gamma(\mathbf{w}')\gamma(\mathbf{w}) = 2(\mathbf{w}, \mathbf{w}')\text{id}.$$

This can be generalized to *any* vector space W with a (possibly degenerate) scalar product: To each such vector space W with scalar product (,), we associate an associative algebra† $C(W)$ called the *Clifford* algebra of W and (,). There should be a linear map $\gamma: W \to C(W)$ such that

(i) Every element of $C(W)$ can be written as a sum of products of the elements $\gamma(W)$ (and of multiples of $\mathbb{1}$)

(ii) $\gamma(\mathbf{w})\gamma(\mathbf{w}') + \gamma(\mathbf{w}')\gamma(\mathbf{w}) = 2(\mathbf{w}, \mathbf{w}')\mathbb{1}$

(When we use the phrase *the* Clifford algebra, this tacitly assumes a theorem – that given W and (,) there exists a $C(W)$ satisfying (i) and (ii) and that $C(W)$ is unique,

† By an associative algebra A we mean that A is a vector space with a bilinear map $A \times A \to A$ called multiplication. One assumes that this multiplication satisfies the associative law and that there exists an identity, $\mathbb{1}$, for multiplication.

up to isomorphism. We briefly sketch the proof of this theorem in the appendix to this chapter.)

Let us work out some examples:

(a) Take W to be one-dimensional with a negative-definite scalar product. So there is a vector $\mathbf{e}$ with

$$(\mathbf{e}, \mathbf{e}) = -1$$

and every element of V is a multiple of $\mathbf{e}$. Condition (ii) says that

$$\gamma(\mathbf{e})^2 = -1$$

and condition (i) then says that every element of $C(W)$ can be written as

$$\mathrm{a}\mathbb{1} + b\gamma(\mathbf{e}).$$

If we call $\gamma(\mathbf{e}) = \mathrm{i}$, we see that $C(W)$ is precisely the algebra of complex numbers.

(b) Take $V = \mathbb{R}^2$ with the negative of the usual scalar product. So

$$(\mathbf{v}, \mathbf{v}) = -x^2 - y^2 \quad \text{if} \quad \mathbf{v} = \begin{pmatrix} x \\ y \end{pmatrix}.$$

Set

$$\mathrm{i} = \gamma\begin{pmatrix} 1 \\ 0 \end{pmatrix} \quad \text{and} \quad \mathrm{j} = \gamma\begin{pmatrix} 0 \\ 1 \end{pmatrix}.$$

Then condition (ii) says that

$$\mathrm{i}^2 = -\mathbb{1}$$
$$\mathrm{j}^2 = -\mathbb{1},$$

and

$$\mathrm{ij} + \mathrm{ji} = 0.$$

Condition (i) then implies that every element of $C(W)$ can be written as

$$a\mathbb{1} + b\mathrm{i} + c\mathrm{j} + d\mathrm{k}$$

where we have set

$$\mathrm{k} = \mathrm{ij}.$$

Then

$$\mathrm{k}^2 = \mathrm{ijij} = -\mathrm{iijj} = -\mathbb{1}$$

so

$$\mathrm{i} = \mathrm{j}^2 = \mathrm{k}^2 = -\mathbb{1}.$$

Also

$$\mathrm{ij} = \mathrm{k},$$
$$\mathrm{jk} = \mathrm{jij} = -\mathrm{ijj} = \mathrm{i},$$

and

$$\mathrm{ki} = \mathrm{iji} = -\mathrm{iij} = \mathrm{j},$$

while

$$\mathrm{kj} = \mathrm{ijj} = -\mathrm{i},$$

and

$$\mathrm{ik} = \mathrm{iij} = -\mathrm{j}.$$

To summarize:

$$\begin{aligned} &\mathrm{i}^2 = \mathrm{j}^2 = \mathrm{k}^2 = -\mathbb{1}, \\ &\mathrm{ij} = \mathrm{k}, \quad \mathrm{jk} = \mathrm{i}, \quad \mathrm{ki} = \mathrm{j}, \\ &\mathrm{ij} = -\mathrm{ji}, \quad \mathrm{jk} = -\mathrm{kj}, \quad \mathrm{ik} = -\mathrm{ki}. \end{aligned}$$

This algebra is known as the algebra of *quaternions*. It was first discovered by Hamilton.

In general, in checking condition (ii), it is enough to check that

$$\gamma(\mathbf{w})^2 = (\mathbf{w}, \mathbf{w})\mathbb{1}.$$

Indeed,

$$\gamma(\mathbf{w}+\mathbf{w}')^2 = [\gamma(\mathbf{w}) + \gamma(\mathbf{w}')]^2 = \gamma(\mathbf{w})^2 + \gamma(\mathbf{w})\gamma(\mathbf{w}') + \gamma(\mathbf{w}')\gamma(\mathbf{w}) + \gamma(\mathbf{w}')^2$$

and

$$\begin{aligned} \gamma(\mathbf{w}+\mathbf{w}')^2 &= (\mathbf{w}+\mathbf{w}', \mathbf{w}+\mathbf{w}')\mathbb{1} \\ &= (\mathbf{w}, \mathbf{w})\mathbb{1} + 2(\mathbf{w}, \mathbf{w}')\mathbb{1} + (\mathbf{w}', \mathbf{w}')\mathbb{1} \end{aligned}$$

so we get condition (ii).

With this in mind, let us take $W = \mathbb{R}^{1,3}$, that is, W is spacetime with its Lorentz metric. Set

$$\mathbf{w} = \begin{pmatrix} t \\ u \\ y \\ z \end{pmatrix} \text{ and } \gamma(\mathbf{w}) = \begin{pmatrix} 0 & 0 & t+z & x+\mathrm{i}y \\ 0 & 0 & x-\mathrm{i}y & t-z \\ t-z & -x-\mathrm{i}y & 0 & 0 \\ -x+\mathrm{i}y & t+z & 0 & 0 \end{pmatrix}.$$

Then

$$\gamma(\mathbf{w})^2 = (t^2 - x^2 - y^2 - z^2)\begin{pmatrix} 1 & 0 & 0 & 0 \\ 0 & 1 & 0 & 0 \\ 0 & 0 & 1 & 0 \\ 0 & 0 & 0 & 1 \end{pmatrix}.$$

So if we call the 4×4 identity matrix $\mathbb{1}$, then we see that (ii) is satisfied. We can, in fact, regard $C(W)$ as a certain subalgebra of the algebra of 4×4 complex matrices. The algebra $C(W)$ is called the *Dirac algebra*. It was developed by Dirac for his study of the electron and plays a central role in the modern theory of elementary particles.

(c) Take W to be any vector space and (,) to be identically zero. Then (ii) says that

$$\gamma(\mathbf{w})\gamma(\mathbf{w}') = -\gamma(\mathbf{w}')\gamma(\mathbf{w})$$

for any $\mathbf{w}$ and $\mathbf{w}'$. In this case, $C(W)$ is exactly the exterior algebra, $\Lambda(W)$.

18.6. The star operator and geometry

We saw in Chapter 16 that the star operator going from $\Lambda^1(\mathbb{R}^3)$ to $\Lambda^2(\mathbb{R}^3)$ determines the Euclidean scalar product on $\mathbb{R}^3$. This is the mathematical expression of the fact that dielectric properties of the vacuum determine the Euclidean geometry of space.

We can thus pose the following question. Let V be a vector space with orientation and let $(\ , \ )$ and $(\ , \ )'$ be two non-degenerate scalar products on V. Suppose that for some $1 \leqslant k < n = \dim V$ they give rise to the *same* star operator from $\Lambda^k(V^*)$ to $\Lambda^{n-k}(V^*)$. Does this imply that $(\ , \ ) = (\ , \ )'$? The answer is 'almost'. Here are some examples.

Take $V = \mathbb{R}^2$ with its usual Euclidean scalar product. We can then identify $\Lambda^1(\mathbb{R}^{2*})$ with $\mathbb{R}^2$. The star operator $\star\colon \Lambda^1(\mathbb{R}^{2*}) \to \Lambda^1(\mathbb{R}^{2*})$ is, as we have seen, in this case just rotation through ninety degrees. Any other scalar product on $\mathbb{R}^2$ (or $\mathbb{R}^{2*}$) is given by

$$(\mathbf{u}, \mathbf{v})' = (A\mathbf{u}, \mathbf{v})$$

for some symmetric matrix A. If $(\ , \ )'$ determines the same operator, we must have

$$(\star\mathbf{u}, \star\mathbf{v})' = (\mathbf{u}, \mathbf{v})'$$

so

$$R^{-1}AR = A.$$

But if $RAR^{-1} = A$ for some rotation R (other than through 0° or 180°), then A must be a scalar multiple of the identity. Thus

$$(\mathbf{u}, \mathbf{v})' = \lambda(\mathbf{u}, \mathbf{v})$$

for some non-zero number λ.

In fact, as we shall check in a moment, any such $(\ , \ )'$ with $\lambda > 0$ determines the same $\star$ operator. Thus in the plane, the $\star$ operator does not determine the Euclidean geometry, but it *does* determine the conformal geometry of the plane. This fact lies at the basis of the theory of functions of a complex variable. We shall present an introduction to this subject in Chapter 20.

Let us see, in general, what the effect of a *scale transformation* – replacing $\mathbf{v}$ by $c\mathbf{v}$ for some non-zero number c – has on the $\star$ operator. Multiplying lengths by c multiplies areas by c^2, volumes by c^3 etc., so we are multiplying by a factor of c^k on $\Lambda^k(V^*)$. Then, if the scale factor c is to have no effect, we must have

$$(c^k\lambda, c^{n-k}\omega)c^n\sigma = c^k\lambda \wedge c^{n-k}\omega$$

or, to get the same linear function of ω, we must have

$$c^{2k-n} = 1.$$

So

For $k = n/2$ a scale transformation has no effect on $\star\colon \Lambda^k(V^*) \to \Lambda^k(V^*)$.

It is not hard to prove – much as the proof we gave in Chapter 16 for $\star\colon \Lambda^1(\mathbb{R}^{3*}) \to \Lambda^2(\mathbb{R}^{3*})$ – that in all other cases the star operator determines the scalar product.

We now turn to another important point that we can only discuss briefly. Recall in our discussion of the dielectric 'constant' we allowed the possibility that ε might be a variable symmetric matrix:

$$\begin{pmatrix} \varepsilon_{xx} & \varepsilon_{xy} & \varepsilon_{xz} \\ \varepsilon_{yx} & \varepsilon_{yy} & \varepsilon_{yz} \\ \varepsilon_{zx} & \varepsilon_{zy} & \varepsilon_{zz} \end{pmatrix}$$

where all the entries are functions on $\mathbb{R}^3$. This suggests the possibility that the very notion of length should be allowed to vary from point to point: that is, that the scalar product of two tangent vectors $\boldsymbol{\xi} = (x, \mathbf{v})$ and $\boldsymbol{\eta} = (x, \mathbf{w})$ at the point x be given by

$$\sum g_{ij}(x) v^i w^j$$

where the v^i and w^j are the coordinates of $\mathbf{v}$ and $\mathbf{w}$. This idea – that the scalar product might be allowed to vary from point to point – was first introduced by Riemann and has been a basic concept in geometry ever since. In particular, it gave rise to the theory of general relativity of Einstein. The theory of the star operator extends without difficulty to this Riemannian geometry.

18.7. The star operator and vector calculus

The existence of a scalar product establishes, as usual, a correspondence between the space V and its dual $V^* = \Lambda^1(V)$, which we can use to identify a differential form with a vector field. As a notational convenience, we shall identify the vector field associated with a differential form by writing an arrow over the symbol for the form: thus, if A is a one-form, $\vec{A}$ is the vector field related to it by

$$\omega[\vec{A}] = (A, \omega) \text{ for all one-forms } \omega.$$

Thus if

$$\vec{A} = A_1 \mathbf{e}_1 + A_2 \mathbf{e}_2 + \cdots$$

the associated one-form is

$$A = \pm A_1 \, \mathrm{d}x^1 \pm A_2 \, \mathrm{d}x^2 + \cdots$$

where for each term the $\pm$ sign is chosen according to whether $(\mathbf{e}_i, \mathbf{e}_i) = \pm 1$. In particular, for $\mathbb{R}^3$ with the Euclidean scalar product, if

$$\vec{A} = A_x \mathbf{e}_x + A_y \mathbf{e}_y + A_z \mathbf{e}_z$$

then

$$A = A_x \, \mathrm{d}x + A_y \, \mathrm{d}y + A_z \, \mathrm{d}z.$$

For four-dimensional spacetime, with the Lorentz scalar product, if

$$\vec{A} = A_t \mathbf{e}_t + A_x \mathbf{e}_x + A_y \mathbf{e}_y + A_z \mathbf{e}_z,$$

then

$$A = A_t \, \mathrm{d}t - A_x \, \mathrm{d}x - A_y \, \mathrm{d}y - A_z \, \mathrm{d}z.$$

We can now express the operations of vector calculus in $\mathbb{R}^3$ in terms of the star operator and the d operator. Then the dot product may be written

$$\vec{A}\cdot\vec{B} = \star(A \wedge \star B) = \star(B \wedge \star A)$$

while the cross product is

$$\vec{A} \times \vec{B} = \overrightarrow{\star(A \wedge B)}.$$

The differential operators are

$$\begin{aligned} \operatorname{grad} f &= \overrightarrow{\mathrm{d}f}, \\ \operatorname{curl} \vec{A} &= \star\overrightarrow{\mathrm{d}A}, \\ \operatorname{div} \vec{A} &= \star\mathrm{d}\star A. \end{aligned}$$

In coordinates other than Cartesian, the association between a vector $\vec{A}$ and a differential form A is most conveniently made by first constructing an orthonormal basis of one-forms at each point. For example, in cylindrical coordinates, $\mathrm{d}r, r\mathrm{d}\theta$, and $\mathrm{d}z$ are orthonormal one-forms, dual to the (unit) basis vectors $\hat{\mathbf{e}}_r$, $\hat{\mathbf{e}}_\theta$, and $\hat{\mathbf{e}}_z$. Thus

$$\vec{A} = A_r\hat{\mathbf{e}}_r + A_\theta\hat{\mathbf{e}}_\theta + A_z\hat{\mathbf{e}}_z$$

is associated with the one-form

$$A = A_r\,\mathrm{d}r + A_\theta(r\,\mathrm{d}\theta) + A_z\,\mathrm{d}z.$$

Similarly, in spherical coordinates, $\mathrm{d}r, r\mathrm{d}\theta$, and $r\sin\theta\mathrm{d}\phi$ are orthonormal, so

$$\vec{A} = A_r\hat{\mathbf{e}}_r + A_\theta\hat{\mathbf{e}}_\theta + A_\phi\hat{\mathbf{e}}_\phi$$

is associated with

$$A = A_r\,\mathrm{d}r + A_\theta r\mathrm{d}\theta + A_\phi r\sin\theta\,\mathrm{d}\phi.$$

It now becomes a simple matter to compute div, grad, and curl in cylindrical or spherical coordinates. The rule for applying d is the same in any coordinate system, while the star operator acts on the orthonormal basis $\{\mathrm{d}r, r\mathrm{d}\theta, \mathrm{d}z\}$ or $\{\mathrm{d}r, r\,\mathrm{d}\theta, r\sin\theta\,\mathrm{d}\phi\}$ exactly as it does on $\{\mathrm{d}x, \mathrm{d}y, \mathrm{d}z\}$. For example, we compute curl $\vec{A}$ in cylindrical coordinates, using curl $\vec{A} = \star\mathrm{d}A$:

$$\begin{aligned} \vec{A} &= A_r\hat{\mathbf{e}}_r + A_\theta\hat{\mathbf{e}}_\theta + A_z\hat{\mathbf{e}}_z, \\ A &= A_r\,\mathrm{d}r + rA_\theta\mathrm{d}\theta + A_z\mathrm{d}z, \end{aligned}$$

$$\begin{aligned} \mathrm{d}A = {} & \left(\frac{\partial A_z}{\partial\theta} - \frac{\partial(rA_\theta)}{\partial z}\right)\frac{1}{r}(r\mathrm{d}\theta \wedge \mathrm{d}z) + \left(\frac{\partial A_r}{\partial z} - \frac{\partial A_z}{\partial r}\right)(\mathrm{d}z \wedge \mathrm{d}r) \\ & + \left(\frac{\partial(rA_\theta)}{\partial r} - \frac{\partial A_r}{\partial\theta}\right)\frac{1}{r}(\mathrm{d}r \wedge r\mathrm{d}\theta), \end{aligned}$$

$$\star\mathrm{d}A = \left(\frac{\partial A_z}{\partial\theta} - \frac{\partial(rA_\theta)}{\partial z}\right)\frac{1}{r}\mathrm{d}r + \left(\frac{\partial A_r}{\partial z} - \frac{\partial A_z}{\partial r}\right)r\,\mathrm{d}\theta + \left(\frac{\partial(rA_\theta)}{\partial r} - \frac{\partial A_r}{\partial\theta}\right)\frac{1}{r}\mathrm{d}z,$$

so that, on replacing $\mathrm{d}r$ by $\hat{\mathbf{e}}_r$, $r\,\mathrm{d}\theta$ by $\hat{\mathbf{e}}_\theta$, $\mathrm{d}z$ by $\hat{\mathbf{e}}_z$,

$$\operatorname{curl} A = \left(\frac{\partial A_z}{\partial\theta} - \frac{\partial(rA_\theta)}{\partial z}\right)\frac{1}{r}\hat{\mathbf{e}}_r + \left(\frac{\partial A_r}{\partial z} - \frac{\partial A_z}{\partial r}\right)\hat{\mathbf{e}}_\theta + \left(\frac{\partial(rA_\theta)}{\partial r} - \frac{\partial A_r}{\partial\theta}\right)\frac{1}{r}\hat{\mathbf{e}}_z.$$

Similarly, we compute in spherical coordinates the divergence of a radial vector field using $\vec{A} = \star \mathrm{d} \star A$:

$$\vec{A} = A_r \hat{\mathbf{e}}_r,$$
$$A = A_r \, \mathrm{d}r,$$
$$\star A = A_r (r \, \mathrm{d}\theta \wedge r \sin\theta \, \mathrm{d}\phi) = r^2 A_r \sin\theta \, \mathrm{d}\theta \wedge \mathrm{d}\phi,$$
$$\mathrm{d}\star A = \frac{\partial (r^2 A_r)}{\partial r} \sin\theta \, \mathrm{d}r \wedge \mathrm{d}\theta \wedge \mathrm{d}\phi = \frac{1}{r^2} \frac{\partial (r^2 A_r)}{\partial r} \mathrm{d}r \wedge r \, \mathrm{d}\theta \wedge r \sin\theta \, \mathrm{d}\phi,$$
$$\star \mathrm{d} \star A = \frac{1}{r^2} \frac{\partial (r^2 A_r)}{\partial r} = \operatorname{div} \vec{A}.$$

In both examples, the secret is to express all differential forms in terms of orthonormal one-forms before applying the star operator.

Appendix: tensor products

The notion of the tensor product of two (or more) vector spaces is essential to much of what we do in this book. It is a basic notion in the theory of vector spaces, but, as it does not appear in many of the elementary linear algebra texts, we give a brief introduction to the subject here.

Let V and W be vector spaces, and let U be a third vector space. A map $f: V \times W \to U$ is called *bilinear* if

$$f(\mathbf{v}_1 + \mathbf{v}_2, \mathbf{w}) = f(\mathbf{v}_1, \mathbf{w}) + f(\mathbf{v}_2, \mathbf{w}),$$
$$f(\mathbf{v}, \mathbf{w}_1 + \mathbf{w}_2) = f(\mathbf{v}, \mathbf{w}_1) + f(\mathbf{v}, \mathbf{w}_2),$$

and

$$f(a\mathbf{v}, \mathbf{w}) = a f(\mathbf{v}, \mathbf{w}) = f(\mathbf{v}, a\mathbf{w}) \quad \text{for scalar } a.$$

In other words, f is called bilinear if it is linear in $\mathbf{v}$ when $\mathbf{w}$ is held fixed, and is linear in $\mathbf{w}$ when $\mathbf{v}$ is held fixed.

A familiar example, where $V = W = U$, is the *vector product* in ordinary three-dimensional space.

Starting with V and W we wish to construct a vector space Z and a bilinear map $b: V \times W \to Z$ which is universal in the following sense. Suppose that $f: V \times W \to U$ is any bilinear map. Then there exists a *unique* linear map $l_f: Z \to U$ such that

$$f = l_f \circ b.$$

Diagrammatically,

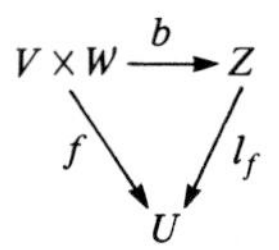

for any f there is a unique linear l_f making the diagram commute i.e. $l_f \circ b = f$.

Notice that *if* b and Z exist, they are unique up to isomorphism. Indeed, suppose b' and Z' were a second *universal* bilinear map. Then taking $U = Z'$ and $f = b'$ in

the above diagram, we get

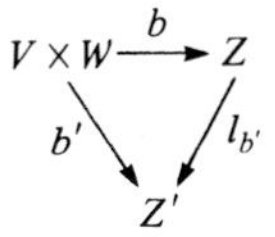

so $l_{b'}: Z \to Z'$ with

$$b' = l_{b'} \circ b.$$

Similarly we get a linear map $l_b: Z' \to Z$ with

$$b = l_b \circ b'.$$

But then

$$b = (l_b \circ l_{b'}) \circ b.$$

But

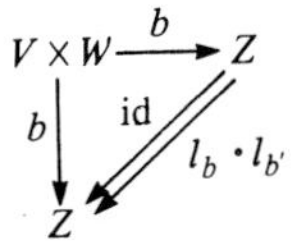

then id and $l_b \circ l_{b'}$ both satisfy the equation

$$b = l \circ b.$$

By the uniqueness, we conclude that

$$l_b \circ l_{b'} = \text{id};$$

and similarly

$$l_{b'} \circ l_b = \text{id}.$$

Thus $l_{b'}$ gives an isomorphism of Z with Z' and $b' = l_{b'} \circ b$. So, up to isomorphism, (b, Z) is unique.

Our problem is thus to show that such a (b, Z) actually exists. We shall give two rather different looking constructions. Of course, we know by uniqueness that they must give the same final answer.

Let Y denote the space of all scalar-valued bilinear functions on $V \times W$. Thus $\alpha \in Y$ is a bilinear map of $V \times W \to \mathbb{R}$. It is clear that the set of all bilinear functions, α, forms a vector space and this is Y. Now take

$$Z = Y^*.$$

So a vector in Z is a linear function on Y. Define the map $b: V \times W \to Z$ by

$$[b(\mathbf{v}, \mathbf{w})](\alpha) = \alpha(\mathbf{v}, \mathbf{w}).$$

That is, $b(\mathbf{v}, \mathbf{w})$ is that linear function that assigns to each bilinear function, α, the value $\alpha(\mathbf{v}, \mathbf{w})$. It is clear that $b(\mathbf{v}, \mathbf{w})$ *is* a linear function of α and that it depends bilinearly on $\mathbf{v}$ and $\mathbf{w}$. We claim that b and Z give a solution to our universal problem.

Indeed, suppose that $f: V \times W \to U$ is a bilinear map, where U is some vector space. For each $v \in U^*$ we get a bilinear function

$$v \circ f \in Y,$$

We have thus defined a linear map, $l^*: U^* \to Y$:

$$l^*(v) = v \circ f.$$

Therefore, we get its adjoint, $l^{**}: Z = Y^* \to U^{**}$. By definition,

$$\begin{aligned} l^{**}(b(\mathbf{v}, \mathbf{w}))(v) &= b(\mathbf{v}, \mathbf{w})(l^*(v)) \\ &= (l^*(v))(\mathbf{v}, \mathbf{w}) \\ &= (v \circ f)(\mathbf{v}, \mathbf{w}) \\ &= v(f(\mathbf{v}, \mathbf{w})). \end{aligned}$$

In other words, $l^{**}(b(\mathbf{v}, \mathbf{w}))$ is that linear function on U^* which assigns to each v the number $v(f(\mathbf{v}, \mathbf{w}))$. Now recall that there is a canonical isomorphism

$$U \overset{i}{\to} U^{**}$$

where each $\mathbf{u} \in U$ is identified with the linear function $i(\mathbf{u})$ on U^* given by

$$i(\mathbf{u})(v) = v(\mathbf{u}) \quad \mathbf{u} \in U, v \in U^*.$$

We can thus write

$$l^{**}(b(\mathbf{v}, \mathbf{w})) = i(f(\mathbf{v}, \mathbf{w})).$$

Using the identification i, we can write $l^{**} = i \circ l$, where l is then defined as a map of $Z \to U$. Thus we have found a map $l: Z \to U$ satisfying

$$l \circ b = f.$$

Suppose there were two such maps. Then the difference, m, would imply

$$(m \circ b)(\mathbf{v}, \mathbf{w}) = 0 \qquad \text{for all} \quad \mathbf{v}, \mathbf{w} \in V.$$

So

$$m^*(v)(b(\mathbf{v}, \mathbf{w})) = 0 \quad \text{for all} \quad v \in U^*.$$

Thus $m^*(v) \in Y$ is that bilinear function on $V \times W$ which assigns 0 to all pairs $(\mathbf{v}, \mathbf{w})$. Thus $m^*(v)$ is the zero bilinear function. Thus m^* sends all of U^* into 0 so is the zero linear map. Thus $m = 0$. This proves the uniqueness of l.

The space Z is called the *tensor product* of the spaces V and W and is denoted by $V \otimes W$. We shall also use the notation

$$b(\mathbf{v}, \mathbf{w}) = \mathbf{v} \otimes \mathbf{w}.$$

Suppose that V and W are finite-dimensional. Let $\{\mathbf{e}_1, \ldots, \mathbf{e}_m\}$ be a basis of V and $\{\mathbf{f}_1, \ldots, \mathbf{f}_n\}$ be a basis of W. Then a bilinear function β on $V \otimes W$ is completely determined by its values $\beta(\mathbf{e}_i, \mathbf{f}_j)$: we can write

$$\beta = \sum \beta(\mathbf{e}_i, \mathbf{f}_j) \varepsilon_{ij}$$

where ε_{ij} is the bilinear function determined by

$$\varepsilon_{ij}(\mathbf{v},\mathbf{w}) = v_i w_j \quad \text{if} \quad \mathbf{v} = v_i\mathbf{e}_1 + \cdots + v_m\mathbf{e}_m, \mathbf{w} = w_1\mathbf{f}_1 + \cdots + w_n\mathbf{f}_n.$$

The ε_{ij} are clearly linearly independent and span Y; they form a basis of Y. The dual basis of $Z = Y^*$ is just $b(\mathbf{e}_i, \mathbf{f}_j) = \mathbf{e}_i \otimes \mathbf{f}_j$. Thus,

If $\{\mathbf{e}_1,\ldots,\mathbf{e}_m\}$ is a basis of V and $\{\mathbf{f}_1,\ldots,\mathbf{f}_n\}$ is a basis of W, then $\{\mathbf{e}_i \otimes \mathbf{f}_j\}_{i=1,\ldots,m;j=1,\ldots,n}$ gives a basis in $V\otimes W$. In particular

$$\dim(V\otimes W) = (\dim V) \times (\dim W).$$

Here is another construction of $V\otimes W$. It is more direct – and abstract – but has the disadvantage of invoking some infinite-dimensional spaces. For any set M, let $F(M)$ denote the vector space of all formal linear combination of elements of M. Thus an element of $F(M)$ is a finite expression of the form

$$a_{m_1}m_1 + \cdots + a_{m_k}m_k$$

where the m_i are elements of M and the a_{m_i} are scalars. We add two such expressions by adding the coefficients. If M were a finite set, $F(M)$ would simply be the space of *all* functions on M, and the function f corresponds to the formal expression

$$\sum_{m\in M} f(m)m.$$

If M is not finite, then $F(M)$ can be thought of as the space of all functions on M which vanish except on a finite number of points. (Here we are identifying $m\in M$ with the function δ_m where $\delta_m(n) = 0$ if $n\in m$ and $\delta_m(m) = 1$. Thus the most general element of $F(M)$ is a finite linear combination of the δ_m. The map $M \to F(M)$ sending $m \to \delta_m$ gives M as a subspace of $F(M)$.)

The space $F(M)$ is *universal* with respect to maps of M into vector spaces: given any map $\phi: M \to U$, where U is a vector space, there is a unique linear map, $L_\phi: F(M) \to U$ such that

$$L_\phi \circ \delta_m = \phi(m).$$

Indeed, this last equation defines L_ϕ on the elements δ_m and hence L_ϕ extends by linearity to all of $F(M)$.

Now let us get back to our problem. Take $M = V\times W$. If ϕ is any map of the *set* $V\times W \to U$, there is an $L_\phi: F(V\times W) \to U$ which is linear and satisfies $L_\phi \circ \delta_m = \phi(m)$ for all $m\in V\times W$. If $f: V\times W \to U$ is bilinear, then L_f must vanish on all the elements

$$\delta_{(\mathbf{v}_1+\mathbf{v}_2,\mathbf{w})} - \delta_{(\mathbf{v}_1,\mathbf{w})} - \delta_{(\mathbf{v}_2,\mathbf{w})},$$
$$\delta_{(\mathbf{v},\mathbf{w}_1+\mathbf{w}_2)} - \delta_{(\mathbf{v},\mathbf{w}_1)} - \delta_{(\mathbf{v},\mathbf{w}_2)},$$
$$\delta_{(r\mathbf{v},\mathbf{w})} - r\delta_{(\mathbf{v},\mathbf{w})},$$
$$\delta_{(\mathbf{v},r\mathbf{w})} - r\delta_{(\mathbf{v},\mathbf{w})}.$$

Thus, L_f must vanish on the subspace B spanned by these elements. We can thus define $V\otimes W$ to be the quotient space

$$V\otimes W = F(V\times W)/B$$

and the map $b: V \times W \to V \otimes W$ by

$$b(\mathbf{v}, \mathbf{w}) = [\delta_{\mathbf{v},\mathbf{w}}]$$

where $[\cdot]$ denotes the equivalence class mod B. It is now clear that this definition of b and $V \otimes W$ fulfills the universal properties.

Let us now draw some consequences of the universal property. Let V' and W' be another pair of vector spaces. Let $A: V \to V'$ and $B: W \to W'$ be linear transformations. Consider the map sending $(\mathbf{v}, \mathbf{w})$ into $A\mathbf{v} \otimes B\mathbf{w}$. This is clearly bilinear. Hence there exists a unique *linear* map from $V \otimes W \to V' \otimes W'$, which we shall denote by $A \otimes B$, such that

$$(A \otimes B)(\mathbf{v} \otimes \mathbf{w}) = A\mathbf{v} \otimes B\mathbf{w}.$$

Suppose that $A': V' \to V''$ and $B': W' \to W''$ are a second pair of vector spaces and linear transformations. Then the map sending $(\mathbf{v}, \mathbf{w})$ into $A'Av \otimes B'Bw$ gives

$$A'A \otimes B'B: V \otimes W \to V'' \otimes W''.$$

But we also have the maps

$$A \otimes B: V \otimes W \to V' \otimes W'$$

and

$$A' \otimes B': V' \otimes W' \to V'' \otimes W''$$

so we get

$$(A' \otimes B') \cdot (A \otimes B): V \otimes W \to V'' \otimes W''.$$

By construction $(A' \otimes B')(A\mathbf{v} \otimes B\mathbf{w}) = AA'\mathbf{v} \otimes BB'\mathbf{w}$. Hence we conclude (from the uniqueness part of the universal property) that

$$(A' \otimes B') \cdot (A \otimes B) = A'A \otimes B'B.$$

Here is another consequence of the defining property of tensor products. Suppose that V and W are equipped with scalar products, $(\ ,\ )_V$ and $(\ ,\ )_W$. Consider the scalar valued function f defined on $(V \times W) \times (V \times W)$ by

$$f((\mathbf{v}, \mathbf{w}), (\mathbf{v}', \mathbf{w}')) = (\mathbf{v}, \mathbf{v}')_V (\mathbf{w}, \mathbf{w}')_W.$$

For fixed $\mathbf{v}'$ and $\mathbf{w}'$, f is bilinear in $\mathbf{v}$ and $\mathbf{w}$, hence defines a linear function $l_{(\mathbf{v}',\mathbf{w}')}$ on $V \otimes W$ satisfying

$$l_{(\mathbf{v}',\mathbf{w}')}(\mathbf{v} \otimes \mathbf{w}) = (\mathbf{v}, \mathbf{v}')_V (\mathbf{w}, \mathbf{w}')_W.$$

For fixed $\mathbf{v}$ and $\mathbf{w}$, this expression is bilinear in $\mathbf{v}'$ and $\mathbf{w}'$. Since every $\alpha \in V \otimes W$ is a finite sum of elements of the form $\mathbf{v} \otimes \mathbf{w}$, we conclude that

$$l_{\mathbf{v}',\mathbf{w}'}(\alpha)$$

is bilinear in $\mathbf{v}'$ and $\mathbf{w}'$, for any fixed α in $V \otimes W$. Thus there is an linear function l^α on $V \otimes W$ such that

$$l^\alpha(\mathbf{v}' \otimes \mathbf{w}') = l_{\mathbf{v}',\mathbf{w}'}(\alpha).$$

Then

$$l^\alpha(\beta)$$

is linear in α and (anti-)linear in β. It is easy to check that

$$(\alpha, \beta)_{V\otimes W} = l^{\alpha}(\beta)$$

defines a scalar product on $V \otimes W$. To summarize:

There is a unique scalar product $(\ ,\ )_{V\otimes W}$ defined on $V \otimes W$ which has the property that

$$(\mathbf{v} \otimes \mathbf{w}, \mathbf{v}' \otimes \mathbf{w}')_{V\otimes W} = (\mathbf{v}, \mathbf{v}')_V (\mathbf{w}, \mathbf{w}')_W.$$

Tensor products and Hom

Here is one further identification involving tensor products which is very useful. Let V and W be vector spaces. We define the vector space $\mathrm{Hom}(W, V)$ to be the space of all linear transformations from W to V. This is a vector space (by the addition of linear transformations) where dimension is $(\dim W)(\dim V)$. For each $\mathbf{v} \in V$ and $\mu \in W^*$, consider the rank 1 linear transformation $T^{\mu}_{\mathbf{v}}: W \to V$ define by

$$T^{\mu}_{\mathbf{v}}(\mathbf{w}) = \langle \mu, \mathbf{w} \rangle \mathbf{v}.$$

Here $\langle \mu, \mathbf{w} \rangle$ denotes the value of the linear function $\mu \in W^*$ on the vector $\mathbf{w} \in W$. The map $T^{\mu}_{\mathbf{v}}$ clearly depends linearly on $\mathbf{v}$ for fixed μ and linearly on μ for fixed $\mathbf{v}$. Thus we have a unique linear map

$$i: V \otimes W^* \to \mathrm{Hom}(W, V)$$

determined by

$$i(\mathbf{v} \otimes \mu)\mathbf{w} = \langle \mu, \mathbf{w} \rangle \mathbf{v}.$$

It is easy to check that the map i is an injection. It maps $V \otimes W^*$ onto the subspace of $\mathrm{Hom}(W, V)$ consisting of those linear transformations of finite rank. If W or V is finite-dimensional, this is all of $\mathrm{Hom}(W, V)$. If V and W are both infinite-dimensional, this is a proper subspace.

Higher order tensor products

There are a number of extensions and modifications of the notion of tensor product which we now briefly describe. Suppose that instead of just two vector spaces V and W we had k vector spaces $V_1, \ldots, V_k$. We can then define a k-linear (or multilinear) map $f: V_1 \times \cdots \times V_k \to U$ to be a map which is linear in any one of the variables when all the others are held fixed. Then there is a universal space $V_1 \otimes \cdots \otimes V_k$ and multilinear map $m: V_1 \times \cdots \times V_k \to V_1 \otimes V_2 \otimes \cdots \otimes V_k$ just as in the other case. It follows from the universal properties that there is an isomorphism of

$$(V_1 \otimes \cdots \otimes V_k) \otimes (V_{k+1} \otimes \cdots \otimes V_{k+l})$$

with

$$V_1 \otimes \cdots \otimes V_{k+l}$$

since they both satisfy the universal property for $(k + l)$-linear maps.

As in the case $k = 2$, if we are given linear maps $A_1: V_1 \to W_1$, $A_2: V_2 \to W_2$, etc., we get a linear map

$$A_1 \otimes \cdots \otimes A_k: V_1 \otimes \cdots \otimes V_k \to W_1 \otimes \cdots \otimes W_k.$$

The generalization of the rule for composition holds as well. Also, under the identification of $(V_1 \otimes \cdots \otimes V_k) \otimes (V_{k+1} \otimes \cdots \otimes V_{k+l})$ with $V_1 \otimes \cdots \otimes V_{k+l}$, we get an identification of $(A_1 \otimes \cdots \otimes A_k) \otimes (A_{k+1} \otimes \cdots \otimes A_{k+l})$ with $A_1 \otimes \cdots \otimes A_{k+l}$.

Now suppose that $V_1 = V_2 = \cdots = V_k$ are the same vector space. We say that a multilinear map $f: V \times \cdots \times V$ (k times) is antisymmetric if

$$f(\mathbf{v}_1, \ldots, \mathbf{v}_i, \mathbf{v}_{i+1}, \ldots, \mathbf{v}_k) = -f(\mathbf{v}_1, \ldots, \mathbf{v}_{i+1}, \mathbf{v}_i, \ldots, \mathbf{v}_k)$$

for all $i = 1, 2, \ldots, k$. In other words, interchanging any two adjacent entries changes the sign of f. It then follows that for any permutation π of the indices $(1, \ldots, k)$

$$f(\mathbf{v}_{\pi(1)}, \ldots, \mathbf{v}_{\pi(k)}) = (\operatorname{sign} \pi) f(\mathbf{v}_1, \ldots, \mathbf{v}_k).$$

We can now look for a universal space and antisymmetric multilinear map. That is, we look for a vector space $\Lambda^k(V)$ and an antisymmetric multilinear map $m_k: V \times \cdots \times V$ (k times) $\to \Lambda^k V$ such that for any antisymmetric multilinear map $f: V \times \cdots \times V \to U$ there is a unique linear map $l_f: \Lambda^k(V) \to U$ such that $l \circ m_k = f$. Diagrammatically,

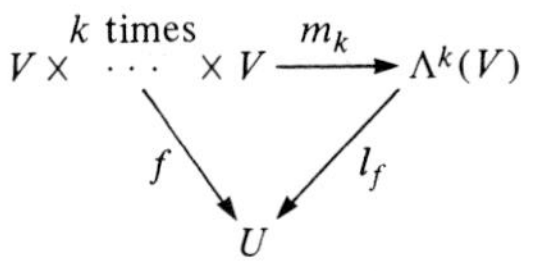

given f, there should exist a unique l_f making the diagram commute.

It follows immediately that, if $(m_k, \Lambda^k(V))$ exist, they are unique up to isomorphism. Also, an examination of either of our two proofs of the existence of the tensor products leads, with minor modifications, to a proof of the existence of $\Lambda^k(V)$ and m_k. In particular, the first proof shows the following:

> Let V^* be the dual space of V. Then $\Lambda^k(V^k)$ can be identified with the space of all k-linear antisymmetric functions on $V \times \cdots \times V$.

Exterior algebra

Consider the map $m_{k+l}: V \times \cdots \times V$($k + l$ times) into $\Lambda^{k+l}(V)$. It is multilinear and antisymmetric in the first k variables and also in the last l variables. Hence it defines a map of

$$\Lambda^k(V) \otimes \Lambda^l(V) \to \Lambda^{k+l}(V)$$

known as exterior multiplication, and denoted by $\wedge$. From the universal properties, it is easy to check that $\wedge$ is associative, that is

$$(\omega \wedge \sigma) \wedge \tau = \omega \wedge (\sigma \wedge \tau)$$

as maps into $\Lambda^{k+l+p}(V)$ where $\omega \in \Lambda^k(V)$, $\sigma \in \Lambda^l(V)$ and $\tau \in \Lambda^p(V)$. Also, by evaluating on basis elements, it can be checked that

$$\omega \wedge \sigma = (-1)^{kl} \sigma \wedge \omega \qquad \omega \in \Lambda^k(V), \sigma \in \Lambda^l(V).$$

Finally, suppose that V has a scalar product. Then the function on $(V \times \cdots \times V) \times (V \times \cdots \times V)$ (k times each factor) given by

$$f(\mathbf{v}_1, \ldots, \mathbf{v}_k; \mathbf{w}_1, \ldots, \mathbf{w}_k) = \operatorname{Det}((\mathbf{v}_i, \mathbf{w}_j)_V)$$

is multilinear and antisymmetric in $\mathbf{v}_1, \ldots, \mathbf{v}_k$ and is multilinear and antisymmetric in the $\mathbf{w}_1, \ldots, \mathbf{w}_k$. Thus, just as we argued for the induced scalar product on $V \otimes W$, it follows from the universal properties that f induces a scalar product on $\Lambda^k(V)$. In other words, we get a unique scalar product $(\ ,\ )_{\Lambda^k(V)}$ which is characterized by

$$(\mathbf{v}_1 \wedge \cdots \wedge \mathbf{v}_k, \mathbf{w}_1 \wedge \cdots \wedge \mathbf{w}_k)_{\Lambda^k(V)} = \mathrm{Det}\,((\mathbf{v}_i, \mathbf{w}_j)_V).$$

We can use the idea of the 'universal construction' to build some of the algebras we have studied in this chapter, such as the exterior algebra and the Clifford algebra.

Here are some of the details. A vector space A is called an *algebra* if there is given a map m (called multiplication)

$$m: A \otimes A \to A.$$

We shall denote $m(a \otimes b)$ by $a \cdot b$ (or just by ab). The algebra is called *associative* if

$$a(bc) = (ab)c$$

for all a, b, and c. It is said to have a unit (denoted by 1_A or just 1 if there is no risk of confusion) if

$$1_A a = a$$

for all elements of A. Unless otherwise specified, all algebras will be assumed associative and to have units.

The tensor algebra

Let V be a vector space. Consider the following universal problem: to find an algebra U and a linear map $i: V \to U$ such that for any algebra A and any linear map $f: V \to A$ there exists a unique *homomorphism* $\phi: U \to A$ such that

$$\phi(1_U) = 1_A$$

and

$$\phi(iv) = f(v) \quad \text{for all } v \in V.$$

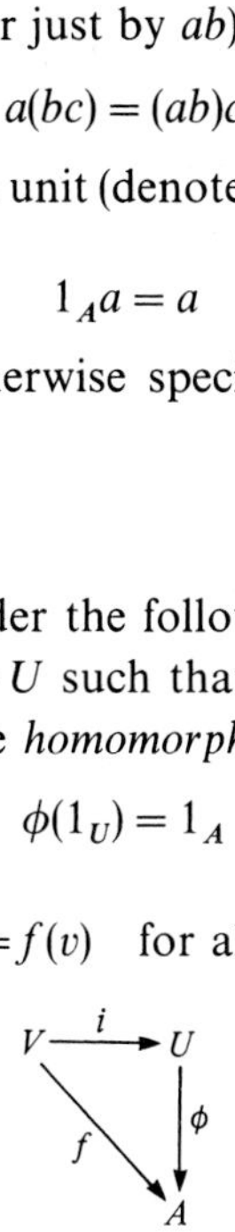

(Here the word homomorphism means that ϕ is a linear map satisfying $\phi(ab) = \phi(a)\phi(b)$.) If U and i exist, then the pair (U, i) is unique up to a 1–1 homomorphism – by the same arguments that apply to all universal constructs. (They are really arguments that belong to category theory.) The problem, as usual, is to construct one such algebra. Consider the (infinite) direct sum

$$T(V) = \mathbb{R} \oplus V \oplus (V \otimes V) \oplus (V \otimes V \otimes V) \otimes \cdots.$$

Make this into an algebra by defining the map

$$m = m_{pq}: (\overbrace{V \otimes \cdots \otimes V}^{p \text{ factors}}) \otimes (\overbrace{V \otimes \cdots \otimes V}^{q \text{ factors}}) \to \overbrace{V \otimes \cdots \otimes V}^{(p+q) \text{ factors}}$$

to be the unique map given by

$$m[(\mathbf{v}_1 \otimes \cdots \otimes \mathbf{v}_p) \otimes (\mathbf{w}_1 \otimes \cdots \otimes \mathbf{w}_q)] = \mathbf{v}_1 \otimes \cdots \otimes \mathbf{v}_p \otimes \mathbf{w}_1 \otimes \cdots \otimes \mathbf{w}_q$$

(just remove the parentheses). Multilinearity guarantees that this is well defined. So we know how to multiply a_p with b_q where

$$a_p \in T_p(V) \overset{\text{def}}{=} \overset{p}{V \otimes \cdots \otimes V}$$

and

$$b_q \in T_q(V).$$

(We define $r \cdot a = ra$ for $r \in \mathbb{R}$). Since every element of

$$T(V) = \bigoplus_{r=0}^{\infty} T_r(V)$$

is a finite sum of $a = a_0 + a_1 + \cdots + a_r$ of elements $a_p \in T_p(V)$, the distributive law (bilinearity) then determines multiplication in $T(V)$. Take $U = T(V)$ and let $i: V \to T(V)$ be the map which simply identifies V as the $r = 1$ piece, $T_1(V)$, of the direct sum, $T(V)$.

Now let $f: V \to A$ be any linear map. Then

$$f(v_1)f(v_2)$$

(multiplication in A) is bilinear in v_1 and v_2, and hence defines a map

$$\phi_2: V \otimes V \to A.$$

Similarly

$$\phi_r: T_r(V) \to A$$

is uniquely defined by

$$\phi_r(\mathbf{v}_1 \otimes \cdots \otimes \mathbf{v}_r) = f(\mathbf{v}_1) \cdots f(\mathbf{v}_r).$$

Finally set

$$\phi(a_0 + \cdots + a_r) = \phi_0(a_0) + \cdots + \phi_r(a_r)$$

(where $\phi_0(a_0) = a_0 \cdot 1_A$). It is easy to check that ϕ is a homomorphism and is the unique one which satisfies

$$\phi(i\mathbf{v}) = f(\mathbf{v}) \quad \text{for all } \mathbf{v} \in V.$$

Thus $T(V)$, i is the 'universal' algebra over V.

If A is an algebra, a subspace $I \subset A$ is called a (two-sided) ideal if

$$a \in A, b \in I \quad \text{implies} \quad ab \in I \quad \text{and} \quad ba \in I.$$

We can then define multiplication on the quotient space

$$B = A/I$$

by

$$[a/I] \cdot [a'/I] = [aa'/I]$$

where we have used the notation $[a/I]$ to denote the equivalence class of a mod I. The point is that the above rule is well defined – it is independent of the particular choice of a or a' mod I.

The Clifford algebra

We can now use this notion of quotient algebra to solve other 'universal construction' problems. For example, let us consider Clifford algebras. Let V be a vector space with a quadratic form Q, and associated scalar product (,). Let A be an associative algebra with unit 1_A. A Clifford map is a linear map

$$f\colon V\to A$$

such that

$$f(\mathbf{v})^2 = Q(\mathbf{v})1_A \quad \text{for all } \mathbf{v}\in\mathbf{V}.$$

We can equally write this condition as

$$f(\mathbf{u})f(\mathbf{v}) + f(\mathbf{v})f(\mathbf{u}) = 2(\mathbf{u},\mathbf{v})1_A.$$

The *Clifford algebra* over V is an algebra $C(V,Q)$ together with a Clifford map $j\colon V\to C(V,Q)$ which is universal in the usual sense: given any Clifford map $f\colon V\to A$ there is a unique *homomorphism* $\phi\colon C(V,Q)\to A$ such that $f=\phi\circ j$.

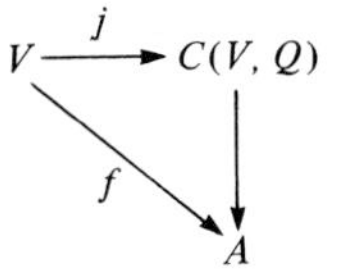

As usual, if a $C(V,Q)$ exists, it must be unique up to isomorphism. The problem is to construct one. The universal property of the tensor algebra forces our hand. If there is to be such a map j of V into the algebra $C(V,Q)$, we must have a unique homomorphism $\phi\colon T(V)\to C(V,Q)$ such that

$$j=\phi\circ i.$$

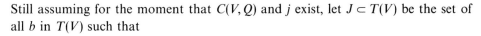

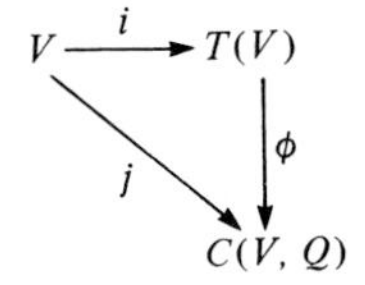

Still assuming for the moment that $C(V,Q)$ and j exist, let $J\subset T(V)$ be the set of all b in $T(V)$ such that

$$\psi(b)=0.$$

Then J is an ideal, and we would get an isomorphism of algebras

$$T(V)/J \xrightarrow{\bar{\psi}} C(V,Q)$$

induced from ψ:

$$\bar{\psi}[a/J]=\psi(a) \quad \text{(independent of the choice of } a \text{ in } [a/J]).$$

Now the elements

$$\mathbf{v}\otimes\mathbf{v} - Q(\mathbf{v})1_{T(V)}$$

would certainly have to lie in J by the defining property of Clifford maps. Hence J would have to contain all sums of right and left multiples of these elements. So let us start afresh. Let I be the ideal in $T(V)$ *generated* by the elements $\mathbf{v}\otimes\mathbf{v}-Q(\mathbf{v})1_{T(V)}$. That is I consists of all (finite) sums of expressions of the form

$$a\cdot(\mathbf{v}\otimes\mathbf{v}-Q(\mathbf{v})1_{T(V)})b$$

where a and b range over $T(V)$ and $\mathbf{v}$ over V. Then I is an ideal. Define

$$C(V,Q)=T(V)/I$$

and

$$j: V\to C(V,Q) \quad j(\mathbf{v})=[i(\mathbf{v})/I]$$
$$=[\mathbf{v}/I].$$

By construction j is a Clifford map and we leave it to you to check that all the universal properties are satisfied. In the special case that $Q\equiv 0$, the Clifford condition becomes

$$f(\mathbf{u})f(\mathbf{v})=-f(\mathbf{v})f(\mathbf{u}).$$

In this case (as you should check), the Clifford algebra is exactly the exterior algebra $\Lambda(V)$. A detailed analysis of the structure of Clifford algebra is given at the end of the Exercises to this chapter.

Summary

A Scalar products and the star operator

Given an orthonormal basis for $\Lambda^1(V^*)$, you should be able to compute scalar products on $\Lambda^k(V^*)$.

Given a vector space with a scalar product (not necessarily positive-definite) and an orientation, you should be able to use the definition of the star operator to compute $*\lambda$ for an arbitrary basis k-form λ.

You should be able to define the Laplace operator and express it in terms of partial derivatives with respect to Cartesian coordinates.

B Vector calculus

You should be able to define div, grad and curl in terms of d and $*$ and to use these definitions to prove identities of vector calculus or to express differential operators in orthogonal coordinate systems.

Exercises

18.1. Two-dimensional spacetime has affine coordinates t and x, with scalar product $(\mathbf{e}_t,\mathbf{e}_t)=+1$, $(\mathbf{e}_x,\mathbf{e}_x)=-1$, $(\mathbf{e}_t,\mathbf{e}_x)=0$. The basis two-form is $\mathrm{d}t\wedge\mathrm{d}x$.

(a) Using the definition of the star operator,

$$\lambda\wedge\omega=(\star\lambda,\omega)\sigma$$

calculate $\star 1$, $\star\mathrm{d}t$, $\star\mathrm{d}x$, and $\star(\mathrm{d}t\wedge\mathrm{d}x)$.

(b) Another pair of affine coordinate functions on this space is

$$u = t - x, v = t + x.$$

Calculate $\mathrm{d}u, \mathrm{d}v, \star\mathrm{d}u, \star\mathrm{d}v$.

(c) Let f be a twice-differentiable function on two-dimensional spacetime. Calculate the Laplacian of f,

$$\Box f = -\mathrm{d}\star\mathrm{d}f$$

in terms of partial derivatives of f with respect to t and x.

(d) Calculate $\Box f$ in terms of partial derivatives of f with respect to u and v.

18.2. When polar coordinates are used in the plane, the vectors $\mathbf{e}_r$ and $\mathbf{e}_\theta$ are an orthonormal basis at every point except the origin, where θ is not defined. The one-forms $\mathrm{d}r$ and $r\mathrm{d}\theta$ are dual to these basis vectors.

(a) Using the definition of the star operator,

$$\omega \wedge \lambda = (\star\omega, \lambda)\sigma,$$

calculate $\star\mathrm{d}r$ and $\star\mathrm{d}\theta$.

(b) Let f be a twice-differentiable function on the plane. Calculate the Laplacian of f

$$\Box f = -\mathrm{d}\star\mathrm{d}f$$

in terms of partial derivatives of f with respect to r and θ.

18.3. Translate the following identities into vector calculus notation (f denotes a function, A a one-form, but your answers will involve a vector $\mathbf{A}$).

(a) $\mathrm{d}(\mathrm{d}f) = 0$.

(b) $\mathrm{d}(\mathrm{d}A) = 0$. (In $\mathbb{R}^3$, $\star\star$ is the identity.)

(c) $\mathrm{d}(fA) = \mathrm{d}f \wedge A + f\mathrm{d}A$. (Apply $\star$ to both sides.)

(d) $\mathrm{d}(A \wedge B) = \mathrm{d}A \wedge B - A \wedge \mathrm{d}B$. (Look at $(\star\mathrm{d}\star)\star(A \wedge B)$.)

(e) $\mathrm{d}(f\star A) = \mathrm{d}f \wedge \star A + f\mathrm{d}\star A$. (Apply $\star$ to both sides.)

18.4. Let $f(t, x, y, z)$ be a twice-differentiable function on four-dimensional spacetime. Calculate the Laplacian $\Box f$ in terms of partial derivatives of f.

18.5. Let $\vec{B}$ be a vector field in spherical coordinates: $\vec{B} = B_r\mathbf{e}_r + B_\theta\mathbf{e}_\theta + B_q\mathbf{e}_q$. By calculating $\star\mathrm{d}\star B$, where B is the associated one-form, develop an expression for div $\vec{B}$.

18.6. (a) Let $\vec{A} = A_x\mathbf{e}_x + A_y\mathbf{e}_y + A_z\mathbf{e}_z$. By evaluating $(\mathrm{dd}^* + \mathrm{d}^*\mathrm{d})\,A$, where A is the one-form associated with $\vec{A}$, find an explicit formula for the *vector Laplacian* of $\vec{A}$ in terms of partial derivatives of A_x, A_y, A_z.

(b) Do the same in cylindrical coordinates, where $\vec{A} = A_r\hat{\mathbf{e}}_r + A_\phi\hat{\mathbf{e}}_\phi + A_z\hat{\mathbf{e}}_z$. Careful: The coefficient of $\mathrm{d}\theta$ will be rA_θ!

18.7. Develop a complete expression for curl $\vec{A}$ in spherical coordinates; i.e., start with

$$\vec{A} = A_r\mathbf{e}_r + A_\theta\mathbf{e}_\theta + A_\phi\mathbf{e}_\phi,$$

so that the corresponding form is

$$A = A_r\,\mathrm{d}r + A_\theta r\,\mathrm{d}\theta + A_\phi r\sin\theta\,\mathrm{d}\phi.$$

Calculate $\star\mathrm{d}A$, and reinterpret as a vector.

The purpose of the next discussion is to describe all (finite-dimensional) Clifford algebras over the real numbers. Recall that the Clifford algebra $C(V, W)$ is completely determined by the vector space V and the quadratic form Q. This has the following implication, as follows immediately from the 'universal' property of the Clifford algebra:

Let $f: V \to V'$ be a linear map which 'preserves length', that is, satisfies

$$Q'(f(v)) = Q(v) \quad \text{for all } v \in V.$$

Then f induces a unique homomorphism, $\phi_f: C(V, Q) \to C(V', Q')$ such that

$$\phi_f(v) = f(v) \quad \text{for} \quad v \in V,$$

when we regard V as a subspace of $C(V, Q)$. In particular, if f is an isometry, that is f is also a linear isomorphism of V with V', then ϕ_f is an isomorphism of $C(V, Q)$ with $C(V', Q)$. Thus the *classification* of Clifford algebras is the same as the classification of vector spaces with quadratic forms. We know the answer to this classification problem: the quadratic form is completely classified by its signature (p, q). More precisely, let $V^\perp \subset V$ denote the subspace of V consisting of all vectors u which satisfy $(u, v)_Q = 0$ for all $v \in V$. Then Q induces a scalar product on the quotient space $V/V^\perp$ which is non-degenerate, and we know from Gram–Schmidt that this induced scalar product is determined up to isomorphism by the numbers p and q of $+$ and $-$ signs in its orthonormal basis. If we choose a vector space complement W to $V^\perp$, so we can write $V = W \oplus V^\perp$, then W is isometric to $V/V^\perp$. If dim $V = n$, then V is isometric to $\mathbb{R}^n$ where the standard basis $\delta_1, \ldots, \delta_n$ is orthogonal with

$$\begin{aligned} Q(\delta_i) &= 1 & 1 \leqslant i \leqslant p, \\ Q(\delta_i) &= -1 & p+1 \leqslant i \leqslant q, \end{aligned}$$

and

$$Q(\delta_i) = 0 \quad p + q < i \leqslant n.$$

Thus the Clifford algebras are completely specified by the integers (n, p, q) with $p + q \leqslant n$. What we wish to do is get some insight into what each of these algebras actually looks like. We begin with an elementary but important consequence of our observation that an isometry $f: V \to V'$ induces an isomorphism ϕ_f of the corresponding Clifford algebra. Consider the map f of V into itself given by

$$f(v) = -v.$$

It is clearly an isometry and $f^2 = \text{id}$. It therefore induces an isomorphism ϕ_f of the Clifford algebra $C = C(V, Q)$ with itself. We shall denote this isomorphism by ω, so

$$\omega = \phi_f.$$

Notice that it follows from general principles that, if $f: V \to V'$ and $g: V' \to V''$ are isometries, then

$$\phi_{g \circ f} = \phi_g \circ \phi_f.$$

Taking $V = V' = V''$ and $g = f$, we get

$$\phi_{f^2} = \omega^2.$$

But $f^2 = \text{id}$ and, by the uniqueness property of ϕ, ϕ_{id} must be the identity isomorphism of C with itself. Thus

$$\omega^2 = \text{id}.$$

We say that ω is an *involution* of C.

The $\mathbb{Z}^2$ grading

Let C_0 be the subspace of C consisting of all c satisfying

$$\omega(c) = c$$

and let C_1 be the subspace consisting of all c satisfying

$$\omega(c) = -c.$$

Then any c can be written as

$$c = c_0 + c_1 \quad c_0 = \tfrac{1}{2}(c + \omega(c)) \quad c_1 = \tfrac{1}{2}(c - \omega(c))$$

with

$$c_0 \in C_0 \quad \text{and} \quad c_1 \in C_1.$$

Thus

$$C = C_0 \oplus C_1$$

as a vector space. Also if

$$\omega(c) = (-1)^i c \quad \text{and} \quad \omega(c') = (-1)^{i'} c'$$

then

$$\omega(cc') = (-1)^{i+i'} cc'$$

so

$$\begin{aligned} c_0 c'_0 \in C_0 \quad & \text{if} \quad c_0 \text{ and } c'_0 \in C_0, \\ c_0 c'_1 \in C_1 \quad & \text{if} \quad c_0 \in C_0 \text{ and } c_1 \in C_1, \\ c_1 c'_0 \in C_1 \quad & \text{if} \quad c_1 \in C_1 \text{ and } c'_0 \in C_0, \end{aligned}$$

and

$$c_1 c'_1 \in C_0 \quad \text{if} \quad c_1 \in C_1 \text{ and } c'_1 \in C_1.$$

We can summarize this fact by writing

$$\begin{aligned} C_0 \cdot C_0 &\subset C_0, \\ C_0 C_1 &\subset C_1, \\ C_1 C_0 &\subset C_1 \end{aligned}$$

and

$$C_1 C_1 \subset C_0.$$

We say that C is a ($\mathbb{Z}_2$-)*graded* algebra. Now $V \subset C_1$ and $\mathbb{1} \in C_0$ and every element of C can be written (perhaps in several ways) as sums of products of ($\mathbb{1}$ and) elements of V. Thus C_0 consists of those elements which can be written as sums of products of elements of V having an even number of factors, while C_1 consists of sums of products with an odd number of factors.

Twisted tensor products

Let $A = A_0 \oplus A_1$ and $B = B_0 \oplus B_1$ be two $\mathbb{Z}_2$-graded algebras. We define their *twisted tensor product* denoted by $\mathrm{A} \hat{\otimes} \mathrm{B}$ as follows. As a vector space, $A \hat{\otimes} B$ is from the Clifford algebraic properties. Now set

$$k = \gamma = ij.$$

Then

$$k^2 = -\mathbb{1}$$

and

$$ki + ik = 0 = kj + jk.$$

Thus $C(-2)$ is spanned by 1, i, j, and k and the above relations are precisely the defining properties of the quanternions. This algebra is frequently denoted by $\mathbb{H}$.

Let A and B be algebras. We define $A \otimes B$ as an algebra to be the vector space $A \otimes B$ with the multiplication law

$$(a \otimes b) \cdot (a' \otimes b') = aa' \otimes bb'.$$

(Notice, in contrast to the twisted tensor product, there is no sign change.)

Periodicity

We wish to prove the following basic formulas

$$C(p, q) \otimes C(2) \cong C(q + 2, p),$$

and

$$C(p, q) \otimes C(1, 1) \cong C(p + 1, q + 1),$$

$$C(p, q) \otimes C(-2) \cong C(q, p + 2),$$

where $\cong$ means isomorphism.

Let us prove the first of these isomorphisms. Let V be a $p + q$ vector space (say $\mathbb{R}^n$) with a quadratic form of type (p, q), and let $W = \mathbb{R}^2$ with its positive-definite inner product. Let γ denote the $\gamma = \mathbf{e}_1 \mathbf{e}_2$ of $C(W) = C(2)$. Consider the map

$$\psi: V \oplus W \to C(p, q) \otimes C(2)$$

given by

$$\psi(\mathbf{v}) = \mathbf{v} \otimes \gamma \quad \mathbf{v} \in V$$

and

$$\psi(\mathbf{w}) = \mathbb{1} \otimes \mathbf{w}.$$

Now

$$(\mathbf{v} \otimes \gamma)(\mathbb{1} \otimes \mathbf{w}) + (\mathbb{1} \otimes \mathbf{w})(\mathbf{v} \otimes \gamma) = \mathbf{v} \otimes (\gamma \mathbf{w} + \mathbf{w} \gamma) = 0$$

so

$$\psi(\mathbf{v})\psi(\mathbf{w}) + \psi(\mathbf{w})\psi(\mathbf{v}) = 0.$$

Also

$$\psi(\mathbf{w})^2 = \mathbb{1} \otimes \mathbf{w}^2 = \|\mathbf{w}\|^2 \mathbb{1}$$

and

$$\begin{aligned} \psi(\mathbf{v})^2 = (\mathbf{v} \otimes \gamma)(\mathbf{v} \otimes \gamma) &= \mathbf{v}^2 \otimes \gamma^2 \\ &= \mathbf{v}^2 \otimes (-\mathbb{1}) \\ &= -Q(\mathbf{v})\mathbb{1} \otimes \mathbb{1}. \end{aligned}$$

Thus, change Q into $-Q$ on V and take the direct sum of $-Q$ on V with $\| \ \|^2$

on W. For this quadratic form on $V \oplus W$, the map ψ satisfies the Clifford identities. This quadratic form has $q+2$ pluses and p minuses. Thus ψ extends to a homomorphism

$$\psi: C(q+2, p) = C(V \oplus W) \to C(p,q) \otimes C(2).$$

We leave it as an exercise to construct the inverse map and verify that ψ is an isomorphism. Exactly the same argument (except that $\|\mathbf{w}\|^2$ is now negative) gives the isomorphisms

$$C(p,q) \otimes C(1,1) \cong C(p+1, q+1)$$

and

$$C(p,q) \otimes C(-2) \cong C(q, p+2).$$

Here are some applications of these isomorphisms:

$$\begin{aligned} \mathbb{H} \otimes \mathbb{H} &= C(-2) \otimes C(-2) \\ &= C(0,2) \otimes C(-2) \\ &\cong C(2,2) \\ &\cong \mathbb{R}(2^2) = \mathbb{R}(4) \end{aligned}$$

so

$$\mathbb{H} \otimes \mathbb{H} \cong \mathbb{R}(4).$$

Now

$$C(1) \cong \mathbb{R} \oplus \mathbb{R}$$

under the isomorphism sending

$$x1 + y\mathbf{e} \to (x-y, x+y)$$

as can be easily checked (exercise). Thus

$$\begin{aligned} C(-3) \sim C(1) \otimes C(-2) &\cong (\mathbb{R} \oplus \mathbb{R}) \otimes \mathbb{H} \\ &\cong \mathbb{H} \oplus \mathbb{H}. \end{aligned}$$

We have already verified that $C(-1) \cong \mathbb{C}$. Hence

$$C(3) \cong C(-1) \otimes C(2) = \mathbb{C} \otimes \mathbb{R}(2)$$

which is just the algebra of complex two-by-two matrices, which we denote by $\mathbb{C}(2)$. Similarly

$$\begin{aligned} \mathbb{C} \otimes \mathbb{H} \cong C(-1) \otimes C(-2) &\cong C(0,1) \otimes C(0,2) \\ &\cong C(1,2) \\ &\cong C(0,1) \otimes C(1,1) \\ &\cong \mathbb{C} \otimes \mathbb{R}(2) = \mathbb{C}(2). \end{aligned}$$

So $\mathbb{C} \otimes \mathbb{H} \sim \mathbb{C}(2)$.

We can use these various isomorphisms to fill in, successively, a table of all the entries of the $C(p,q)$:

just the ordinary tensor product, $A\otimes B$, of A and B. Thus, as a vector space,

$$\begin{aligned}A\hat{\otimes} B &= (A_0\oplus A_1)\otimes(B_0\oplus B_1)\\ &= A\otimes B_0\oplus A_1\otimes B_0\oplus A_0\otimes B_1\oplus A_1\otimes B_1.\end{aligned}$$

We define multiplication:

$$\begin{aligned}(a\otimes b_0)\cdot(a'\otimes b') &= aa'\otimes b_0 b,\\ (a\otimes b)\cdot(a'_0\otimes b') &= aa'_0\otimes bb',\\ (a\otimes b_1)\cdot(a_1\otimes b') &= -aa'_1\otimes b_1 b',\end{aligned}$$

where subscript denotes whether the element is even or odd. A lack of subscript means that the equation is true regardless of whether the element is even or odd. More succinctly

$$(a\otimes b_i)(a'_j\otimes b') = (-1)^{ij}aa'_j\otimes b_i b'.$$

In other words, one must multiply by $(-1)^{ij}$ in order to move the a'_j past the b_i. The utility of the concept of twisted tensor product for us lies in the following. Let V and W be vector spaces equipped with quadratic forms Q_V and Q_W respectively. On the direct sum, $V\oplus W$, put the direct sum quadratic form. That is, V and W are to be orthogonal, and the quadratic form $Q_{V\oplus W}$ is to coincide with Q_V on V and with Q_W on W. Then we have an isomorphism

$$C(V\oplus W, Q_{V\oplus W}) \cong C(V, Q_V)\hat{\otimes} C(W, Q_W).$$

More precisely: the maps $f: V\to V\oplus W, f(v) = v\oplus 0$ and $g: W\to V\oplus W$, $g(w) = 0\oplus w$ 'preserve length' and so induce homomorphisms

$$\phi_f: C(V, Q_V)\to C(V\oplus W, Q_{V\oplus W})$$

and

$$\phi_g: C(W, Q_W)\to C(V\oplus W, Q_{V\oplus W}).$$

Define the map

$$\Phi: C(V, Q_V)\hat{\otimes} C(W, Q_W)\to C(V\oplus W, Q_{V\oplus W})$$

by

$$\Phi(a\hat{\otimes} b) = \phi_f(a)\cdot\phi_g(b).$$

We claim that ϕ is an isomorphism. First we must show that Φ is a homomorphism, i.e., that

$$\phi((a\otimes b)\cdot(a'\otimes b')) = \phi(a\otimes b)\cdot\phi(a'\otimes b').$$

It is enough to check this for

$$b = w_1\dots w_i$$

and

$$a' = v_1\dots v_j.$$

Now for $v\in V$ and $w\in W$ we have $(v, w) = 0$ in $V\oplus W$. (We write v instead of $f(v) = v\oplus 0$ and similarly for w.) So in the Clifford algebra $C(V\oplus W, Q_{V\oplus W})$ we have

$$v\cdot w + w\cdot v = 0$$

and hence

$$ba' = (-1)^{ij}a'b.$$

This is just what is required to prove that Φ is a homomorphism. To see that ϕ is an isomorphism, consider the map $h: V \oplus W \to C(V, Q_V) \hat{\otimes} C(W, Q_W)$ given by

$$h(v \oplus w) = 1_V \hat{\otimes} w + v \hat{\otimes} 1_W$$

where 1_V is the $\mathbb{1}$ of $C(V, Q_V)$ and 1_W is the $\mathbb{1}$ of $C(W, Q_W)$. It follows immediately from the definition of $\hat{\otimes}$ that

$$h(v \oplus 0)h(0 \oplus w) + h(0 \oplus w)h(v \oplus 0) = 0$$

while $h(v \oplus 0)^2 = Q(v)1_V$ and $h(0 + w)^2 = Q(w)1_W$. Thus by the universal properties of $C(V \oplus W, Q_{V \oplus W})$ the map h induces a homomorphism

$$\phi_h: C(V \oplus W, Q_{V \oplus W}) \to C(V, Q_V) \hat{\otimes} C(W, Q_W).$$

and

$$\begin{aligned} \phi \circ \phi_h(v \oplus w) &= \phi(v \hat{\otimes} 1_W + 1_V \hat{\otimes} w) \\ &= \phi_f(v \hat{\otimes} 1_W) + \phi_g(1_W \hat{\otimes} w) \\ &= v + w. \end{aligned}$$

Thus $\phi \circ \phi_h = \text{id}$ on $V + W$, hence, by uniqueness,

$$\phi \circ \phi_h = \text{id}.$$

Thus ϕ_h is a (right inverse) to ϕ. A similar argument shows that $\phi_h \circ \phi = \text{id}$. Thus ϕ is an isomorphism.

For example, if U is a vector space with the quadratic form $Q \equiv 0$, its Clifford algebra is just $\Lambda(U)$ (as follows from the definitions). Now let V be any vector space with quadratic form Q. Recall that we can write

$$V = V^{\perp} \oplus W$$

where the restriction, Q_W, of Q to the subspace W is non-degenerate. Then ϕ establishes an isomorphism of $C(V, Q)$ with $\Lambda(V^{\perp}) \hat{\otimes} C(W, Q_W)$. So, up to throwing in a twisted tensor product by an exterior algebra, we may restrict attention to quadratic forms which are non-degenerate. Let us introduce the notation

$$C(p, q) = C(\mathbb{R}^n, Q_{p,q})$$

where $Q_{p,q}$ denotes the standard quadratic form on $\mathbb{R}^n (p + q = n)$ with p +'s and q −'s. We wish to understand the structure of $C(p, q)$.

The split case

For example, in the case $p = q$, as a possible model for $C(p, p)$ we could take $V = U + U^*$ where U is a p-dimensional vector space and we take $(u, u') = 0$, $(u^*, u'^*) = 0$ for $u, u' \in U$ and $u^*, u'^* \in U^*$ and

$$(u^*, u) = \langle u^*, u \rangle$$

(the evaluation map). As discussed in the text we have a description the corresponding Clifford algebra in terms of creation and annihilation operators on $\Lambda(U)$.

An examination of the discussion there will show that the Clifford algebra becomes identified with the algebras of *all* linear transformations on $\Lambda(U)$. Since dim $\Lambda(U) = 2^p$ we can thus identify the Clifford algebra

$$C(p,p) \quad \text{with} \quad \mathbb{R}(2^p)$$

where we have used the notation $\mathbb{R}(n)$ to denote the algebra of all real $n \times n$ matrices.

For $p \neq q$ observe that we can write $\mathbb{R}^n$ as the direct sum of $\mathbb{R}^p$ with a positive-definite form and $\mathbb{R}^q$ with a negative-definite form. Hence, by our direct sum theorem, we have

$$C(p,q) = C(p,0) \mathbin{\hat{\otimes}} C(0,q).$$

If we use the notation $C(p) = C(p,0)$ and $C(-q) = C(0,q)$ we can write the preceding equation as

$$C(p,q) = C(p) \mathbin{\hat{\otimes}} C(-q).$$

We thus wish to understand $C(p)$ and $C(-q)$.

The nondegenerate case

For the moment, however, we study the general case of a vector space V with a non-degenerate quadratic form Q and let

$$C = C(V,Q).$$

Let $\{\mathbf{e}_1,\ldots,\mathbf{e}_n\}$ be an orthonormal basis of V (with $Q(\mathbf{e}_1) = \cdots = Q(\mathbf{e}_p) = 1$ and $Q(\mathbf{e}_i) = -1$ for $i = p+1,\ldots,n$). Consider the element

$$\gamma = \mathbf{e}_1 \cdots \mathbf{e}_n \quad \text{in } C.$$

If

$$x_i = \sum_{j=1}^{n} a_{ij}\mathbf{e}_i \quad i = 1,\ldots,n$$

are any elements of V, it follows from the orthogonality of the $\mathbf{e}_j$ that

$$x_1 \cdots x_n = \operatorname{Det}(a_{ij})\mathbf{e}_1 \cdots \mathbf{e}_n = \operatorname{Det}(a_{ij})\gamma.$$

In particular, if $\{\mathbf{f}_1,\ldots,\mathbf{f}_n\}$ is some other orthogonal basis, then

$$\mathbf{f}_1 \cdots \mathbf{f}_n = \pm\, \mathbf{e}_1 \cdots \mathbf{e}_n.$$

Thus, up to a sign, the element γ is independent of the choice of basis. (For the Dirac algebra, $C(1,3)$, the element γ is usually denoted in the physics literature by γ_5. The subscript 5 refers to the fact that in this four-dimensional case γ is the product of four elements.) Notice that

$$\gamma^2 = (-1)^{n(n-1)/2+q}\gamma.$$

The $n(n-1)/2$ comes from moving $\mathbf{e}_1$ past $\mathbf{e}_2 \cdots \mathbf{e}_n$, then $\mathbf{e}_2$ past $\mathbf{e}_3 \cdots \mathbf{e}_n$, etc. The q comes from the fact that $\mathbf{e}_{p+1}^2 = \cdots = \mathbf{e}_n^2 = -1$ where $n - p = q$.

Notice that for any i with $1 \leqslant i \leqslant n$

$$\gamma\mathbf{e}_i = \mathbf{e}_1 \cdots \mathbf{e}_n\mathbf{e}_i = (-1)^{n-i}\mathbf{e}_1 \cdots \mathbf{e}_i\mathbf{e}_i\mathbf{e}_{i+1} \cdots \mathbf{e}_n$$

while

$$\mathbf{e}_i\gamma = \mathbf{e}_i\mathbf{e}_1 \cdots \mathbf{e}_n = (-1)^{i-1}\mathbf{e}_1 \cdots \mathbf{e}_i\mathbf{e}_i\mathbf{e}_{i+1} \cdots \mathbf{e}_n.$$

Thus

$$\gamma \mathbf{e}_i = (-1)^{n-1}\mathbf{e}_i\gamma.$$

Since the $\mathbf{e}_i$ form a basis of V, it follows that

$$\gamma \mathbf{v} = (-1)^{n-1}\mathbf{v}\gamma \quad \text{for all} \quad \mathbf{v}\in V.$$

Hence

$$\gamma a = \omega(a)^{n-1}a\gamma \quad \text{for any} \quad a\in C.$$

Low-dimensional examples

Here are two important low-dimensional examples. Take $V = \mathbb{R}^2$. We claim that

$$C(2) \cong M(2) = \text{the algebra of all } 2\times 2 \text{ matrices},$$

when $\cong$ means 'is isomorphic to'. To prove this, let

$$\phi\colon \mathbb{R}^2 \to M(2)$$

be given by

$$\phi\left(\begin{pmatrix} x \\ y \end{pmatrix}\right) = \begin{pmatrix} x & y \\ y & -x \end{pmatrix}.$$

Then

$$\left(\phi\left(\begin{pmatrix} x \\ y \end{pmatrix}\right)\right)^2 = \begin{pmatrix} x^2+y^2 & 0 \\ 0 & x^2+y^2 \end{pmatrix} = (x^2+y^2)\begin{pmatrix} 1 & 0 \\ 0 & 1 \end{pmatrix}$$

so

$$Q(\mathbf{u})^2 = Q(\mathbf{v})\begin{pmatrix} 1 & 0 \\ 0 & 1 \end{pmatrix}$$

and $M(2)$ thus has all the desired properties of the Clifford algebra $C(2)$. If we take the orthogonal basis $\begin{pmatrix} 1 \\ 0 \end{pmatrix}, \begin{pmatrix} 0 \\ 1 \end{pmatrix}$ then the corresponding elements of the Clifford algebra $M(2)$ are

$$\mathbf{e}_1 = \begin{pmatrix} 1 & 0 \\ 0 & -1 \end{pmatrix} \quad \text{and} \quad \mathbf{e}_2 = \begin{pmatrix} 0 & 1 \\ 1 & 0 \end{pmatrix}$$

so

$$\gamma = \mathbf{e}_1\mathbf{e}_2 = \begin{pmatrix} 0 & 1 \\ -1 & 0 \end{pmatrix}.$$

Notice that $\gamma^2 = -1$ as required. The even piece, C_0, of C consists of all 2×2 matrices which commute with γ. This is just the set of conformal matrices. The subspace C_1, the odd component of C, consists, in this special case, of the subspace $\mathbb{R}^2$ itself, identified with the space of symmetric matrices of trace zero.

We have already investigated the case of $C(-2)$ and found that it is isomorphic to the quaternions. We review the construction. We define i to be the element $\begin{pmatrix} 1 \\ 0 \end{pmatrix}$ in $C(-2)$ and j to be the element $\begin{pmatrix} 0 \\ 1 \end{pmatrix}$. Then

$$\mathrm{i}^2 = -\mathbb{1}, \quad \mathrm{j}^2 = -\mathbb{1} \quad \mathrm{ij} + \mathrm{ji} = 0$$

	0	1	2	3	4 ...
0	$\mathbb{R}$	$\mathbb{C}$	$\mathbb{H}$	$\mathbb{H}\oplus\mathbb{H}$	
1	$\mathbb{R}+\mathbb{R}$	$\mathbb{R}(2)$			
2	$\mathbb{R}(2)$		$\mathbb{R}(4)$		
3	$\mathbb{C}(2)$				
4					
$\vdots$					

For example, to find $C(1,2)$ we tensor the entry, $\mathbb{C}$, for $C(0,1)$ by $\mathbb{R}(2) \cong C(1,1)$ to find that $C(1,2) \cong \mathbb{C}(2)$. Similarly, to find $C(3,1)$ we tensor the entry $\mathbb{R}(2) \cong C(2,0)$ by $\mathbb{R}(2)$ to find that $C(3,1) \cong \mathbb{R}(2) \otimes \mathbb{R}(2)$. But, by the definition of matrix algebras, $\mathbb{R}(R) \otimes \mathbb{R}(l) \cong \mathbb{R}(Rl)$. So $\mathbb{R}(2) \otimes \mathbb{R}(2) \cong \mathbb{R}(4)$. Similarly,

$$C(1,3) \cong \mathbb{H} \otimes \mathbb{R}(2) = \mathbb{H}(2),$$

the algebra of all 2×2 matrices with quaternionic entries. (Notice that $C(1,3)$ and $C(3,1)$ are *not* isomorphic.) Similarly

$$C(4,0) \cong C(0,2) \otimes C(2,0) = \mathbb{H} \otimes \mathbb{R}(2) \cong \mathbb{H}(2)$$

and

$$C(0,4) \cong C(2,0) \otimes C(0,2) \cong \mathbb{H}(2).$$

Thus by working back and forth you will find

	q								
	0	1	2	3	4	5	6	7	8
0	$\mathbb{R}$	$\mathbb{C}$	$\mathbb{H}$	$\mathbb{H}\oplus\mathbb{H}$	$\mathbb{H}(2)$	$\mathbb{C}(4)$	$\mathbb{R}(8)$	$\mathbb{R}(8)+\mathbb{R}(8)$	$\mathbb{R}(16)$
1	$\mathbb{R}\oplus\mathbb{R}$								
2	$\mathbb{R}(2)$								
3	$\mathbb{C}(2)$								
4	$\mathbb{H}(2)$								
5	$\mathbb{H}(2)\oplus\mathbb{H}(2)$								
6	$\mathbb{H}(6)$								
7	$\mathbb{C}(8)$								
8	$\mathbb{R}(16)$								

The rest of the table can be filled in by moving along the diagonal, tensoring by $\mathbb{R}(2)$ for each step. This eight by eight block then determines *all* the Clifford algebras, as $C(n+8,0) \cong C(n,0) \otimes \mathbb{R}(16)$ and $C(0,n+8) \cong C(0,n) \otimes \mathbb{R}(16)$ as follows from a fourfold application of our basic isomorphisms: For example

To compute $C(9)$ we can start from $C(1)$ and successively tensor by $C(-2)$,

$C(2)$, $C(-2)$, $C(2)$. But this is the same as tensoring by

$$\begin{aligned} &C(2)\otimes C(-2)\otimes C(2)\otimes C(-2)\\ &\cong C(-4)\otimes C(2)\otimes C(-2)\\ &\cong C(6)\otimes C(-2)\\ &\cong C(8)\cong C(-8)\cong \mathbb{R}(16). \end{aligned}$$

This same argument shows that

$$C(k+8)\cong C(k)\otimes \mathbb{R}(16)$$

and

$$C(-k-8)\cong C(-k)\otimes \mathbb{R}(16).$$

Thus

$$C(8r+k)\cong C(k)\otimes \mathbb{R}(16^r) \quad \text{(Bott periodicity)}$$

and

$$C(-kr-k)\cong C(-k)\otimes \mathbb{R}(16^r).$$

This completely determines the structure of all finite-dimensional Clifford algebras.

19

Maxwell's equations

Chapter 19 can be thought of as the culmination of the course. It applies the results of the preceding chapters to the study of Maxwell's equations and the associated wave equations.

19.1. The equations

Electrostatics and magnetostatics are only approximately correct. For time-varying fields they must be replaced by Maxwell's theory. We begin with *Faraday's law of induction* which says that for any surface S spanning a curve γ the time derivative of B is related to E by

$$\frac{\mathrm{d}}{\mathrm{d}t}\int_S B = -\int_\gamma E.$$

We now rewrite this in a form suitable to four dimensions. Consider an interval $[a, b]$ in time and the three-dimensional cylinder $S \times [a, b]$ whose boundary is the two-dimensional cylinder $\gamma \times [a, b]$ together with the top and the bottom of this two-dimensional cylinder. (See figure 19.1.) Integrating Faraday's law of induction

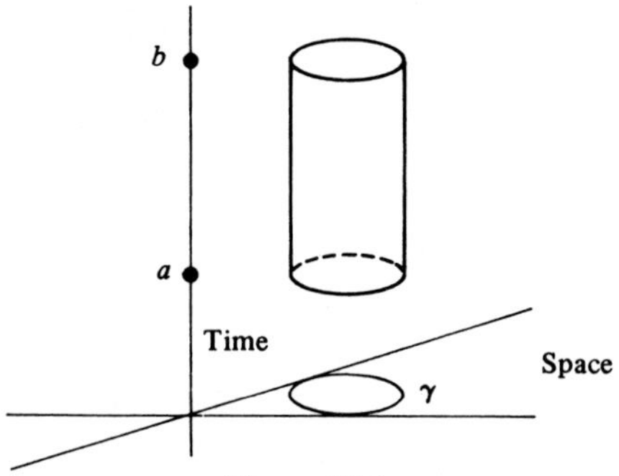

Figure 19.1

with respect to t from a to b gives the equation

$$\int_{S\times\{b\}} B - \int_{S\times\{a\}} B + \int_{\gamma\times\{a,b\}} E \wedge \mathrm{d}t = 0.$$

Let us set

$$F = B + E \wedge \mathrm{d}t$$

so that F is a two-form defined on four-dimensional space. Let C denote the three-dimensional cylinder, $C = S \times [a, b]$, so that

$$\partial C = S \times \{b\} - S \times \{a\} + \gamma \times [a, b].$$

Now, B is a two-form involving just the spatial differentials and, therefore, must vanish when restricted to the side $\gamma \times [a, b]$ of the cylinder, while $\mathrm{d}t$ and hence $E \wedge \mathrm{d}t$ must vanish on the top and the bottom. Thus, we can write Faraday's law of induction as

$$\int_{\partial C} F = 0.$$

We can also consider a three-dimensional region C lying entirely in space at one fixed constant time. In this case, ∂C will be a surface on which $\mathrm{d}t = 0$ so that

$$\int_{\partial C} F = \int_{\partial C} B$$

and, by the absence of true magnetism, the surface integral must vanish. Thus, $\int_{\partial C} F = 0$ for all three-dimensional cubes whose sides are parallel to any three of the four coordinate axes. This is enough to imply that $\mathrm{d}F = 0$, where now, of course, d stands for the exterior derivative in four-space. Since

$$F = B_x \mathrm{d}y \wedge \mathrm{d}z - B_y \mathrm{d}x \wedge \mathrm{d}z + B_z \mathrm{d}x \wedge \mathrm{d}y + E_x \mathrm{d}x \wedge \mathrm{d}t + E_y \mathrm{d}y \wedge \mathrm{d}t + E_z \mathrm{d}z \wedge \mathrm{d}t,$$

the equation $\mathrm{d}F = 0$ is equivalent to the four equations

$$\frac{\partial B_x}{\partial x} + \frac{\partial B_y}{\partial y} + \frac{\partial B_z}{\partial z} = 0, \qquad \frac{\partial B_y}{\partial t} - \frac{\partial E_z}{\partial x} + \frac{\partial E_x}{\partial z} = 0,$$

$$\frac{\partial B_x}{\partial t} - \frac{\partial E_y}{\partial z} + \frac{\partial E_z}{\partial y} = 0, \qquad \frac{\partial B_z}{\partial t} - \frac{\partial E_x}{\partial y} + \frac{\partial E_y}{\partial x} = 0.$$

We will use Faraday's law to define B so that the units of B are

$$\frac{\text{voltage}\cdot\text{time}}{\text{area}} = \frac{\text{energy}\cdot\text{time}}{\text{charge}\cdot(\text{length})^2}.$$

Ampère's law relates current to magnetism. It says that the electric current flux through a surface S whose boundary is γ equals the magnetic loop tension around γ. According to Maxwell's great discovery, we must write the electric current flux as the sum of two terms $\partial D/\partial t + 4\pi J$, where D is the dielectric displacement and J is the current density of moving charges. (For slowly varying fields, the first term is negligible in comparison to the second and did not appear in Ampère's original formulation.) The magnetic loop tension is obtained by integrating a linear differential form H, called the *magnetic field strength*, around γ. Thus Maxwell's modification of Ampère's law says

$$\int_S\left(\frac{\partial D}{\partial t}+4\pi J\right)=\int_\gamma H.$$

Consider the three-dimensional cylinder $C=S\times[a,b]$ as before and set

$$G=D-H\wedge \mathrm{d}t.$$

We integrate Ampère's law from a to b with respect to t and get

$$\int_{\partial C} G=-4\pi\int_C J\wedge \mathrm{d}t.$$

If we consider a three-dimensional region R at constant time, then $\mathrm{d}t$ vanishes on ∂R and the integral of G over ∂R is the same as the integral of D. This equals $4\pi\times$ (the total charge in R) which is $4\pi\int\rho\mathrm{d}x\wedge \mathrm{d}y\wedge \mathrm{d}z$. Thus, for regions which lie in constant time, we have

$$\int_{\partial R} G=4\pi\int_R \rho\mathrm{d}x\wedge \mathrm{d}y\wedge \mathrm{d}z.$$

Let us set

$$j=\rho\mathrm{d}x\wedge \mathrm{d}y\wedge \mathrm{d}z-J\wedge \mathrm{d}t$$

and we see that

$$\int_{\partial C} G=4\pi\int_C j$$

for any three-dimensional cube whose sides are parallel to the coordinate axes. Thus

$$\mathrm{d}G=4\pi j,$$

from which it follows that $\mathrm{d}j=0$.

We summarize:

Set	then Maxwell's equations say
$F=B+E\wedge \mathrm{d}t,$	$\mathrm{d}F=0,$
$G=D-H\wedge \mathrm{d}t,$	$\mathrm{d}G=4\pi j.$

Note that Maxwell's equations are invariant under smooth changes of coordinates.

We will use Ampère's law to define H. D has units charge/area so $\partial D/\partial t$ has units charge/(area·time) and J has units current/area = charge/(area·time). Thus H has units charge/(time·length). *In vacuo* we have the constitutive relations

$$D=\varepsilon_0\star E \quad\text{and}\quad B=\mu_0\star H.$$

ε_0 has units:

$$\frac{\text{charge}}{\text{area}}\times\frac{\text{length}}{\text{voltage}}=\frac{(\text{charge})^2}{\text{energy}\cdot\text{length}}.$$

μ_0 has units

$$\frac{\text{energy} \cdot \text{time}}{\text{charge} \cdot (\text{length})^2} \times \frac{\text{time} \cdot \text{length}}{\text{charge}} = \frac{\text{energy} \cdot (\text{time})^2}{(\text{charge})^2 \cdot \text{length}}.$$

Thus $1/\varepsilon_0\mu_0$ has units $(\text{length})^2/(\text{time})^2 = (\text{velocity})^2$. Thus the theory of electromagnetism has a fundamental velocity built into it. It was Maxwell's great discovery that this velocity is exactly c – the velocity of light! So introduce cdt instead of dt and the four-dimensional $\star$ operator becomes

$$\star(dx \wedge dy) = -cdt \wedge dz,$$
$$\star(dx \wedge dz) = cdt \wedge dy,$$
$$\star(dy \wedge dz) = -cdt \wedge dx,$$
$$\star(dx \wedge cdt) = -dy \wedge dz,$$
$$\star(dy \wedge cdt) = dx \wedge dz,$$
$$\star(dz \wedge cdt) = -dx \wedge dy.$$

Then

$$\star F = c(B_z dz \wedge dt + B_y dy \wedge dt + B_z dx \wedge dt)$$
$$- (1/c)(E_x dy \wedge dz - E_y dx \wedge dz + E_z dx \wedge dy).$$

The constitutive relations can now be written as

$$G = -\frac{\varepsilon_0}{\mu_0} \star F.$$

Finally, in vacuum, where ε_0/μ_0 is a constant, we can absorb all constants into a redefinition of j and write Maxwell's equations as

$$dF = 0 \quad \text{and} \quad d \star F = 4\pi j.$$

From the last chapter, we know a procedure for dealing with this system of equations: Assuming the appropriate topological condition, we look for a one-form A with

$$dA = F, \quad d \star A = 0$$

and

$$-\square A = 4\pi \star j.$$

Thus the solutions of Maxwell's equations are closely related to the study of the *wave operator*

$$-\square = \frac{\partial^2}{\partial t^2} - \frac{\partial^2}{\partial x^2} - \frac{\partial^2}{\partial y^2} - \frac{\partial^2}{\partial z^2}$$

if we use coordinates where $c = 1$.

19.2. The homogeneous wave equation in one dimension

In this section, as a warm-up to the study of Maxwell's equations, and of interest in its own right, let us study the equation

$$\frac{\partial^2 u}{\partial t^2} - \frac{\partial^2 u}{\partial x^2} = 0.$$

If we make the change of variables $p = x + t, q = x - t$, this equation becomes

$$\frac{\partial^2 u}{\partial p\,\partial q} = 0$$

so, by integration,

$$u = u_1(p) + u_2(q)$$

or

$$u(x, t) = u_1(x + t) + u_2(x - t). \tag{19.1}$$

Any such function is clearly a solution. The function $u_2(x - t)$ can be thought of dynamically. At each instant of time t, the graph of $u_2(x - t)$ is just given by the graph of $u_2(x)$ displaced t units to the right. We say that $u_2(x - t)$ represents a wave moving without distortion to the right. Thus, the most general solution of the homogeneous wave equation in two dimensions is the sum of two undistorted waves, one moving to the left and the other moving to the right.

We are usually interested in describing the wave motion corresponding to some initial disturbance – say at time $t = 0$. Since the equation is of degree two, we expect to be able to specify the initial values u_0 and v_0 where

$$u_0(x) = u(x, 0)$$

and

$$v_0(x) = \frac{\partial u}{\partial t}(x, 0).$$

Indeed, let us denote by $A(x, t)$ the operation of averaging over an interval of length $2t$ (or radius t) centered at x. Thus for any function w, by definition,

$$A(x, t)w = \frac{1}{2}\int_{x-t}^{x+t} w(s)ds.$$

We claim that

$$u(x, t) = A(x, t)v_0 + \frac{\partial}{\partial t}[A(x, t)u_0]$$

is the unique solution of the wave equation with the given initial conditions. Indeed, writing $u(x, t) = u_1(x + t) + u_2(x - t)$, we see that

$$u(x, 0) = u_1(x) + u_2(x)$$

and

$$\frac{\partial u}{\partial t}(x, 0) = u_1'(x) - u_2'(x).$$

Substituting $v_0 = \partial u/\partial t$ and integrating this last equation give

$$u_1(x) - u_2(x) = \int_0^x v_0(s)ds + c$$

where c is some constant. So adding and subtracting the equation $u_0(x) = u_1(x) + u_2(x)$, we see that

$$2u_1(x) = u_0(x) + \int_0^x v_0(s)ds + c,$$

$$2u_2(x) = u_0(x) - \int_0^x v_0(s)ds - c.$$

So, from (19.1), we see that

$$u(x,t) = \frac{1}{2}\left(u_0(x+t) + u_0(x-t) + \int_{x-t}^{x+t} v_0(s)ds \right).$$

This is just another way of writing our formula.

Notice that we have proved an existence and uniqueness theorem for our initial-value problem. We also see from the explicit formula that small changes in the initial values cause small changes in the solution. We say that the problem is *well-posed*. Notice that the value of u at (x_0, t_0) does not depend on all the values of u and $\partial u/\partial t$ at

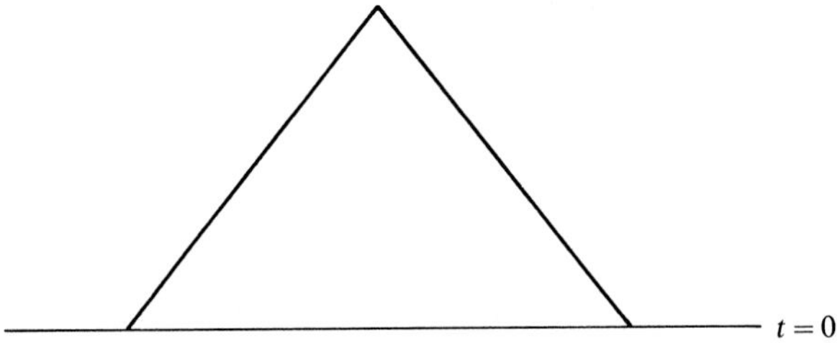

Figure 19.2. Domain of dependence

$t = 0$, but only on the values in the interval $[x_0 - t_0, x_0 + t_0]$. Put another way, the values at $(x, 0)$ only influence those spacetime points which lie in the *forward cone*

$$\|x_0\| = \|t_0\|.$$

This is known as the *principle of causality*.

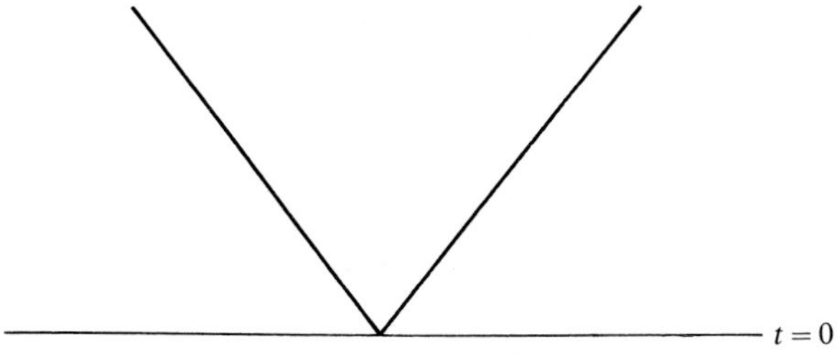

Figure 19.3. Domain of influence

19.3. The homogeneous wave equation in $\mathbb{R}^3$

We wish to solve the initial-value problem

$$\frac{\partial^2 u}{\partial t^2} - \frac{\partial^2 u}{\partial x^2} - \frac{\partial^2 u}{\partial y^2} - \frac{\partial^2 u}{\partial z^2} = 0,$$

$$u(\mathbf{x}, 0) = u_0(\mathbf{x}),$$

$$\frac{\partial u}{\partial t}(\mathbf{x}, 0) = v_0(\mathbf{x}).$$

Let $A(t, \mathbf{x})$ denote the process of averaging over a sphere of radius $|t|$ centered at $\mathbf{x}$. Thus

$$A(t, \mathbf{x})f = \int_S f(\mathbf{x} + t\cdot)\tau.$$

Here $f(\mathbf{x} + t\cdot)$ denotes the function on the unit sphere S whose value at any vector $\mathbf{v}$ with $\|\mathbf{v}\| = 1$ is given by $f(\mathbf{x} + t\mathbf{v})$ and τ is $1/4\pi$ times the solid angle form. We claim that the function u given by

$$u(\mathbf{x}, t) = tA(t, \mathbf{x})v_0 + \frac{\partial}{\partial t}[tA(t, \mathbf{x})u_0]$$

is the unique solution to our problem.

We first show that it is *a* solution and begin by showing that it satisfies the initial conditions. For small values of t we can expect

$$u(\mathbf{x}, t) = A(t, \mathbf{x})u_0 + \mathrm{O}(t)$$

and the limit of $A(t, \mathbf{x})$ is clearly $u_0(\mathbf{x})$ as we are averaging over smaller and smaller spheres centered at $\mathbf{x}$. Thus

$$u(\mathbf{x}, 0) = u_0(\mathbf{x}).$$

Also

$$\frac{\partial u}{\partial t}(\mathbf{x}, 0) = \lim_{t \to 0} \left[A(t, \mathbf{x})v_0 + \frac{\partial A}{\partial t}(t, \mathbf{x})u_0 \right]$$

since all remaining terms are $\mathrm{O}(t)$. But

$$\frac{\partial A}{\partial t}(0, \mathbf{x})u_0 = \int_S \sum \frac{\partial u}{\partial x_i}(\mathbf{x})x_i\tau$$

$$= \sum \frac{\partial u}{\partial x_i}(x) \int_S x_i\tau = 0$$

since the average of an odd function – like the coordinate x_i – over the unit sphere must vanish.

To prove that u satisfies the wave equation, it is clearly enough to prove that, for any function ω of $\mathbf{x}$, the function

$$W(\mathbf{x}, t) = tA(t, \mathbf{x})\omega$$

satisfies the wave equation. This is because our expression for u is a sum of two terms – one of the above form and the other of the form $\partial W/\partial t$. And if W is a solution of the wave equation, so is $\partial W/\partial t$.

For any twice-differentiable function ω, we have

$$\int_{\|\mathbf{y}-\mathbf{x}\|\leqslant t}\left(\frac{\partial^2\omega}{\partial x^2}+\frac{\partial^2\omega}{\partial y^2}+\frac{\partial^2\omega}{\partial z^2}\right)\mathrm{d}x\,\mathrm{d}y\,\mathrm{d}z=\int_{\|\mathbf{y}-\mathbf{x}\|\leqslant t}\mathrm{d}\star\mathrm{d}\omega=\int_{S_{\mathbf{x},t}}\star\mathrm{d}\omega$$

where $S_{\mathbf{x},t}$ denotes the sphere of radius $|t|$ centered at $\mathbf{x}$.

Now on the surface of a sphere of radius r we can write

$$(\star\mathrm{d}\omega)\wedge\mathrm{d}r=\frac{\partial\omega}{\partial r}\mathrm{d}x\wedge\mathrm{d}y\wedge\mathrm{d}z$$

and

$$\mathrm{d}x\wedge\mathrm{d}y\wedge\mathrm{d}z=\tau\wedge\mathrm{d}r$$

so

$$\frac{1}{4\pi t^2}\int_{S_{\mathbf{x},t}}\star\mathrm{d}\omega=\frac{\partial}{\partial t}A(t,\mathbf{x})\omega.$$

Thus

$$\begin{aligned}\frac{\partial}{\partial t}A(t,\mathbf{x})\omega&=\frac{1}{4\pi t^2}\int_{\|\mathbf{y}-\mathbf{x}\|\leqslant|t|}(\Delta\omega)(\mathbf{y})\mathrm{d}\mathbf{y}\\&=\frac{1}{4\pi t^2}\int_0^t\mathrm{d}\rho\int_{\|\mathbf{y}-\mathbf{x}\|=\rho}(\Delta\omega)(\mathbf{y})\mathrm{d}S.\end{aligned}$$

Differentiating with respect to t, we see that

$$\frac{\partial^2A}{\partial t^2}=\frac{-1}{2\pi t^3}\int_0^t\mathrm{d}\rho\int_{\|\mathbf{y}-\mathbf{x}\|=\rho}(\Delta\omega)(\mathbf{y})\mathrm{d}S+\frac{1}{4\pi t^2}\int_{\|\mathbf{y}-\mathbf{x}\|=\rho}\Delta\omega(\mathbf{y})\mathrm{d}S.$$

Now set $W(\mathbf{x},t)=tA(t,\mathbf{x})\omega$. Then

$$\begin{aligned}\frac{\partial^2W}{\partial t^2}&=2\frac{\partial}{\partial t}A(t,\mathbf{x})\omega+t\frac{\partial^2A}{\partial t^2}(t,\mathbf{x})\omega\\&=\frac{1}{4\pi t}\int_{\|\mathbf{y}-\mathbf{x}\|=t}\Delta\omega(\mathbf{y})\mathrm{d}S.\end{aligned}$$

But

$$\begin{aligned}\Delta W&=\frac{t}{4\pi}\int_{\|\mathbf{v}\|=1}\Delta_{\mathbf{x}}\omega(\mathbf{x}+t\mathbf{v})\mathrm{d}S\\&=\frac{1}{4\pi t}\int_{\|\mathbf{y}-\mathbf{x}\|=t}\Delta\omega(\mathbf{y})\mathrm{d}S\end{aligned}$$

so W is a solution of the wave equation.

We can turn this argument around and prove the uniqueness of our solution by reducing the three-dimensional homogeneous wave equation to the one-dimensional equation. We have proved the integral formula

$$\int_{\|\mathbf{y}-\mathbf{x}\|\leqslant r} \Delta u(\mathbf{y})\mathrm{d}\mathbf{y} = r^2 \int_{\|\mathbf{v}\|=1} \frac{\partial}{\partial r} u(\mathbf{x}+\mathbf{v}r)\mathrm{d}\omega.$$

So, for a function $u(\mathbf{x}, t)$, we have

$$\int_{\|\mathbf{y}-\mathbf{x}\|\leqslant r} \Delta u(\mathbf{y}, t) = 4\pi r^2 \frac{\partial}{\partial r} A(r, \mathbf{x})u.$$

If u satisfies the equation

$$\frac{\partial^2 u}{\partial t^2} = \Delta u,$$

then we can substitute into the above equation to obtain

$$4\pi r^2 A(r, \mathbf{x})u = \int_{\|\mathbf{y}-\mathbf{x}\|\leqslant r} \frac{\partial^2 u}{\partial t^2}(\mathbf{y}, t) = \int_0^r \mathrm{d}\rho \int_{\|\mathbf{y}-\mathbf{x}\|=r} \frac{\partial^2 u}{\partial t^2}(\mathbf{y}, t)\mathrm{d}S.$$

Differentiation with respect to r now shows that the function

$$Z(r, t) = rA(r, \mathbf{x})u(\mathbf{x}, v)$$

satisfies the one-dimensional wave equation

$$\frac{\partial^2 Z}{\partial r^2} - \frac{\partial^2 Z}{\partial t^2} = 0.$$

Thus

$$Z(r, t) = rA(r, \mathbf{x})u(\mathbf{x}, v)$$

Since $Z(0, t) = 0$, we see that

$$w_1(t) = -w_2(-t).$$

So writing w for w_1

$$Z(r, t) = w(r+t) - w(t-r).$$

Differentiating this with respect to r and letting $r \to 0$, we get

$$u(\mathbf{x}, t) = A(0, \mathbf{x})u(\mathbf{x}, t) = 2w'(t).$$

Also

$$\frac{\partial Z}{\partial r} + \frac{\partial Z}{\partial t} = 2w'(r+t)$$

so

$$u(\mathbf{x}, r) = 2w'(r) = \lim_{t\to 0} 2w'(r+t) = \lim_{t\to 0} \frac{\partial Z}{\partial r} + \lim_{t\to 0} \frac{\partial Z}{\partial t}.$$

Substituting the expression Z into this last equation gives

$$u(\mathbf{x}, r) = \frac{\partial}{\partial r}\left(\frac{r}{4\pi}\int_{\|\gamma\|=1} u(\mathbf{x}+\gamma r, 0)\mathrm{d}S + \frac{r}{4\pi}\int_{\|\gamma\|=1} \frac{\partial u}{\partial t}(\mathbf{x}+\gamma r, 0)\mathrm{d}S\right).$$

But

$$u(\mathbf{x}+\gamma r, 0) = u_0(\mathbf{x}+\gamma r)$$

and

$$\frac{\partial u}{\partial t}(\mathbf{x}+\gamma r) = v_0(\mathbf{x}+\gamma r).$$

Putting $r = t$ in the last equation gives us back our original solution. This proves the uniqueness.

19.4. The inhomogeneous wave equation in $\mathbb{R}^3$

We can solve Maxwell's equations by solving a succession of equations of the form

$$\Box u = f, \quad \Box = \frac{\partial^2}{\partial t^2} - \Delta.$$

Let us first show how to solve this equation with the initial conditions

$$u(\mathbf{x}, 0) \equiv 0, \quad \frac{\partial u}{\partial t}(\mathbf{x}, 0) \equiv 0.$$

We can solve the inhomogeneous wave equation by reducing it to solving the homogeneous wave equation with a series of initial conditions as follows. Let $v(\mathbf{x}, t, \tau)$ be the solution to the initial value problem

$$\Box v = 0, \quad v(\mathbf{x}, \tau) = 0, \quad \frac{\partial v}{\partial t}(\mathbf{x}, \tau) = f(\mathbf{x}, \tau).$$

We set

$$u(\mathbf{x}, t) = -\int_0^t v(\mathbf{x}, t, \tau)\mathrm{d}\tau.$$

Then it is obvious that $u(\mathbf{x}, 0) = 0$ and

$$\frac{\partial u}{\partial t}(\mathbf{x}, t) = -v(\mathbf{x}, t, t) - \int_0^t \frac{\partial v}{\partial t}(\mathbf{x}, t, \tau)\mathrm{d}\tau.$$

But $v(\mathbf{x}, t, t) = 0$ by the initial conditions on v. So

$$\frac{\partial u}{\partial t}(\mathbf{x}, t) = -\int_0^t \frac{\partial v}{\partial t}(\mathbf{x}, t, \tau)\mathrm{d}\tau$$

and therefore

$$\frac{\partial u}{\partial t}(\mathbf{x}, 0) = 0.$$

Also, differentiating the equation for $\partial u/\partial t$ gives

$$-\frac{\partial^2 u}{\partial t^2}(\mathbf{x}, t) = \frac{\partial v}{\partial t}(\mathbf{x}, t, t) + \int_0^t \frac{\partial^2 v}{\partial t^2}(\mathbf{x}, t, \tau)\mathrm{d}\tau$$

while

$$\Delta u = -\int_0^t v(\mathbf{x}, t, \tau)\mathrm{d}\tau$$

so

$$\left(\frac{\partial^2}{\partial t^2} - \Delta\right)u = f.$$

Now the function v is given by

$$v(\mathbf{x}, t, \tau) = (t-\tau)A_{\mathbf{x},t-\tau}f = \frac{1}{4\pi}(t-\tau)\int_{\|\gamma\|=1} f(\mathbf{x}+(t-\tau)\gamma,\tau)\mathrm{d}\gamma$$

so

$$u(\mathbf{x}, t) = -\frac{1}{4\pi}\int_0^t (t-\tau)\int_{\|\gamma\|=1} f(\mathbf{x}+(t-\tau)\gamma, \tau)\mathrm{d}\gamma\mathrm{d}z$$

$$= -\frac{1}{4\pi}\int_0^t \tau' \int_{\|\gamma\|=1} f(\mathbf{x}+\tau'\gamma, t-\tau')\mathrm{d}\gamma\mathrm{d}\tau'.$$

Setting $\mathbf{y} = \mathbf{x} + \gamma\tau'$ this last integral becomes

$$-\frac{1}{4\pi}\int_0^t \frac{1}{\tau'}\int_{\|\mathbf{y}-\mathbf{x}\|=t} f(\mathbf{y}, t-\tau')\mathrm{d}S\mathrm{d}\tau'$$

$$= -\frac{1}{4\pi}\int_{\|\mathbf{y}-\mathbf{x}\|=\tau'} \frac{f(\mathbf{y}, t-\|\mathbf{y}-\mathbf{x}\|)}{\|\mathbf{y}-\mathbf{x}\|}\mathrm{d}\mathbf{y}.$$

This is the formula of *retarded potentials*:

$$u(x, t) = -\frac{1}{4\pi}\int \frac{f(\mathbf{y}, t-\|\mathbf{y}-\mathbf{x}\|)}{\|\mathbf{y}-\mathbf{x}\|}\mathrm{d}\mathbf{y}.$$

It looks like Poisson's formula for the potential, but the contribution from the sphere of radius a about x is given by $1/a$ times the value of f at time a units.

Once we have a solution for the inhomogeneous equation with zero initial conditions, we can then reduce the inhomogeneous equation with arbitrary initial conditions, by subtracting off our given solution, to solving the homogeneous equation.

Se can now apply the preceding formula for the solution of the inhomogeneous wave equation to each of the components of the *four-potential* A in Maxwell's equations. We obtain

$$A(\mathbf{x}) = -\int \frac{\star j(\mathbf{x}, t-|\mathbf{x}-\mathbf{x}'|)}{|\mathbf{x}-\mathbf{x}'|}\mathrm{d}\mathbf{x}.$$

This solution is known as the *Lienard–Wiechert potential* for the given charge-current density j. The general solution to Maxwell's equations is obtained by adding to A a potential whose components satisfy the homogeneous wave equation.

19.5. The electromagnetic Lagrangian and the energy-momentum tensor

Let A be a one-form and j a three-form on Minkowski space, $\mathbb{R}^{1,3}$. Consider the four-form

$$\mathscr{L} = \mathscr{L}(A, J) = -\tfrac{1}{2}\mathrm{d}A \wedge \star\mathrm{d}A - 4\pi A \wedge \star j.$$

We can integrate $\mathscr{L}$ over any bounded four-dimensional region N and so obtain a function

$$L_{N,j}(A) = \int_N \mathscr{L}.$$

Here we are regarding N and j as fixed and given, so that L is considered as a function of A. We say that A is an *extremal* of L if

$$\frac{\mathrm{d}}{\mathrm{d}s} L_{N,J}(A_s)|_{s=0} = 0$$

for any differentiable family of one-forms A_s with $A_0 = A$ and $A_s = A$ on ∂N.

Write $\left.\dfrac{\mathrm{d}A_s}{\mathrm{d}s}\right|_{s=0} = B$; then

$$\frac{\mathrm{d}}{\mathrm{d}s}(A_s, j)|_{s=0} = -\tfrac{1}{2}(\mathrm{d}B \wedge \star\mathrm{d}A + \mathrm{d}A \wedge \star\mathrm{d}B) - 4\pi B \wedge j.$$

Now

$$\mathrm{d}(B \wedge \star\mathrm{d}A) = \mathrm{d}B \wedge \star\mathrm{d}A + B \wedge \mathrm{d}\star\mathrm{d}A$$

and

$$\mathrm{d}B \wedge \star\mathrm{d}A = \mathrm{d}A \wedge \star\mathrm{d}B$$

so

$$\left.\frac{\mathrm{d}}{\mathrm{d}s}\right|_{s=0} = B \wedge [\mathrm{d}\star\mathrm{d}A - 4\pi j] = \mathrm{d}(B \wedge \star\mathrm{d}A).$$

Now the boundary integral $\int_{\partial N} B \wedge \star\mathrm{d}A$ vanishes since $B = 0$ on the boundary. Thus

$$\left.\frac{\mathrm{d}L_{N,j}(A_s)}{\mathrm{d}s}\right|_{s=0} = \int_N B \wedge [\mathrm{d}\star\mathrm{d}A - 4\pi j].$$

If this vanishes for *all* B vanishing at the boundary, we must have

$$\mathrm{d}\star\mathrm{d}A = 4\pi j.$$

But, on a region where $H^2 = 0$ this is equivalent to Maxwell's equation $F = \mathrm{d}A$. Notice that as a necessary condition to be able to solve these equations, we must have

$$\mathrm{d}j = 0.$$

We have thus derived Maxwell's equations from a *variational principle*: Maxwell's equations are precisely the equations for the *critical one-forms A* of the *Lagrangian L*. There are several advantages to a Lagrangian formulation. One is that, by modifying L, we get a procedure for modifying the equations for its extremals, and thus exploring various alternative laws of nature. A second is that the Lagrangian formulation lends itself to one of the procedures for passing from classical mechanics to quantum mechanics. The function L plays a key role in quantum electrodynamics. A third advantage of a Lagrangian formulation is that every one-parameter group of symmetries of L gives a conservation law by a method known as *Noether's theorem*. We explain the method in our special case of electromagnetism.

Let ξ be a vector field on $\mathbb{R}^{1,3}$ with the property that

$$D_\xi \star\omega = \star D_\xi\omega \quad \text{for all two-forms } \omega.$$

As we have seen, any conformal transformation of spacetime preserves the star operator on two-forms. So ξ can be the vector field associated with any one-parameter group of conformal transformations. In particular, ξ can be any constant vector field – corresponding to translations, or any infinitesimal Lorentz transformation.

Now

$$D_\xi\mathscr{L} = -\tfrac{1}{2}\mathrm{d}D_\xi A \wedge \star\mathrm{d}A - \tfrac{1}{2}\mathrm{d}A\, D_\xi\star\mathrm{d}A - D_\xi A \wedge 4\pi j - A \wedge 4\pi D_\xi j.$$

Since

$$D_\xi \star\mathrm{d}A = \star D_\xi \mathrm{d}A = \star\mathrm{d}D_\xi A$$

and

$$\mathrm{d}D_\xi A \wedge \star\mathrm{d}A = \mathrm{d}A \wedge \star\mathrm{d}D_\xi A$$

the first and second terms are equal, so they add up to

$$-\mathrm{d}D_\xi A \wedge \star\mathrm{d}A = -\mathrm{d}(D_\xi A \wedge \star\mathrm{d}A) + D_\xi A \wedge \mathrm{d}\star\mathrm{d}A.$$

Thus

$$D_\xi\mathscr{L} = -\mathrm{d}(D_\xi A \wedge \star\mathrm{d}A) + D_\xi A \wedge (\mathrm{d}\star\mathrm{d}A - 4\pi j) - A \wedge 4\pi D_\xi j.$$

But $\mathrm{d}\star\mathrm{d}A = 4\pi j$ by Maxwell's equations. So

$$\begin{aligned} D_\xi\mathscr{L} &= -\mathrm{d}(D_\xi A \wedge \star\mathrm{d}A) - A \wedge 4\pi D_\xi j \\ &= -\mathrm{d}((i(\xi)\mathrm{d}A + \mathrm{d}i(\xi)A) \wedge \star\mathrm{d}A) - A \wedge 4\pi D_\xi j \\ &= -\mathrm{d}[i(\xi)F \wedge \star F] - \mathrm{d}i(\xi)A \wedge \mathrm{d}\star\mathrm{d}A - A \wedge 4\pi D_\xi j \\ &= -\mathrm{d}[i(\xi)F \wedge \star F] - \mathrm{d}i(\xi)A \wedge 4\pi j - A \wedge 4\pi D_\xi j \end{aligned}$$

using $\mathrm{d}A = F$ and $\mathrm{d}\star\mathrm{d}A = j$. On the other hand,

$$\begin{aligned} D_\xi A &= i(\xi)\mathrm{d}A + \mathrm{d}A + \mathrm{d}i(\xi)A \\ &= i(\xi)F + \mathrm{d}i(\xi)A \end{aligned}$$

while

$$\begin{aligned} D_\xi\mathscr{L} &= -\tfrac{1}{2}D_\xi(F \wedge \star F) = D_\xi(A \wedge 4\pi j) \\ &= -\tfrac{1}{2}\mathrm{d}i(\xi)(F \wedge \star F) - (i(\xi)F + \mathrm{d}i(\xi)A) \wedge j - A \wedge D_\xi j \end{aligned}$$

since $d(F \wedge \star F) = 0$ because $F \wedge \star F$ is a four-form. Setting these two expressions for $D_\xi \mathscr{L}$ equal, we get

$$\begin{aligned} &-d[i(\xi)F \wedge \star F] - di(\xi)A \wedge 4\pi j - A \wedge D_\xi j \\ &= -\tfrac{1}{2}d[i(\xi)F \wedge \star F + F \wedge i(\xi)\star F] - i(\xi)F \wedge j - di(\xi)A \wedge 4\pi j - A \wedge 4\pi D_\xi j \end{aligned}$$

or

$$d\tfrac{1}{2}[F \wedge i(\xi)\star F - i(\xi)F \wedge \star F] = -i(\xi)F \wedge 4\pi j.$$

In the absence of currents, i.e., if $j = 0$, we would find that the three-form

$$C(\xi) = \tfrac{1}{2}[F \wedge i(\xi)\star F - i(\xi)F \wedge \star F]$$

would be closed. This is interpreted as a conservation law: the integral of $C(\xi)$ over two three-dimensional regions which bound some four-dimensional region will give the same answer.

If F vanishes sufficiently rapidly as we go out to infinity in space-like directions, then the integrals of $C(\xi)$ over any two space-like three-surfaces will be equal.

For example, in some coordinate system, we take the three-surfaces to be given by $t = t_1$ and $t = t_2$. So the total amount of $C(\xi)$ at time $t = t_1$ equals the total amount

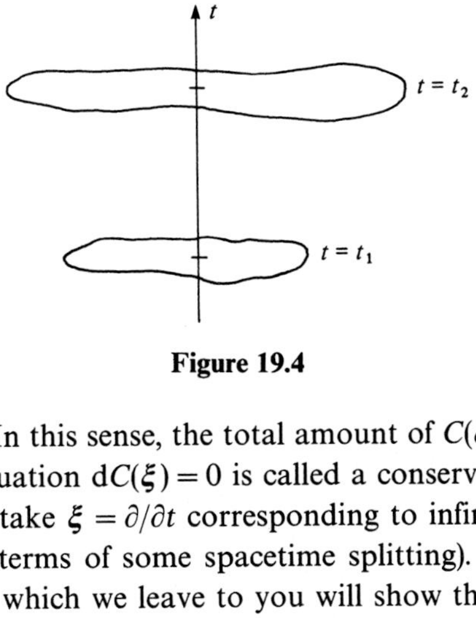

Figure 19.4

of $C(\xi)$ at time $t = t_2$. In this sense, the total amount of $C(\xi)$ is conserved in time. For this reason the equation $dC(\xi) = 0$ is called a conservation law.

For example, let us take $\xi = \partial/\partial t$ corresponding to infinitesimal translation in the time direction (in terms of some spacetime splitting). A straightforward but instructive calculation which we leave to you will show that

$$\begin{aligned} C(\xi) = \tfrac{1}{2}(\|\mathbf{E}\|^2 + \|\mathbf{B}\|^2)dx \wedge dy \wedge dz \\ + P_1 dy \wedge dz \wedge dt - P_2 dx \wedge dz \wedge dt + P_3 dx \wedge dy \wedge dt \end{aligned}$$

where

$$\begin{aligned} P_1 &= E_y B_z - B_y E_z, \\ P_2 &= E_x B_z - B_x E_z, \\ P_3 &= E_x B_y - B_x E_y, \end{aligned}$$

or more succinctly

$$\vec{\mathbf{P}} = \vec{\mathbf{E}} \times \vec{\mathbf{B}}.$$

(The three-vector **P** is called the Poynting vector.) Thus on a surface $t = \text{constant}$, the integral of $C(\xi)$ gives the total energy of the electric and magnetic fields. Our conservation law thus becomes the conservation of energy. The remaining components of $C(\xi)$ in this case describe the flux of (local energy density) across any surface.

Just as energy is associated with time translation, the x-component of momentum is associated with infinitesimal translation in the x-direction. This is a general principle of mechanics – see for example Loomis & Sternberg, Chapter 13, or, for a more sophisticated version, Guillemin & Sternberg, Chapter 2. Thus $C(\partial/\partial x)$ will be a closed three-form whose integral over the space-like three-surface $t = \text{constant}$ gives the total momentum of the electromagnetic field in the x-direction. Over surfaces of the form $[a, b] \times S$, where S is a surface in $\mathbb{R}^3$ and $[a, b]$ an interval of time, the integral of $C(\partial/\partial x)$ gives the flux of x-momentum across S in the time interval $[a, b]$.

For each constant vector ξ, $C(\xi)$ is a three-form. Thus $\star C(\xi)$ is a one-form. Thus $T(\xi, \boldsymbol{\eta}) = C(\xi)(\boldsymbol{\eta})$ makes sense for some other constant vector $\boldsymbol{\eta}$. The function $T = T(\xi, \boldsymbol{\eta})$ is called the *energy-momentum tensor* because coded into it are all components of energy or momentum density and flux.

In the presence of a source term j, the forms $C(\xi)$ are not closed, but satisfy

$$dC(\xi) = i(\xi)F \wedge 4\pi j.$$

This represents an exchange of energy-momentum from the electromagnetic field to the charge-current. It is the *Lorentz force*.

19.6. Wave forms and Huyghens' principle

A solution of the wave equation

$$\frac{\partial^2 u}{\partial t^2} - \Delta u = 0$$

of the form

$$u(\mathbf{x}, t) = f(\mathbf{x})g(h(\mathbf{x}) \pm t)$$

is called a *progressing wave*. The expression $h(\mathbf{x}) \pm t$ is called the *phase* of the wave and the level surfaces $h = \text{constant}$ are called *wave fronts*.

In one dimension, we have seen that taking $f \equiv 1$, $h(\mathbf{x}) = x$ and g arbitrary gives a solution (and that the general solution can be written as the sum of two such expressions, one with a $+$ and the other with a $-$).

In three dimensions we can take h to be a linear function of $\mathbf{x}$

$$h(\mathbf{x}) = \mathbf{k} \cdot \mathbf{x}$$

and $f \equiv 1, g$ again arbitrary. The only condition to solve the equation is that $\|\mathbf{k}\|^2 = 1$. (In fact, we have reduced the problem to a one-dimensional problem: By rotation we may assume that $\mathbf{k} = (1, 0, 0)$. So it is really just a function of $\mathbf{x}$ and t and we are reduced to the wave equation in one spatial dimension.)

In particular, if we take $g(\theta) = e^{i\theta}$, we get the functions

$$e^{i(\mathbf{k}\cdot\mathbf{x}-t)}, \ \|\mathbf{k}\|^2 = 1.$$

The importance of these sinusoidal phase wave solutions is the following: It is a theorem that every smooth function on $\mathbb{R}^{1,3}$ (or more generally on $\mathbb{R}^n$) which vanishes sufficiently rapidly at infinity can be written as a superposition of functions of the form $e^{i\mathbf{l}\cdot\mathbf{v}}$. That is, we can write any such function as

$$f(\mathbf{x}) = \frac{1}{(2\pi)^{3/2}} \int \hat{f}(\xi) e^{i\xi\cdot\mathbf{x}} d\mathbf{x}$$

for a suitable function, $\hat{f}$. We shall give a rapid proof of this fact in Chapter 21. It turns out that this *Fourier inversion formula* is true in greater generality, and that (in our case), the most general solution of the wave equation can be represented as a superposition of the functions $e^{i(\mathbf{k}\cdot\mathbf{x}-t)}$, $\|\mathbf{k}\|^2 = 1$.

As a second example of a wave form, consider spherically symmetric solutions of the wave equation (defined outside the line $r = 0$). In terms of polar coordinates, we have

$$\Delta u = \frac{1}{r^2}\frac{\partial}{\partial r} r^2 \frac{\partial u}{\partial r} + \frac{1}{r^2 \sin\theta}\frac{\partial}{\partial\theta}\sin\theta\frac{\partial u}{\partial\theta} + \frac{1}{r^2\sin^2\theta}\frac{\partial^2 u}{\partial\phi^2}$$

so that if u is spherically symmetric, $u = u(r, t)$, the wave equation becomes

$$\frac{\partial^2 u}{\partial t^2} = \frac{1}{r^2}\frac{\partial}{\partial r} r^2 \frac{\partial u}{\partial r} = \frac{1}{r}\left[2\frac{\partial u}{\partial r} + \frac{\partial^2 u}{\partial r^2}\right] = \frac{1}{r}\frac{\partial^2}{\partial r^2}(ru).$$

Thus $v = ru$ satisfies the one-dimensional wave equation

$$\left(\frac{\partial^2}{\partial t} - \frac{\partial^2}{\partial r^2}\right) v = 0.$$

The general solution of this equation is given by

$$v(r, t) = f(r + t) + g(r - t)$$

and so the general solution of the symmetric wave equation is given by

$$u(r, t) = \frac{f(r+t)}{r} + \frac{g(r-t)}{r}.$$

Here the first term represents an incoming wave and the second term represents an outgoing wave. In particular, if we take $f = 0$ and $g(s) = e^{iks}$ then

$$w_k(t, r) = \frac{e^{ik(r-t)}}{r}$$

represents an outgoing (sinusoidal) wave of frequency k. Indeed, up to normalizing constants, it is easy to check that

$$E_k(r) = \frac{e^{ikr}}{r}$$

is the 'fundamental solution' to the reduced wave operator $\Delta + k^2$, i.e.,

$$(\Delta + k^2)E_k = 0$$

for $r \neq 0$ while

$$-\frac{1}{\pi}\int E_k(\Delta + k^2)\phi \mathrm{d}\mathbf{x} = \phi(0)$$

for any smooth function ϕ. We proved this result for the case $k = 0$ in Chapter 15 from Green's formula. The identical proof works here.

Furthermore, if D is any bounded region in $\mathbb{R}^3$ and u satisfies the reduced wave (or Helmholtz) equation

$$\Delta u + k^2 u = 0$$

then the argument from Green's formula shows that the following formula due to Helmholtz is true:

$$\frac{1}{4\pi}\iint_{\partial D}\left[\frac{\mathrm{e}^{ikr_P}}{r_P}\star \mathrm{d}u - u\star \mathrm{d}\left(\frac{\mathrm{e}^{ikr_P}}{r_P}\right)\right] = \begin{cases} u(P) & \text{if } P\in D, \\ 0 & \text{if } P\notin D, \end{cases}$$

where r_P denotes the distance from P.

In many applications we are interested in the situation where D, instead of being bounded, represents the exterior to some surface S. Let us first apply the formula to the bounded region D_R, consisting of the intersection of D with a ball of radius R centered at P.

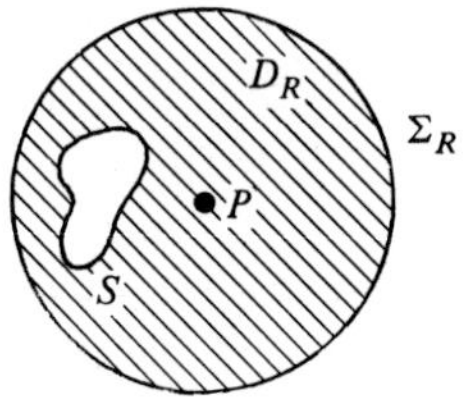

Figure 19.5

If R is taken large enough the previous integral becomes a sum of two surface integrals, yielding

$$\frac{1}{4\pi}\iint_S + \frac{1}{4\pi}\int_{\Sigma_R}\int = \begin{cases} u(P) & \text{if } P\in D, \\ 0 & P\notin D. \end{cases}$$

where Σ_R is the sphere of radius R. Now

$$\mathrm{d}\left(\frac{\mathrm{e}^{ikr}}{r}\right) = \frac{\mathrm{e}^{ikr}}{r}\left[ik - \frac{1}{r}\right]\mathrm{d}r$$

and $\star\mathrm{d}r = R^2\mathrm{d}\omega$ on Σ_R, where $\mathrm{d}\omega$ is the element of solid angle on Σ_R. Thus the second integral becomes

$$\iint \mathrm{e}^{ikr}\left[r\left(\frac{\partial u}{\partial r} - iku\right) + u\right]_{r=R}\mathrm{d}\omega.$$

Thus the integral over the sphere will go to zero as $R \to \infty$ if

$$\iint |u|\,\mathrm{d}\omega = \mathrm{o}(1) \text{ and } \iint \left|\frac{\partial u}{\partial r} - iku\right| \mathrm{d}\omega = \mathrm{o}(R^{-1}).$$

where the integrals are evaluated for $r = R$. These conditions are known as the Sommerfeld radiation conditions. Their significance is that they represent the condition that u consists of expanding waves radiating outward and no incoming waves*. Let us assume that this condition is satisfied. Then the value of u outside some surface S is given by

$$u(P) = \frac{1}{4\pi}\int_S \left[\frac{\mathrm{e}^{ikr}}{r}\star\mathrm{d}u - u\star\mathrm{d}\left(\frac{\mathrm{e}^{ikr}}{r}\right)\right]. \tag{19.2}$$

In this way, the solution exterior to S is described in terms of radiation emitted from S. It was Huyghens who originally had the idea that propagated disturbances in the wave theory could be represented as the superposition of *secondary disturbances* along an intermediate surface such as S; but he did not have an adequate explanation of why there was no *backward wave*, i.e. why the propagation was only in the outward direction. The idea that the backward waves would cancel one another out because of phase differences was due to Fresnel. Fresnel believed that if all the sources were inside S, the *secondary radiation* (i.e. the integrand in Helmholtz formula) from each separate surface element would produce a null effect at each interior point due to interference. The above argument, due essentially to Helmholtz, was the first rigorous mathematical treatment of the problem, and shows that the internal cancellation is due to the total effect of the boundary. Nevertheless, as we shall see in Chapter 21, Fresnel was right, up to terms of order $1/k$. This is due to a method for giving an asymptotic evaluation of the integrals in Helmholtz formula.

So far, we have dealt with *monochromatic radiation u*, corresponding to the time-dependent function v where $v(x, y, z, t) = u(x, y, z)\mathrm{e}^{-ikt}$. For a fixed point, P, let v_P denote the function $v_P(x, y, z, t) = v(x, y, z, t - r)$, where r is the distance from P to (x, y, z). The substitution into Helmholtz' formula shows that

$$v(P, t) = \frac{1}{4\pi}\int_S (v_P\star\mathrm{d}(1/r) - (1/r)(\partial u/\partial t_P)\star\mathrm{d}r - (1/r)\star\mathrm{d}v_P).$$

* For a precise mathematical explanation of the Sommerfeld radiation conditions see the book by Lax & Phillips, pp. 120–128. The gist of what they prove is the following. Let $f = \{f_1, f_2\}$ be Cauchy data for the wave equation; we thus seek a solution of the wave equation

$$\frac{\partial^2 w}{\partial t^2} - \Delta w = 0$$

with $w(x, 0) = f_1$ and $(\partial w/\partial t)(x, 0) = f_2$. We say f is *eventually outgoing* if there is some constant c such that $w = 0$ for $|x| < t - c$. If we seek a solution of fixed frequency, then the appropriate Cauchy data are $\{w, ikw\}$. Suppose that w is a solution of the reduced wave equation outside some bounded domain. Then $\{w, ikw\}$ is eventually outgoing if and only if the Sommerfeld radiation conditions are satisfied.

This is Kirchhoff's formula. Since it is linear in v, and does not explicitly involve the frequency, it is true for any superposition of monochromatic waves of varying frequencies, and hence for an arbitrary solution of the wave equation. In this form, the relation with Huygens' principle is very apparent.

Summary

A Maxwell's equations

You should be able to state Faraday's law and Ampère's law in both integral and differential form and to show that the two-forms

$$F = B + E \wedge \mathrm{d}t \quad \text{and} \quad G = D - H \wedge \mathrm{d}t$$

satisfy $G = \star F$ in vacuum.

You should be able to convert Maxwell's equations to the form $\Box A = 4\pi j$ and to describe the solutions to this wave equation.

You should know how to formulate energy-momentum conservation for electromagnetism in terms of differential forms.

Exercises

19.1. In four-dimensional spacetime, if we use units where $c = 1$, $(\mathrm{d}t, \mathrm{d}t) = 1$, $(\mathrm{d}x, \mathrm{d}x) = (\mathrm{d}y, \mathrm{d}y) = (\mathrm{d}z, \mathrm{d}z) = -1$ and $\sigma = \mathrm{d}t \wedge \mathrm{d}x \wedge \mathrm{d}y \wedge \mathrm{d}z$. A two-form F, which represents the electromagnetic field, may be expressed in terms of coordinates x, y, z, t as $F = B_z(x, y, t)\mathrm{d}x \wedge \mathrm{d}y - E_x(x, y, t)\mathrm{d}t \wedge \mathrm{d}x - E_y(x, y, t)\mathrm{d}t \wedge \mathrm{d}y$. Its other terms are all zero.
(a) What relation among the partial derivatives of B_z, E_x, and E_y must hold in order that $\mathrm{d}F = 0$?
(b) Express $\star\mathrm{d}\star F$ in terms of partial derivatives of B_z, E_x, and E_y.

19.2. A Lorentz transformation α corresponding to velocity v along the x-axis may be described in terms of pullback as follows (for convenience, we set $c = 1$):

$$\alpha^* t' = \gamma(t - vx),$$
$$\alpha^* x' = \gamma(x - vt),$$
$$\alpha^* y' = y \quad \alpha^* z' = z,$$

where $\gamma = 1/\sqrt{(1 - v^2)}$. If x, y, z, t are coordinates used by platform observers and x', y', z', t' are used by observers on board a train moving with velocity v along the x-axis, then α^* pulls back the train coordinates x', y', z', t' of an event into the coordinates of that even in the platform frame of reference.
(a) Show explicitly that α^* preserves the star operator; i.e. that $\alpha^*\star\omega = \star(\alpha^*\omega)$. Using the star operator for the coordinates x, y, z, t; i.e., $\star\mathrm{d}t = \mathrm{d}x \wedge \mathrm{d}y \wedge \mathrm{d}z$, etc., check the following:

$$\alpha^*\star\mathrm{d}x' \wedge \mathrm{d}y' \wedge \mathrm{d}z' = \star\alpha^*(\mathrm{d}x' \wedge \mathrm{d}y' \wedge \mathrm{d}z'),$$
$$\alpha^*\star\mathrm{d}t' \wedge \mathrm{d}y' \wedge \mathrm{d}z' = \star\alpha^*(\mathrm{d}t' \wedge \mathrm{d}y' \wedge \mathrm{d}z'),$$

$$\alpha^* \star dt' \wedge dx' = \star\alpha^*(dt' \wedge dx'),$$

$$\alpha^* \star dx' \wedge dy' = \star\alpha^*(dx' \wedge dy').$$

(b) In terms of the train coordinates, the potential **A** may be expressed as

$$\mathbf{A} = A'_t dt' + A'_x dx' + A'_y dy' + A'_z dz'.$$

Calculate $\alpha^*\mathbf{A}$ and equate it to $A_t dt + A_x dx + A_y dy + A_z dz$, thereby obtaining expressions for A_t, A_x, etc., in terms of $A'_t, A'_x, \ldots$.

19.3. Carry out the same program as in Exercise 19.2 for the two-form

$$F = B' + E' \wedge dt',$$

$$F = B'_x dy' \wedge dz' + \cdots + E'_z dz' \wedge dt'.$$

Calculate $\star F$, compare the result with

$$F = B_x dy \wedge dz + \cdots + E_z dz \wedge dt,$$

and thereby obtain expressions for the field components $B_x, B_y, B_z; E_x, E_y, E_z$ in terms of $B'_x, B'_y, B'_z, E'_x, E'_y, E'_z$.

19.4. For a point charge at rest at the origin in the train frame, we know that

$$F = B' + E' \wedge dt' = \frac{x'dx' + y'dy' + z'dz'}{(x'^2 + y'^2 + z'^2)^{3/2}} \wedge dt'.$$

Calculate $\alpha^* F$, thereby obtaining the description of the fields of a charge in uniform motion along the x-axis.

19.5. From the two-form F it is possible to create two different zero-forms:

$$f_1 = \star(F \wedge F), \quad f_2 = \star(F \wedge \star F).$$

Express these Lorentz invariant quantities in terms of the components of F; i.e., in terms of $B_x, B_y \ldots E_z$.

20

Complex analysis

The material in Chapter 20 is a relatively standard treatment of the theory of functions of a complex variable.

Introduction

This chapter is devoted to the study of the theory of functions of a complex variable. This theory, in its various ramifications, represents the major achievements of nineteenth-century mathematics. We shall touch on the key results here. Recall from Chapters 7–8 the basic facts of the calculus of several variables, in particular of two variables: Let x and y be the standard coordinates on $\mathbb{R}^2$. The basic objects in the calculus are functions, linear differential forms, and forms of degree 2. Each of these may only be defined on an open subset of $\mathbb{R}^2$. A function is a rule, f, which assigns the number $f(x, y)$ to the point (x, y). Unless otherwise specified, all functions are assumed to be differentiable. A linear differential form is an expression like

$$\omega = a\mathrm{d}x + b\mathrm{d}y$$

where a and b are functions. If γ is an oriented curve (which is piecewise differentiable and lying in the domain of definition of ω), then we can form the line integral $\int_\gamma \omega$. If $\gamma = \gamma_1 \ldots \gamma_n$ where the γ_i are differentiable curves and γ_i is parameterized by $x_i(t)$, $y_i(t)$ $\alpha_i \leqslant t \leqslant \beta_i$ then

$$\int_\gamma \omega = \int_{\gamma_1} \omega + \cdots + \int_{\gamma_n} \omega$$

where

$$\int_{\gamma_i} \omega = \int_{\alpha_i}^{\beta_i} [a(x_i(t), y_i(t))x_i'(t) + b(x_i(t), y_i(t))y_i'(t)]\mathrm{d}t.$$

Given any function f its differential, $\mathrm{d}f$, is the linear differential form given by

$$\mathrm{d}f = \frac{\partial f}{\partial x}\mathrm{d}x + \frac{\partial f}{\partial y}\mathrm{d}y.$$

Stokes' theorem for functions (the fundamental theorem of the calculus) says that, if γ is any curve going from p to q, then $\int_\gamma \mathrm{d}f = f(q) - f(p)$.

A two-form is an expression like

$$\Omega = c\mathrm{d}x \wedge \mathrm{d}y$$

where c is a function. If D is a bounded region in the plane (contained in the domain of definition of Ω), then we can form the integral $\int_D \Omega$. If D is given the standard orientation, then $\int_D \Omega$ is just the double integral

$$\int_D \omega = \int_D c(x, y)\mathrm{d}x\,\mathrm{d}y.$$

(If D is given the opposite orientation, then we must reverse the sign.) All regions will be assumed to have a piecewise differentiable boundary which gets an induced orientation. This boundary (with orientation) is denoted by ∂D. If $\omega = a\mathrm{d}x + b\mathrm{d}y$, then $\mathrm{d}\omega$ is the two-form given by

$$\mathrm{d}\omega = \left(\frac{\partial a}{\partial y} - \frac{\partial b}{\partial x}\right)\mathrm{d}x \wedge \mathrm{d}y$$

and Stokes' theorem (for one-forms) says that

$$\int_{\partial D} \omega = \int_D \mathrm{d}\omega.$$

We assume that the reader is familiar with the basic facts about the complex numbers. Every complex number c can be written as $c = r + \mathrm{i}s$, where r and s are real numbers and $\mathrm{i}^2 = -1$. Here r is called the *real part* and s the *imaginary part* of c. The number $\bar{c} = r - \mathrm{i}s$ is called the *complex conjugate* of c and $c\bar{c} = |c|^2 = r^2 + s^2$. All the usual algebraic laws such as the commutative and associative laws for addition and multiplication and the distributive law hold.

20.1. Complex-valued functions

We can now allow the functions (and differential forms) that we have been considering to take on complex values. Thus a *complex-valued function* f assigns a complex-number $f(x, y) = u(x, y) + \mathrm{i}v(x, y)$ to each point in the plane, where u and v are real-valued functions. In other words, giving a complex-valued function is the same as giving a pair of real-valued functions, i.e., the same as giving a map of (an open subset of) $\mathbb{R}^2$ into $\mathbb{R}^2$. We say that f is differentiable if this map is differentiable, that is if both functions u and v are differentiable. Thus $\partial f/\partial x = \partial u/\partial x + \mathrm{i}\partial v/\partial x$ and $\partial f/\partial y = \partial u/\partial y + \mathrm{i}\partial v/\partial y$ give the partial derivatives of f. We can consider various linear combinations of these partial derivatives with real or complex coefficients.

For example, we can consider the expression $\frac{1}{2}\frac{\partial f}{\partial x}+\frac{i}{2}\frac{\partial f}{\partial y}$. We can regard this expression as the result of applying the *differential operator* $\frac{1}{2}(\partial/\partial x + i\partial/\partial y)$ to the function f. This operator will be of crucial importance to us in all that follows and we shall denote it by $\partial/\partial\bar{z}$. Thus

$$\frac{\partial}{\partial \bar{z}}=\frac{1}{2}\left(\frac{\partial}{\partial x}+i\frac{\partial}{\partial y}\right) \tag{20.1}$$

and, similarly, we define

$$\frac{\partial}{\partial z}=\frac{1}{2}\left(\frac{\partial}{\partial x}-i\frac{\partial}{\partial y}\right). \tag{20.2}$$

The reasons for this notation will become clear in a little while. Notice that if $f=u+iv$ then

$$\begin{aligned}\frac{\partial f}{\partial \bar{z}}&=\frac{\partial u}{\partial \bar{z}}+i\frac{\partial v}{\partial \bar{z}}\\&=\frac{1}{2}\left(\frac{\partial u}{\partial x}+i\frac{\partial u}{\partial y}\right)+\frac{i}{2}\left(\frac{\partial v}{\partial x}+i\frac{\partial v}{\partial y}\right),\end{aligned}$$

so, collecting real and imaginary parts,

$$\frac{\partial f}{\partial \bar{z}}=\frac{1}{2}\left(\frac{\partial u}{\partial x}-\frac{\partial v}{\partial y}\right)+\frac{i}{2}\left(\frac{\partial u}{\partial y}+\frac{\partial v}{\partial x}\right). \tag{20.3}$$

Notice that for any pair of differentiable functions f and g we have

$$\frac{\partial}{\partial \bar{z}}(fg)=\frac{\partial f}{\partial \bar{z}}g+f\frac{\partial g}{\partial \bar{z}} \tag{20.4}$$

with a similar equation for $\partial/\partial z$.

We define the particular complex function z as $z(x, y)=x+iy$. We write this as

$$z=x+iy \tag{20.5}$$

and similarly

$$\bar{z}=x-iy.$$

Then the function z^2 is given by

$$z^2=(x^2-y^2)+i2xy$$

while

$$\bar{z}^2=(x^2-y^2)-i2xy$$

and

$$z\bar{z}=|z|^2=x^2+y^2.$$

Similarly we can form polynomials in z alone such as $17z^3-5z^2+2z-3$ or polynomials in $\bar{z}$ or mixed polynomials like $3z^2\bar{z}-2z\bar{z}^3+z^2-5\bar{z}+1$, etc. All these are examples of complex-valued functions. If we substitute z for f in (20.3) where now $u(x, y)\equiv x$ and $v(x, y)\equiv y$, we see that

$$\frac{\partial z}{\partial \bar{z}}\equiv 0 \tag{20.6}$$

and, similarly,

$$\frac{\partial z}{\partial z} \equiv 1.$$

By (20.4) and the linearity of $\partial/\partial z$ we see that, if $P = P(z)$ is any polynomial in z alone, we have

$$\frac{\partial P}{\partial \bar{z}} \equiv 0.$$

If $P(z) = a_0 + a_1 z + \cdots + a_n z^n$, then

$$\frac{\partial P}{\partial z} = a_1 + 2a_2 z + \cdots + n a_n z^{n-1}.$$

This is just the polynomial $P'(z)$, the ordinary derivative of P in the sense of polynomials. Thus

$$\frac{\partial P}{\partial \bar{z}} = 0 \quad \text{and} \quad \frac{\partial P}{\partial z} = P'.$$

20.2. Complex-valued differential forms

We can also consider complex-valued differential forms. A linear differential form is an expression

$$a\,\mathrm{d}x + b\,\mathrm{d}y$$

where now a and b are complex-valued functions. The notion of line integral and the (zero-dimensional) Stokes' theorem are as before, with complex numbers as values. We define the linear differential forms $\mathrm{d}z$ and $\mathrm{d}\bar{z}$ by

$$\mathrm{d}z = \mathrm{d}x + \mathrm{i}\,\mathrm{d}y, \quad \text{and} \quad \mathrm{d}\bar{z} = \mathrm{d}x - \mathrm{i}\,\mathrm{d}y.$$

Notice that these differential forms are linearly independent at each point and in fact

$$\mathrm{d}x = \frac{1}{2}(\mathrm{d}z + \mathrm{d}\bar{z}), \quad \mathrm{d}y = \frac{1}{2\mathrm{i}}(\mathrm{d}z - \mathrm{d}\bar{z}).$$

We can thus write any linear differential form as

$$a\,\mathrm{d}x + b\,\mathrm{d}y = A\,\mathrm{d}z + B\,\mathrm{d}\bar{z}$$

where

$$A = \tfrac{1}{2}(a - ib) \quad \text{and} \quad B = \tfrac{1}{2}(a + ib).$$

In particular, for any (complex-valued) differential function f, we have

$$\begin{aligned} \mathrm{d}f &= \frac{\partial f}{\partial x}\mathrm{d}x + \frac{\partial f}{\partial y}\mathrm{d}y \\ &= \frac{1}{2}\left(\frac{\partial f}{\partial x} - \mathrm{i}\frac{\partial f}{\partial y}\right)\mathrm{d}z + \frac{1}{2}\left(\frac{\partial f}{\partial x} + \mathrm{i}\frac{\partial f}{\partial y}\right)\mathrm{d}\bar{z}. \end{aligned}$$

Going back to the definitions (20.1) and (10.2), we can write this as

$$\mathrm{d}f = \frac{\partial f}{\partial z}\mathrm{d}z + \frac{\partial f}{\partial \bar{z}}\mathrm{d}\bar{z}. \tag{20.7}$$

Let us work out some examples. If we take $f = z$, then (20.5) and (20.6) say that $\mathrm{d}(z) = \mathrm{d}z$. Thus our notation is consistent (and this is the reason for the choice of the $\partial/\partial z$ and $\partial/\partial \bar{z}$ notations). If $f = P$ is a polynomial in z, then

$$\mathrm{d}P = P'\mathrm{d}z$$

with no $\mathrm{d}\bar{z}$ component. If we take $f = |z|^2 = z\bar{z}$, we get

$$\mathrm{d}|z|^2 = \bar{z}\mathrm{d}z + z\,\mathrm{d}\bar{z}.$$

Let us define the complex exponential function e^z by

$$\mathrm{e}^z = \mathrm{e}^x(\cos y + \mathrm{i}\sin y).$$

Then

$$\begin{aligned}\mathrm{d}\mathrm{e}^z &= \mathrm{e}^x(\cos y + \mathrm{i}\sin y)\,\mathrm{d}x + \mathrm{e}^x(-\sin y + \mathrm{i}\cos y)\,\mathrm{d}y \\ &= \mathrm{e}^x(\cos y + \mathrm{i}\sin y)\,(\mathrm{d}x + \mathrm{i}\,\mathrm{d}y)\end{aligned}$$

so

$$\mathrm{d}\mathrm{e}^z = \mathrm{e}^z\mathrm{d}z.$$

From this we can conclude that

$$\frac{\partial \mathrm{e}^z}{\partial \bar{z}} \equiv 0.$$

A (complex-valued) two-form is an expression like

$$c\mathrm{d}x \wedge \mathrm{d}y$$

where now c is a complex-valued function. We can form the exterior product, $\omega_1 \wedge \omega_2$, of two complex-valued linear differential forms ω_1 and ω_2. Thus, for example

$$\begin{aligned}\mathrm{d}z \wedge \mathrm{d}\bar{z} &= (\mathrm{d}x + \mathrm{i}\,\mathrm{d}y) \wedge (\mathrm{d}x - \mathrm{i}\,\mathrm{d}y) \\ &= -2\mathrm{i}\,\mathrm{d}x \wedge \mathrm{d}y\end{aligned}$$

or

$$\mathrm{d}x \wedge \mathrm{d}y = \frac{\mathrm{i}}{2}\mathrm{d}z \wedge \mathrm{d}\bar{z}.$$

In particular,

$$\mathrm{d}(A\,\mathrm{d}z) \equiv 0 \quad \text{if and only if} \quad \frac{\partial A}{\partial \bar{z}} \equiv 0.$$

Stokes' theorem remains true without change except that now ω is a complex-valued linear differential form. Thus, for example, if $A\,\mathrm{d}z$ satisfies $\mathrm{d}(A\,\mathrm{d}z) = 0$, then Stokes' theorem says that for any region D contained in the domain of definition of A, we have

$$\boxed{\int_{\partial D} A\,\mathrm{d}z = 0.}$$

This important result is known as *Cauchy's theorem.*

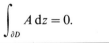

20.3. Holomorphic functions

A differentiable function f (defined on some domain D of $\mathbb{R}^2$) is called *holomorphic* if it satisfies the identity

$$\frac{\partial f}{\partial \bar{z}} \equiv 0. \tag{20.8}$$

Since this definition is basic to the whole subject, the rest of this section will be devoted to discussing equivalent formulations and elementary consequences of the notion of holomorphicity. Since $\partial/\partial\bar{z}$ is a differential operator, we have the equations

$$\frac{\partial}{\partial \bar{z}}(f+g) = \frac{\partial f}{\partial \bar{z}} + \frac{\partial g}{\partial \bar{z}}, \quad \frac{\partial}{\partial \bar{z}}(fg) = \frac{\partial f}{\partial \bar{z}}g + f\frac{\partial g}{\partial \bar{z}},$$

$$\frac{\partial}{\partial \bar{z}}\left(\frac{f}{g}\right) = \frac{1}{g^2}\left(\frac{\partial f}{\partial \bar{z}}g - f\frac{\partial g}{\partial \bar{z}}\right), \quad g \neq 0.$$

From these we see that the sum, product and quotient of two holomorphic functions are holomorphic (in the case of quotient only if the denominator is nowhere zero). We have already used this remark in the preceding section to conclude that $(\partial/\partial\bar{z})P(z) \equiv 0$, i.e., that any polynomial (in z alone) is holomorphic. Similarly any rational function, i.e., the quotient of two polynomials, P/Q, is holomorphic in a region D, provided Q does not vanish anywhere in D.

In view of (20.7), a function f is holomorphic if and only if

$$\mathrm{d}f = h\,\mathrm{d}z \tag{20.9}$$

that is, if and only if $\mathrm{d}f$ has no $\mathrm{d}\bar{z}$ component. Here, of course, $h = \partial f/\partial z$. Equivalently, we can write this condition as

$$\mathrm{d}(f\,\mathrm{d}z) = 0. \tag{20.10}$$

Let us write $f = u + \mathrm{i}v$ where the real-valued functions u and v are the real and imaginary parts of f. Then, setting the real and imaginary parts of $\partial f/\partial\bar{z}$ equal to zero, we see by (20.3) that

$$\boxed{\frac{\partial u}{\partial x} = \frac{\partial v}{\partial y}, \quad \frac{\partial u}{\partial y} = \frac{-\partial v}{\partial x}.} \tag{20.11}$$

These equations are known as the *Cauchy–Riemann equations.* They are a system of first-order partial differential equations for u and v. Again, f is holomorphic if and only if the Cauchy–Riemann equations hold.

Here is another way of writing the Cauchy–Riemann equations. Consider the differential form

$$u\,\mathrm{d}x + v\,\mathrm{d}y.$$

Then relative to the Euclidean metric in the plane

$$\star(u\,\mathrm{d}x + v\,\mathrm{d}y) = -v\,\mathrm{d}x + u\,\mathrm{d}y.$$

Thus the Cauchy–Riemann equations are equivalent to

$$\mathrm{d}(u\,\mathrm{d}x + v\,\mathrm{d}y) = 0$$

and

$$\mathrm{d}\star(u\,\mathrm{d}x + v\,\mathrm{d}y) = 0.$$

Notice that they are the analogues, in $\mathbb{R}^2$, of the Maxwell equations in $\mathbb{R}^{1,3}$. In Maxwell's equation we dealt with a two-form on four-space. The Cauchy–Riemann equations are for a one-form on two-space. As we remarked in Chapter 18, the star operator: $\Lambda^{n/2} \to \Lambda^{n/2}$ (n even) depends only on the conformal structure.

Let us think of f as giving a map of (a subset of) $\mathbb{R}^2$ into $\mathbb{R}^2$, sending (x, y) into $(u(x, y), v(x, y))$. The Jacobian matrix of this map is

$$\begin{pmatrix} \dfrac{\partial u}{\partial x} & \dfrac{\partial u}{\partial y} \\ \dfrac{\partial v}{\partial x} & \dfrac{\partial v}{\partial y} \end{pmatrix}.$$

Equation (20.11) says that this matrix has the form $\begin{pmatrix} a & -b \\ b & a \end{pmatrix}$. Now a matrix in $\mathbb{R}^2$ is of this form, with $a^2 + b^2 \neq 0$, if and only if it is conformal (i.e., preserves angles) and orientation-preserving. Thus we can rewrite (20.11) as

$$\begin{pmatrix} \dfrac{\partial u}{\partial x} & \dfrac{\partial u}{\partial y} \\ \dfrac{\partial v}{\partial x} & \dfrac{\partial v}{\partial y} \end{pmatrix} \text{ either is zero or is conformal and orientation-preserving.} \tag{20.12}$$

Let $f = u + \mathrm{i}v$ and $g = r + \mathrm{i}s$ be complex functions. We think of f and g as maps of (subsets of) $\mathbb{R}^2$ into $\mathbb{R}^2$. If the image of g lies in the domain of f, then we can form the composite function $f \circ g$. By the chain rule, the Jacobian matrix of $f \circ g$ (thought of as a map of $\mathbb{R}^2$ into $\mathbb{R}^2$) when evaluated at (x, y) is the product of the Jacobian matrix of f, evaluated at $(r(x, y), s(x, y))$ with the Jacobian matrix of g, evaluated at (x, y). Now the product of two conformal, orientation-preserving matrices is again conformal and orientation-preserving. Also the product of the zero matrix with any matrix is zero. Thus it follows from (20.12) that the composition of two holomorphic functions is again holomorphic. The expression that we use for the composition of two holomorphic functions is standard: if, for example, f is given by $f(z) = \mathrm{e}^z$ and g is given by $g(z) = 3z^2 - 2$, then $f \circ g(z) = \mathrm{e}^{3z^2 - 2}$. Also, since $\partial/\partial z$ is a linear combination of the operators $\partial/\partial x$ and $\partial/\partial y$ (with constant coefficients), the chain rule implies that

$$\frac{\partial}{\partial z}(f \circ g) = \left(\frac{\partial f}{\partial z} \circ g\right)\frac{\partial g}{\partial z}. \tag{20.13}$$

So, in the case of the preceding example,

$$\frac{\partial}{\partial z}\mathrm{e}^{3z^2 - 2} = 6z\mathrm{e}^{3z^2 - 2}.$$

In other words, for holomorphic functions, we can compute with $\partial/\partial z$ as with the ordinary derivative of a function of one variable. We shall presently see the reason why this is so.

Condition (20.12) is a bit awkward in that there are two separate cases according to whether the Jacobian matrix is or is not the zero matrix. Here is a way of writing (20.12) as one single condition. Consider the matrix

$$J = \begin{pmatrix} 0 & -1 \\ 1 & 0 \end{pmatrix}.$$

Then a matrix

$$A = \begin{pmatrix} a & c \\ b & d \end{pmatrix}$$

commutes with J, i.e., satisfies

$$AJ = JA$$

if and only if $a = d$ and $c = -b$, as can be seen by multiplying out both sides. Thus $AJ = JA$ if and only if A has the form

$$A = \begin{pmatrix} a & -b \\ b & a \end{pmatrix}$$

(where now a and b can both be zero). Thus we can write (20.12) as

$$\begin{pmatrix} \dfrac{\partial u}{\partial x} & \dfrac{\partial u}{\partial y} \\ \dfrac{\partial y}{\partial x} & \dfrac{\partial v}{\partial y} \end{pmatrix} \begin{pmatrix} 0 & -1 \\ 1 & 0 \end{pmatrix} = \begin{pmatrix} 0 & -1 \\ 1 & 0 \end{pmatrix} \begin{pmatrix} \dfrac{\partial u}{\partial x} & \dfrac{\partial u}{\partial y} \\ \dfrac{\partial v}{\partial x} & \dfrac{\partial v}{\partial y} \end{pmatrix} \tag{20.14}$$

We can give an interesting interpretation to the condition $AJ = JA$ and thus to (20.14). The matrix A is a 2×2 matrix. It gives a linear transformation of the real two-dimensional vector space $\mathbb{R}^2$ into itself. Now we can identify $\mathbb{R}^2$ with the one-dimensional complex vector space $\mathbb{C}$ by identifying the vector $\begin{pmatrix} x \\ y \end{pmatrix}$ with the complex number $x + \mathrm{i}y$. Multiplication by i sends $x + \mathrm{i}y$ into $-y + \mathrm{i}x$. But the complex number $-y + \mathrm{i}x$ corresponds to the vector $\begin{pmatrix} -y \\ x \end{pmatrix} = J\begin{pmatrix} x \\ y \end{pmatrix}$. Thus J is the matrix of multiplication by i from the real point of view. Now any linear transformation of the one-dimensional complex vector space $\mathbb{C}$ is a linear transformation of $\mathbb{R}^2$. But not every linear transformation A of $\mathbb{R}^2$ over the *reals* corresponds to a linear transformation of $\mathbb{C}$ over the complex numbers. A complex linear transformation must commute with multiplication by complex numbers, in particular by i. Thus for A to correspond to a complex linear transformation, it must satisfy $AJ = JA$. If A satisfies this equation, so that $A = \begin{pmatrix} a & -b \\ b & a \end{pmatrix}$, it is immediate to check that A corresponds to multiplication by the complex number $a + ib$, i.e., that

$A\begin{pmatrix} x \\ y \end{pmatrix}$ corresponds to $(a+ib)(x+iy)$. Multiplication by a complex number on $\mathbb{C}$ is obviously complex linear. Thus we can write (20.14) as

$$\text{The transformation } \begin{pmatrix} \dfrac{\partial u}{\partial x} & \dfrac{\partial u}{\partial y} \\ \dfrac{\partial v}{\partial y} & \dfrac{\partial v}{\partial y} \end{pmatrix} \text{ is complex linear.} \tag{20.15}$$

So far none of the above equivalent formulations, with the exception of the Cauchy–Riemann equations (20.11), would be familiar to a nineteenth-century mathematician. We now turn to the classical formulation, one which is closely related to (20.15). Let f be a differentiable function, not necessarily holomorphic. Let $\begin{pmatrix} x \\ y \end{pmatrix}$ be a point in the domain of definition of F. Let $\begin{pmatrix} k \\ l \end{pmatrix}$ be a small vector. Then we have

$$f\left(\begin{pmatrix} x+k \\ y+l \end{pmatrix}\right) - f\left(\begin{pmatrix} x \\ y \end{pmatrix}\right) = \frac{\partial f}{\partial x}k + \frac{\partial f}{\partial y}l + o((\|k^2+l^2\|)^{1/2})$$

by the definition of differentiability. $\left(\text{Here the partial derivatives are evaluated at } \begin{pmatrix} x \\ y \end{pmatrix}\right)$. Let us write $z=x+iy$ and $h=k+il$ and write $f(z)$ instead of $f\left(\begin{pmatrix} x \\ y \end{pmatrix}\right)$. So we may write the preceding equation as

$$f(z+h)-f(z) = \frac{\partial f}{\partial x}k + \frac{\partial f}{\partial y}l + o(|h|).$$

By (20.7) we can write this as

$$f(z+h)-f(z) = \frac{\partial f}{\partial z}h + \frac{\partial f}{\partial \bar{z}}\bar{h} + o(|h|).$$

This is an equation involving complex numbers, so, if $h \neq 0$, we can divide both sides of the equation by h so as to obtain

$$\frac{f(z+h)-f(z)}{h} = \frac{\partial f}{\partial z} + \frac{\partial f}{\partial \bar{z}}\frac{\bar{h}}{h} + o(1).$$

Notice that the left-hand side looks like an ordinary difference quotient, but relative to the complex numbers. Suppose we let $|h| \to 0$. The value of $\bar{h}/h$ can be any complex number of absolute value one. (For instance, if h is real, then $\bar{h}/h = 1$, while, if $h = si$ is pure imaginary, then $\bar{h}/h = -1$.) So the limiting expression on the right may depend on the angle at which $h \to 0$. If we want this limit to exist, that is, to be independent of the angle of approach, we must have $\partial f/\partial \bar{z} = 0$ at the point $\begin{pmatrix} x \\ y \end{pmatrix}$. Thus if f is holomorphic, we conclude that the limit on the left-hand side exists at all points in the domain of definition of f and this limit equals

$\partial f/\partial z$ which is a continuous function. Conversely, suppose that at each point the limit

$$\lim_{h\to 0}\frac{f(z+h)-f(z)}{h}=f'(z)$$

exists (independently of how h approaches zero) and is a continuous function. Then taking h real shows that $\partial f/\partial x$ exists and is continuous and taking h purely imaginary shows that $\partial f/\partial y$ exists and is continuous. Thus f is continuously differentiable. The preceding argument then shows that $\partial f/\partial \bar{z}=0$ and $f'(z)=\partial f/\partial z$. Thus we have proved that f is holomorphic if and only if

$$\lim_{h\to 0}\frac{f(z+h)-f(z)}{h}=f'(z) \tag{20.16}$$

exists (independently of how h approaches zero) and is a continuous function. In short, we can formulate (20.16) as saying that f is continuously differentiable from the complex point of view. This is the approach that was mainly taken in nineteenth-century mathematics. It is somewhat deceptive in that the condition of complex differentiability is so much more restrictive than the standard notion of differentiability. Nevertheless we shall adopt $f'(z)$ as a more convenient notation that $\partial f/\partial z$.

We have observed that Stokes' theorem and (20.10) imply that

$$\int_{\partial D} f(z)\,\mathrm{d}z=0, \tag{20.17}$$

which is Cauchy's integral theorem, for any domain D contained in the domain of definition of a holomorphic function f. Conversely, if f is any function, then $\int_{\partial D} f(z)\,\mathrm{d}z=\int_D \mathrm{d}(f\,\mathrm{d}z)$. This integral can vanish for all D only if the integrand vanishes identically, i.e., if f is holomorphic.

We have thus seen that the conditions (20.8)–(20.17) are all equivalent. A function that satisfies any one (and hence all) of them is *holomorphic.*

20.4. The calculus of residues

We begin by calculating the line integral $\int_\gamma z^n\,\mathrm{d}z$ where γ is a circle centered at the origin. If $n\geqslant 0$, the function z^n is holomorphic in the entire plane and it follows from Cauchy's theorem that the integral is zero. For negative values of n, the function z^n is not holomorphic (and not defined) at the origin and hence we must evaluate the integral directly. If γ has radius r, we can use polar coordinates to write $z=r\mathrm{e}^{\mathrm{i}\theta}$ and thus, along γ, $\mathrm{d}z=\mathrm{i}r\mathrm{e}^{\mathrm{i}\theta}\,\mathrm{d}\theta$. Thus

$$z^n\mathrm{d}z=\mathrm{i}r^{n+1}\mathrm{e}^{\mathrm{i}(n+1)\theta}\,\mathrm{d}\theta$$

and

$$\int_\gamma z^n\,\mathrm{d}z=\mathrm{i}r^{n+1}\int_0^{2\pi}\mathrm{e}^{\mathrm{i}(n+1)\theta}\,\mathrm{d}\theta.$$

This last integral vanishes except when $n = -1$, when it equals 2π. In this case, $r^{n+1} = 1$ and so

$$\int_\gamma z^n \, dz = \begin{cases} 0 & \text{if } n \neq -1, \\ 2\pi i & \text{if } n = -1. \end{cases}$$

Let D be some region containing the origin and let f be a function which is holomorphic in $D - \{0\}$. Suppose, further, that near the origin we can write

$$f = a_{-n}z^{-n} + a_{-n+1}z^{-(n-1)} + \cdots + a_{-1}z^{-1} + g,$$

where g is holomorphic near the origin.

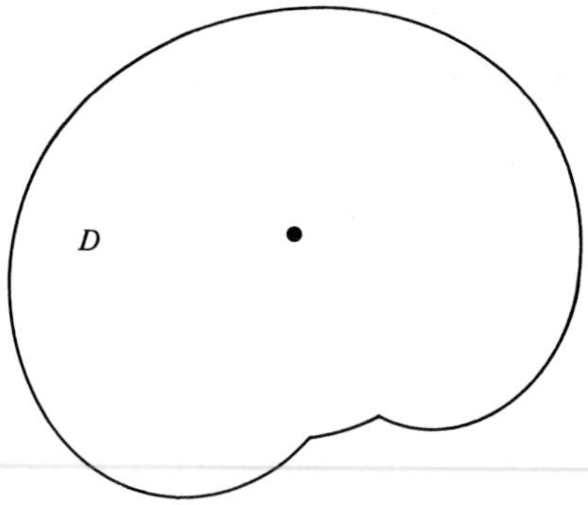

Figure 20.1

Let us evaluate $\int_{\partial D} f(z)\,dz$. Let us draw a small circle γ about the origin (lying entirely in D) and in a region where the above expansion holds. Then f is holomorphic in that region of D which is exterior to γ. Therefore, by the Cauchy integral theorem,

$$\int_{\partial D} f \, dz = \int_\gamma f \, dz.$$

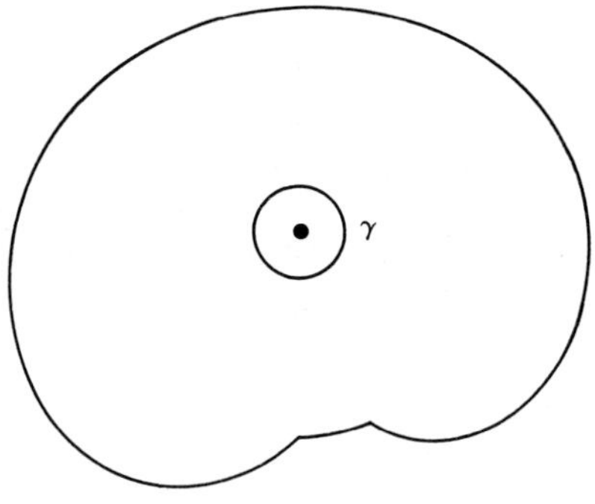

Figure 20.2

If we now substitute the above expansion for f, and integrate term by term, all the negative powers give zero except $a_{-1}z^{-1}$, while $\int_\gamma g \, \mathrm{d}z = 0$ since g is holomorphic. Thus

$$\int_{\partial D} f(z) \, \mathrm{d}z = 2\pi \mathrm{i} a_{-1}.$$

Now there is nothing special about the origin, which we can replace by any other point in D, or by a finite number of points. A function f is said to have a *pole* at the point α if it is holomorphic in $U - \{\alpha\}$ where U is some neighborhood of α and if (near α) it has an expansion

$$f(z) = a_{-n}(z - \alpha)^{-n} + \cdots + a_{-1}(z - \alpha)^{-1} + g$$

where g is holomorphic near α. The number a_{-1} is called the *residue* of f at α, and it follows from the preceding argument that the integral of $f(z) \, \mathrm{d}z$ around a small circle about α equals $2\pi i a_{-1} = 2\pi \mathrm{i} \, \mathrm{res}_\alpha(f)$. A function which is holomorphic in a domain D except for poles is called *meromorphic*.

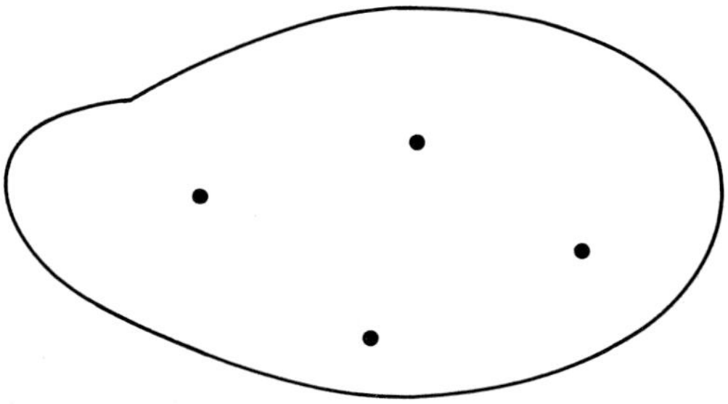

Figure 20.3

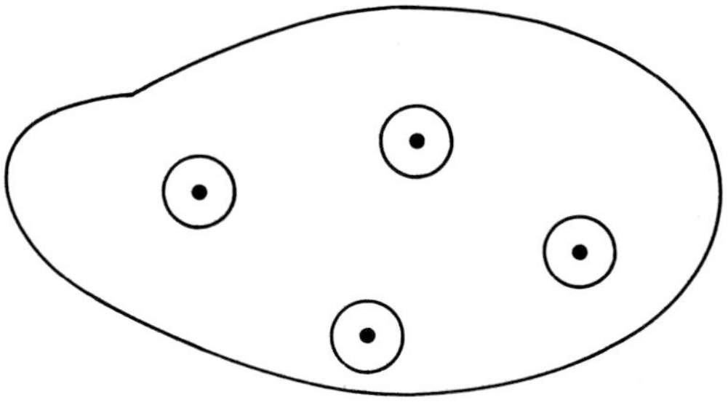

Figure 20.4

Suppose that f is meromorphic in D with a finite number of poles. Then we can draw little circles about each of the poles, and apply the Cauchy integral theorem to conclude that

$$\int_{\partial D} f(z) \, \mathrm{d}z = 2\pi \mathrm{i} \sum_{\text{poles } \alpha} \mathrm{res}_\alpha(f).$$

This is known as *Cauchy's residue theorem*. It is very useful in the computation of definite integrals. For example, suppose we wish to evaluate the definite integral

$$\int_0^{\pi} \frac{d\theta}{a+\cos\theta},$$

where $a > 1$. We write this as

$$\frac{1}{2}\int_0^{2\pi} \frac{d\theta}{a+\cos\theta}.$$

Let us set $z = e^{i\theta}$ so that, on the unit circle, $dz/iz = d\theta$. Also, $2\cos\theta = e^{i\theta} + e^{-i\theta} = z + z^{-1}$. Thus the preceding integral becomes

$$\frac{1}{2}\int_{\gamma} \frac{dz}{iz(a+\frac{1}{2}(z+z^{-1}))} = \frac{1}{i}\int_{\gamma} \frac{dz}{(z^2+2az+1)}$$

where γ is the unit circle. But

$$z^2 + 2az + 1 = (z-\alpha_1)(z-\alpha_2),$$

where

$$\alpha_1 = -a + \sqrt{(a^2-1)}$$

and

$$\alpha_2 = -a - \sqrt{(a^2-1)}$$

and we have the partial fraction expansion

$$\frac{1}{z^2+2az+1} = \frac{1}{\alpha_1-\alpha_2}\left(\frac{1}{z-\alpha_1} - \frac{1}{z-\alpha_2}\right).$$

Now, since $a > 1$, it follows that $\alpha_2 < -1$ and so α_2 lies outside the unit circle. Since $\alpha_1\alpha_2 = 1$, it follows that α_1 lies inside the unit circle and the above partial fraction expansion shows (since $1/(z-\alpha_2)$ is holomorphic inside the unit circle) that the residue is

$$\frac{1}{\alpha_1-\alpha_2} = \frac{1}{2\sqrt{(a^2-1)}}.$$

Thus, by Cauchy's residue theorem,

$$\int_0^{\pi} \frac{d\theta}{a+\cos\theta} = \frac{1}{2i}\int_{\gamma} \frac{dz}{z^2+2az+1} = \frac{\pi}{\sqrt{(a^2-1)}}.$$

We shall soon see that any rational function $R(z) = P(z)/Q(z)$ is meromorphic with poles located at the zeros of q. (This follows immediately from the partial fraction expansion for rational functions. But we will not assume this here.) We shall also get an effective way of evaluating the residues. This will, for example, allow us to evaluate integrals of the form $\int_0^{2\pi} R(\cos\theta, \sin\theta)\, d\theta$ by setting $z = e^{i\theta}$, $\cos\theta = \frac{1}{2}(z+z^{-1})$, and $\sin z = (1/2i)(z-z^{-1})$ and proceed as above.

We begin with a lemma. Suppose that g is holomorphic in $\bar{D}$† except for a finite

† $\bar{D}$ is the *closure* of D: it includes all points of D plus points on the boundary.

number of points $\alpha_1, \ldots, \alpha_k$. Suppose that at each one of these points the function g satisfies

$$\lim_{z \to \alpha} |z - \alpha| |g(z)| = 0.$$

Then

$$\int_{\partial D} g \, \mathrm{d}z = 0.$$

Indeed, by putting little circles γ_i around each α_i, we conclude as before that $\int_{\partial D} g \, \mathrm{d}z = \sum \int_{\gamma_1} g \, \mathrm{d}z$. Now

$$\left| \int_{\gamma_i} g(z) \, \mathrm{d}z \right| \leqslant 2\pi r M_r$$

where r is the radius of γ_i and M_r is the maximum of $|g|$ on r_i. By assumption, $rM_r \to 0$ as $r \to 0$. Thus, by shrinking the radius of the circles, we conclude that $\int_{\partial D} g \, \mathrm{d}z = 0$.

Now let f be holomorphic in $\bar{D}$ and let a be a point of D. We apply the preceding lemma to

$$g(z) = \frac{f(z) - f(a)}{z - a}.$$

The function g is holomorphic in $\bar{D}$ except at the single point a. Furthermore, $(z - a)g(z) = f(z) - f(a)$ and this tends to zero as $t \to a$. Thus

$$\int_{\partial D} \frac{f(z) - f(a)}{z - a} \mathrm{d}z = 0.$$

But $f(a)$ is a constant and

$$\int_{\partial D} \frac{\mathrm{d}z}{z - a} = 2\pi \mathrm{i}.$$

Thus

$$\int_{\partial D} \frac{f(a) \, \mathrm{d}z}{z - a} = 2\pi \mathrm{i} f(a),$$

and we can write the preceding equation as

$$f(a) = \frac{1}{2\pi \mathrm{i}} \int_{\partial D} \frac{f(z)}{z - a} \mathrm{d}z.$$

For convenience we shall replace a by z and z by ξ and write *Cauchy's integral formula*

$$f(z) = \frac{1}{2\pi \mathrm{i}} \int_{\partial D} \frac{f(\xi)}{\xi - z} \mathrm{d}\xi \qquad (20.18)$$

This formula is the cornerstone of all the theoretical developments in the subject of complex analysis.

We use the following lemma. Let ϕ be a complex-valued function which is defined and continuous on the boundary, ∂D, of some domain D. Then the functions

$$F_n(z) = \int \frac{\phi(\xi)}{(\xi - z)^n} \, d\xi$$

are all holomorphic in D and satisfy

$$F_n'(z) = nF_{n+1}(z).$$

We prove this by induction on n. First of all we prove that F_1 is continuous in D. Let z_0 be some point of D and choose a so small that the neighborhoods $|z'' - z_0| < 2a$ and $|z - z_0| < a$ lie entirely in D. Thus, the distance of z' to all points of ∂D is at least a and hence, for $\xi \in \partial D$,

$$\frac{1}{|\xi - z'||\xi - z_0|} < \frac{1}{2a^2} < \frac{1}{a^2}$$

for all such z. Now, suppressing the prime,

$$\frac{1}{\xi - z} - \frac{1}{\xi - z_0} = \frac{1}{(\xi - z)(\xi - z_0)}(z - z_0)$$

so

$$|F_1(z) - F_1(z_0)| = |z - z_0| \left| \int_{\partial D} \frac{\phi(\xi)\, d\xi}{(\xi - z)(\xi - z_0)} \right| \leqslant |z - z_0| \frac{1}{a^2} \int_{\partial D} |\phi(\xi)|\, d\xi.$$

This shows that F_1 is continuous. Furthermore,

$$\frac{F_1(z) - F_1(z_0)}{z - z_0} = \int_{\partial D} \frac{\phi(\xi)\, d\xi}{(\xi - z)(\xi - z_0)}.$$

Keeping z_0 fixed, the function $\phi(\xi)/(\xi - z_0)$ is continuous on ∂D. Hence, the right-hand side of the above equation is a continuous function of z, and equals the left-hand side at all $z \neq z_0$. Letting $z \to z_0$ shows that F_1 is differentiable in the complex sense and that

$$F_1'(z_0) = \int \frac{\phi(\xi)}{(\xi - z_0)^2} = F_2(z_0).$$

Now

$$\frac{1}{(\xi - z)(\xi - z_0)} + \frac{(z - z_0)}{(\xi - z)^2(\xi - z_0)} = \frac{1}{(\xi - z)^2}$$

so

$$F_2(z) - F_2(z_0) = \int \frac{\phi(\xi)\, d\xi}{(\xi - z)(\xi - z_0)} - \int \frac{\phi(\xi)\, d\xi}{(\xi - z_0)^2} + (z - z_0) \int \frac{\phi(\xi)\, d\xi}{(\xi - z)^2(\xi - z_0)},$$

showing by the preceding argument that F_2 is continuous, and hence that F_1 is holomorphic. Also

$$\frac{F_2(z) - F_2(z_0)}{z - z_0} = \frac{R_1(z) - R_1(z_0)}{z - z_0} + \int \frac{\phi(\xi)}{(\xi - z)^2(\xi - z_0)} \, d\xi$$

with

$$R_1(z) = \int \frac{\phi(\xi)\,\mathrm{d}\xi}{(\xi - z)(\xi - z_0)}.$$

Since the function $\phi(\xi)/(\xi - z_0)$ is continuous on ∂D, we conclude (by the preceding result) that R_1 is differentiable and $R_1'(z_0) = \int(\phi(\xi)/(\xi - z_0)^3)\,\mathrm{d}\xi$. Thus, passing to the limit in the above equation, we conclude that F_2 is differentiable in the complex sense and that

$$F_2' = 2F_3.$$

It is now clear how to proceed in the induction argument. Suppose that we have proved that $F_{n-1}' = (n-1)F_n$ *for all continuous functions* $\boldsymbol{\phi}$. Using the identity

$$\frac{1}{(\xi - z)^{n-1}(\xi - z_0)} + \frac{z - z_0}{(\xi - z)^n(\xi - z_0)} = \frac{1}{(\xi - z_0)^n},$$

we see that

$$F_n(z) - F_n(z_0) = \int \frac{\phi(\xi)}{(\xi - z)^{n-1}(\xi - z_0)}\,\mathrm{d}\xi - \int \frac{\phi(\xi)\,\mathrm{d}\xi}{(\xi - z_0)^n} + (z - z_0)\int \frac{\phi(\xi)\,\mathrm{d}\xi}{(\xi - z)^n(\xi - z_0)}$$

$$= R_{n-1}(z) - R_{n-1}(z_0) + (z - z_0)\int \frac{\phi(\xi)\,\mathrm{d}\xi}{(\xi - z)^n(\xi - z_0)},$$

where the functions

$$R_k(z) = \int \frac{\phi(\xi)}{(\xi - z)^k(\xi - z_0)}\,\mathrm{d}\xi$$

are associated to the continuous function $\phi(f)/(\xi - z_0)$. We see that F_n is continuous.

By induction $F_n'(z_0) = nR_n(z_0) = nF_{n+1}(z_0)$, so dividing the preceding equation for $F_n(z) - F_n(z_0)$ by $z - z_0$ and letting $z \to z_0$ show that $F_n' = nF_{n+1}$, completing the induction.

We now apply this result to the Cauchy integral formula, (20.18). Since f is continuous, we conclude the remarkable fact that if f is holomorphic in $\bar{D}$, it must be infinitely differentiable there, each of its (complex) derivatives $f^{(k)}$ is holomorphic, and that

$$f^{(n)}(z) = \frac{n!}{2\pi \mathrm{i}} \int_{\partial D} \frac{f(\xi)\mathrm{d}\xi}{(\xi - z)^{n+1}}. \qquad (20.19)$$

There are many important consequences of (20.18) and (20.19) which we will derive in the next few sections. We begin with the *principle of removable singularities*. Suppose that a function f is defined and holomorphic in $D - \{a\}$ where $a \in D$; i.e., in all of D except at a single point a. A necessary and sufficient condition that there exist a holomorphic function defined throughout D which coincides with f on $D - \{a\}$ is that $\lim_{z \to a}(z - a)f(z) = 0$. The extended function (if it exists) is uniquely determined.

Proof. Let $f(a)$ denote the value of the extended function at a. Since the extended function is holomorphic, it is continuous at a, and $\lim_{z\to a} f(z) = f(a)$. Therefore $\lim_{z\to a}(z-a)f(z) = \lim_{z\to a}(z-a)f(a) = 0$, which proves the necessity. To prove the sufficiency, draw a circle γ about a which lies entirely within D. Consider the function

$$\frac{1}{2\pi i}\int_\gamma \frac{f(\xi)d\xi}{\xi - z}.$$

This function is holomorphic inside γ. Suppose we draw a small circle γ_1 about a lying inside γ. Then, at all points in the annulus between γ_1 and γ, Cauchy's integral formula (20.18) says that

$$f(z) = \frac{1}{2\pi i}\int_\gamma \frac{f(\xi)}{\xi - z}d\xi - \frac{1}{2\pi i}\int_{\gamma_1} \frac{f(\xi)}{\xi - z}d\xi.$$

By hypothesis, if we hold z fixed and shrink γ_1 to a, the second integral goes to zero. So we have

$$f(z) = \frac{1}{2\pi i}\int_\gamma \frac{f(\xi)}{\xi - z}d\xi$$

for all $z \neq a$ inside γ. This shows that the right-hand side gives a holomorphic function, defined throughout the interior of γ, which coincides with f at $z \neq a$. This extension is clearly unique since Cauchy's integral formula must hold for it. We shall continue to denote this extended function by f.

Let f be holomorphic in a domain D and let a be a point of D. Let us apply the principle of removable singularities to the difference quotient

$$f_1(z) = \frac{f(z) - f(a)}{z - a}.$$

The function f_1 is clearly defined and holomorphic in $D - \{a\}$ and satisfies $\lim_{z\to a}(z-a)f_1(z) = 0$. Hence it extends to a holomorphic function defined throughout D and

$$f_1(z) = \frac{1}{2\pi i}\int_\gamma \frac{f(\xi) - f(a)}{(\xi - a)(\xi - z)}d\xi.$$

But

$$\frac{1}{(\xi - a)(\xi - z)} = (a - z)^{-1}\left(\frac{1}{\xi - a} - \frac{1}{\xi - z}\right),$$

and so

$$(a - z)f_1(z) = \frac{1}{2\pi i}\frac{f(\xi) - f(a)}{\xi - a} - \frac{1}{2\pi i}\frac{f(\xi) - f(a)}{\xi - z} = f(a) - f(z).$$

Thus

$$f(z) = f(a) + (z - a)f_1(z) \quad \text{where} \quad f_1(a) = \frac{1}{2\pi i}\int_\gamma \frac{f(z)}{(\xi - z)(\xi - a)}d\xi.$$

Of course

$$f_1(a) = f'(a).$$

We can apply the same result to f_1 so as to define the function f_2 by writing $f_1(z)=f_1(a)+(z-a)f_2(z)$ and in fact take the method n steps:

$$\begin{aligned} f(z) &= f(a)+(z-a)f_1(z), \\ f_1(z) &= f_1(a)+(z-a)f_2(z), \\ &\vdots \\ f_{n-1}(z) &= f_{n-1}(a)+(z-a)f_n(z), \end{aligned}$$

so that

$$f(z)=f(a)+(z-a)f_1(a)+\cdots+(z-a)^{n-1}f_{n-1}(a)+(z-a)^n f_n(z).$$

Differentiating $k \leqslant n$ times, we see that $f_k(a)=(1/k!)f^{(k)}(a)$, so we get the Taylor expansion with remainder

$$f(z)=f(a)+(z-a)f'(a)+\frac{(z-a)^2}{2!}f''(a)+\cdots+(z-a)^n f_n(z), \tag{20.20}$$

where, as above,

$$f_n(z)=\frac{1}{2\pi i}\int_\gamma \frac{f(\xi)}{(\xi-a)^n(\xi-z)}\mathrm{d}\xi \tag{20.21}$$

for all z inside γ.

We can derive a useful estimate on the remainder term from (20.21). Suppose that γ is a circle of radius R centered at a and that M is the maximum value of $|f(\zeta)|$ for ζ on γ. Substituting into (20.21) gives

$$|f_n(z)| \leqslant \frac{M}{R^{n-1}(R-|z-s|)}. \tag{20.22}$$

Since $f_n(a)=(1/n!)f^{(n)}(a)$ we obtain, from (20.21) and (20.22), the formula and estimate

$$f^{(n)}(a)=\frac{n!}{2\pi i}\int_\gamma \frac{f(\zeta)\mathrm{d}\zeta}{(\zeta-z)^{n+1}} \tag{20.23}$$

and

$$|f^{(n)}(a)| \leqslant M\, n!\, R^{-n}, \tag{20.24}$$

if we take γ to be a circle of radius R centered at a and where M is the maximum of f on this circle.

Notice the remarkable logic of the situation. We started out with the assumption that the function f possesses a continuous derivative of the first order which satisfies the Cauchy–Riemann partial differential equations. These equations imply that f has derivatives of all orders and that the nth derivative of f, in the complex sense, exists and is given by the simple expression (20.23) in terms of the values of f itself.

We shall draw some consequences of these facts in the next section.

20.5. Applications and consequences

As a first illustration of an application of (20.24), we prove *Liouville's theorem* which asserts that

> A function which is holomorphic and bounded in the whole plane must be a constant.

Proof. Let $|f(z)| \leqslant M$ for all z. In (20.24) take $n = 1$. We may choose R arbitrarily large, which shows that $f'(a) = 0$ for all a. Thus both partial derivatives of f vanish identically and so f is a constant.

An immediate consequence of Liouville's theorem is the *fundamental theorem of algebra* which says that:

> Any polynomial P of positive degree has at least one root.

Proof. Write $P(z) = a_n z^n + a_{n-1} z^{n-1} + \cdots + a_0$, where $n > 0$, and $a_n \neq 0$. Then

$$|P(z)| \geqslant [|a_n| - (|a_{n-1}|\,|z|^{-1} + \cdots + |a_0|\,|z|^{-n})]|z|^n.$$

For $|z|$ large enough, the expression in square brackets is $> \frac{1}{2}|a_n|$. Thus $|P(z)| \to \infty$ as $|z| \to \infty$ and hence the function $1/P$ is holomorphic and bounded outside some large circle. If $P(z)$ were never zero, the function $1/P$ would be holomorphic and bounded in the entire plane, and hence would have to be a constant. Since P is not a constant, this is impossible. Thus there must be at least one zero of P, proving the theorem.

If $P(a) = 0$, then $(z - a)[P(z)/(z - a)] \to 0$, and so we would apply the principle of removable singularities to conclude that $P(z)/(z - a)$ is again a holomorphic function. In fact, since $z^j - a^j = (z - a)(z^{j-1} + \cdots + a^{j-1})$, it follows that $P(z) - P(a)$ is divisible by $z - a$, and since $P(a) = 0$, we see that $P(z)/(z - a)$ is a polynomial of degree $n - 1$. If $n > 1$, we can apply the fundamental theorem of algebra once again. We conclude:

> A polynomial of degree n has exactly n roots (counted with multiplicity).

As our next application we show that a holomorphic function, defined in some connected region D, cannot vanish together with all its derivatives at some point a inside D unless it is identically zero. Indeed, suppose that $f(a) = 0$ and $f^{(n)}(a) = 0$ for all n. Let us choose R so that the disk of radius R centered at a lies entirely within the region D. We first show that f is identically zero in this disk. By (20.20)

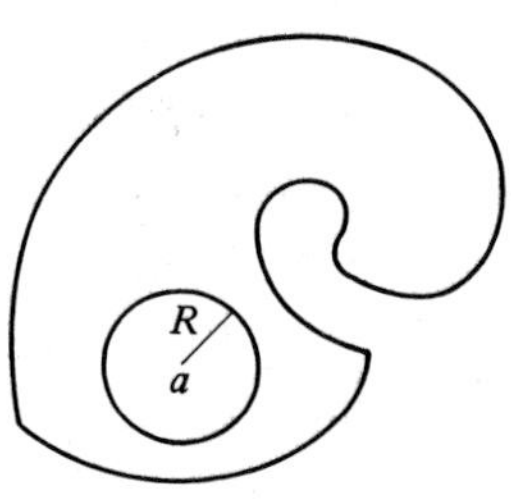

Figure 20.5

and (20.22) we know that

$$f(z) = (z-a)^{n+1} f_{n+1}(z) \quad \text{where} \quad |f_{n+1}(z)| \leqslant \frac{M}{R^n(R-|z-a|)}$$

so

$$|f(z)| \leqslant \frac{M}{(R-|z-a|)} \frac{|z-a|^{n+1}}{R^n}.$$

Since $|z-a| < R$ and n can be taken as large as we like, the preceding inequality implies that $f(z)=0$. Now we can take any point in the disk as a new center, and so enlarge the region on which we know that $f(z)=0$. If b is any other

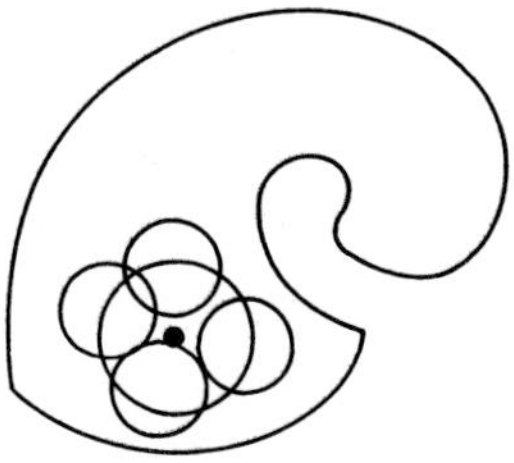

Figure 20.6

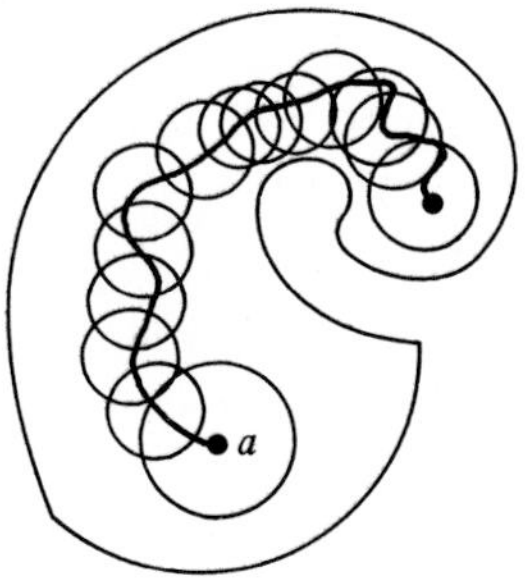

Figure 20.7

point in D, we can join a to b by a curve which lies entirely in D and hence is at a finite distance from the boundary. We can then draw a finite number of disks starting with one centered at a and ending with one containing b to conclude that $f(b)=0$. We have thus proved:

> Suppose that f is holomorphic in some connected region D and f vanishes with all its derivatives at some a inside D. Then f is identically zero.

Suppose that a_n is a sequence of points in D such that $a_n \to a$ where a is also a point inside D. Suppose that $f(a_n)=0$ for all n. Then $f(a) = \lim f(a_n) = 0$ and $f'(a) = \lim(f(a) - f(a_n))/(a - a_n) = 0$. We claim that in fact all the derivatives $f^{(k)}(a) = 0$ and hence $f \equiv 0$. Indeed, suppose that k is the first positive integer with $f^{(k)}(a) \neq 0$. Then by (20.20), $f(z) = (z-a)^k f_k(z)$ with $f_k(a) \neq 0$, and hence $f_k(z) \neq 0$ for z near a. But since $z - a \neq 0$ for $z \neq a$, we conclude that $f(z) \neq 0$ for all z close enough to but not equal to a. This contradicts the hypothesis that $f(a_n)=0$ with $a_n \to a$. We express this by saying that the zeros of a (non-trivial) holomorphic function must be isolated. If f and g are two holomorphic functions, we can apply the preceding result to the holomorphic function $f-g$ to conclude that:

> If f and g are holomorphic in a connection region D, and if $f(a_n)=g(a_n)$ at a sequence of points a_n having a limit $a=\lim a_n$ lying in D, then $f=g$ throughout D.

This result shows how strikingly different the theory of holomorphic functions is from the theory of C^∞ functions of a real variable. The function f defined by

$$f(x)=\begin{cases} e^{-1/x} & \text{for } x>0, \\ 0 & \text{for } x \leqslant 0, \end{cases}$$

is a C^∞ function on $\mathbb{R}$, which coincides with the identically zero function for all negative x. Thus the behavior of a C^∞ function in one region of its domain of definition has no effect on its behavior on some other region at some finite distance. This is not the case for holomorphic functions. Once we are given its behavior on some small portion of its domain of definition, it is completely determined throughout.

We have already seen that a holomorphic function, g, cannot vanish to infinite order at any point interior to its domain of definition. So, if $g(\alpha)=0$ at some such point α, then there is some smallest k with $g^{(k)}(\alpha)\neq 0$ and thus

$$g(z)=(z-\alpha)^k g_k(z)\neq 0 \quad \text{for } z \text{ near } \alpha, z\neq\alpha,$$

and where g_k is a holomorphic function. Suppose that f is also holomorphic in the same region and let us consider the quotient function $h=f/g$ which is defined for z near $\alpha, z\neq\alpha$. Let us use the expansion (20.20) of f about α:

$$f(z)=f(\alpha)+(z-\alpha)f'(\alpha)+\cdots+(z-\alpha)^{k-1}\frac{f^{(k-1)}(\alpha)}{(k-1)!}+(z-\alpha)^k f_k(z).$$

Then

$$h(z)=f(z)/g(z)=\frac{f(\alpha)}{(z-\alpha)^k}\frac{1}{g_k(z)}+\frac{f'(\alpha)}{(z-\alpha)^{k-1}}\frac{1}{g_k(z)}+\cdots+\frac{f^{(k-1)}(\alpha)}{(z-\alpha)(k-1)!}\frac{1}{g_k(z)}+\frac{f_k(z)}{g_k(z)}.$$

Since $1/g_k$ is holomorphic near α, we can expand it about α as well:

$$1/g_k(z)=b_0+b_1(z-\alpha)+\cdots+(z-\alpha)^k b_k(z)$$

where

$$b_0=1/g_k(\alpha)=1/g^{(k)}(\alpha) \quad \text{etc.}$$

If we substitute this expansion into the preceding expression for h, we see that

$$h(z)=a_{-k}(z-\alpha)^{-k}+a_{-k+1}(z-\alpha)^{-(k-1)}+\cdots+a_{-1}(z-\alpha)^{-1}+h_0(z)$$

where

$$\left.\begin{aligned} a_{-k}&=f(\alpha)b_0=f(\alpha)/g_k(\alpha), \\ a_{-k+1}&=f(\alpha)b_1+f'(\alpha)b_0 \end{aligned}\right\} h_0 \text{ holomorphic near } \alpha,$$

etc. We have thus proved that:

The quotient of two holomorphic functions is meromorphic.

The preceding formulas simplify if g has a *simple* zero at α, i.e., if $k=1$. In that case, we can assert that:

If $g(\alpha)=0$ and $g'(\alpha)\neq 0$, then f/g has a simple pole at α with residue $f(\alpha)/g'(\alpha)$, i.e., $f(z)/g(z)=a_{-1}(z-\alpha)^{-1}+h_0(z)$ near α with $a_{-1}=f(\alpha)/g'(\alpha)$ and $h_0(z)$ holomorphic near α. (20.25)

The number k that we have been using in the preceding analysis is called the *order* of the zero. Thus a holomorphic function g has a zero of first order at α if $g(\alpha)=0$ and $g'(\alpha)\neq 0$. It has a zero of order k at α if $g(\alpha)=\cdots=g^{(k-1)}(\alpha)=0$, while $g^{(k)}(\alpha)\neq 0$.

Let the function f be holomorphic in a bounded region D and suppose that f has a finite number of zeros in D. That is, we suppose that there are a finite number of distinct points $\alpha_1,\ldots,\alpha_r$ at which f vanishes and let k_j be the order of the zero of f at α_j. Then $f(z)=(z-\alpha_1)^{k_1}g(z)$, where g vanishes only at $\alpha_2,\ldots,\alpha_r$ and has the same order at each of these points as f. We can write $g(z)=(z-\alpha_2)^{k_2}h(z)$ where h vanishes only at $\alpha_3,\ldots,\alpha_r$. Proceeding in this way, we conclude that we can write

$$f(z)=(z-\alpha_1)^{k_1}(z-\alpha_2)^{k_2}\ldots(z-\alpha_r)^{k_r}F(z)$$

where F is holomorphic in D and does not vanish anywhere in D. Differentiating and dividing by f shows that

$$\frac{f'(z)}{f(z)}=\frac{k_1}{z-\alpha_1}+\cdots+\frac{k_r}{z-\alpha_r}+\frac{F'(z)}{F(z)}$$

where F'/F is holomorphic in D. In particular, if g is some other function holomorphic in D, then gf'/f has only simple poles located at the α_j with residues $k_jg(\alpha_j)$. Thus by Cauchy's formula we get the following result:

Suppose that f and g are holomorphic in a bounded region D and that f has only finitely many zeros in D located at $\alpha_1,\ldots,\alpha_r$ with orders $k_1,\ldots,k_r$ and $f\neq 0$ on ∂D.

$$\frac{1}{2\pi i}\int_{\partial D}g(\zeta)\frac{f'(\zeta)}{f(\zeta)}\,d\zeta=k_1g(\alpha_1)+\cdots+k_rg(\alpha_r). \tag{20.26}$$

In particular, taking $g\equiv 1$, we get

$$\frac{1}{2\pi i}\int_{\partial D}\frac{f'(\zeta)}{f(\zeta)}\,d\zeta=k_1+\cdots+k_r=\text{the number of zeros of } f \text{ counted with multiplicity.} \tag{20.27}$$

Notice that if f is holomorphic on a larger region, E, such that $\bar{D}$ is contained in E, then f can only have finitely many zeros in D, for otherwise f would vanish at a sequence of points converging to a limit in E, which would imply that f is identically zero. So we can readily guarantee that the hypotheses of (20.26) and (20.27) are satisfied.

Rouché's theorem says that:

> If f and h are both holomorphic in a bounded region D and continuous on $\bar{D} = D \cup \partial D$, and if $|f(z) - h(z)| < |f(z)|$ on D, then f and h have the same number of zeros (counted with multiplicity) in D. (20.28)

Notice that the hypothesis implies that f does not vanish on ∂D and hence does not vanish in some small strip about ∂D, so f has finitely many zeros in D and the same is true for h. In fact, the hypothesis implies that for any $0 \leqslant t \leqslant 1$, the function h_t defined by

$$h_t(z) = f(z) + t(h(z) - f(z))$$

is holomorphic in D, does not vanish on ∂D, and has a finite number of zeros in D which we can compute by (20.27). The formula (20.26) depends continuously on t and is an integer. Hence it must be a constant. But $h_0 = f$ and $h_1 = h$, so f and h must have the same number of zeros, proving Rouché's theorem.

Suppose that a function f is holomorphic at all points $z \neq \beta$ but near β, and that $|f(z)| \to \infty$ as $z \to \beta$. Then $1/f(z) \neq 0$ for z near enough to β and so $1/f$ is holomorphic for z near $\beta, z \neq \beta$ and $1/f(z) \to 0$ as $z \to \beta$. This means that $1/f$ has a removable singularity at β so that $1/f$ becomes holomorphic at β if we assign the value zero there. Since $1/f$ is now holomorphic near and including β, and is not identically zero, it must vanish to some finite order at β, so $1/f(z) = (z - \beta)^j g_j(z)$, where g_j is holomorphic near β and $g_j(\beta) \neq 0$. But then

$$f(z) = (z - \beta)^{-j} h(z) \quad \text{where } h = 1/g_j \text{ is holomorphic near } \beta.$$

If we use the Taylor expansion of h about β, we see that f has a pole of order j at β. We have proved that:

> If f is holomorphic for z near β, $z \neq \beta$ and $|f(z)| \to \infty$ as $z \to \beta$, then f has a pole of finite order at β.

Suppose that f is meromorphic in a bounded region D and continuous (and nowhere zero) on ∂D. Then f can have only a finite number of zeros and poles in D. Dealing with each zero or pole one at a time as above, we conclude that

$$f(z) = \frac{(z - \alpha_1)^{k_1} \ldots (z - \alpha_r)^{k_r}}{(z - \beta_1)^{j_1} \ldots (z - \beta_p)^{j_p}} F(z)$$

where F is holomorphic with no zeros in D. (Here the α_i are the zeros with order k_i and the β_i are the poles with order j_i.) We can then proceed as in the proof of (20.25) and (20.26). We conclude:

> Suppose that f is meromorphic in a bounded region $\bar{D}$, continuous in D near ∂D and nowhere zero on ∂D. Suppose that g is holomorphic in $\bar{D}$, and continuous on $\bar{D}$. Then
>
> $$\frac{1}{2\pi i} \int_{\partial D} g(\zeta) \frac{f'(\zeta)}{f(\zeta)} d\zeta = k_1 g(\alpha_1) + \cdots + k_r g(\alpha_r) - (j_1 g(\beta_1) + \cdots + j_p g(\beta_p)).$$
>
> Here $\alpha_1, \ldots, \alpha_r$ are the zeros of f with orders $k_1, \ldots, k_r$ and $\beta_1, \ldots, \beta_p$ are the

poles of f with orders $j_1, \ldots, j_p$. In particular, taking $g \equiv 1$, we get

$$\frac{1}{2\pi i}\int_{\partial D}\frac{f'(\zeta)}{f(\zeta)}d\zeta$$

$= \text{number of zeros of } f - \text{number of poles of } f$ (counted with multiplicities).

Suppose that f is holomorphic for $z \neq \gamma$, z near γ. Suppose that γ is not a removable singularity of f and also that γ is not a pole of f (so that we do *not* have $|f(z)| \to \infty$ as $z \to \gamma$). Then γ is called an *essential singularity* of f. Some idea of the complicated behavior of a function near an essential singularity is expressed by the following result:

> Suppose that f is holomorphic near γ except at γ and has an essential singularity at γ. Then given any complex number c whatsoever, we can find a sequence of points $a_i \to \gamma$ such that $f(a_i) \to c$.

Indeed, suppose that this did not hold for some c. This means that we can find some neighborhood, U, of γ such that $f(z)$ stays a finite distance away from c for all z in U, $z \neq \gamma$. Say $|f(z) - c| > 1/M$ for some M. Let g be defined by $g(z) = 1/(f(z) - c)$. Then g is holomorphic in U (except at γ) and $|g(z)| < M$. So g has a removable singularity at γ. But then $f = 1/(g - c)$ is meromorphic (with at worst a pole) at γ contradicting the assumption. An example of a function with an essential singularity at 0 is the function $e^{1/z}$. Let us show explicitly that the conclusion of (20.29) holds for this function. We must show that we can make $e^{1/z}$ as close as we please to any complex number, with z arbitrarily close to the origin. Let us set $w = 1/z$. We want to show that we can make e^w as close as we like to any complex number c with $|w| > R$ for any R. Write $c = re^{i\theta}$ and $w - u + iv$. Then we can arrange that $|w| > R$ by simply choosing $v > R$. Now $e^w = e^u \cdot e^{iv}$, so if $r > 0$, we can choose $u = \log r$ and $v = \theta + 2\pi n$, where $2\pi n$ is large enough so that $v > R$. So we have exactly solved the equation $e^w = c$, and infinitely often by choosing larger and larger values of n, which amounts to the corresponding $z = 1/w$ getting closer and closer to 0. The only value of c that we cannot hit exactly is $c = 0$. But we can choose a sequence of $c_k \to 0$, $c_k \neq 0$ and corresponding w_k with $|w_k| \to \infty$ and $e^{w_k} = c_k$. Then $z_k = 1/w_k$ is the desired sequence of points approaching 0 with $e^{1/z_k} \to 0$.

It is a somewhat deeper fact, which we will not prove here, that the above behavior of $e^{1/z}$ is typical of a holomorphic function near an essential singularity. That is, we can actually solve $f(z) = c$ exactly, infinitely often near $z = \gamma$ for all values of c with perhaps one possible exception. (For $e^{1/z}$ the exceptional value was $c = 0$.)

20.6. The local mapping

In this section we study the local properties of holomorphic functions a little more closely. Our first result is an easy consequence of Rouché's theorem:

Suppose that f is holomorphic and not constant near a and $f(a)=b$. Then there is an $\varepsilon>0$, such that for all w satisfying $|w-b|<\varepsilon$ there is a z near a with $f(z)=w$. (20.30)

Proof. Set $g(z)=f(z)-b$. Then g has a zero of some finite order at a. We can find some small enough r such that $g(z)\neq 0$ for $|z-a|\leqslant r$ except at a. Take D to be this disk of radius r centered at a and choose $\varepsilon>0$ so that $|g(z)|>\varepsilon$ for $z\in\partial D$. Suppose that $|w-b|<\varepsilon$ and define g_w by

$$g_w(z)=f(z)-w=g(z)+(b-w).$$

By Rouché's theorem, (20.28), g_w has the same number of zeros (counted with multiplicity) in D as does g. Since g has at least one, so does g_w, so we can find at least one value of z in D with $f(z)=w$.

(20.30) is sometimes expressed by saying that a holomorphic function defines an *open* map; that is, if a point is in the image, then a whole neighborhood of the point is in the image.

(20.30) shows another striking difference between the theory of holomorphic functions of a complex variable and the theory of smooth functions of a real variable. Consider the map of $\mathbb{R}\to\mathbb{R}$ sending x into $u=x^2$. Then u takes on only

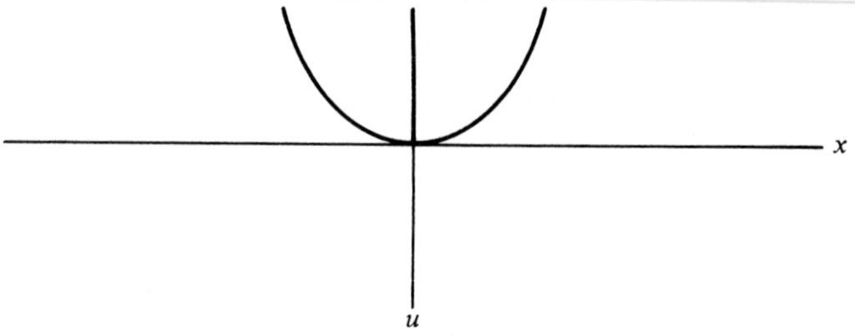

Figure 20.8

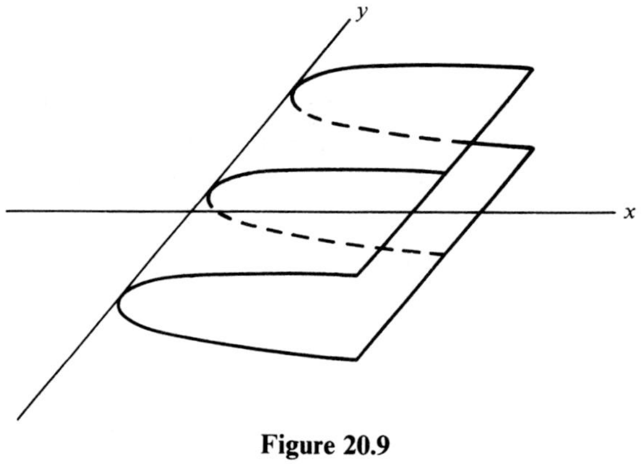

Figure 20.9

non-negativevalues. So 0 is in the image of this map but no neighborhood of 0 is in the image. We can easily jack this example up to a map of the plane into the plane by sending (x, y) into (u, v) where $u = x^2$ and $v = y$. In this map the plane is folded over along the y-axis so that no points (u, v) with u negative are in the image.

(20.30) says that this cannot happen for maps corresponding to holomorphic functions. For instance, if we take $f(z) = z^2$, then for $w = re^{i\theta} \neq 0$, we can find two values of z, namely $z = r^{1/2}e^{i\theta/2}$ and $z = r^{1/2}e^{i(\pi+\theta/2)}$, with $z^2 = w$. So instead of finding no solutions of $z^2 = w$ for some range of w, we find two solutions.

This last property of the function $f(z) = z^2$ is also typical. Suppose that f is a holomorphic function near a with $f(a) = b$. Let us go back to the proof of (20.30). The function g defined there has a zero of finite order at a. Let k be the order of the zero. If $k > 1$, then $g'(a) = 0$. We can choose the r (possibly smaller than in the proof of (20.30)) so that not only g, but also g', has no other zeros in D. Thus in all cases, $k = 1$ or $k > 1$, g has no zeros in D except at a and g' has no zeros in D except possibly at a. Notice that g_w, defined in the proof of (20.30), differs from g by a constant. Thus $g'_w = g'$ and so $g'_w(z) \neq 0$ in D for $z \neq a$. But, for $w \neq b$, the zeros of g_w are not located at a. This means that each of the zeros of g_w must be simple, and hence there must be k distinct zeros. We have thus proved:

> Let f be holomorphic near a with $f(a) = b$ and suppose that the function g defined by $g(z) = f(z) - b$ has a zero of order k at a. Then we can find an $r > 0$ and an $\varepsilon > 0$ such that for any $w \neq b$ with $|w - b| < \varepsilon$ there are exactly k distinct z values, $z_1, \ldots, z_k$, with $|z_i - a| < r$ and $f(z_i) = w$. (20.31)

(20.31) shows the true meaning of a zero of order k. It is where k distinct roots coalesce.

The case $k = 1$ of (20.31) is sufficiently important for us to record it separately:

> Let f be holomorphic near a with $f(a) = b$ and $f'(a) \neq 0$. Then we can find an $r > 0$ and an $\varepsilon > 0$ so that for each w with $|w - b| < \varepsilon$ there is a unique z with $|z - a| < r$ satisfying $f(z) = w$. In other words, there is a unique function g defined for $|w - b| < \varepsilon$ satisfying $|g(w) - b| < r$ and $f(g(w)) = w$. (20.32)

We claim that:

> The function g defined in (20.32) is holomorphic and $g'(w) = 1/f'(g(w))$. (20.33)

We have chosen the r in (20.32) so that $f'(z) \neq 0$ for $|z - a| > r$. So to prove (20.33), it is enough for us to prove that g is differentiable in the complex sense at the point b and $g'(b) = 1/f'(a)$. The same argument would then apply to any $z = g(w)$ and the formula for g' would show that it is continuous. Suppose that $f'(a) = c \neq 0$. Let $\eta < \frac{1}{2}|c|$. By the definition of the derivative of f, we can find a $\delta > 0$ such that

$$\left|\frac{f(z) - f(a)}{z - a} - f'(a)\right| < \eta \quad \text{for } |z - a| < \delta. \tag{20.34}$$

Taking δ as a new r (if necessary) in (20.32) shows that, for all w close enough to

b, we have

$$\left|\frac{w-b}{z-a}-c\right|<\tfrac{1}{2}|c| \quad \text{so} \quad |w-b|>\tfrac{1}{2}|c(z-a)| \quad \text{where} \quad z=g(w),$$

or

$$\left|\frac{z-a}{(w-b)c}\right|<\frac{2}{|c|^2}.$$

If we now multiply (20.34) by $(z-a)/c(w-b)$, we get

$$\left|\frac{1}{f'(a)}-\frac{g(w)-a}{w-b}\right|<2|c|^{-2}\eta.$$

We can make η as small as we like by choosing δ and hence ε small enough. This shows that g is differentiable in the complex sense at b with $g'(b)=1/f'(a)$, proving (20.33).The two assertions, (20.32) and (20.33), constitute the implicit function theorem for holomorphic functions. We must emphasize that the implicit function theorem is a local theorem. Let us look again at the function $f(z)=z^2$. We know that for any $w\neq 0$ there are two roots of $f(z)=w$. Suppose that we specify the square root of a positive number by demanding that it be positive; for example, take $a=1$ with $f(a)=1$. The (20.32) implies that for w near to one, we will have specified a unique square root by demanding that it be close to one, and (20.33) implies this function $g(w)=w^{1/2}$ will be holomorphic near $w=1$. Indeed, it is not hard to see that we can take the ε in (20.33) to be any number less than 1. Of course, we can take any point w in the disc of radius ε about $w=1$ as a new choice of b, with its $g(w)$ as a new choice of a and apply the implicit function theorem once again. Suppose we make such a succession of choices with $|w|=1$ as indicated in figure 20.10.

As we come back full circle to the positive w-axis, we end up with the opposite choice of the square root, which is not surprising. Thus, although (20.32) guarantees the existence of a local inverse for a function f, a succession of applications of (20.32) may lead to a global inconsistency. In the case of $f(z)=z^2$ there simply is no globally well-defined function $w^{1/2}$ on the w-plane.

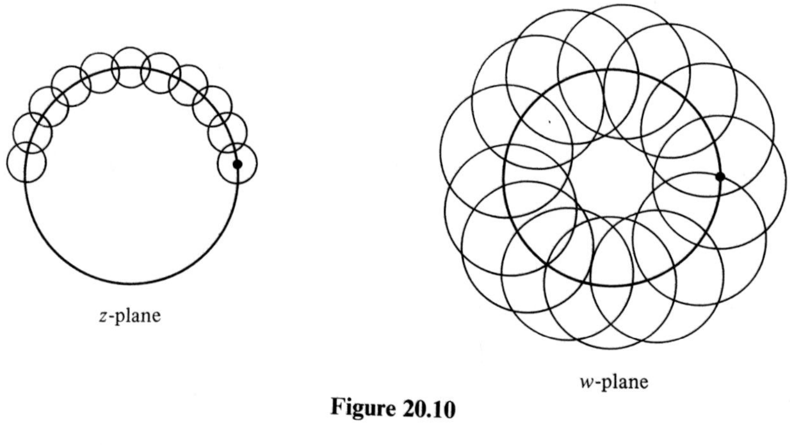

Figure 20.10

There are two ways that are used to deal with this problem of ambiguity and we shall illustrate each of them for the special case of the square-root function. One is simply to live with the ambiguity. We think of the z-plane as being a *two-sheeted covering* of the w-plane with a *ramification point* at the origin where only one z-value corresponds to the w-value. Whenever we are given a functional expression such as $\cos(w^{3/2}+1)$, we understand that this is not really a function defined on the w-plane but is a function of z; i.e., $\cos(z^3+1)$. The z-plane is the *Riemann surface* associated to all functions of $w^{1/2}$ in the sense that they are all defined as functions of z. This was the point of view espoused by Riemann. The detailed study of the structure of such Riemann surfaces for various other kinds of holomorphic functions had a profound effect on the development of geometry, topology and algebra well up to the present time. We will not go into these matters here.

An alternative, more mundane, approach to the problem of defining $w^{1/2}$ is to specify some curve extending from the origin out to infinity along which we decide that $w^{1/2}$ is not to be defined, so as to make its specification unique everywhere else. For example, suppose we agree to *cut* the w-plane along the negative w-axis. Thus every w not on this *cut line* can be written as $w = re^{i\theta}$ with $-\pi < \theta < \pi$. Then choosing $w^{1/2} = r^{1/2}$ for $\theta = 0$ (i.e., a positive square root for positive w), and continuing from this choice into the first and fourth quadrants and throughout the range of θ indicated above, gives

$$w^{1/2} = r^{1/2}e^{i\theta/2} \quad \text{for} \quad -\pi < \theta < \pi.$$

So, for example, with this choice of square root,

$$i^{1/2} = e^{\pi i/4}$$

while

$$(-i)^{1/2} = e^{-\pi i/4}.$$

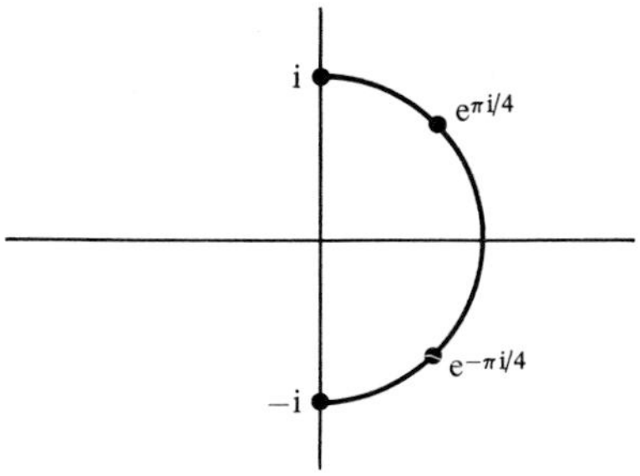

Figure 20.11

Where to draw the cut line is mostly determined by convenience to the application one has in mind. The above choice is frequently convenient and is sometimes known as the *principal branch* of the square root.

Let us illustrate an application of such a choice to the computation of some integrals which will be of extreme importance to us later on. Consider the integral

$$\int_{-\infty}^{\infty} \mathrm{e}^{-\lambda x^2/2}\,\mathrm{d}x.$$

When $\operatorname{Re}\lambda > 0$, the function $\mathrm{e}^{-\lambda x^2/2}$ vanishes rapidly at $\pm\infty$ and the above integral converges absolutely. When $\operatorname{Re}\lambda < 0$ the function $\mathrm{e}^{-\lambda x^2/2}$ blows up at infinity, so the integral makes no sense. When $\operatorname{Re}\lambda = 0$, the function $\mathrm{e}^{-\lambda x^2/2}$ has absolute value one for all x, so the integral certainly does not converge absolutely. Nevertheless, we claim that it does converge. In fact, we claim:

The integral $\int_{-\infty}^{\infty}\mathrm{e}^{-\lambda x^2/2}\,\mathrm{d}x$ converges uniformly when λ satisfies $\operatorname{Re}\lambda \geqslant 0$, $|\lambda| \geqslant 1$ and its value is therefore a continuous function of λ on this range.

To establish uniform convergence, we must show that for any $\varepsilon > 0$, we can find some R_ε (depending on ε but not on λ) so that

$$\left|\int_R^S \mathrm{e}^{-\lambda x^2/2}\,\mathrm{d}x\right| < \varepsilon \quad\text{and}\quad \left|\int_{-S}^{-R} \mathrm{e}^{-\lambda x^2/2}\,\mathrm{d}x\right| < \varepsilon \quad \text{for all } S > R > R_\varepsilon.$$

Now

$$\frac{\mathrm{d}}{\mathrm{d}x}\mathrm{e}^{-\lambda x^2/2} = -\lambda x \mathrm{e}^{-\lambda x^2/2},$$

so integration by parts gives

$$\int_R^S \mathrm{e}^{-\lambda x^2/2}\,\mathrm{d}x = -\int_R^S \frac{1}{\lambda x}\frac{\mathrm{d}}{\mathrm{d}x}\mathrm{e}^{-\lambda x^2/2}\,\mathrm{d}x = \frac{1}{\lambda R}\mathrm{e}^{-\lambda R^2/2} - \frac{1}{\lambda S}\mathrm{e}^{-\lambda S^2/2} - \frac{1}{\lambda}\int_R^S \frac{1}{x^2}\mathrm{e}^{-\lambda x^2/2}\,\mathrm{d}x.$$

Since $|\mathrm{e}^{-\lambda x^2/2}| \leqslant 1$ for all x and λ under consideration, we can estimate the integral on the right-hand side by $\int_R^S x^{-2}\,\mathrm{d}x$ and thus (since $S > R$) the whole expression by

$$\left|\int_R^S \mathrm{e}^{-\lambda x^2/2}\,\mathrm{d}x\right| \leqslant \frac{4}{|\lambda| R},$$

which will be less than ε if $R > 4|\lambda|^{-1}\varepsilon^{-1}$. Since $|\lambda| \geqslant 1$, we can take $R_\varepsilon = 4/\varepsilon$, with exactly the same argument working for $\int_{-S}^{-R}$. (Notice that, instead of $|\lambda| \geqslant 1$, we could take $|\lambda| \geqslant c$ for any fixed $c > 0$ and have uniformity of convergence.) Since $\mathrm{e}^{-\lambda x^2/2}$ is a continuous function of λ and x, the uniform convergence of the integral guarantees that the value of the integral is a continuous function of λ. We shall evaluate this integral for $\operatorname{Re}\lambda > 0$, where it converges absolutely, by a trick. In fact, we compute the square of the integral by passing to polar coordinates:

$$\left[\int_{-\infty}^{\infty} \mathrm{e}^{-\lambda x^2/2}\,\mathrm{d}x\right]^2 = \int_{-\infty}^{\infty}\int_{-\infty}^{\infty} \mathrm{e}^{-\lambda x^2/2}\cdot\mathrm{e}^{-\lambda y^2/2}\,\mathrm{d}x\,\mathrm{d}y = \int_{-\infty}^{\infty}\int_{-\infty}^{\infty} \mathrm{e}^{-\lambda(x^2+y^2)/2}\,\mathrm{d}x\,\mathrm{d}y$$

$$= \int_0^{2\pi}\int_0^{\infty} \mathrm{e}^{-\lambda r^2/2} r\,\mathrm{d}r\,\mathrm{d}\theta = 2\pi\int_0^{\infty} \mathrm{e}^{-\lambda r^2/2} r\,\mathrm{d}r = 2\pi/\lambda.$$

Thus

$$\int_{-\infty}^{\infty} \mathrm{e}^{-\lambda x^2/2} = (2\pi/\lambda)^{1/2}. \tag{20.35}$$

Which square root do we take? When λ is real and positive, the integrand, $e^{-\lambda x^2/2}$, is a positive function and so the integral must be a positive number. Hence, for positive real λ, we must choose the positive square root. Since the integral is a continuous function of λ, we must take the principal branch of the square root. This is a typical illustration of how the problem at hand tells us which determination of the square root to use.

Formula (20.35) is valid, by continuity, for all $\operatorname{Re}\lambda \geqslant 0$. In particular, setting $\lambda = -it$ and $\lambda = it$, we obtain the important formulas

$$\int_{-\infty}^{\infty} e^{itx^2/2}dx = e^{\pi i/4}(2\pi/t)^{1/2} \quad \text{and} \quad \int_{-\infty}^{\infty} e^{-itx^2/2}dx = e^{-\pi i/4}(2\pi/t)^{1/2}$$

valid for all real $t > 0$, and where the positive square root is meant. These integrals are known as *Fresnel integrals* and are of importance in optics. We shall have much need for them later on.

All that we have said about the square root goes over with minor change for the nth root, for any positive integer n. In fact, it also goes over for the exponential function. The function e^z maps the z-plane onto the w-plane infinitely often; each strip of width 2π parallel to the x-axis is mapped onto the entire w-plane. Thus the logarithm function, $\log w$, is not well defined; it is only defined up to adding an arbitrary integer multiple of 2π. Again there are two approaches to this ambiguity. One is to think of the z-plane as being an *infinitely sheeted covering* of the w-plane, and any functional expression involving $\log w$ is to be thought of as a function of z. The alternative is to cut the w-plane, say along the negative real axis and then, writing

$$w = re^{i\theta}, \quad -\pi < \theta < \pi,$$

define the *principal branch* of the logarithm function by

$$\log w = \log r + i\theta.$$

By (20.32) and (20.33) we know that $\log w$ is a holomorphic function whose derivative is $1/w$. Having defined the logarithm, we then can define the complex power of any complex number. For any complex number c, define

$$w^c = e^{c \log w}.$$

As mentioned above, there are two possible interpretations of this definition. One is to consider w^c as a *multiply valued* function of w; i.e., not a function of w but a function of z. The other is to pick a branch, say the principal branch of $\log w$, so then w^c becomes defined at all points except for w lying on the non-positive real axis.

20.7. Contour integrals

One of the applications of the Cauchy residue theorem is to the calculation of definite integrals. We have indicated a typical application in section 20.4. We have also given, in section 5, especially formula (20.25), a convenient tool for the evaluation of residues. In this section, we give some more illustrative examples, some of which will be of use to us later on. In all cases, a certain amount of ingenuity is required in the choice of contour. One is given a definite integral to evaluate. This usually translates into an integral over a curve in the complex plane. One then adds other curves so as to get the boundary of some region and apply Cauchy's residue theorem. One must choose the other pieces so that their contribution to the integral either is known, or is some multiple of the desired integral, or becomes vanishingly small when an appropriate limit is taken.

(a) Let $R(x) = P(x)/Q(x)$ be a rational function, so P and Q are polynomials. Suppose that Q has no real zeros. If the degree of Q is at least two more than the degree of P, the integral $\int_{-\infty}^{\infty} R(x)\,\mathrm{d}x$ converges, and is the limit of $\int_{-r}^{r} R(x)\,\mathrm{d}x$ as $r \to \infty$. We think of this as the integral of the complex function $R(z)$ along the line segment from $-r$ to r on the real axis. Adjoin the semicircle of radius r centered at the origin in the upper half-plane.

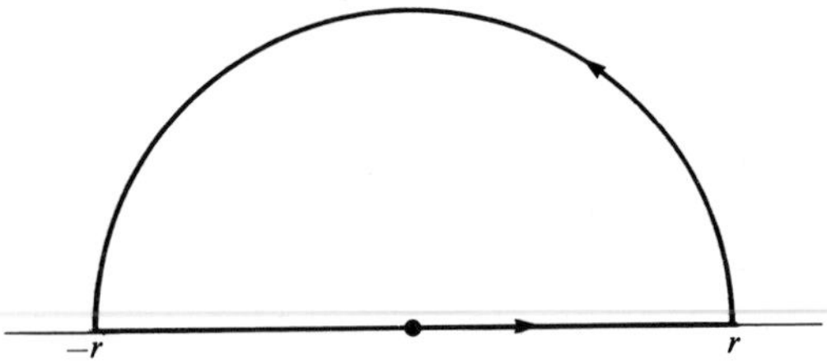

Figure 20.12

We now have a closed path which is the boundary of the half-disk in the upper half-plane, and we can apply the Cauchy residue theorem. On the other hand, $|R(z)| \leqslant Kr^{-2}$ for z on the semicircle and K some suitable constant. Since the semicircle has length πr, this means that the integral over the semicircle tends to zero as $r \to \infty$. Thus

$$\int_{-\infty}^{\infty} R(x)\,\mathrm{d}x = 2\pi\mathrm{i} \sum_{\operatorname{Im} z > 0} \operatorname{res} R(z),$$

the sum being taken over all poles in the upper half-plane.

(b) Consider the integral $\int_{-\infty}^{\infty} \mathrm{e}^{\mathrm{i}x} R(x)\,\mathrm{d}x$ where $R = P/Q$ is a rational function with no poles on the real axis. If $\deg Q \geqslant \operatorname{deq} P + 2$, we can proceed exactly as before. If $\deg Q \geqslant \deg P + 1$, so all we know is that $R(x)$ vanishes like $1/x$ at infinity, we still can prove that the integral is convergent. In fact, we can use an integration by parts argument as in section 20.6. Since $\mathrm{e}^{\mathrm{i}x} = -\mathrm{i}(\mathrm{d}/\mathrm{d}x)\,\mathrm{e}^{\mathrm{i}x}$, we have $\int_r^s \mathrm{e}^{\mathrm{i}x} R(x)\,\mathrm{d}x = -\mathrm{i}\mathrm{e}^{\mathrm{i}s} R(s) + \mathrm{i}\mathrm{e}^{\mathrm{i}r} R(r) + \int_r^s \mathrm{e}^{\mathrm{i}x} R'(x)\,\mathrm{d}x$, and R' vanishes to order $1/x^2$ at infinity. To evaluate the integral, we again take the limit of the integral from $-r$ to r and evaluate this integral by adjoining the integral over a semicircle in the upper half-plane. Here we must be slightly more careful in estimating the integral over the semicircle. If $z = r\mathrm{e}^{\mathrm{i}\theta} = r(\cos\theta + \mathrm{i}\sin\theta)$, then

$$\mathrm{e}^{\mathrm{i}z} = \mathrm{e}^{\mathrm{i}r\cos\theta}\mathrm{e}^{-r\sin\theta}$$

and so the integral over the semicircle can be estimated as follows:

$$\left|\int_{\substack{\text{semi}\\\text{circle}}} e^{iz}R(z)\,dz\right| = \left|\int_0^{\pi} e^{ir\cos\theta}e^{-r\sin\theta}R(re^{i\theta})r\,d\theta\right| \leqslant \int_0^{\pi} e^{-r\sin\theta}r|R(re^{i\theta})|\,d\theta.$$

Now $|rR(re^{i\theta})|$ is bounded by some constant, say K. So our problem becomes one of estimating the integral $\int_0^{\pi} e^{-r\sin\theta}\,d\theta = 2\int_0^{\pi/2} e^{-r\sin\theta}\,d\theta$. Now $\sin\theta \geqslant 2\theta/\pi$ for $0 \leqslant \theta \leqslant \pi/2$. (Indeed, both sides are equal at the end point and the difference has one maximum in the interior.) So

$$\int_0^{\pi/2} e^{-r\sin\theta}\,d\theta \leqslant \int_0^{\pi/2} e^{-2r\theta/\pi}\,d\theta = (\pi/2r)(1-e^{-r}) \to 0.$$

Thus the integral over the semicircle goes to zero and we get

$$\int_{-\infty}^{\infty} e^{ix}R(x)\,dx = 2\pi i \sum_{\operatorname{Im} z>0} e^{iz}\operatorname{res} R(z),$$

the sum being taken over all poles in the upper half-plane. If Q has a zero on the real axis, the above integral makes no sense as it stands. Still, it might make sense in some case as a special kind of limit known as the *Cauchy principal value*, as is illustrated in the following example. Suppose that R has a simple pole at $z = 0$ with residue A, so that $R(a) = A/z + R_0(z)$ with R_0 holomorphic near the origin. Let us consider the same line integral as before, except that we make a detour along a small semicircle of radius ε in the lower half-plane to avoid the origin.

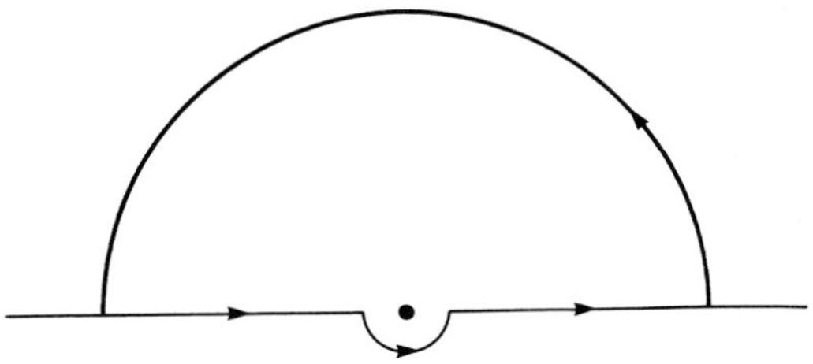

Figure 20.13

The enclosed region now contains all the poles in the upper half-plane together with the pole at the origin. The integral of A/z around the little semicircle gives πiA, canceling out half the residue from the origin. Thus

$$P\left(\int_{-\infty}^{\infty} e^{ix}R(x)\,dx\right) = \lim_{\varepsilon\to 0}\left[\int_{-\infty}^{-\varepsilon} + \int_{\varepsilon}^{\infty}\right]e^{ix}R(x)\,dx = \pi iA + 2\pi i\sum_{\operatorname{Im} z>0}\operatorname{res}(e^{iz}R(z)).$$

where $P(\;)$ means the Cauchy principal value. For example, taking $R(x) = 1/x$ gives

$$P\left(\int_{-\infty}^{\infty}\frac{e^{ix}}{x}\right) = \pi i.$$

Taking the real and imaginary parts of this equation gives $P(\int_{-\infty}^{\infty}(\cos x/x)\,dx) = 0$, which is obvious since $\cos x/x$ is an odd function, and $P(\int_{-\infty}^{\infty}(\sin x/x)\,dx) = \pi$. Now $\sin x/x$ has no singularity at the origin, so we do not have to worry about taking the principal value for this last integral and $\sin x/x$ is an even function. So we get

$$\int_0^\infty \frac{\sin x}{x}\,dx = \pi/2.$$

We shall have occasion to use this formula in the theory of Fourier series.

(c) Suppose we consider integrals of the form $\int_0^\infty x^\alpha R(x)\,dx$, where α is some real number with $0 < \alpha < 1$, and where R is a rational function vanishing to second order at infinity and with at worst a pole of first order at the origin. Here the trick is to first make the change of variables $x = t^2$, so the integral now becomes $2\int_0^\infty t^{2\alpha+1}R(t^2)\,dt$. For evaluating this integral, it is convenient to choose the branch of $z^{2\alpha}$ which is obtained by cutting the z-plane along the negative imaginary axis so that we write $z = re^{i\theta}$ with $-\pi/2 < \theta < 3\pi/2$, so that $z^{2\alpha} = r^{2\alpha}e^{i2\alpha\theta}$ with $-\pi\alpha < 2\alpha\theta < 3\pi\alpha$. We must now choose our contour so that it does not come in contact with the negative imaginary axis. Choose a contour to consist of two line segments, along the negative and positive real axis, a large semicircle and a small

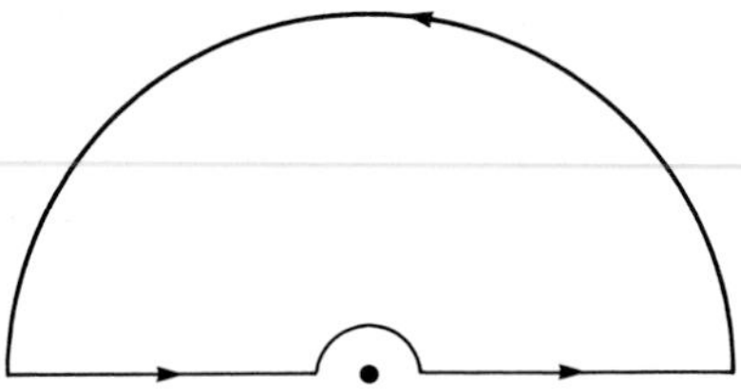

Figure 20.14

semicircle in the upper half-plane. Our assumptions clearly imply that the integrals over the semicircles tend to zero as the large semicircle expands out and the small semicircle shrinks down to zero. Thus

$$2\pi i \sum \text{residues} = \int_{-\infty}^{\infty} z^{2\alpha+1}R(z^2)\,dz = \int_0^\infty (z^{2\alpha+1} + (-z)^{2\alpha+1})R(z^2)\,dz.$$

But our choice of branch of z^2 is such that $(-z)^{2\alpha} = e^{2\pi i\alpha}z^{2\alpha}$ for z positive, and so the integral on the right becomes

$$(1 - e^{2\pi i\alpha})\int_0^\infty t^{2\alpha+1}R(t^2)\,dt.$$

Since $(1 - e^{2\pi i\alpha}) \neq 0$, we can divide by it to give an evaluation of our original integral. The residues to be summed are the residues of $z^{2\alpha+1}R(z^2)$ in the upper half-plane. These are the same as the residues of $z^\alpha R(z)$ in the whole plane cut along the positive real axis. So another way of solving our problem is not to make the initial substitution $x = t^2$ but to cut the z-plane along the positive real axis and use the branch of z^α determined by writing $z = re^{i\theta}$ with $0 < \theta < 2$. Notice that we have cut precisely along the path of our original integral! So we cannot get our original integral as a precise component of a closed contour. Instead, integrate along a path parallel to the positive real axis but deformed slightly into the upper half-plane, with the idea of then passing to the limit to get our original integral. So we consider a closed contour consisting of two line segments, one just above and the other just below the positive real axis, together with most of a small and a large circle. The circle contributions vanish in the limit. The two line segment

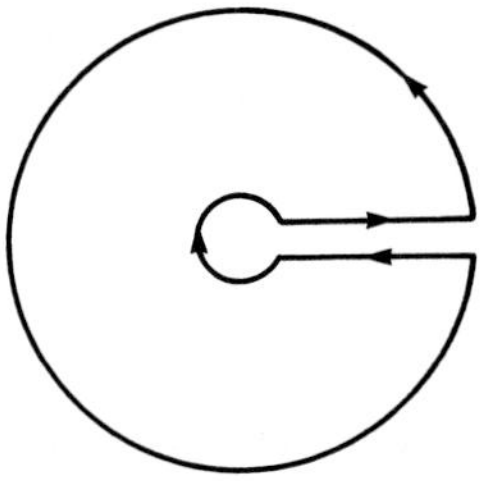

Figure 20.15

contributions do not cancel each other out precisely because the determinations of z^α just above and just below the positive real axis differ by a factor of $e^{2\pi i\alpha}$. So it is precisely the fact that z^α is not well defined that allows us to evaluate the integral.

(d) We can use contour integration to sum various infinite series. The idea is to apply Cauchy's residue theorem to larger and larger domains where the boundary integrals tend to zero. We are then left with an (infinite) sum of residues which must vanish. Bringing one summand to the other side of the equation gives us a sum of an infinite series. We illustrate this with some sums that are useful in the study of trigonometric functions. For this purpose we first define the trigonometric functions of a complex variable. We set

$$\cos z = \tfrac{1}{2}(e^{iz} + e^{-iz}) \quad \text{and} \quad \sin z = -(\tfrac{1}{2}i)(e^{iz} - e^{-iz}).$$

These functions are clearly holomorphic in the entire complex plane and coincide with their usual definitions for real values of z. Notice that $\sin z = 0$ if and only if $e^{2iz} = 1$, and this can only happen for the real values $z = n\pi$. Similarly the only complex zeros of $\cos z$ are at the real points $(n + \frac{1}{2})\pi$. We then define

$$\tan z = \frac{\sin z}{\cos z} \quad \text{and} \quad \cot z = \frac{\cos z}{\sin z}$$

which are meromorphic with poles located at the zeros of their denominators. We can use (20.25) to compute the residues at these poles since all the zeros are simple. So, for example, $\cot z$ has its poles located at the points $n\pi$ and all its residues equal 1. Notice that

$$\cot z = i\frac{e^{iz} + e^{-iz}}{e^{iz} - e^{-iz}}$$

will be bounded in absolute value so long as z stays a finite distance away from the points $n\pi$. In particular, $|\cot z|$ will be uniformly bounded on all circles of radius $(n+\frac{1}{2})\pi$ centered at the origin. Now consider the integral

$$\int_C \frac{\cot\zeta}{\zeta^2 - z^2}\,d\zeta$$

over these circles, where z is some fixed complex number $\neq n\pi$. As we let the circles expand out to infinity the integrals go to zero since $|\cot\zeta|$ is bounded, and we have a polynomial in ζ of degree two in the denominator. The poles of the integrand are located at $\pm z$ and at $n\pi$. The residues at $\pm z$ are each equal to $\frac{1}{2}\cot z/z$ and the residue at $n\pi$ is $(n^2\pi^2 - z^2)^{-1}$. So

$$\frac{\cot z}{z} + \sum_1^\infty (n^2\pi^2 - z^2)^{-1} = 0.$$

Bringing the first term over to the other side, multiplying by z and separating off the summand corresponding to $n = 0$, give the formula

$$\cot z = \frac{1}{z} + 2z\sum_1^\infty \frac{1}{z^2 - n^2\pi^2}. \tag{20.36}$$

The series on the right converges, and converges uniformly in z so long as z stays a fixed distance away from all the poles. If $f(z) = \sin z$, then $f'(z)/f(z) = \cot z$. We can also write $f'(z)/f(z) = (d/dz)(\log f(z))$ and so write (20.36) as

$$\frac{d}{dz}\left(\log\frac{\sin z}{z}\right) = \sum_1^\infty \frac{d}{dz}\log(z^2 - n^2\pi^2).$$

The function $\sin z/z$ does not vanish at $z = 0$. Let us integrate the above equation along any path joining 0 to z and staying a finite distance away from all the points $n\pi$. We get the equation

$$\log(\sin z/z) - \log 1 = \sum_1^\infty (\log(z^2 - n^2\pi^2) - \log(-n^2\pi^2)).$$

This equation is somewhat ambiguous as it stands, since the values of the logarithm, defined as $\log g$ = the integral of g'/g, depend on the path chosen. But exponentiating both sides of the equation eliminates all these ambiguities and gives the famous formula

$$\sin z = z\prod_1^\infty\left(1 - \frac{z^2}{n^2\pi^2}\right)$$

representing $\sin z$ as an infinite product.

20.8. Limits and series

We begin with a straightforward application of Cauchy's integral formula.

> Let f_n be a sequence of functions all holomorphic in a domain D and suppose that $f_n(z) \to f(z)$ uniformly in D. Then f is holomorphic in D and the $f'_n(z)$ converge uniformly to $f'(z)$ on any subset of D which is a finite distance from ∂D. (20.37)

Proof. For any $z \in D$ choose a circle C lying with its interior inside D with z inside C. Then

$$f_n(z) = \frac{1}{2\pi i} \int_C \frac{f_n(\zeta)}{\zeta - z} \, d\zeta.$$

Passing to the limit gives

$$f(z) = \frac{1}{2\pi i} \int_C \frac{f(\zeta)}{\zeta - z} \, d\zeta$$

and we know from the discussion immediately following (20.18) that this implies that f is holomorphic. By formula (20.19) we have

$$f'_n(z) = \frac{1}{2\pi i} \int_C \frac{f_n(\zeta)}{(\zeta - z)^2} \, d\zeta$$

which clearly converges to

$$f'(z) = \frac{1}{2\pi i} \int_C \frac{f(\zeta)}{(\zeta - z)^2} \, d\zeta$$

and the uniformity of the convergence is clear. We may frequently want to apply (20.37) to the case where D is a bounded subregion of some larger domain on which the f_ns are defined and converge. That is, the f_ns are defined and holomorphic on some large region E and converge uniformly on each of a sequence of subregions D_k with $\bigcup D_k = E$. By applying (20.37) to each D_k, we conclude that the limit function f is holomorphic in all of E.

Applying (20.37) to the partial sums of a series gives

> If a series with holomorphic terms $f(z) = f_1(z) + f_2(z) + \cdots$ converges uniformly in D, then the sum f is holomorphic and the series can be differentiated term by term. (20.38)

The most important series associated with a holomorphic function is its Taylor series. Suppose that f is holomorphic in a domain D and the disk of radius R centered at a point a lies entirely inside D. Let z be any point interior to this disk so that $|z - a| < R$. Then we can use (20.22) to estimate the last term in (20.20) as having absolute value at most

$$\frac{M|z-a|}{R - |z-a|} \frac{|z-a|^{n-1}}{R^{n-1}}$$

which tends uniformly to zero on any slightly smaller disk. Thus by (20.38) we get

> If f is holomorphic in a domain D and a is any point of D, then the series
>
> $$f(a) + f'(a)(z-a) + \frac{f''(a)}{2!}(z-a)^2 + \cdots + \frac{f^{(n)}(a)}{n!}(z-a)^n + \cdots$$
>
> converges to $f(z)$ in the largest open disk centered at a lying entirely in D. (20.39)

The series in (20.39) is known as the *Taylor series* of f at a. Many operations on holomorphic functions correspond to simple operations on their Taylor series. For ease in exposition let us take $a = 0$. Then if

$$f(z) = a_0 + a_1 z + a_2 z^2 + \cdots$$

and

$$g(z) = b_0 + b_1 z + b_2 z^2 + \cdots$$

then

$$f'(z) = a_1 + 2a_2 z + 3a_3 z^2 + \cdots,$$
$$f(z) + g(z) = (a_0 + b_0) + (a_1 + b_1)z + (a_2 + b_2)z^2 + \cdots,$$
$$f(z)g(z) = a_0 b_0 + (a_1 b_0 + a_0 b_1)z + (a_2 b_0 + a_1 b_1 + a_0 b_2)z^2 + \cdots.$$

If $a_0 = b_0 = 0$, then

$$\begin{aligned} f(g(z)) &= a_1(b_1 z + b_2 z^2 + \cdots) + a_2(b_1 z + b_2 z^2 + \cdots)^2 + \cdots \\ &= a_1 b_1 z + (a_1 b_2 + a_2 b_1^2)z^2 + (a_1 b_3 + 2a_2 b_1 b_2 + a_3 b_1^3)z^3 + \cdots. \end{aligned}$$

From this last equation, we can recursively determine the power series of the inverse of a function, f, if we know the power series of f. If we know the as and we want to find the bs such that $f(g(z)) = z$, we get

$$a_1 b_1 = 1 \quad \text{so} \quad b_1 = a_1^{-1}$$

(we must assume $f'(0) \neq 0$ to be able to invert the function),

$$\begin{aligned} a_1 b_2 + a_2 b_1^2 = 0 \quad &\text{so} \quad b_2 = -a_1^{-1}(a_2 b_1^2), \\ a_1 b_3 + 2a_2 b_1 b_2 + a_3 b_1^3 = 0 \quad &\text{so} \quad b_3 = -a_1^{-1}(2a_2 b_1 b_2 + a_3 b_1^3), \end{aligned}$$

and so on. In general, the coefficient of z^n will be an expression of the form $a_1 b_n +$ (terms involving bs of order less than n) and so we can solve recursively for each b_n.

It is an elementary fact that any power series has a radius of convergence, R (possibly zero), such that the series converges for $|z| < R$, and for no z with $|z| > R$. It follows from (20.38) that the series represents a holomorphic function inside its circle of convergence. It follows from (20.39) that, conversely, the Taylor series of a holomorphic function has, as its radius of convergence, the radius of the largest disk contained in the true domain of definition of f.

A series of the form $a_0 + a_{-1}z^{-1} + a_{-2}z^{-2} + \cdots$ can be thought of as a power series in the variable $1/z$. So it will converge outside some circle of radius R (possibly $R = \infty$). A series of the form $\sum_{-\infty}^{\infty} a_n z^n$ is said to converge if and only if its positive and negative parts converge. The positive part will converge for $|z| < R_2$ for some R_2 and the negative part will converge for $|z| > R_1$ for some R_1. So there will be some non-empty region of convergence for the double series if and only if $R_1 < R_2$, in which case, by (20.38), it represents a holomorphic function on the annulus $R_1 < |z| < R_2$. Conversely, suppose that f is holomorphic in such an annulus. We shall show that f has such a double series expansion. In fact, to prove this, all we have to do is show that f can be written as the sum $f = f_1 + f_2$, where f_1 is holomorphic for $|z| < R_2$ and f_1 is holomorphic for $|z| > R_1$ and is such that $f_2(1/z)$ has a removable singularity at 0, so that $f_2(1/z)$ is holomorphic for

$|1/z| < 1/R_1$. To do this, define

$$f_1(z) = \frac{1}{2\pi i}\int_{|\zeta|=r_2} \frac{f(\zeta)}{\zeta - z}\,d\zeta$$

where the integration is over any circle of radius r_2 with $|z| < r_2 < R_2$ and

$$f_2(z) = -\frac{1}{2\pi i}\int_{|\zeta|=r_1} \frac{f(\zeta)}{\zeta - z}\,d\zeta$$

where this integral is over any circle of radius r_1 satisfying $R_1 < r_1 < |z|$. It is clear that f_1 is holomorphic for $|z| < R_2$ and f_2 is holomorphic for $|z| > R_1$. By Cauchy's integral formula, $f_1(z) + f_2(z) = f(z)$. If we set $z' = 1/z$ and substitute $\zeta' = 1/\zeta$ in the integral defining f_2, we get

$$f_2(1/z') = \frac{1}{2\pi i}\int_{|\zeta'|=1/r_1} \frac{f(1/\zeta')}{\zeta' - z'}\frac{z'}{\zeta'}\,d\zeta'$$

which shows that $f_2(1/z')$ is holomorphic in z' for $|z'| < 1/R_1$. We can expand f_1 in a power series

$$f_1(z) = \sum_0^\infty a_n z^n \quad \text{with} \quad a_n = \frac{1}{2\pi i}\int_{|\zeta|=r} \frac{f(\zeta)}{\zeta^{n+1}}\,d\zeta$$

and we can expand $f_2(1/z') = \sum_1^\infty b_n z'^n$ with

$$b_n = \frac{1}{2\pi i}\int_{|\zeta'|=1/r} \frac{f(1/\zeta')}{\zeta'^{n+1}}\,d\zeta' = \frac{1}{2\pi i}\int_{|\zeta|=r} f(\zeta)\zeta^{n-1}\,d\zeta$$

where r is any value between R_1 and R_2. Setting $a_{-n} = b_n$, we get the desired expansion, known as the *Laurent expansion*

$$f(z) = \sum_{-\infty}^{\infty} a_n z^n \quad \text{where} \quad a_n = \frac{1}{2\pi i}\int_{|\zeta|=r} \frac{f(\zeta)}{\zeta^{n+1}}\,d\zeta. \tag{20.40}$$

It will be very important for us later on to write out this formula explicitly in the case where $R_1 < 1 < R_2$. Thus we assume that f is defined and holomorphic in some neighborhood of the unit circle, $|z| = 1$. Let us write $F(\theta) = f(e^{i\theta})$ and take $r = 1$ in (20.40). Then we get

$$f(z) = \sum a_n z^n$$

where

$$a_n = \frac{1}{2\pi}\int_0^{2\pi} F(\theta)e^{in\theta}\,d\theta. \tag{20.41}$$

In particular, taking $z = e^{i\theta}$ to be on the unit circle and substituting into the Laurent expansion of f, we get

$$F(\theta) = \sum_{-\infty}^{\infty} a_n e^{in\theta} \tag{20.42}$$

with the a_ns given by (20.41). The series (20.42) is known as the *Fourier series* for F. If f is holomorphic on some region $R < |z| < 1/R$ where R is some number,

$R < 1$, then, since the a_n are coefficients of convergent power series, it follows from the convergence criterion for power series that

$$|a_n| \leqslant cr^{-|n|} \quad \text{for any } r > R$$

for some suitable constant $c = c_r$.

Summary

A Holomorphic functions

You should be able to define a holomorphic function in terms of two-forms and show that its Jacobian matrix represents a conformal transformation.

You should be able to derive the Cauchy integral theorem and residue formula from Stokes' Theorem.

You should be able to calculate the residue of a function at a simple or multiple pole.

B Contour integration

You should be able to evaluate integrals by applying the residue theorem to a specified contour, and in a few standard cases you should be able to construct the appropriate contour.

C Power series

You should know how to expand a holomorphic function in a Taylor or Laurent series and be able to calculate the radius of convergence of such a series.

Exercises

20.1. (a) Suppose that $f(z)$ is a holomorphic function. Let $z = x + \mathrm{i}y$, $f(z) = u + \mathrm{i}v$. Show that the families of curves $u(x, y) = \text{constant}$ and $v(x, y) = \text{constant}$ are orthogonal; i.e. they cross at right angles.

(b) Verify this property explicitly for the case where $f(z) = z^2$. Sketch a few curves. Do the same for $f(z) = 1/z$. What happens at the origin in this case?

20.2. (a) Suppose that $f(z)$ is a holomorphic function whose real part is $3x^2y - y^3$. Determine the imaginary part of $f(z)$.

(b) Can there be a holomorphic function whose real part is xy^2? If so, find one. If not, explain why not.

20.3. We can define holomorphic functions $\cos z$ and $\sin z$ by using the identities $\cos z = \frac{1}{2}(\mathrm{e}^{\mathrm{i}z} + \mathrm{e}^{-\mathrm{i}z})$, $\sin z = (\mathrm{e}^{\mathrm{i}z} - \mathrm{e}^{-\mathrm{i}z})/2\mathrm{i}$ and the definition of the complex exponential function.

(a) Express $\cos z$ and $\sin z$ in terms of trigonometric and hyperbolic functions of x and y. Calculate $\mathrm{d}(\cos z)$ and $\mathrm{d}(\sin z)$ by using these expressions.

(b) Prove that the *addition formulas* $\sin(z + w) = \sin z \cos w + \cos z \sin w$, $\cos(z + w) = \cos z \cos w - \sin z \sin w$ hold even for complex z and w.

(c) Using the addition formulas, show explicitly that the functions $\sin z$ and $\cos z$ are continuously differentiable from the complex point of view, and calculate the derivatives of these functions by evaluating the appropriate limits.

20.4. Use the technique presented on pp. 736–8 to establish the following definite integrals:

(a) $$\int_0^\pi \sin^6\theta\,\mathrm{d}\theta = \frac{5\pi}{16};$$

(b) $$\int_0^\pi \frac{\cos\theta\,\mathrm{d}\theta}{1-2a\cos\theta+a^2} = \frac{\pi a}{1-a^2}, \quad a^2<1.$$

20.5. Use technique (a) on p. 736 to establish the following definite integrals:

(a) $$\int_{-1}^\infty \frac{\mathrm{d}x}{(x+b)^2+a^2} = \frac{\pi}{a}, \quad a>0;$$

(b) $$\int_0^\infty \frac{\mathrm{d}x}{(x^4+4a^4)} = \frac{\pi}{8a^3}, \quad a>0;$$

(c) $$\int_0^\infty \frac{x^2\,\mathrm{d}x}{x^6+1} = \frac{\pi}{6}.$$

20.6. (a) Suppose that $f(z)$ is holomorphic in a region except at $z=a$, where it has a pole of *second* order. Prove that the residue at the pole is given by the formula

$$\operatorname{res} = \lim_{z\to a}\frac{\mathrm{d}}{\mathrm{d}z}((z-a)^2 f(z)).$$

(b) Use the above result to evaluate

$$\int_{-\infty}^\infty \frac{\mathrm{d}x}{(x^2+a^2)^2}.$$

(c) Generalize the above technique to evaluate

$$\int_{-\infty}^\infty \frac{\mathrm{d}x}{(x^2+a^2)^k}$$

for an arbitrary positive integer k.

20.7. (a) Here is an alternative approach to technique (b) on p. 736. Let.

$$R(z) = P(z)/Q(z)$$

where P and Q are polynomials with deg $Q \geqslant \deg P + 1$ so that

$$|R(z)| \leqslant A/|z|$$

for some constant A for sufficiently large $|z|$. Suppose that Q has no real zeros. The purpose of this exercise is to prove that the

$$\lim_{\substack{X_2\to\infty\\ X_1\to\infty}} \int_{-X_1}^{X_2} R(x)\mathrm{e}^{\mathrm{i}x}\,\mathrm{d}x$$

exists and its value is

$$\int_{-\infty}^\infty R(x)\mathrm{e}^{\mathrm{i}x}\,\mathrm{d}x = 2\pi\mathrm{i}\sum \operatorname{res}(R(z)\mathrm{e}^{\mathrm{i}z})$$

where the sum is taken over all the poles lying in the upper half-plane.

Step 1. Choose X_1, X_2 and Y large enough so that the rectangle in figure 20.16 contains all the poles in the upper half-plane.

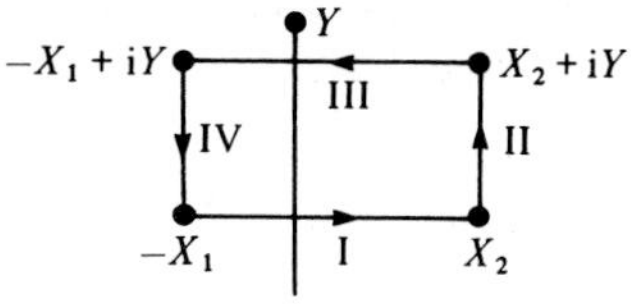

Figure 20.16

Step 2. Show that

$$\left|\int_{\mathrm{II}} R(z)\mathrm{e}^{\mathrm{i}z}\,\mathrm{d}z\right| \leqslant A\int_0^Y \frac{\mathrm{e}^{-y}}{|z|}\,\mathrm{d}y \leqslant \frac{A}{X_2}\int_0^Y \mathrm{e}^{-y}\,\mathrm{d}y \leqslant \frac{A}{X_2}$$

for X_2 large enough.

Step 2′. Show that

$$\left|\int_{\mathrm{IV}} R(z)\mathrm{e}^{\mathrm{i}z}\,\mathrm{d}z\right| \leqslant \frac{A}{X_1}.$$

Step 3. Show that

$$\left|\int_{\mathrm{III}} R(z)\mathrm{e}^{\mathrm{i}z}\,\mathrm{d}z\right| \leqslant A\mathrm{e}^{-Y}(X_1+X_2)/Y.$$

Step 4. Let $Y \to \infty$. Conclude that

$$\left|\int_{-X_1}^{X_2} R(x)\mathrm{e}^{\mathrm{i}x}\,\mathrm{d}x - 2\pi\mathrm{i}\sum_{z\to 0} \mathrm{res}\,(R(z)\mathrm{e}^{\mathrm{i}z}) < A\left(\frac{1}{X_2}+\frac{1}{X_1}\right).\right.$$

Complete the proof.

(b) Evaluate $\int_0^\infty (\cos x/(x^2+a^2))\,\mathrm{d}x$.

20.8. Suppose that $Q(0)=0$ while $Q'(0)\neq 0$ so that $R(z)\mathrm{e}^{\mathrm{i}z}$ has a simple pole at 0. Let B be the residue at 0. Consider the region of integration which is the same as before except that we avoid the origin by following a small semicircle of radius δ in the lower half-plane.

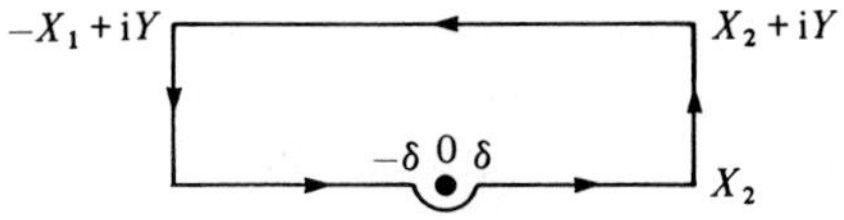

Figure 20.17

(a) Show that the limit as $\delta \to 0$ of the integral around the semicircle is $\pi\mathrm{i}B$.

(b) Conclude that the Cauchy principal value

$$\lim_{\delta\to 0}\left(\left[\int_{-\infty}^{-\delta} + \int_{\delta}^{\infty}\right] R(x)\mathrm{e}^{\mathrm{i}x}\,\mathrm{d}x\right) = 2\pi\mathrm{i}\sum_{z\to 0}\mathrm{res}\,(R(z)\mathrm{e}^{\mathrm{i}z}) + \tfrac{1}{2}B.$$

(c) Show that

$$\int_0^\infty \frac{\sin x}{x}\,\mathrm{d}x = \pi/2.$$

20.9. Let $f(z)=(f_{ij}(z))$ be a matrix-valued function. We say that f is *holomorphic* in a region D if each of the matrix entries f_{ij} is a holomorphic function on D. If γ is any curve, define $\int_\gamma f(z)\,dz$ to be the matrix whose ijth entry is $\int_\gamma f_{ij}(z)\,dz$. Notice that if B and C are *constant* matrices, then if f is a holomorphic matrix-valued function so is BfC, and for any curve γ we have

$$\int_\gamma Bf(z)C\,dz = B\left(\int_\gamma f(z)\,dz\right)C.$$

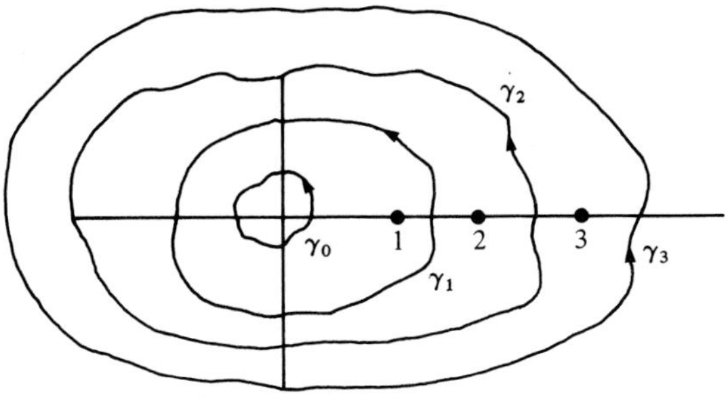

Figure 20.18

(a) Let $A=\begin{pmatrix}1&0&0\\0&2&0\\0&0&3\end{pmatrix}$ and set $f(z)=(zI-A)^{-1}$, where $I=\begin{pmatrix}1&0&0\\0&1&0\\0&0&1\end{pmatrix}$.
Show that f is holomorphic everywhere except at $z=1$, $z=2$, and $z=3$.
Let $P_{\gamma_i}=(1/2\pi i)\int_{\gamma_i} f(z)\,dz$. Evaluate P_{γ_i} for $i=0,1,2$ and 3.

(b) Let D be a 3×3 matrix whose eigenvalues are 1, 2 and 3. Let $g(z)=(zI-D)^{-1}$. Show that g is holomorphic everywhere except at $z=1$, 2 or 3. Let

$$Q_{\gamma_i}=\frac{1}{2\pi i}\int_{\gamma_i} g(z)\,dz.$$

Describe $\operatorname{im} Q_{\gamma_1}$ and $\ker Q_{\gamma_1}$. What is $Q_{\gamma_1}^2$? Describe $\operatorname{im} Q_{\gamma_2}$ and $\ker Q_{\gamma_2}$. What is $Q_{\gamma_2}^2$?

(c) Formulate a general theorem. Let A be an $n\times n$ matrix with distinct eigenvalues $\lambda_1,\dots,\lambda_n$. Let $f(z)=(zI-A)^{-1}$. Where is f holomorphic? Let γ be a curve that does not pass through any of the eigenvalues and is the boundary of a region which contains $\lambda_1,\dots,\lambda_k$ but none of the remaining eigenvalues. Set

$$P_\gamma=\frac{1}{2\pi i}\int_\gamma f(z)\,dz=\frac{1}{2\pi i}\int(zI-A)^{-1}\,dz.$$

What is $\operatorname{im} P_\gamma$? What is $\ker P_\gamma$? What is P_γ^2?

20.10. (a) Let $f(z)$ denote the branch of the function $z^{1/4}$, defined everywhere except on the positive real axis, with the property that

$$\lim_{\varepsilon\to 0} f(x+i\varepsilon)=|x|^{1/4} \text{ (a real quantity).}$$

Evaluate $f(-x)$ and $\lim_{\varepsilon\to 0} f(x-i\varepsilon)$ for real x.

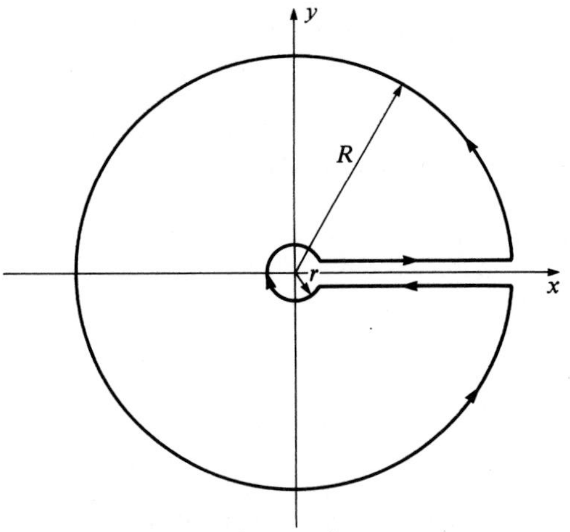

Figure 20.19

(b) By using the contour shown in figure 20.19 evaluate the definite integral

$$I = \int_0^\infty \frac{t^{1/4}\,dt}{(t+a)^2}.$$

By sure to discuss the contributions from the circles.

20.11. Use technique (c) on pp. 738–9 to establish the following integrals by integrating around the contour shown on p. 738.

(a) $$\int_0^\infty \frac{x^{\alpha-1}}{x+1}\,dx = \frac{\pi}{\sin \alpha\pi} \quad 0 < \alpha < 1,$$

(b) $$\int_0^\infty \frac{x^{\alpha-1}}{(x+1)^2}\,dx = \frac{(1-\alpha)\pi}{\sin \alpha\pi} \quad 0 < \alpha < 2.$$

20.12. (a) By evaluating $\int f(z)\log z\,dz$ around the contour on p. 739 using the branch of $\log z$ defined by

$$\log z = \log r + i\theta \quad 0 < \theta < 2\pi,$$

establish the formula

$$\int_0^\infty f(x)\,dx = -\sum \operatorname{res}\,(f(z)\log z),$$

valid whenever $f(z)$ tends to zero as $|z| \to \infty$ fast enough so that the contribution from the large circle vanishes.

(b) Show that

$$\int_0^\infty \frac{dx}{(x+a)(x^2+b^2)} = \frac{\pi a + 2b\log(b/a)}{2b(a^2+b^2)} \quad a, b > 0.$$

(c) By considering $\int f(z)(\log z)^2\,dz$, devise a similar technique for evaluating $\int_0^\infty f(x)\log x\,dx$, and use it to establish the formula

$$\int_0^\infty \frac{\log x}{x^2+a^2}\,dx = \frac{\pi \log a}{2a}.$$

20.13. (a) Determine the Laurent expansion of

$$f(z)=\frac{5}{z^2-\frac{5}{2}z+1}$$

by contour integration around the unit circle. Where does this expansion converge?

(b) Find the same expansion by doing a partial fraction decomposition of $f(z)$, then expanding each term in powers of z or $1/z$ as appropriate so that the series converges when $|z|=1$.

(c) Find the Fourier expansion of

$$F(\theta)=(\tfrac{5}{2}+2\cos\theta)^{-1}.$$

20.14. Find the Fourier expansion of the function

$$F(\theta)=\theta \qquad -\pi<\theta<\pi.$$

Can this be obtained as the Laurent expansion of a holomorphic function?

21

Asymptotic evaluation of integrals

Chapter 21 discusses some of the more elementary aspects of asymptotics.

Introduction

Many important laws of physics can be understood in terms of the asymptotic behavior of integrals. In this chapter we give some elementary illustrations of the mathematical techniques involved and present some important physical and mathematical applications.

One of the earliest, and still most important, *asymptotic formulas* is

$$\text{Stirling's formula: } n! \sim (2\pi)^{1/2} n^{n+1/2} e^{-n}$$

discovered by James Stirling in the first third of the eighteenth century. Here the sign $\sim$ may be taken to mean that the ratio of the two sides (which are both tending to ∞ with n) tends to 1 as $n \to \infty$. We shall soon state and prove a more accurate version of his formula together with an estimate of the error.

Although Stirling's formula only makes an assertion about large n, it gives a remarkably accurate approximation (in terms of percentage error) even for small n. Thus for $n = 1$ we have $n! = 1$ while the right-hand side of Stirling's formula gives 0.92. For $n = 10$ we have $10! = 3\,628\,800$ and the right-hand side gives $3\,598\,600$, an error of less than one percent. This is a recurrent phenomenon. In this chapter we will give estimates valid for a large value of a parameter, but the formula seems to work well for small values as well.

21.1. Laplace's method

We illustrate the method by giving a refined version of Stirling's formula. Integration by parts shows that the Γ *function*, defined by

$$\Gamma(x) = \int_0^\infty e^{-t} t^{x-1} dt \quad x > 0,$$

satisfies

$$\Gamma(x+1) = x\Gamma(x), \quad \Gamma(1) = 1$$

so that

$$\Gamma(n+1) = n!.$$

We can thus rewrite Stirling's formula as

$$\Gamma(x) \sim \mathrm{e}^{-x}x^x\left(\frac{2\pi}{x}\right)^{1/2}.$$

In fact, we shall prove a more precise result. We shall show that

$$\Gamma(x) \sim \mathrm{e}^{-x}x^x\left(\frac{2\pi}{x}\right)^{1/2}\left(1 + \frac{1}{12x} + \frac{1}{288x^2} + \cdots\right).$$

An asymptotic series of this form means, for example, that

$$\left|\Gamma(x) - \mathrm{e}^{-x}x^x\left(\frac{2\pi}{x}\right)^{1/2}\right| < C_0\left(\mathrm{e}^{-x}x^x\left(\frac{2\pi}{x}\right)x^{-1}\right),$$

that

$$\left|\Gamma(x) - \mathrm{e}^{-x}x^x\left(\frac{2\pi}{x}\right)^{1/2}\left(1 + \frac{1}{12x}\right)\right| < C_1\left(\mathrm{e}^{-x}x^x\left(\frac{2\pi}{x}\right)^{1/2}x^{-2}\right),$$

that

$$\left|\Gamma(x) - \mathrm{e}^{-x}x^x\left(\frac{2\pi}{x}\right)^{1/2}\left(1 + \frac{1}{12x} - \frac{1}{288x^2}\right)\right| < C_2\left(\mathrm{e}^{-x}x^x\left(\frac{2\pi}{x}\right)^{1/2}x^{-3}\right),$$

etc., where the C_i are constants. In other words, whenever we break off the series, we can estimate the error in terms of the next higher power of x^{-1}. We begin by writing $\Gamma(x) = \Gamma(x+1)/x$ so

$$\begin{aligned}\Gamma(x) &= \frac{1}{x}\int_0^\infty \mathrm{e}^{-t}t^x\,\mathrm{d}t \\ &= \mathrm{e}^{-x}x^x\int_{-1}^\infty \mathrm{e}^{-x[w-\log(1+w)]}\mathrm{d}w \\ &= \mathrm{e}^{-x}x^x\int_{-1}^\infty \mathrm{e}^{-xp(w)}\,\mathrm{d}w\end{aligned}$$

where $t = (1+w)x$, so that $\mathrm{d}t = x\mathrm{d}w$ and $t = 0 \Leftrightarrow w = -1$, $t = \infty \Leftrightarrow w = \infty$, and where ·

$$p(w) = w - \log(1+w) = +\tfrac{1}{2}w^2 - \tfrac{1}{3}w^3 + \cdots$$

has the above Taylor expansion at the origin. Notice that $p'(w) = 0$ only at $w = 0$ and, in fact, p has an absolute minimum at the origin. The factor $\mathrm{e}^{-x}x^x$ occurs on the right-hand side of Stirling's formula. So we need only concentrate our attention on the asymptotic properties of the integral

$$\int_{-1}^\infty \mathrm{e}^{-xp(w)}\mathrm{d}w.$$

Our first observation is that, for large x, the only substantial contribution to the integral comes from w near 0, the minimum of p. For example, if ε is any fixed positive number

$$\int_{-1}^{-\varepsilon} e^{-xp(w)}dw < \int_{-1}^{-\varepsilon} e^{-x\cdot\frac{1}{4}w^2}dw < \int_{-\infty}^{-\varepsilon} e^{-xw^2/2}dw.$$

We can estimate this last integral very crudely by writing

$$e^{-xw^2/2} = e^{-xw^2/4}\cdot e^{-xw^2/4} \leqslant e^{-\varepsilon^2 x/4}e^{-xw^2/4}$$

for $-\infty < w < -\varepsilon$. So

$$\int_{-\infty}^{-\varepsilon} e^{-xw^2/2}dw \leqslant C(x)e^{-\varepsilon^2 x/4}, \quad \text{where} \quad C(x) = \int_{-\infty}^{0} e^{-xw^2/4}dw$$

is a convergent integral, and, in fact $C(x)\to 0$ as $x\to\infty$. So $\int_{-1}^{-\varepsilon}$ decreases exponentially with x. Similarly, for $w > 0$, $p(w)$ is strictly increasing so that

$$\int_{\varepsilon}^{\infty} e^{-xp(w)}dw < e^{-p(\varepsilon)x/2}\int_{\varepsilon}^{\infty} e^{-xp(w)/2}dw.$$

This second integral converges absolutely and uniformly in $x > 0$ (and more and more rapidly as $x\to\infty$). Thus we can control both $\int_{-1}^{-\varepsilon}$ and $\int_{\varepsilon}^{\infty}$ and concentrate attention on $\int_{-\varepsilon}^{\varepsilon} e^{-xp(w)}dw$ for any $\varepsilon > 0$. Now $p(0) = 0$, $p'(0) = 0$ and $p''(0) = 1$. We can therefore make a change of variables $w = w(s)$ near the origin so that p goes over into $\frac{1}{2}s^2$, i.e.,

$$p(w(s)) = s^2/2.$$

For convenience, we remind the reader how this is done. Since $p(0) = 0$, we have

$$p(w) = \int_0^w p'(r)dr = w\int_0^1 p'(wu)du$$

where we have set $r = wu$ to get from the first integral to the second. Similarly,

$$p'(w) = w\int_0^1 p''(wu)du$$

so that

$$p(w) = \tfrac{1}{2}w^2 q(w)$$

where

$$q(w) = 2\int_0^1\int_0^1 up''(uvw)du\,dv.$$

Notice that q is a differentiable function of w and $q(0) = 1$. Thus $q(w) > 0$ for w sufficiently small. Thus we may make the differentiable change of variables

$$s = w\sqrt{q(w)}$$

and

$$p(w(s)) = \tfrac{1}{2}s^2.$$

Let us choose our $\varepsilon > 0$ small enough that this change of variables is valid for

$-\varepsilon < w < \varepsilon$. (Thus, so that $q(w)$ is positive on this range.) We can then make this change of variables in the integral $\int_{-\varepsilon}^{\varepsilon} e^{-xp(w)}dw$ to get

$$\int_{-\varepsilon}^{\varepsilon} e^{-xp(w)}dw = \int_{-\varepsilon_1}^{\varepsilon_2} e^{-xs^2/2}\left(\frac{dw}{ds}\right)ds$$

where $\varepsilon_2 = \varepsilon\sqrt{q(\varepsilon)}$ and $-\varepsilon_1 = -\varepsilon\sqrt{q(-\varepsilon)}$.

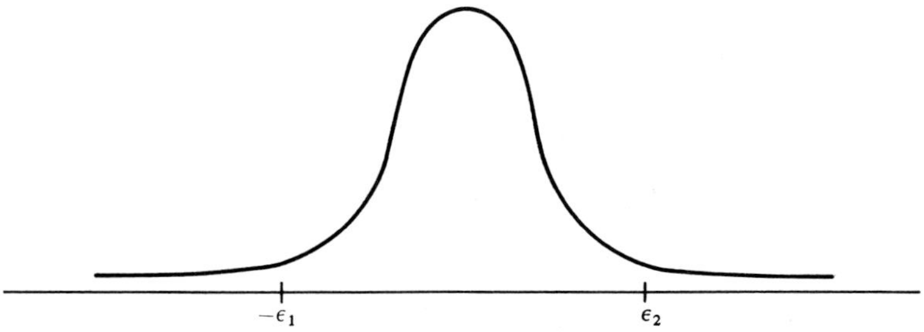

Figure 21.1

In estimating the integral, the only region that matters is the neighborhood of $s = 0$. For example, suppose we choose some function ρ, differentiable to all orders and such that $\rho \equiv 1$ for $|s| < \eta$ and $\rho \equiv 0$ for $s > 2\eta$ and $-\varepsilon_1 < -2\eta, 2\eta < \varepsilon$. Then set

$$b(s) = \rho(s)\frac{dw}{ds} \quad \text{so} \quad b(s) \equiv \frac{dw}{ds}(s) \quad \text{for} \quad |s| < \eta.$$

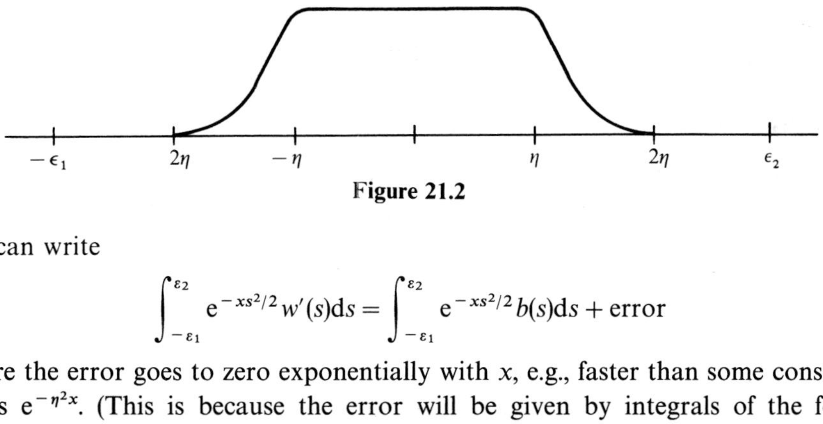

Figure 21.2

We can write

$$\int_{-\varepsilon_1}^{\varepsilon_2} e^{-xs^2/2} w'(s)ds = \int_{-\varepsilon_1}^{\varepsilon_2} e^{-xs^2/2} b(s)ds + \text{error}$$

where the error goes to zero exponentially with x, e.g., faster than some constant times $e^{-\eta^2 x}$. (This is because the error will be given by integrals of the form $\int_{-\varepsilon_1}^{\eta} e^{-xs^2/2}(b(s) - w'(s))\,ds$ and $\int_{\eta}^{\varepsilon_2} e^{-xs^2/2}(b(s) - w'(s))\,ds$ and the integrand goes to zero faster than some constant times $e^{-\eta^2 x}$.)

We may replace $\int_{-\varepsilon_1}^{\varepsilon_2} e^{-s^2x/2} b(s)ds$ by

$$\int_{-\infty}^{\infty} e^{-s^2x/2} b(s)ds$$

since $b \equiv 0$ for $s < -\varepsilon_1$ and $s > \varepsilon_2$. We are thus reduced to estimating this last integral.

We now apply Taylor's formula with remainder to write

$$b(s) = b_0 + b_1 s + b_2 s^2 + \cdots + b_{N-1} s^{N-1} + b_N(s) s^N$$

where $b_N(s)$ is a smooth bounded function. Then

$$\int_{-\infty}^{\infty} e^{-xs^2/2} b(s) ds = b_0 \int_{-\infty}^{\infty} e^{-xs^2/2} ds + b_1 \int_{-\infty}^{\infty} e^{-xs^2/2} s ds + b_2 \int_{-\infty}^{\infty} e^{-x^2 s/2} s^2 \, ds + \cdots + R_N(s)$$

where

$$R_N(s) = \int_{-\infty}^{\infty} e^{-xs^2/2} b_N(s) s^N \, ds$$

so

$$|R_N(s)| \leqslant D \int_{-\infty}^{\infty} e^{-xs^2/2} |s|^N \, ds$$

where

$$D_N = \sup |b_N(s)|.$$

Now, by symmetry,

$$\int_{-\infty}^{\infty} e^{-xs^2/2} s^k ds = 0 \quad \text{if } k \text{ is odd,}$$

while, for even k, make the substitution

$$x^{1/2} s = y$$

so

$$ds = \frac{dy}{x^{1/2}}$$

and

$$\int_{-\infty}^{\infty} e^{-xs^2/2} s^k ds = x^{-1/2} x^{-k} \int_{-\infty}^{\infty} e^{-y^2/2} y^k dy$$
$$= C_R x^{-1/2} x^{-k/2}$$

where

$$C_R = \int_{-\infty}^{\infty} e^{-y^2/2} y^k dy$$

is independent of x (with $C_0 = C_1 = \sqrt{(2\pi)}$). Notice that

$$\int_{-\infty}^{\infty} e^{-y^2/2} y^k \, dy = \int_{-\infty}^{\infty} e^{-u} (2u)^{k-1/2} \cdot \tfrac{1}{2} du = 2 \frac{k-3}{2} \Gamma\left(\frac{k+1}{2}\right)$$

so

$$C_k = 2 \frac{k-1}{2} \Gamma\left(\frac{k+1}{2}\right).$$

Similarly, the remainder term can be estimated by some constant times $x^{-1/2}x^{-N/2}$. Thus, taking $N = 2n$ even

$$\int_{-\infty}^{\infty} e^{-xs^2/2} b(s) \mathrm{d}s = \left(\frac{2\pi}{x}\right)^{1/2} (a_0 + a_1 x^{-1} + \cdots + a_{n-1} x^{-n+1} + \varepsilon_n(x))$$

where $\varepsilon_n(x) = o(x^{-n})$. Here $a_0 = b_0, a_1 = b_2, a_2 = b_4/\sqrt{(2\pi)}$, etc. Notice also that b and $\mathrm{d}w/\mathrm{d}s$ have the *same* Taylor expansion at the origin. Substituting the Taylor expansion for $p(w) = w - \log(1 + w)$ gives the refined Stirling approximation.

More generally, suppose that we are given a smooth positive function p with a single absolute minimum at a point t_0 interior to its domain of definition and such that $p''(t_0) > 0$. Let $a(t)$ be some other smooth function and consider the integral

$$\int a(t) e^{-kp(t)} \mathrm{d}t.$$

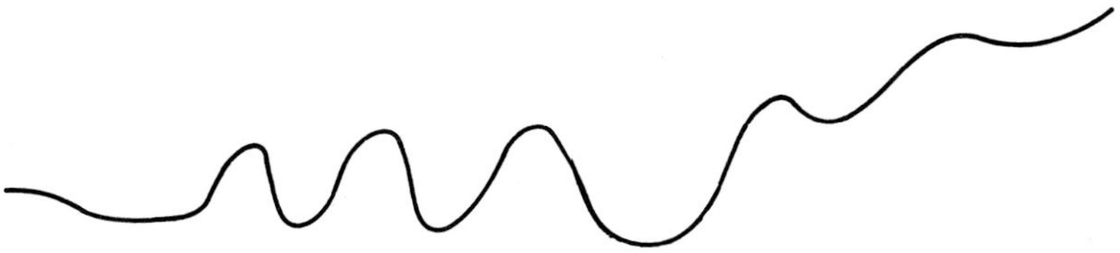

Figure 21.3

We can write this as

$$e^{-kp(t_0)} \int a(t) e^{-k(p(t) - p(t_0))} \mathrm{d}t$$

where now $p(t) - p(t_0)$ takes the minimum value 0 at $t = t_0$. We can thus apply the preceding method to conclude the *principle of Laplace* (1820):

$$\int a(t) e^{-kp(t)} \mathrm{d}t \sim \frac{e^{-kp(t_0)}}{\sqrt{p''(t_0)}} \left(\frac{2\pi}{k}\right)^{1/2} \left(a_0 + \frac{a_1}{k} + \frac{a_2}{k^2} + \cdots\right)$$

where $a_0 = a(t_0)$ and the higher-order coefficients can be expressed in terms of a, p and their derivatives at t_0.

21.2. The method of stationary phase

Let us now examine what happens in Laplace's method when we replace $-kp$ by ikp in the exponential. We will find that a similar formula holds, but that this time *all* the critical points of p enter, not just the absolute minimum. In Laplace's method, the factor $e^{-p(w)}$ integrand was decaying exponentially relative to its value near the minimum of p. For this reason all the asymptotic information was located at the minimum. When we consider an integral of the form

$$\int e^{ikp(y)} a(y) \mathrm{d}y$$

the factor $e^{ikp(y)}$ is not going to zero as $k \to \infty$. But it is oscillating rapidly as a function of y, at points where $p'(y) \neq 0$. These rapid oscillations mean that there will be a mutual cancellation and for this reason the important contribution will come from near the critical points. To illustrate, let us consider the case where $p'(y)$ does not vanish at all on the region where $a(y) \neq 0$. More precisely, let us assume that $a(y)$ has derivatives of all orders and has compact support, that is, that $a(y)$ vanishes outside some finite interval. Let us also assume that

$$|p'(y)| > C > 0 \quad \text{at all } y \text{ where} \quad a(y) \neq 0.$$

We then claim that for any N there is a constant C_N such that

$$\left| \int e^{ikp(y)} a(y) dy \right| < \frac{C_N}{k^N}.$$

In other words, we claim that the integral tends to zero faster than any inverse power of k. Indeed, consider the differential operator

$$D = \frac{1}{p'(y)} \frac{\partial}{\partial y}$$

which (by virtue of our assumptions) is well-defined on the range where $a(y) \neq 0$, and

$$De^{ikp(y)} = ike^{ikp(y)}$$

or

$$e^{ikp(y)} = \frac{1}{ik}(De^{ikp(y)}).$$

Then

$$\begin{aligned} \int e^{ikp(y)} a(y) dy &= \int \frac{1}{ik} (De^{ikp(y)}) a(y) dy \\ &= \int \frac{1}{ik} \left(\frac{\partial}{\partial y} e^{ikp(y)} \right) \frac{a(y)}{p'(y)} dy \\ &= \frac{1}{k} \int e^{ikp} c(y) dy \end{aligned}$$

where

$$c(y) = \frac{-1}{i} \frac{\partial}{\partial y} \left(\frac{a(y)}{p'(y)} \right)$$

and we have used integration by parts and the fact that a vanishes (with its derivatives) at the end points of integration. The function c has all the properties of a so we can repeat the process to conclude that

$$\int b(y) e^{ikp(y)} dy = \frac{1}{k^N} \int c_N(y) e^{ikp(y)} dy$$

for some suitable function $c_N(y)$. Letting

$$c_N = \left| \int c_N(y) e^{ikp(y)} dy \right|$$

proves our claim.

Let us now consider the general case. We will still suppose that a is smooth with compact support. We will no longer assume that $p'(y)$ has no zeros. But we will assume that p has a finite number of critical points, $y_1,\ldots,y_r$ (points at which $p'(y)=0$) and that $p''(y)\neq 0$ at each of the critical points. We then claim that

$$a(y)e^{ikp}(y)dy \sim \left(\frac{2\pi}{k}\right)\sum_{\substack{\text{critical}\\ \text{points, } y_j}} \frac{e^{\pm\pi i/4}e^{ikp(y_j)}}{\sqrt{|p''(y_j)|}}\left(a_{0,j}+\frac{a_{1,j}}{k}+\cdots\right)$$

where $a_{0,j}=a(y_j)$ and the higher coefficients can be expressed in terms of the values of a, p and their derivatives at y_j. The $+$ sign in $e^{\pm\pi i/4}$ is used at minima, where $p''(y_j)>0$ and the $-$ sign at maxima where $p''(y_j)<0$. In the above expression, p is a real-valued, and a may be a complex-valued function of y. This is known as the formula of stationary phase. We shall prove this formula as follows. About each critical point, y_i, we can find coordinates s such that

$$p(y(s)) = \pm s^2/2$$

(the $\pm$ sign depending on the sign of $p''(y_j)$). In making this change of variables, we will have

$$\frac{dy}{ds}=\frac{1}{\sqrt{|p''(y_j)|}} \quad \text{at} \quad y_j.$$

This accounts for the factor $1/\sqrt{|p''(y_j)|}$ occurring in the stationary phase formula.

About each point y_j choose a function ϕ_j such that $\phi_j\equiv 1$ near y_j and $\phi_j\equiv 0$ outside some small neighborhood where the changes of variables are defined. Then

$$b = a-(a_1\phi_1+a_2\phi_2+\cdots+a_r\phi_r)$$

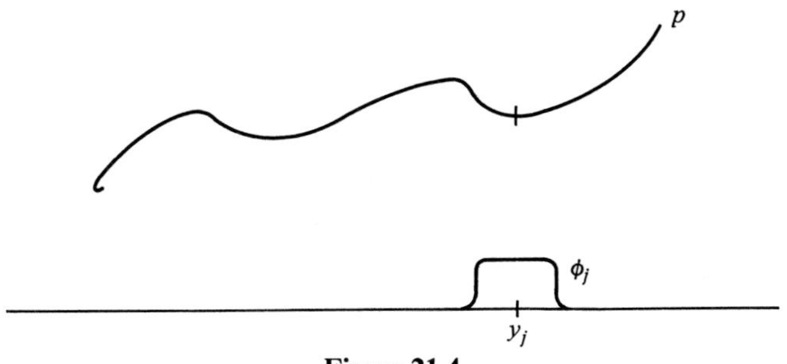

Figure 21.4

Figure 21.5

has the property that it has compact support and $p'(y)\neq 0$ anywhere that $a(y)\neq 0$. Thus the preceding result applies to b, so we have

$$\int e^{ikp}a\,dy=\sum\int e^{ikp}(a_j\phi_j)dy+O(k^{-N})$$

for any N. We may thus consider each summand separately. We may make the change of variables in each summand. So the jth summand looks like

$$\int e^{\pm iks^2/2}\left(a_j\phi_j\frac{dy}{ds}\right)ds.$$

Let us define the function ψ by

$$\psi = a_j\phi_j\frac{dy}{ds}.$$

Then ψ has compact support and we are reduced to an evaluation of

$$\int_{-a}^{\infty} e^{\pm iks^2/2}\psi(s)ds.$$

For this purpose, we pause to collect some facts about Gaussian integrals. We already touched upon this subject in the previous chapter.

21.3. Gaussian integrals

Everything in this section stems from the basic fact that

$$\frac{1}{\sqrt{(2\pi)}}\int_{-\infty}^{\infty} e^{-x^2/2}dx = 1.$$

This is proved by taking the square of the left-hand side:

$$\left[\frac{1}{\sqrt{(2\pi)}}\int_{-\infty}^{\infty} e^{-x^2/2}\,dx\right]^2 = \frac{1}{2\pi}\int_{-\infty}^{\infty}\int_{-\infty}^{\infty} e^{-(x^2+y^2)/2}\,dx\,dy$$
$$= \frac{1}{2\pi}\int_{0}^{2\pi}\int_{-\infty}^{\infty} e^{-r^2/2}r dr\,d\theta$$
$$= \int_{0}^{\infty} e^{-r^2/2}r dr = 1.$$

Now

$$\frac{1}{\sqrt{(2\pi)}}\int_{-\infty}^{\infty} e^{-x^2/2-\eta x}dx$$

converges for all complex values of η and uniformly in any compact region. Hence it defines an analytic function that may be evaluated by taking η to be real and then using analytic continuation. For real η, we simply complete the square and make a change of variables:

$$\frac{1}{\sqrt{(2\pi)}}\int \exp(-x^2/2 - x\eta)dx = \frac{1}{\sqrt{(2\pi)}}\int \exp[-(x+\eta)^2/2 + \eta^2/2]dx$$
$$= \exp(\eta^2/2)\frac{1}{\sqrt{(2\pi)}}\int \exp[-(x+\eta)^2/2]dx$$
$$= e^{\eta^2/2}.$$

This is true for all complex values of η. In particular, taking $\eta = \mathrm{i}\zeta$, we get

$$\frac{1}{\sqrt{(2\pi)}}\int \exp(-x^2/2 - \mathrm{i}\zeta x)\mathrm{d}x = \exp(-\zeta^2/2).$$

In other words, the Fourier transform of $\exp(-x^2/2)$ is $\exp(-\zeta^2/2)$. Now set $x = \sqrt{\lambda}u$, where $\sqrt{\lambda} > 0$ is some positive real number. Then $\mathrm{d}x = \sqrt{\lambda}\mathrm{d}u$ and setting $\xi = \sqrt{\lambda}\zeta$, we get

$$\frac{1}{\sqrt{(2\pi)}}\int \mathrm{e}^{-\lambda u^2/2}\mathrm{e}^{-\mathrm{i}\xi u}\mathrm{d}u = \frac{1}{\sqrt{(2\pi)}}\mathrm{e}^{-\xi^2/2}.$$

Now this last integral converges for any λ with $\mathrm{Re}\,\lambda > 0$ and $\lambda \neq 0$. Indeed, for $S > R > 0$,

$$\begin{aligned}\int_R^S \mathrm{e}^{-\lambda x^2/2}\mathrm{e}^{-\mathrm{i}\xi x}\mathrm{d}x &= -\int_R^S \frac{1}{\lambda x}\frac{\mathrm{d}}{\mathrm{d}x}(\mathrm{e}^{-\lambda x^2/2})\mathrm{e}^{-\mathrm{i}\xi x}\mathrm{d}x \\ &= \frac{\mathrm{e}^{-\lambda R^2/2}\mathrm{e}^{-\mathrm{i}\xi R}}{\lambda R} - \frac{\mathrm{e}^{-\lambda S^2/2}\mathrm{e}^{-\mathrm{i}\xi S}}{\lambda S} + \int_R^S \mathrm{e}^{-\lambda x^2/2}\frac{\mathrm{d}}{\mathrm{d}x}\left(\frac{\mathrm{e}^{\mathrm{i}\xi x}}{\lambda x}\right)\mathrm{d}x.\end{aligned}$$

In this last integral we can write

$$\mathrm{e}^{-\lambda x^2/2} = -\frac{1}{\lambda x}\frac{\mathrm{d}}{\mathrm{d}x}\mathrm{e}^{-\lambda x^2/2}$$

and integrate by parts once again. We thus see that

$$\left|\int_R^S \mathrm{e}^{-\lambda x^2/2}\mathrm{e}^{-\mathrm{i}\xi x}\mathrm{d}x\right| = \mathrm{O}\left(\frac{1}{|\lambda|R}\right).$$

We thus see that

$$\frac{1}{\sqrt{(2\pi)}}\mathrm{e}^{-\lambda x^2/2}\mathrm{e}^{-\mathrm{i}\xi x}\mathrm{d}x$$

converges uniformly (but not absolutely) for λ in any region of the form

$$\begin{gathered}\mathrm{Re}\,\lambda > 0, \\ |\lambda| > \delta \quad \text{any } \delta > 0.\end{gathered}$$

By analytic continuation, its value is given by the same formula as for λ real and positive, where the square root must now be taken to be the analytic continuation of the positive square root. In particular, if we set $\lambda = -\mathrm{i}r$, $r > 0$, then

$$\lambda^{1/2} = |\lambda|^{1/2}\mathrm{e}^{-\pi\mathrm{i}/4}$$

and

$$\frac{1}{\sqrt{(2\pi)}}\int \mathrm{e}^{\mathrm{i}rx^2/2}\mathrm{e}^{-\mathrm{i}\xi x}\mathrm{d}x = \mathrm{e}^{\pi\mathrm{i}/4}r^{-1/2}\mathrm{e}^{-\mathrm{i}\eta^2/2r}$$

and, similarly with $\lambda = \mathrm{i}r$,

$$\frac{1}{\sqrt{(2\pi)}}\int \mathrm{e}^{-\mathrm{i}rx^2/2}\mathrm{e}^{-\mathrm{i}\xi x}\,\mathrm{d}x = \mathrm{e}^{-\pi\mathrm{i}/4}r^{-1/2}\mathrm{e}^{\mathrm{i}\eta^2/2r}.$$

We can now complete the proof of the formula of stationary phase. We wish to find an asymptotic evaluation of an integral of the form

$$\int_{-\infty}^{\infty} e^{\pm iks^2/2}\psi(s)ds$$

where ψ is a function of compact support:

$$\psi(s) = \psi(0) + sx(s).$$

Multiply this equation by some function ρ which is identically one where $\psi \neq 0$ and vanishes for large values of $|s|$. Then

$$\psi(s) = \psi_0(s) + s\rho(s)x(s)$$

where

$$\psi_0(s) = \rho(s)\psi(0) \equiv \psi(0) \text{ near } s = 0.$$

Thus

$$\int \psi_0(s)e^{\pm iks^2/2}ds = \frac{2\pi}{k}e^{\pm i/4} + \text{error}$$

where the error term vanishes to infinite order in k. But

$$\int s[\rho(s)x(s)]e^{iks^2/2} = \frac{1}{\pm ik}\int [\rho(s)x(s)]\frac{d}{ds}e^{\pm iks^2/2}$$
$$= \frac{-1}{ik}\int \frac{d}{ds}[\rho(s)x(s)]\cdot e^{\pm iks^2/2}ds$$

and $(d/ds)[\rho(s)x(s)]$ vanishes for large $|s|$. We can thus repeat the process to yield the derived asymptotic series, completing the proof of the formula.

We can give an n-dimensional version of the Gaussian integral. Take

$$\begin{pmatrix} x_1 \\ \vdots \\ x_n \end{pmatrix}$$

to be a vector variable and let

$$Q = \begin{pmatrix} \pm r_1 & & 0 \\ & \ddots & \\ 0 & & \pm r_n \end{pmatrix}$$

be an $n \times n$ matrix with all the $r_i > 0$ and $\operatorname{sgn} Q$ = number of $+$s minus number of $-$s. Then $|r_1|\cdots|r_n| = |\operatorname{Det}(Q)|$, and, by multiplying the formulas for the one-dimensional integrals, we get

$$\frac{1}{(2\pi)^{n/2}}\int \exp\left(\frac{i}{2}Q\mathbf{x}\cdot\mathbf{x}\right)\exp(-i\xi\cdot\mathbf{x})d\mathbf{x} = |\operatorname{Det}(Q)|^{-1/2} - \exp\left(\frac{\pi i}{4}\operatorname{sgn} Q\right)$$
$$\times \exp\left(-\frac{i}{2}Q^{-1}\xi\cdot\xi\right),$$

where now ξ is also a vector and Q^{-1} is the inverse matrix. We have proved this for diagonal Q. But given any nonsingular symmetric matrix Q, we can find an orthogonal matrix O such that OQO^{-1} is diagonal. This proves that the above formula is true for all such Q.

We can use the n-dimensional version of the Gaussian integral to obtain an n-dimensional version of the stationary phase formula. Let p be a function with a finite number of critical points. We assume that each of the critical points of p is non-degenerate. Thus the Hessian, the matrix

$$H(y_j) = \left(\frac{\partial^2 p}{\partial y_k \partial y_l}\right)(y_j),$$

is non-degenerate at each critical point, y_j. Let $\sigma = \sigma_j$ denote the signature of the jth critical point. Then

$$\int a(y)e^{ikp(y)}\,dy \sim \left(\frac{2\pi}{k}\right)^{n/2} \sum \frac{e^{\pi i\sigma/4}}{\sqrt{|\mathrm{Det}(H(y_j))|}}\left(a(y_j) + \frac{a_1(y_j)}{k} + \cdots\right)$$

where the sum extends over all critical points. Here the terms $a_1(y_j), a_2(y_j)$, etc., can be expressed in terms of a, p and their derivatives of various orders at y_j.

21.4. Group velocity

As an application of the method of stationary phase in one variable, let us consider the following situation. Suppose we have a family of traveling waves

$$e^{-(1/h)(E(p)t - px)}$$

where h is a small number so that $1/h$ plays the role of our large parameter, k. The wave number of the *space variation* is p/h at each fixed time, t. Since we allow E to depend on p, each of these waves is traveling with a different velocity. Now suppose we superimpose such a family of traveling waves so that we consider an integral of the form

$$\int a(p)e^{-(i/h)(E(p)t - px)}dp.$$

Let us further assume that the function $a(p)$ is concentrated about some fixed value p_0. In other words, we assume that a vanishes except for values of p close to p_0. Now the method of stationary phase says that the only non-negligible contributions to the integral come from values of p for which the derivative of the exponential term with respect to p vanishes, i.e., for which

$$E'(p)t - x = 0.$$

Since $a(p)$ vanishes unless p is close to p_0, this equation is really a constraint on x and t: it says that the integral is essentially zero except for those values of x and t such that

$$x = E'(p_0)t$$

holds approximately. In other words, the integral looks like a little blip, when thought of as a function of x, and this blip moves in time with velocity $E'(p)$. (This blip is called a *wave packet* and the velocity $E'(p)$ is called the *group velocity*.)

Let us examine what kind of function E can be of p if we demand that the expression $E \cdot t - p \cdot x$ be invariant under Lorentz transformations. Under a Lorentz transformation, we can move (t, x) into any other point (t', x') so long as

$$c^2t^2 - x^2 = c^2t'^2 - x'^2.$$

Thus (E, p) can be transformed into any other (E', p') with

$$E^2 - c^2p^2 = E'^2 - c^2p'^2.$$

Thus the only invariant relation between E and p is of the form

$$E^2 - (pc)^2 = \text{constant}.$$

Let us call this constant m^2c^4 so that

$$E^2 - (pc)^2 = m^2c^4$$

or

$$E(p) = ((pc)^2 + m^2c^4)^{1/2}.$$

Then

$$E'(p) = \frac{pc^2}{E(p)} = \frac{p}{M}$$

where

$$E(p) = Mc^2 \quad \text{or} \quad M = (m^2 + (pc)^2)^{1/2}$$

(so that, if p/c is small in comparison to m, then $M \doteq m$). If we think of M as a *mass*, then the relationship between the *group velocity* $E'(p)$ and p is precisely the relationship between velocity and momentum. In this way we have associated a wave number $k = p/h$ to the momentum p and the associated wavelength is

$$\lambda = \frac{1}{k} = \frac{h}{p} \quad \text{(de Broglie's formula).}$$

If we think of E as energy, then the relation between the energy and the frequency, ν, of the time variation is given by

$$E = h\nu \quad \text{(Einstein's formula).}$$

In these formulas we have been thinking of h as a small parameter which we have been letting tend to zero. The great discovery of quantum physics is that h should not tend to zero, but rather is a fundamental constant of nature (known as Planck's constant). In Einstein's formula it occurs as a conversion factor from inverse time to energy, and hence has units energy × time. It is given by

$$h = 6.626 \times 10^{-34}\ \text{J s}.$$

21.5. The Fourier inversion formula

Let us give an important application of the stationary phase formula. Let us take $n = 2$ with coordinates x and ξ. Consider the function

$$p(x, \xi) = p_\eta(x, \xi) = x(\xi - \eta)$$

where η is a fixed number. This function has only one critical point, at

$$x = 0, \quad \xi = \eta.$$

This critical point is non-degenerate with signature $\sigma = 0$.

Thus

$$\frac{1}{2\pi}\int e^{ikx(\xi-\eta)}a(x,\xi)\mathrm{d}x\,\mathrm{d}\xi = \frac{1}{k}\left(a(0,\eta) + \frac{a_1}{k} + \cdots\right).$$

This formula was proved for $a(x, \xi)$ of compact support. But an expansion of the proof shows that it works as well for functions which vanish at infinity, together with their derivatives at any order faster than any inverse power of k. In particular, we shall take

$$a(x,\xi) = f(x)g(\xi)$$

where f and g are smooth functions of one variable, which vanish rapidly at infinity with all their derivatives. So we have

$$\frac{1}{2\pi}\iint f(x)g(\xi)e^{-ikx(\xi-\eta)}\mathrm{d}x\,\mathrm{d}\xi = \frac{1}{k}f(0)g(\eta) + O\left(\frac{1}{k^2}\right).$$

Let us make the change of variable $u = kx$ in the integral. So

$$\frac{1}{k}f(0)g(\eta) + O\left(\frac{1}{k^2}\right) = \frac{1}{2\pi k}\int f\left(\frac{u}{k}\right)g(\xi)e^{-iu(\xi-\eta)}\mathrm{d}u\,\mathrm{d}\xi$$

$$= \frac{1}{k}\frac{1}{\sqrt{(2\pi)}}\int f\left(\frac{u}{k}\right)e^{iu\eta}\left(\frac{1}{\sqrt{(2\pi)}}\int g(\xi)e^{-iu\xi}\mathrm{d}\xi\right)\mathrm{d}u.$$

Define $\hat{g}$, the *Fourier transform* of g, by the formula

$$\hat{g}(u) = \frac{1}{\sqrt{(2\pi)}}\int g(\xi)e^{-i\xi u}\mathrm{d}\xi.$$

Thus, substituting into the preceding formula, we have

$$\frac{1}{k}\frac{1}{\sqrt{(2\pi)}}\int f\left(\frac{u}{k}\right)\hat{g}(u)e^{iu\eta}\,\mathrm{d}u = \frac{1}{k}f(0)g(\eta) + O\left(\frac{1}{k^2}\right).$$

Suppose we choose f with $f(0) = 1$, so

$$f\left(\frac{u}{k}\right) = 1 + O\left(\frac{1}{k}\right)$$

and

$$\frac{1}{\sqrt{(2\pi)}}\int f\left(\frac{u}{k}\right)g(u)e^{iu\eta}\mathrm{d}u = \frac{1}{\sqrt{(2\pi)}}\int g(u)e^{iu\eta}\,\mathrm{d}u + O\left(\frac{1}{k}\right).$$

Letting $k \to \infty$ we see that

$$g(\eta) = \frac{1}{\sqrt{(2\pi)}}\int \hat{g}(u)e^{iu\eta}\mathrm{d}u.$$

This formula is known as the *Fourier inversion formula*. It shows how to recover a function from its Fourier transform.

Of course the same proof works in n dimensions to prove that

$$g(\eta) = \frac{1}{(2\pi)^{n/2}} \int \hat{g}(u) e^{iu\eta} \, du.$$

21.6. Asymptotic evaluation of Helmholtz's formula

Let u be a solution of the reduced wave equation $(\Delta + k^2)u = 0$ in three-space. Recall from section 19.6 that, if u satisfies the reduced wave equation, then outside some closed surface S, we have Helmholtz's formula

$$u(P) = \frac{1}{4\pi} \int_S \left(\frac{e^{ikr}}{r} \star du - u \star d\left(\frac{e^{ikr}}{r} \right) \right), \tag{19.2}$$

for P outside S, while the integral vanishes for P inside S. As we indicated there, this was Helmholtz explanation of Huyghens' principle – and the absence of backward waves. However, this vanishing inside S depends on contributions from the whole surface. Fresnel believed that if all the waves were inside S, the integral from

grad $\phi(y)$ = $-$grad $r(y)$

$\star d\phi(y) = - \star dr(y)$

$P = y + r \operatorname{grad} \phi(y)$

(a)

case (a)

grad $\phi(y)$ = 2(grad $\phi(y)$, n)n $-$ grad $r(y)$

$\star d\phi(y) = \star dr(y)$

$P = y - r(2(\operatorname{grad} \phi(y), n)n - \operatorname{grad} \phi(y))$

(b)

case (b)

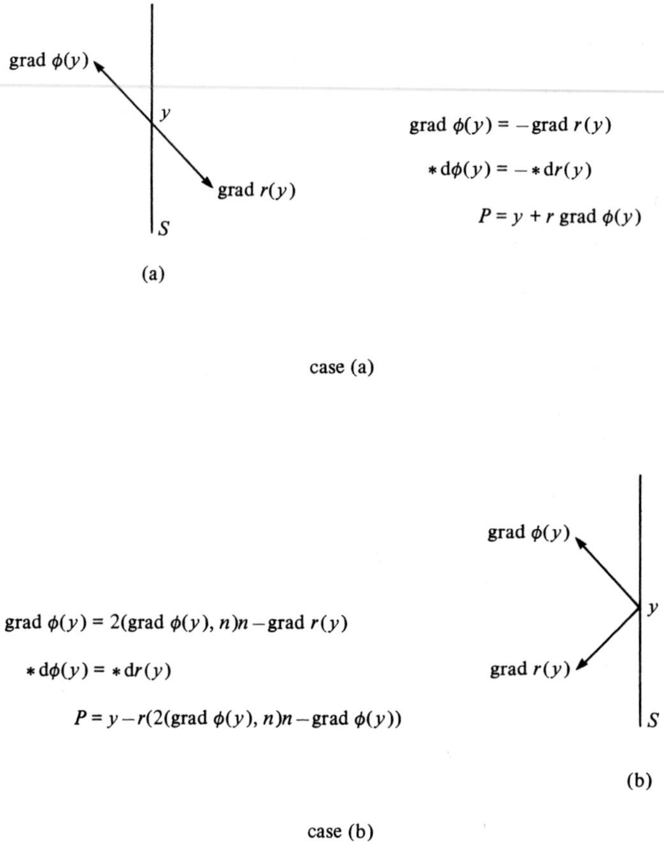

Figure 21.6

each small surface element would produce a null effect at interior points due to interference. We shall now show, using stationary phase, that Fresnel was right, up to terms of order $1/k$. In applying stationary phase, the function u, which occurs on the right-hand side of the formula will itself be oscillatory, and we must make some assumptions about its form before we can proceed. We shall assume that, near $S, u = a\mathrm{e}^{\mathrm{i}k\phi}$ where a and ϕ are smooth, and that $\|\operatorname{grad}\phi\| = 1$. This would be the case, for example, if u represented radiation from a single point, Q, lying inside S, where $\phi(y) = y - Q$.

We shall assume that we are sufficiently far from S so that $1/r^2$ is negligible in comparison with k, and that a and $\mathrm{d}a$ are also negligible in comparison with k, Substituting into (19.2) we find that the top-order term (relative to powers of k) is

$$\frac{\mathrm{i}k}{4\pi}\iint_S (a/r)\mathrm{e}^{\mathrm{i}k(\phi+r)}(\star\mathrm{d}\phi - \star\mathrm{d}r).$$

Now the points of stationary phase are those points y, on S where $\operatorname{grad}\phi(y) + \operatorname{grad} r(y)$ is normal to S. There are two possible situations.

Let us suppose for the moment that y is a non-degenerate critical point of type (b). The top-order term in the stationary phase formula will vanish, and the total contribution coming from y in Helmholtz formula will be of order $1/k$. (Notice that, if S were convex and grad ϕ pointed outward, then, for any P inside S, all the critical points would be of type (b). This, in a sense, justifies Fresnel's view that there is local cancellation of the backward wave.) For non-degenerate critical points of type (a) since $\star\mathrm{d}\phi(y) = -\star\mathrm{d}r(y)$, we may, in computing the highest-order contribution to the stationary phase formula, replace the above integral by

$$\frac{\mathrm{i}k}{2\pi}\iint (a/r)\mathrm{e}^{\mathrm{i}k(\phi+y)}\mathrm{d}\star r.$$

This shows, that (up to order $1/k$) the induced *secondary radiation* along S behaves as if it

(i) has an amplitude equal to $1/\lambda$ times the amplitude of the primary wave where $\lambda = 2\pi/k$ is the wavelength, and

(ii) has phase one-quarter of a period ahead of the primary wave. (This is a way of interpreting the factor i.)

Fresnel made these two assumptions directly in his formulation of Huyghens' principle, and this led many to regard his theory as being *ad hoc*. As we have seen, they are a consequence of the method of stationary phase and Helmholtz's formula.

We still must discuss the question of when the critical points are non-degenerate. We shall treat points of type (a); the points of type (b) can be treated in an identical manner. Actually, the discussion is almost the same as in our treatment of emitted radiation. Let us define the *exponential map* $E: S \times \mathbb{R}^+ \to \mathbb{R}^3$ by

$$E(y, r) = y + r \operatorname{grad}\phi(y).$$

Then the critical points on S associated with a point P consist precisely of those y such that $E(y,r)=P$, where $r=\|y-P\|$. If grad $\phi(y)$ is not tangent to S, then E is a diffeomorphism near $(y,0)$.

It is not hard to show that y is a degenerate critical point for $P=E(y,r)$ if and only if (y,r) is a point at which the map E is singular. In this case we call P a focal point of the map E at y. If P is not a focal point, then the index of the Hessian of $\phi+r$ at y is the number of focal points on the ray segment from y to P (counted with multiplicity). We leave the details to the reader.

Summary

A Asymptotic expansion of integrals

Given a function defined by an integral of the form

$$\int \mathrm{e}^{-xp(\omega)}\,\mathrm{d}\omega,$$

you should be able to identify what range of values of ω makes the major contribution to the integral when x is large and to develop the first two terms of the asymptotic expansion of the integral.

B Stationary phase

Given an integral of the form

$$\int a(y)\mathrm{e}^{ikp(y)}\,\mathrm{d}y,$$

you should be able to develop and apply a formula for the first two terms in the asymptotic expansion for large k.

You should be able to develop the Fourier inversion formula by the method of stationary phase.

Exercises

21.1. The *modified Bessel function of the third kind* may be defined by the integral

$$k_0(x)=\frac{1}{2}\int_{-\infty}^{\infty}\mathrm{e}^{-x\cosh t}\,\mathrm{d}t.$$

Determine the first two terms in the asymptotic expansion of $k_0(x)$.

21.2. The Bessel function $J_0(x)$ may be defined by the integral

$$J_0(x)=\frac{1}{2\pi}\int_0^{2\pi}\mathrm{e}^{\mathrm{i}x\sin t}\,\mathrm{d}t.$$

Use stationary phase to find the first two terms in the asymptotic expansion of $J_0(x)$.

21.3. The beta function $B(x+1, x+1)$ may be represented by the integral

$$I = \int_0^1 [t(1-t)]^x \, dt$$
$$= \int_0^1 e^{x[\log t + \log(1-t)]} \, dt.$$

(a) For very large x, where does the principal contribution to the integral come from?

(b) Find the leading term in the asymptotic expansion of $B(x+1, x+1)$.

(c) Describe carefully, and as explicitly as you can, how you would calculate more terms in the asymptotic expansion. Calculate one more term.

21.4. (a) Describe the general strategy for obtaining an asymptotic expansion of an integral of the form

$$I = \int_{\theta_1}^{\theta_2} e^{-ixp(\theta)} q(\theta) \, d\theta.$$

Assume that $p(\theta)$ has only one critical point on $[\theta_1, \theta_2]$, at $\theta = \alpha$.

(b) Apply this strategy to obtain the leading term in the asymptotic expansion of the Bessel function

$$J_n(x) = \frac{1}{\pi}\left(\operatorname{Re} \int_0^\pi e^{ix \sin\theta} e^{-in\theta} \, d\theta \right)$$

where n is an integer.

22

Thermodynamics

Chapter 22 shows how the exterior calculus can be used in classical thermodynamyics, following the ideas of Born and Carathéodory.

The subject matter of this chapter is equilibrium thermodynamics. This subject has several branches. The first of these – 'classical thermodynamics' – deals with the notions of heat and work and gives rise to the concepts of entropy and temperature and the celebrated 'second law'. It deals with general principles but does not attempt to describe the behavior of substances or compute measurable quantities in terms of a microscopic model. A second branch – 'equilibrium statistical mechanics' – whether in its classical or quantum versions, provides very explicit formulas for all 'macroscopically observable quantities' of equilibrium states in terms of a model of the atomic or molecular interactions. These formulas are of a general nature. That is, they all have a common underlying form, and are associated with the names of Maxwell, Boltzmann and especially Gibbs. A third branch deals with explaining why these formulas work, and studies this problem from the point of view of probability theory or dynamics or both, and also deals with the question of approach to equilibrium. We shall have very little to say about this third branch. The first few sections will deal with classical thermodynamics. However, we will begin with a purely mathematical theorem due to Carathéodory which is geometrically quite plausible – although we will postpone some of the details of the proof to the appendix to this chapter. This theorem serves as the mathematical tool which implies the existence of the entropy function according to the approach to thermodynamics given by Born and Carathéodory. We then derive some physical consequences of the theory and devote the remainder of the chapter to 'equilibrium statistical mechanics'.

22.1. Carathéodory's theorem

Let $\alpha = A_1 \mathrm{d}x^1 + \cdots + A_n \mathrm{d}x^n$ be a linear differential form in $\mathbb{R}^n$. Here the A_i are all functions on $\mathbb{R}^n$. If we look at some fixed point $\mathbf{P}$ then the 'value of α at $\mathbf{P}$' can be thought of as the row vector $(A_1(\mathbf{P}), \ldots, A_n(\mathbf{P}))$. If this row vector is not zero, its null space is the $(n-1)$-dimensional space of all $\mathbf{X} = \begin{pmatrix} X^1 \\ \vdots \\ X^n \end{pmatrix}$ such that

$$A_1(\mathbf{P})X^1 + \cdots + A_n(\mathbf{P})X^n = 0.$$

Let $\gamma = x(t)$ be a piecewise differentiable curve. Then

$$\int_\gamma \alpha = \int [A_1(x(t))\dot{x}^1(t) + \cdots + A_n(x(t))\dot{x}^n(t)]\mathrm{d}t.$$

In particular, if for every t the tangent vector $\dot{x}(t) = \begin{pmatrix} \dot{x}^1(t) \\ \vdots \\ \dot{x}^n(t) \end{pmatrix}$ lies in the null space of $(A_1(x(t)), \ldots, A_n(x(t)))$, then the integral $\int_\gamma \alpha = 0$. A curve γ is called a *null curve* of α if γ is continuous and piecewise differentiable and if, at every t for which $\dot{x}(t)$ is defined, $\dot{x}(t)$ lies in the null space of $(A_1(x(t)), \ldots, A_n(x(t)))$. We want to consider the following geometrical problem.

Suppose we start at some point $\mathbf{P}$. What are the points $\mathbf{Q}$ which we can join to $\mathbf{P}$ by null curves? For example, suppose that

$$\alpha = \mathrm{d}f$$

for some function f. Then if γ is a null curve joining $\mathbf{P}$ to $\mathbf{Q}$, we have

$$0 = \int_\gamma \alpha = f(\mathbf{Q}) - f(\mathbf{P}).$$

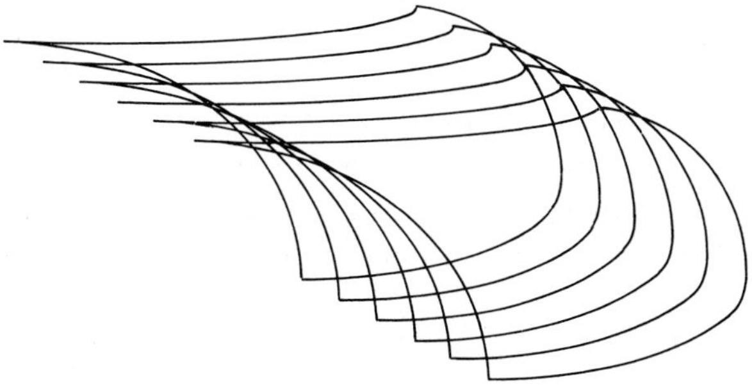

f = const

Figure 22.1

Thus we must have $f(\mathbf{Q})=f(\mathbf{P})$. If $\mathrm{d}f\neq 0$, the set $f(\mathbf{P})=\text{const}$ is an $(n-1)$-dimensional surface passing through $\mathbf{P}$, and the condition on $\mathbf{Q}$ is that it must lie on this surface.

In particular, there will be points arbitrarily close to $\mathbf{P}$ which can *not* be joined to $\mathbf{P}$ by null curves.

The same is true if $\alpha = g\mathrm{d}f$ where g is some non-zero function and $\mathrm{d}f\neq 0$. This is because γ is a null curve for α if and only if it is a null curve for $g^{-1}\alpha=\mathrm{d}f$. The conditions to be a null curve are the same. So again $\mathbf{Q}$ must lie on the $(n-1)$-dimensional surface $f=f(\mathbf{P})$ if we are to be able to join $\mathbf{Q}$ to $\mathbf{P}$ by a null curve of α. Notice that if $\alpha = g\mathrm{d}f$ then $\mathrm{d}\alpha=\mathrm{d}g\wedge\mathrm{d}f$ so

$$\alpha\wedge\mathrm{d}\alpha = g\mathrm{d}f\wedge\mathrm{d}g\wedge\mathrm{d}f=0.$$

On the other hand, consider the form

$$\alpha=\mathrm{d}z+x\mathrm{d}y$$

defined on $\mathbb{R}^3$. Here we have $\mathrm{d}\alpha=\mathrm{d}x\wedge\mathrm{d}y$ so

$$\alpha\wedge\mathrm{d}\alpha=\mathrm{d}x\wedge\mathrm{d}y\wedge\mathrm{d}z$$

is nowhere zero. We wish to show that one can get from any point to any other point by a null curve of $\mathrm{d}z+x\mathrm{d}y$. It is enough to show that we can get from the origin, $\mathbf{0}$, to any other point $\mathbf{Q}$. Let us write

$$\mathbf{Q}=\begin{pmatrix}a\\b\\c\end{pmatrix}.$$

Let us first assume that $b\neq 0$. Now the x-axis is a null curve for α since both $\mathrm{d}z$ and $\mathrm{d}y$ vanish along the line $z=y=0$, More generally any line parallel to the x-axis is a null curve for α for the same reason. So let us first move along the x-axis from $\mathbf{0}$ to the point whose x coordinate is $-c/b$ during time $0\leqslant t\leqslant 1$. Then go from

$$\begin{pmatrix}-c/b\\0\\0\end{pmatrix}\text{ to }\begin{pmatrix}-c/b\\b\\c\end{pmatrix}\text{ along }\mathbf{x}(t)=\begin{pmatrix}-c/b\\0\\0\end{pmatrix}+\begin{pmatrix}0\\(t-1)b\\(t-1)c\end{pmatrix},\quad 1\leqslant t\leqslant 2.$$

Since

$$\dot{\mathbf{x}}^1(t)=\begin{pmatrix}0\\b\\c\end{pmatrix}\text{ and }(A_1(\mathbf{x}(t)),A_2(\mathbf{x}(t)),A_3(\mathbf{x}(t)))=\left(0,-\frac{c}{b},1\right)$$

we see that this is also a null curve. Finally, go from

$$\begin{pmatrix}-c/b\\b\\c\end{pmatrix}\text{ to }\begin{pmatrix}a\\b\\c\end{pmatrix}\text{ along }\mathbf{x}(t)=\begin{pmatrix}-c/b\\b\\c\end{pmatrix}+\begin{pmatrix}(t-2)(a-c/b)\\0\\0\end{pmatrix},\quad 2\leqslant t\leqslant 3.$$

Since this is a line parallel to the x-axis it is also a null curve. We have thus gotten from $\mathbf{0}$ to $\mathbf{Q}$ along a continuous, broken path having three pieces, each a straight line segment which is a differentiable null curve.

If $b=0$ and $c\neq 0$ we make a slight detour. First go to the point

$$\begin{pmatrix}1\\c\\0\end{pmatrix}$$

along a null curve. Since $c\neq 0$ we can get to this point by the above procedure. Then from

$$\begin{pmatrix}1\\c\\0\end{pmatrix}\text{ to }\begin{pmatrix}1\\0\\c\end{pmatrix}\text{ along }\mathbf{x}(t)=\begin{pmatrix}1\\c\\0\end{pmatrix}+\begin{pmatrix}0\\-(t-3)c\\(t-3)c\end{pmatrix},\qquad 3\leqslant t\leqslant 4.$$

Since along this curve $\mathrm{d}z=c\mathrm{d}t$ and $\mathrm{d}y=-c\mathrm{d}t$ and $x\equiv 1$, we see that $\mathrm{d}z+x\mathrm{d}y$ vanishes identically along this curve and so it is a null curve. We can then go to the point $\mathbf{Q}$ along a line parallel to the x-axis. That is, keep the y and z coordinates fixed and move the x coordinate from 1 to a.

Of course, if $b=c=0$ we can go straight from $\mathbf{0}$ to $\mathbf{Q}$ along the x-axis. We have thus proved that any two points of $\mathbb{R}^n$ can be joined by a null curve of $\mathrm{d}z+x\mathrm{d}y$.

We can thus consider three types of one-forms on $\mathbb{R}^3$:

Type	*Example*
(i) α with $\alpha\neq 0$ but $\mathrm{d}\alpha\equiv 0$	$\alpha=\mathrm{d}z$
(ii) α with $\mathrm{d}\alpha\neq 0$ but $\alpha\wedge\mathrm{d}\alpha\equiv 0$	$\alpha=x\mathrm{d}y$
(iii) α with $\alpha\wedge\mathrm{d}\alpha\neq 0$	$\alpha=\mathrm{d}z+x\mathrm{d}y.$

In case (i) we know that if $\mathrm{d}\alpha\equiv 0$, then, locally, we can find a function f such that $\alpha=\mathrm{d}f$. (We can do so in any star-shaped region contained in the domain of definition of α.) If α does not vanish at some point $\mathbf{P}$ then the implicit function theorem implies that we can make a change of variables so the function f becomes one of the coordinates, say the coordinate z. In other words, up to a change of variables, in the first row, the example represents the general case. The mathematical theorem that we wish to quote says that the same holds in the remaining two cases. We will give a statement and indicate the proof of this theorem later on in this section, but defer the details to an appendix. Our statement and proof will be valid in n dimensions, not just three. Now we have seen that in case (iii) we can join any point to any nearby point by a null curve of α. Hence, if the form α has the property that for any point $\mathbf{P}$ there are points $\mathbf{Q}$ arbitrarily close which can not be joined to $\mathbf{P}$ by a null curve of α then we must be in case (i) or (ii). This assertion, or rather its n-dimensional generalization, is the content of Carathéodory's theorem.

Carathéodory's theorem. Let α be a linear differential form with the property that for any point $\mathbf{P}$ *there are points* $\mathbf{Q}$ *arbitrarily close to* $\mathbf{P}$ *which can not be joined to* $\mathbf{P}$ *by a null curve of α. Then (locally) there exist functions f and g such that*

$$\alpha=f\,\mathrm{d}g.$$

The expression for α is not quite unique. Indeed, if H is any differentiable function of one variable with nowhere vanishing derivative, and

$$G=H\circ g$$

then $dG = H'(g)dg$ by the chain rule, so we can also write α as

$$\alpha = FdG \quad \text{where} \quad G = H \circ g \quad \text{and} \quad F = f/H'(g).$$

Before continuing our purely mathematical discussion of Carathéodory's theorem, let us sketch how it is actually used in thermodynamics. The detailed discussion with the precise definitions and axioms will be given in the next section.

The quantity of heat given off by a chemical reaction *performed in a specified way* such as at constant *volume* or at constant *pressure* is readily measurable. (It can, for example be measured in a clever device known as an ice calorimeter which makes use of the fact that as ice melts it contracts, and it takes a specified amount of heat to melt a certain amount of ice under standard conditions.) The amount of heat needed to effect some small change (such as to raise the temperature of one gram of water by one degree on a standard thermometer, or to expand its volume) when performed in some specified way is (approximately) linearly related to these small changes. So the 'quantity of heat added' to a system in equilibrium is a linear differential form, α. For a long period of time it was thought that this form was exact. That is, it was thought that there was a function, C (called the caloric), representing the 'total amount of heat in the system' such that $\alpha = dC$. In other words, it was thought that the 'caloric' in the system was changed by the quantity of heat added. It was only gradually realized that the form α is not closed and therefore cannot represent the infinitesimal change in any function.

On the other hand it was realized quite early that the work done on a system is given by a linear differential form ω which is not closed. Indeed, if we compress a piston at high temperature, where the pressure is high, we do more work than compressing it at a low temperature where the pressure is low. Therefore, we can compress a piston at high temperature, then cool it off in its compressed state (doing no work since the piston is stationary), expand it when cool and then heat it up in its expanded condition. We will have gone around a 'cycle'. That is, we will have traversed a closed curve around which the integral of ω is not zero. (Of course to check whether or not a form is closed, an effective way is to integrate it around closed curves. This is why the 'cycles' play such an important role in the development of the ideas of thermodynamics.)

The *first law of thermodynamics* asserts that although neither α nor ω are closed, the sum is closed – that $d(\alpha + \omega) = 0$. In other words, the first law states that (locally)

$$\alpha + \omega = dU$$

where U is a well defined function on the system (determined up to an additive constant) known as the *internal energy*. The existence of U is a version of the physical principle of conservation of energy.

The second law of thermodynamics derives from the behavior of a system when no heat exchange is allowed. If we put a system in an enclosure (known as an adiabatic enclosure) where no heat is exchanged, this means that we are constraining the system to change according to paths which are null curves of α. It is an observed fact of nature that then near every state of the system there will be nearby inaccessible states. By Carathéodory's theorem, this implies that there are functions f and g such that

$$\alpha = f\mathrm{d}g.$$

Of course this does not determine f or g completely, as we mentioned in our discussion above. However an analysis of exchange between systems in thermal contact shows that we can choose f to be a universal function, T, of all systems known as the *absolute temperature* which is determined up to a multiplicative constant (a scale factor). Having fixed the scale of the absolute temperature, this means that the function g is determined up to an additive constant for each fixed system. It is usually denoted by S and is called the *entropy* of the system. Thus the *second law of thermodynamics* asserts that there exists a universal temperature scale T (called the absolute temperature and determined up to a multiplicative constant) and a function S on each system (called the entropy and determined up to an additive constant once the temperature scale is fixed) such that

$$\alpha = T\mathrm{d}S.$$

The temperature function can be chosen to be always positive. With this convention, the change of entropy has the same sign as the amount of heat added along any tangent vector (i.e. infinitesimal change) to the system. Since the net change of S around any closed path must be zero, this means that around any cycle, if heat is added in one portion, heat must be extracted in some other portion. This is Kelvin's famous formulation of the second law – that 'no cycle can exist whose net effect is a total conversion of heat into work', work can only be produced by a heat engine by dumping a certain amount of the heat into a cold reservoir.

We will come back to give a precise formulation of all of these laws in the next section. For the moment let us return to a purely mathematical discussion surrounding Carathéodory's theorem. Suppose we consider the form

$$\alpha = \mathrm{d}z + x\mathrm{d}y$$

in a six-dimensional space, $\mathbb{R}^6$, with coordinates x, y, z, u, v, w. Obviously any curve along which y and z are constant is a null curve for α. So we can move from $\mathbf{0}$ to any point $\mathbf{Q}$ in $\mathbb{R}^6$ along a null curve of α by first moving in the three-dimensional space $u = v = w = 0$ to the point whose first three coordinates are the same as those of $\mathbf{Q}$ and then along the straight line from there to $\mathbf{Q}$. (Along this line x, y and z are constant.) Now consider the form

$$\alpha = \mathrm{d}z + x\mathrm{d}y + u\mathrm{d}v.$$

Once again, we can move from the origin to the point whose first three coordinates are the same as those of $\mathbf{Q}$ in the $u = v = w = 0$ space as before. Then, holding x and y constant, we can adjust the u and v coordinates, because along any curve with $y = \text{const}$ the form α takes the same value as the form $\mathrm{d}z + u\mathrm{d}v$, so we are back in the three-dimensional situation. Then we can adjust w while keeping the other five variables constant. Next consider the form

$$\alpha = w\mathrm{d}z + x\mathrm{d}y + u\mathrm{d}v.$$

Suppose the point **Q** has its w coordinate $\neq 0$, say

$$w(\mathbf{Q}) = r \neq 0.$$

We can first move from **0** along the line $x = y = z = u = v = 0$ to the point whose w coordinate is r. We will then move entirely in the hyperplane $w = r$. Now on any curve in the hyperplane $w = r$, the form α restricts to the same values as the form

$$\alpha' = r\mathrm{d}z' + x\mathrm{d}y + u\mathrm{d}v.$$

If we make a change of coordinates replacing z by $z' = rz$ then

$$\alpha' = \mathrm{d}z' + x\mathrm{d}y + u\mathrm{d}v.$$

We can now adjust the remaining five variables using the three-dimensional case as before. Similarly, if either the x coordinate or the u coordinate of **Q** does not vanish the same argument goes through: simply interchange the role of the various coordinates.

On the other hand, if **Q** lies in the three-dimensional subspace given by the equations $x = u = w = 0$, then any curve in this subspace is a null curve, so once again we can join the origin to **Q**. Thus in $\mathbb{R}^6$ we can consider six types of one-forms:

Type	*Example*
(i) $\alpha \neq 0$ but $\mathrm{d}\alpha \equiv 0$	$\mathrm{d}z$
(ii) $\mathrm{d}\alpha \neq 0$, but $\alpha \wedge \mathrm{d}\alpha \equiv 0$	$x\mathrm{d}y$
(iii) $\alpha \wedge \mathrm{d}\alpha \neq 0$ but $\mathrm{d}\alpha \wedge \mathrm{d}\alpha \equiv 0$	$\mathrm{d}z + x\mathrm{d}y$
(iv) $\mathrm{d}\alpha \wedge \mathrm{d}\alpha \neq 0$ but $\alpha \wedge \mathrm{d}\alpha \wedge \mathrm{d}\alpha \equiv 0$	$x\mathrm{d}y + u\mathrm{d}v$
(v) $\alpha \wedge \mathrm{d}\alpha \wedge \mathrm{d}\alpha \neq 0$ but $\mathrm{d}\alpha \wedge \mathrm{d}\alpha \wedge \mathrm{d}\alpha \equiv 0$	$\mathrm{d}z + x\mathrm{d}y + u\mathrm{d}v$
(vi) $\mathrm{d}\alpha \wedge \mathrm{d}\alpha \wedge \mathrm{d}\alpha \neq 0$	$w\mathrm{d}z + x\mathrm{d}y + u\mathrm{d}v$.

Our arguments show that for examples (iii)–(vi) we can move from any point to any other by a null curve. It is a mathematical theorem that for each of the cases on the left, we can make a local change of coordinates so that in the new coordinates α has the form given on the right. Granted this fact (whose proof we shall sketch in the appendix) we can prove Carathéodory's theorem (in six dimensions) as follows. We begin by showing that we must have

$$\mathrm{d}\alpha \wedge \mathrm{d}\alpha \wedge \mathrm{d}\alpha \equiv 0.$$

Indeed, suppose the contrary – that α is such that $\mathrm{d}\alpha \wedge \mathrm{d}\alpha \wedge \mathrm{d}\alpha$ is not identically zero. Then at some point **P** (and therefore in an entire neighborhood of **P**) we are in case (vi). Hence we can make a change of coordinates near **P** so that α has the form indicated in the right hand column. But then we can join **P** to any nearby point by a null curve of α. However our hypothesis is that *every* point has nearby points which are unreachable by null curves. Contradiction. So we must have $\mathrm{d}\alpha \wedge \mathrm{d}\alpha \wedge \mathrm{d}\alpha \equiv 0$. But now the same argument shows that we must have

$$\alpha \wedge \mathrm{d}\alpha \wedge \mathrm{d}\alpha \equiv 0$$

for otherwise we will be in case (v) near some point etc. So we can work our way down to prove that $\alpha \wedge \mathrm{d}\alpha \equiv 0$ and thus conclude Carathéodory's theorem. A similar argument works in n dimensions.

22.2. Classical thermodynamics according to Born and Carathéodory

In this section we present the basics of thermodynamics from the point of view of Born and Carathéodory. This approach has the advantage that it avoids the intricate arguments involving Carnot cycles and the non-existence of perpetual motion machines. Rather it is based on ideas abstracted from everyday experience and some simply stated physical laws. In this approach 'heat' is a derived concept. The theory is formulated in terms of standard concepts of elementary mechanics. But although the concepts of mechanics enter into the theory, the laws of elementary mechanics must be modified. It is the modification of these laws that involves such notions as temperature and heat. Then two simply stated physical laws together with an application of Carathéodory's theorem lead to the concepts of absolute temperature and entropy.

The first basic assumption, common to almost all physical theories, is that we can isolate a portion of the universe that we are interested in studying from the rest of the universe. This portion of the universe is called a *system.* The system can exist in various *states.* The description of all possible states of the system might be enormously complicated. For instance, if our system consisted of a gas in some enclosure, the gas might be in turbulent motion in which case we would have to specify the local velocity of each small portion of the gas as part of the description of its state. Thus we would expect that the collection of *all* states might be infinite-dimensional in an appropriate sense. In any event, we shall deal with 'system' and 'state' as undefined terms in our theory. Later on, when we develop a 'model' for thermodynamics, such as (classical) statistical mechanics, the undefined terms of thermodynamics will be given mathematical definitions in terms of the model. Another term which is intuitively clear but which will be left undefined is *interaction.* We assume that we can specify the various interactions of our system with the rest of the universe or with other systems.

The next basic assumption is that among all the possible states of the system there is a distinguished class of states called *equilibrium* states. If all interactions of the system with the rest of the universe are held constant then the system will pass through various states but tend to a definite equilibrium state (determined by the initial state of the system at the moment the interactions are held constant). Although the general states of the system are very complicated, the class of equilibrium states are relatively simple to describe and can be parameterized by a finite number of variables. Thus the set of equilibrium states has the structure of a submanifold* of a finite-dimensional vector space. This 'manifold of equilibrium states' is a subset of the set of all states of the system.

The next assumptions single out certain types of interactions: There exists a special form of interaction between two systems called *diathermal contact.* In this

* See the discussion of submanifold in section 10.9.

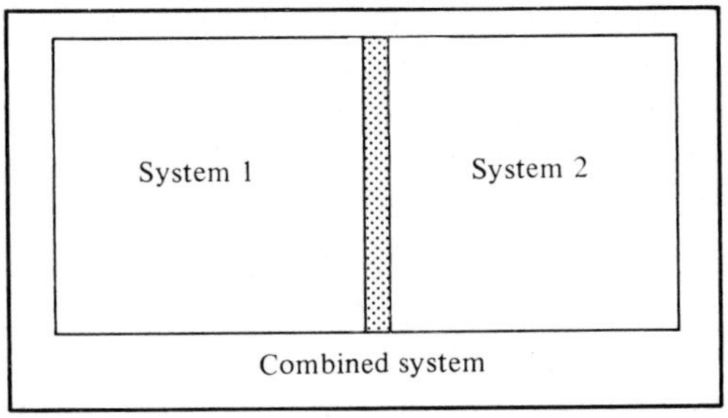

Figure 22.2

form of interaction there is no observed macroscopic motion nor any exchange of material. The changes which do occur can be described as follows. Consider the combined system of the two original systems as a new system. This combined system is the direct product of the two original systems in the sense that the states of the new system are pairs (p_1, p_2) where p_1 is a state of the first system and p_2 is a state of the second. In *diathermal contact* the equilibrium states of the combined system constitute a subset of the set of pairs of equilibrium states of the original system. In other words, if p_1 is an equilibrium state of the first system and p_2 is an equilibrium state of the second, and the two system are brought into diathermal contact, the combined system will not necessarily be in equilibrium but will tend to a definite

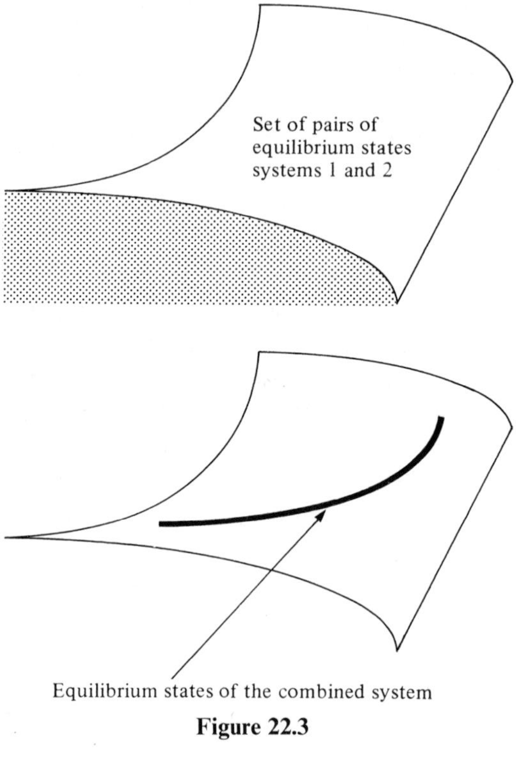

Figure 22.3

equilibrium state (q_1, q_2) where q_1 is a new equilibrium state of the first system and q_2 is a new equilibrium state of the second system. We say that the first system in state q_1 is in *thermal equilibrium* with the second system in state q_2.

It is an observed law of nature (sometimes called the *zeroth law of thermodynamics*) that *thermal equilibrium is an equivalence relation* in the sense that if we have three systems and if p_1, p_2, and p_3 are equilibrium states of these systems such that p_1 would be in thermal equilibrium with p_2 if the first two systems were to be brought into diathermal contact and p_2 and p_3 would be in thermal equilibrium if the second and the third system were to be brought into diathermal contact, then the first and third system would be in thermal equilibrium if they were brought into equilibrium at the states p_1 and p_3.

The equivalence class of all systems at thermal equilibrium is called an *abstract temperature*. Suppose that we fix one definite system and some numerical function of the equilibrium states of this system which takes distinct values on states belonging to differing abstract temperatures. We thus obtain a numerical function, θ, defined on all equilibrium states of all systems called an *empirical temperature*. The chosen system with its numerical function is called a *thermometer*.

It is part of the assertion of the 'zeroth law of thermodynamics' that we can choose θ to be a differentiable function on the set of equilibrium states of any system. Furthermore we can find some definite system and choose θ so that $d\theta \neq 0$ any where on the manifold of equilibrium states of that particular system. We shall only use such functions θ as empirical temperatures. Of course, if θ' is some other empirical temperature, then θ' is a strictly monotonic differentiable function of θ. Conversely, given any strictly monotonic differentiable function, we can apply it to θ to obtain a new empirical temperature, θ'. In any event, two states have the same abstract temperature if and only if they have the same value of θ, for any choice of empirical temperature.

In the description of each system there are certain functions that play an important role. Thus, for example, if the system consists of gas in a container, the total volume occupied would be such a function, as would the total mass of each of the various chemical constituents of the gas. These functions are called *configurational variables*. Thus each system has n functions defined on the set of its states called configurational variables. Here the number n of configurational variables may depend on the system but for any system $n \geqslant 1$. The configurational variables, when restricted to the equilibrium states, are differentiable functions. The configurational variables together with θ form a system of local coordinates on the manifold of equilibrium states. Thus the manifold of equilibrium states is $(n+1)$-dimensional. If $x_1, x_2, \ldots, x_n$ denote the configurational variables, then $\theta, x_1, \ldots, x_n$ will be a local system of coordinates on the manifold of equilibrium states. In many examples, the total volume of the system, denoted by V, is one of the configurational variables. We will then usually take $x_1 = V$ so that $\theta, V, x_2, \ldots, x_n$ will be local coordinates, where $x_2, \ldots, x_n$ are the remaining configurational variables.

Since $d\theta$ does not vanish anywhere, the sets $\theta = \text{const}$ form n-dimensional submanifolds by the implicit function theorem. These submanifolds $\theta = \text{const}$ are called *isotherms* and the configurational variables can be used as local coordinates on the isotherms.

There exists another important class of interactions of the system with the rest of the universe, called *adiabatic interactions.* These interactions have the property that for such interactions the equilibrium is not disturbed unless there is some change in the configurational variables. If, for a period of time, all the interactions of the system are adiabatic, the system is said to be in an *adiabatic enclosure.* A familiar, everyday (approximation to an) adiabatic enclosure is a thermos bottle.

A curve γ joining two states ρ and ρ' while the system is in an adiabatic enclosure is called an *adiabatic curve.* The *first law of thermodynamics* is a generalization from experience which asserts that if a system in an adiabatic enclosure is brought from one state ρ to another state ρ' by applying external work, the amount of this work is always the same no matter how this work is applied. There is therefore a function U on the space of states of the system called the *internal energy* such that $W = U(\rho') - U(\rho)$ represents the amount of work done on the system when the system is moved from ρ to ρ' along any adiabatic curve.

If the system is not in an adiabatic enclosure and is brought from ρ to ρ' by some curve of interactions, then the total external work applied need no longer equal $U(\rho') - U(\rho)$. The difference $U(\rho') - U(\rho) - W = Q$ is called the *heat* supplied to the system by the process. Here the work done and the heat supplied depend on the process and not merely on the initial and final states ρ and ρ'.

(This is the formulation of the first law of thermodynamics in the Born–Carathéodory approach. Notice that in this approach heat is *defined* as the difference between the change in internal energy and the work applied to this system. This idea, to regard heat as a derived rather than a fundamental quantity, is both the strength and the weakness of the Born–Carathéodory approach. It is its strength because it reflects the fundamental view in physics for more than a century – that the basic object is energy which is conserved when all its forms are taken into account. But its weakness is that it does not reflect the experimental or historical realities. In fact, it is usually the heat added to a system which is easiest to measure. Joule's experiments in the 1840s which inaugurated this whole subject were to derive the *mechanical equivalent of heat.* That is if a system in an adiabatic enclosure was changed by adding a specific amount of heat, this same change could be effected by doing work on the system. But the total work done is a definite amount, always the same, no matter how the work was done. So in the actual experiments it is the heat added which is the measured quantity, along with various types of work. It is a *deduction* from the existence of a unique mechanical equivalent of heat that allows us to conclude that the total work in bringing the system from one state to another is independent of the path if the work is done adiabatically.)

The next basic notion in the theory is that of a reversible curve. A curve γ is called (almost) *reversible* if for every t the state $\gamma(t)$ is (sufficiently close to) an equilibrium state.

The intuitive meaning of reversibility is that the interactions are proceeding very slowly relative to the 'relaxation time' of the system, i.e., the time it takes the system to reach its equilibrium state. For example, suppose the interaction takes place through some piston and rod arrangement which changes the volume of a gas. If the gas is in an equilibrium state, and we suddenly push in the piston, the new state of the gas will not be an equilibrium state; there will be all sorts of eddies or even shock waves set up in the gas. Of course, if we hold the piston in its new position, the gas will settle down to a new equilibrium state. We can reckon that the rate of return to equilibrium is quite rapid, so that if we move the piston sufficiently slowly, the states of the system are (approximately) equilibrium states for all time. Such a curve of states is called *reversible*. For the case of volume changes, it is a familiar observation that there exists a function, p, called the pressure, so that the work done on the system along any reversible curve γ is given by $-\int_\gamma p\mathrm{d}V$.

Notice that up until now we have been measuring work in terms of change external to the system. The pressure is a function on the equilibrium states which allows us to compute the work done by a reversible change of volume. More generally, it is observed that there are similar functions corresponding to the other configurational variables. That is, there exist functions $v_1, \ldots, v_n$ on the space of equilibrium states of a system such that the work done on the system along any (almost) reversible curve γ is (approximately) equal to the integral along γ of the linear differential form

$$\omega = v_1 \mathrm{d}x_1 + \cdots + v_n \mathrm{d}x_n$$

so

$$W(\gamma) = \int_\gamma \omega.$$

Therefore the heat supplied along the path γ is given by the integral

$$Q(\gamma) = \int_\gamma \alpha$$

where α is the linear differential form

$$\alpha = \mathrm{d}U - \omega.$$

In case $x_1 = V$ is the volume, it is customary to call $p = -v_1$ the pressure so that

$$\omega = -p\mathrm{d}V + v_2 \mathrm{d}x_2 + \cdots + v_n \mathrm{d}x_n$$

and

$$\alpha = \mathrm{d}U - \omega = \mathrm{d}U p\mathrm{d}V - v_2 \mathrm{d}x_2 - \cdots - v_n \mathrm{d}x_n.$$

In particular, by definition, an *adiabatic reversible curve* is a curve on the manifold of equilibrium states which is a null curve for the linear differential form α.

We now come to the celebrated 'second law of thermodynamics'. It is an everyday experience that certain types of work done on a system cannot be recovered. More precisely, given a system in an adiabatic enclosure, there are certain types of work that can be done on the system, such as violent shaking or stirring, which are such that you cannot get your work back by any reversible adiabatic curve. The assertion is that this can happen near any equilibrium state. Thus the *second law of thermodynamics* asserts that:

> Near any equilibrium state ρ of any system there exist arbitrarily close equilibrium states which can not be joined to ρ by reversible adiabatic curves.

This simple assertion has far-reaching consequences as we shall now see. The law asserts that near any point of the manifold of equilibrium states there are arbitrarily close points which cannot be reached by null curves of α. It follows from Carathéodory's theorem that for any system there exist functions λ and ϕ defined on the manifold of equilibrium states such that $\alpha = \lambda \mathrm{d}\phi$.

Of course, for any given system, neither the function λ nor the function ϕ is determined completely by the equation $\alpha = \lambda \mathrm{d}\phi$. However, once we have fixed one such choice of λ, then ϕ will be determined up to an additive constant. In the next section we shall show that by considering how α must behave when we combine systems, we can conclude that there is a preferred choice of temperature function, T, (determined up to a constant factor) so that we can take $\lambda = T$ for all systems. This function T is called the *absolute temperature.* In terms of some chosen empirical temperature θ, we are asserting that there is some universal function $T = T(\theta)$ such that we may choose $\lambda = T(\theta)$ for all systems.

Once we decide to use the absolute temperature for our temperature scale, then for any system we may write

$$\alpha = T\mathrm{d}S$$

where the function S is then determined up to an additive constant. The function S is called the *entropy.*

22.3. Entropy and absolute temperature

Suppose that two systems are in diathermal contact. We may assume that the energy involved in bringing the two systems together is negligible. There are no moving walls between the systems and no direct exchange of matter. Thus the total internal energy U at any state of the combined system is the sum of the energies of the component states

$$U(\rho_1, \rho_2) = U_1(\rho_1) + U_2(\rho_2)$$

for any state (ρ_1, ρ_2) of the combined system. Here U denotes the internal energy function of the combined system and U_1 the internal energy function of the first system while U_2 denotes the internal energy function of the second system.

Similarly, the work forms are additive:

$$\omega = \omega_1 + \omega_2.$$

This equation is to be understood in the following sense. ω_1 is a linear differential form defined on the set of equilibrium states of the first system. It can be considered as a linear differential form defined on the product space consisting of all pairs of equilibrium states. It simply does not involve the equilibrium states of the second system. Similarly ω_2 can be defined on the set of such pairs of equilibrium states. Hence $\omega_1 + \omega_2$ is defined on the space of all pairs of equilibrium states. The space of equilibrium states of the combined system in thermal contact is a submanifold of the space of all pairs (in fact the submanifold specified by the equation $\theta_1 = \theta_2$). Then the above equation asserts that ω is the restriction of $\omega_1 + \omega_2$ to this submanifold.

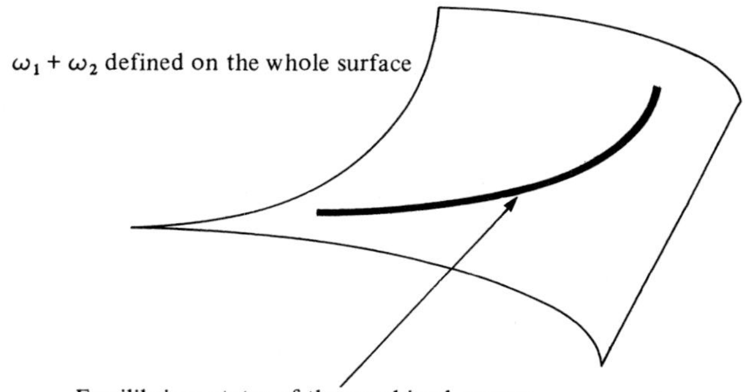

Figure 22.4

It follows that (in the same sense)

$$\alpha = \alpha_1 + \alpha_2.$$

Writing

$$\alpha = \lambda \mathrm{d}\phi, \quad \alpha_1 = \lambda_1 \mathrm{d}\phi_1, \quad \text{and} \quad \alpha_2 = \lambda_2 \mathrm{d}\phi_2$$

we obtain

$$\lambda \mathrm{d}\phi = \lambda_1 \mathrm{d}\phi_1 + \lambda_2 \mathrm{d}\phi_2. \tag{$*$}$$

For any system, we can change the temperature (for example by changing V at constant pressure) adiabatically. This means that for system 1 the forms α_1 and $\mathrm{d}\theta_1$ are linearly independent, and similarly for system 2. As $\alpha_1 = \lambda_1 \mathrm{d}\phi_1$ this implies that the differential forms $\mathrm{d}\phi_1$ and $\mathrm{d}\theta_1$ are independent. By the implicit function theorem, this means that we can make a change of variables so that θ_1 and ϕ_1 are the first two coordinates, i.e. that the local coordinates on the manifold of equilibrium states of system 1 are $(\theta_1, \phi_1, y_1, \ldots)$ and similarly for system 2. So local coordinates on the set of pairs of equilibrium states can be taken as

$$(\theta_1, \theta_2, \phi_1, \phi_2; y_1, \ldots; z_2, \ldots)$$

while the equilibrium states of the combined system (in thermal contact) are described by the condition $\theta_1 = \theta_2$. So we can use

$$(\theta, \phi_1, \phi_2; y_1, \ldots; z_2, \ldots)$$

as coordinates on the set of equilibrium states of the combined system. Equation (*) says that $d\phi$ is a linear combination of $d\phi_1$ and $d\phi_2$ in these coordinates, i.e. $\phi = F(\phi_1, \phi_2)$. In other words, the function ϕ of the combined system depends only on the values of the functions ϕ_1 and ϕ_2 and is independent of all the other coordinates. Now we can always add heat to any system at any state, so the form α and hence the function λ does not vanish anywhere. So we may divide (*) by λ to write it as

$$d\phi = (\lambda_1/\lambda)d\phi_1 + (\lambda_2/\lambda)d\phi_2. \qquad (**)$$

This means that

$$\lambda_1/\lambda = f_1(\phi_1, \phi_2) \quad \text{where} \quad f_1 = \partial F/\partial\phi_1$$

is a function only of ϕ_1 and ϕ_2 and similarly

$$\lambda_2/\lambda = f_2(\phi_1, \phi_2) \quad \text{where} \quad f_2 = \partial F/\partial\phi_2.$$

We may divide the first of these equations by the second since none of the functions λ_1, λ_2 or λ vanishes. We get

$$\lambda_1/\lambda_2 = f_1/f_2$$

or

$$\log \lambda_1 - \log \lambda_2 = \log (f_1/f_2)$$

where the right hand side depends only on ϕ_1 and ϕ_2. In particular, if we compute the partial derivative of both sides with respect to θ the right hand side vanishes and we get

$$\frac{\partial \log \lambda_1}{\partial\theta} = \frac{\partial \log \lambda_2}{\partial\theta}.$$

The left hand side of this equation is a function of the variables describing system 1, that is, it is a function of $(\theta, \phi_1, y_1, \ldots)$ and the right hand side is a function of $(\theta, \phi_2, z_2, \ldots)$. The only way that they can agree for all values of the variables is for each side to depend only on θ and to be the same function of θ. In other words, there is some universal function g such that

$$\partial(\log \lambda)/\partial\theta = g(\theta)$$

for all systems. By examining some specific systems one verifies that this function g is nowhere equal to zero.

Now suppose that we make a change of variables in the empirical temperature – that is, replace θ by $T(\theta)$. Then the coordinates are now $(T, \phi, \ldots)$ and we have

$$\partial(\log \lambda)/\partial T = [\partial(\log \lambda)/\partial\theta](d\theta/dT) = [\partial(\log \lambda)/\partial\theta](dT/d\theta)^{-1}.$$

In particular, if we choose T so that

$$dT/d\theta = Tg(\theta) \qquad (***)$$

then

$$\partial(\log \lambda)/\partial T = 1/T = \partial(\log T)/\partial T.$$

Equation (***) determines T up to a multiplicative constant. Indeed $T(\theta) = ce^{G(\theta)}$ where G is an indefinite integral of g and c is a constant. Any such choice of T is called an *absolute temperature*. Let us now go back over our entire discussion in this section but now use an absolute temperature, T, for the θ. Then for each system we have

$$\log \lambda_i = \log T + \log G_i$$

where G_i is independent of T. Thus

$$\lambda_1 = TG_1, \quad \lambda_2 = TG_2 \quad \text{and} \quad \lambda = TG$$

where the functions G_1, G_2, and G are independent of T. Also we know that λ_1/λ_2 is a function only of ϕ_1 and ϕ_2. But λ_1 is a function of the first system, hence independent of ϕ_2 and similarly λ_2 is independent of ϕ_1. Hence G_i is a function of ϕ_i alone for each system.

Let us drop the subscripts once again. What we have shown is that for any system we have

$$\alpha = TG(\phi)\mathrm{d}\phi,$$

where G is a function (depending on the system) which does not vanish anywhere. But now we can solve the equation

$$\mathrm{d}S = G\mathrm{d}\phi$$

which determines S up to an arbitrary additive constant. In other words, S is any indefinite integral of G. Then we get

$$\alpha = T\mathrm{d}S.$$

To summarize, we have proved that

> There exists a universal *absolute temperature* scale T determined up to a multiplicative constant. Fixing this constant, and thus choosing T, determines a function S on the set of equilibrium states of every system. This function S is determined up to an additive constant (one for each system) and is called the *entropy*. The heat form α is given by
>
> $$\alpha = T\mathrm{d}S.$$

As we have already indicated, the existence of the function S and the equation $\alpha = T\mathrm{d}S$ imply Kelvin's formulation of the second law – around any closed curve one must have $\int \mathrm{d}S = 0$ and T has a constant sign (which may be chosen to be positive). If heat is added along one portion of the cycle, heat must be extracted along some other portion and hence 'no cycle can exist whose net effect is a total conversion of heat into work'. Thus 'perpetual motion machines of the second kind' cannot exist.

A simple type of cycle, C, is a curve built up out of four portions, two of which, say C_3 and C_4, are on surfaces where S is constant (and so no heat is exchanged) along one of which, say C_1, the system is in thermal contact with a large system (called a heat reservoir) at a high temperature T_1, and along the fourth, C_2, the system is in thermal contact with a cold reservoir at a low temperature, T_2. Let $Q_1 = \int_{C_1} \alpha$ be the total heat *absorbed* by the system along C_1 and let $Q_2 = -\int_{C_2} \alpha$ be the heat *emitted*

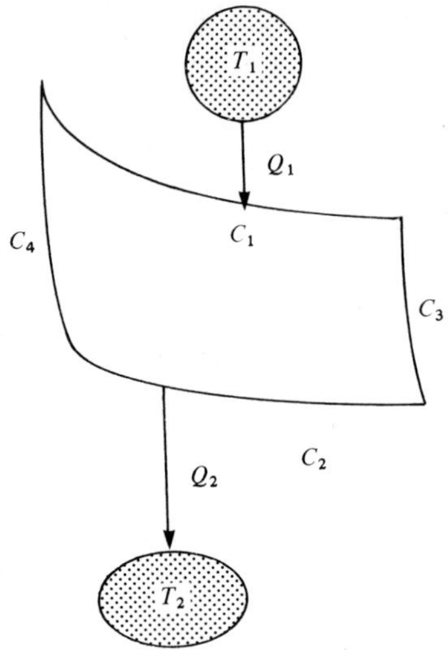

Figure 22.5

along C_2. Then $Q_1 = \int_{C_1} T\mathrm{d}S = T_1 \int_{C_1} \mathrm{d}S$ and $Q_2 = -T_2 \int_{C_2} \mathrm{d}S$. But

$$0 = \int_C \mathrm{d}S = \int_{C_1} \mathrm{d}S + \int_{C_2} \mathrm{d}S$$

so

$$\frac{Q_1}{Q_2} = \frac{T_1}{T_2}.$$

Now we have

$$0 = \int_C \mathrm{d}U = \int_C \omega + \int_C \alpha = \int_C \omega + Q_1 - Q_2.$$

Thus if W denotes the work done *by* the system so that

$$W = -\int_C \omega$$

then

$$W = Q_1(1 - T_2/T_1).$$

This is, of course, the famous formula found in all the textbooks for the work done by a Carnot cycle operating between the temperatures T_2 and T_1. The actual statement of Carnot is that this is the most efficient way of extracting work from heat between these two temperatures. That is, that for any machine extracting the amount of heat Q_1 at T_1 and giving up heat at T_2 the above formula represents the maximum possible output of work. This stronger statement has to do with irreversible processes, and relates to the changes in the function S in the course of irreversible adiabatic processes. Let us formulate the general statement, and leave the applications to work efficiency to the reader (or to the standard texts).

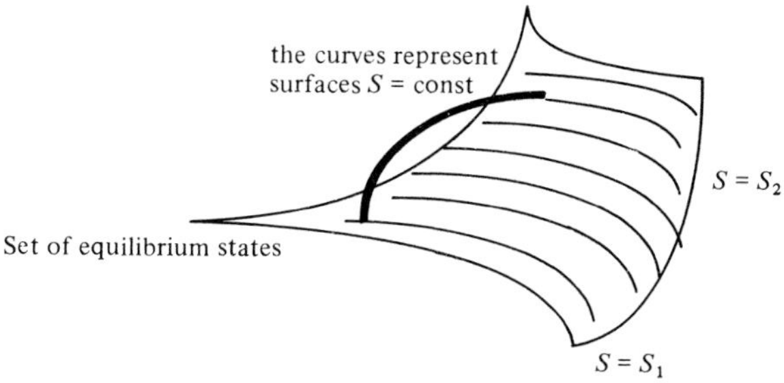

Figure 22.6

If two equilibrium states, σ_1 and σ_2, can be joined by a reversible adiabatic curve then we know that they lie on the same $S = \text{const}$ surface, i.e. $S(\sigma_1) = S(\sigma_2)$. But suppose that we start with the equilibrium state σ_1 and apply an adiabatic process which is not necessarily reversible, and end up at an equilibrium state σ_2. Thus the curve joining σ_1 to σ_2 might not lie in the submanifold of equilibrium states, although the end points do. For example, we might keep a fluid in a thermos bottle and stir a propeller in the fluid. After the stirring ends, the fluid in the thermos bottle comes to an equilibrium state, with a different value of S. Experience shows that then we cannot get back to the original state unless we extract heat from the system. Thus (if we have chosen the multiplicative constant so that T is positive) we have $S(\sigma_1) < S(\sigma_2)$ for such an irreversible adiabatic curve. We can thus state this stronger version of the second law of thermodynamics as:

> if the equilibrium states σ_1 and σ_2 can be joined by an adiabatic curve (that is if a process can lead from σ_1 to σ_2 while the system is in an adiabatic enclosure) then
>
> $$S(\sigma_1) \leqslant S(\sigma_2)$$
>
> with $S(\sigma_1) = S(\sigma_2)$ if and only if σ_1 and σ_2 can be joined by a reversible adiabatic curve.

22.4. Systems with one configurational variable

In this section we shall study the important special case where there is one configurational variable, and hence the manifold of equilibrium states is two-dimensional. Throughout this section we shall assume that the configurational variable in question is volume. The corresponding function entering into the expression for the work form is pressure and the work done on the system is given by the differential form

$$\omega = -p\mathrm{d}V.$$

However, the considerations equally apply to many other interesting physical systems. In many of the general theorems, all that has to be done is to replace volume by the appropriate configurational variable and the pressure by the appropriate

Table 22.1 *Systems with one configurational variable*

System	Configurational variable	Generalized force	Work from ω in joules = newton meters
Gas	volume, V, in m^3	pressure p in N/m^2	$\omega = -p\mathrm{d}V$
Wire	length L in m	tension T in N	$\omega = T\mathrm{d}L$
Surface area	area A in m^3	surface tension S in N/m	$\omega = S\mathrm{d}A$
Electrical cell	charge Z in coulombs	voltage E in volts	$\omega = E\mathrm{d}Z$
Magnetic material	total magnetic moment (total magnetization) M in amp/meter	magnetic field H in amp/meter^2	$\omega = \mu_0 H\mathrm{d}M$

'generalized force' in all the formulas of this section. Table 22.1 lists some of the important physical systems. We will make no further mention of them in this section.

Before embarking on our general discussion, some comments are in order about the history and literature of our subject. Many of the most basic mathematical concepts that we use in this book such as 'set', 'function', 'manifold', 'coordinates', 'linear differential form' were introduced (in their modern form), or became commonly known, *after* the fundamental discoveries in thermodynamics. Furthermore, most of the early heroes of our story (with the notable exceptions of Kelvin and Helmholtz) were not mathematically trained, or (Carnot) chose not to express themselves in mathematical language. As a result, most of the standard texts to this very day use a mathematically obscure language reflecting the early formulations of the subject, with a plethora of formulas involving relations between partial derivatives. Here is an example: the notion of a linear differential form and its line integral together with a version of Stokes' theorem (in the plane) were introduced to the world by Ampère in the early 1920s in a breathtakingly original series of papers combining important new mathematics with ingenious physical experiments. The more general version of Stokes' theorem for a line integral of a one-form in three-space first occurs in a letter from Kelvin to Stokes in 1850. (Stokes published it as an examination question (!) some years later.) Yet in all his papers on thermodynamics Kelvin never once uses Stokes' theorem, although it would have greatly simplified his arguments.

Now we have expressed the heat added to a system as a linear differential form, α, on the manifold of equilibrium states. The functions T and V have independent differentials and so if the manifold is two-dimensional we can use T and V as local coordinates and write

$$\alpha = \Lambda_V \mathrm{d}V + C_V \mathrm{d}T, \tag{22.1}$$

where Λ_V and C_V are functions on the manifold of equilibrium states. Similarly, the functions p and T are independent so we can use them as coordinates and write

$$\alpha = \Lambda_p \mathrm{d}p + C_p \mathrm{d}T \tag{22.2}$$

where Λ_p and C_p are two other functions. (Strictly speaking we should, from the historical point of view, use an empirical temperature, θ, instead of T since we are discussing the subject prior to the discovery of the second law of thermodynamics. But then we would have to rewrite all of the formulas with T when we return to the modern period. So we will let the reader make the various mental substitutions.) Of course the four functions are related to one another by the change of variables formula going from the V, T to the p, T coordinates, and we shall return to this point shortly. The underlying physical law expressed by the two preceding assertions was formulated as the 'doctrine of latent and specific heats'. The function Λ_V was known as the 'latent heat with respect to volume' and the function C_V was called 'the specific heat at constant volume'. Similarly the function Λ_p was known as 'the latent heat with respect to pressure' and C_p was called 'the specific heat at constant pressure'. The meaning of the word 'latent' was as follows. For a long time there was a confusion between the concepts of heat and temperature, and it was thought that the temperature of a body reflects the 'total amount of heat that it contains' (a meaningless concept, of course, once we know that α is not closed). If we add heat we raise the temperature. But if some of the heat added goes into expanding the volume or increasing the pressure then it becomes 'hidden' or 'latent'. Of course equations (22.1) and (22.2) give clear mathematical formulations to rather mysterious sounding physical 'doctrines'.

In two dimensions many computations simplify because of the fact that if Ω_1 and Ω_2 are two-forms with $\Omega_2 \neq 0$ then $\Omega_1 = f\Omega_2$ where f is a function. In other words, we can take the 'quotient' of two two-forms to obtain a function, provided that the denominator does not vanish, and we would write $f = \Omega_1/\Omega_2$.

For example, consider the function

$$\frac{\alpha \wedge \mathrm{d}p}{\alpha \wedge \mathrm{d}V} = f.$$

What does it represent? At any point $\mathbf{x}$ let ξ be a vector tangent to the adiabatic curve passing through $\mathbf{x}$ so that $\alpha(\xi) = 0$, and η some tangent vector independent of ξ so that $\alpha(\eta) \neq 0$. Then evaluating the forms $\alpha \wedge \mathrm{d}p$ and $\alpha \wedge \mathrm{d}V$ on the parallelogram spanned by ξ and η shows that

$$f(\mathbf{x}) = \frac{\mathrm{d}p(\xi)}{\mathrm{d}V(\xi)}.$$

We can write the right hand side of this equation as

$$f = (\mathrm{d}p/\mathrm{d}V)_{\text{adiabatic}}$$

where the right hand side means the ratio of the linear differential forms $\mathrm{d}p$ and $\mathrm{d}V$ when evaluated on vectors tangent to adiabatics. Similarly, and with similar meaning

$$\frac{dT \wedge dp}{dT \wedge dV} = (dp/dV)_{\text{isothermal}}.$$

If we take the exterior product of (22.2) with dp we get

$$\alpha \wedge dp = C_p dT \wedge dp$$

and the exterior product of (22.1) with dV gives

$$\alpha \wedge dV = C_V dT \wedge dV.$$

Dividing the first equation by the second gives the important result

$$(dp/dV)_{\text{adiabatic}} = \gamma (dp/dV)_{\text{isothermal}} \text{ where } \gamma = C_p/C_V. \tag{22.3}$$

If we introduce the density

$$\rho = m/V$$

we can also write (22.3) as

$$(dp/d\rho)_{\text{adiabatic}} = \gamma (dp/d\rho)_{\text{isothermal}}, \quad \gamma = C_p/C_V. \tag{22.4}$$

This result is important for the following physical and historical reasons. One of Newton's famous results was to show that his laws implied that the speed of sound, c, in a gas is given by

$$c^2 = dp/d\rho \tag{22.5}$$

where Newton assumed that p and ρ are related by the equation

$$p = N\rho \text{ where } N \text{ is a constant,} \tag{22.6}$$

i.e.

$$pV = \text{const.} \tag{22.7}$$

Now the pressure, density, and speed of sound (in air for example can' be independently measured, and Newton's assertion that $c^2 = p/\rho$ was repeatedly contradicted by experiment for the next 100 years. The speed of sound was found to be greater than that predicted by Newton's formula by a factor of about 1.4. On the other hand, experiments by Gay-Lussac and others showed that equation (22.7) holds at constant temperature, but that the constant on the right hand side of (22.7) varies with the temperature. Thus, if we use the independent functions p and V as coordinates on the manifold of equilibrium states, equation (22.7) for differing values of the constant are the equations for isothermals, and we can write Newton's formula as

$$c^2 = (dp/d\rho)_{\text{isothermal}}.$$

Laplace (in 1816) argued on physical grounds (essentially that the speed of propagation of the disturbance is far greater than the time it takes for heat to be transferred) that one should replace Newton's formula by

$$c^2 = (dp/d\rho)_{\text{adiabatic}} = \gamma (dp/d\rho)_{\text{isothermal}}. \tag{22.8}$$

To quote Laplace: 'The true speed of sound equals the product of the speed according to Newton's formula with the square root of the ratio of the specific heat of air subject to the constant pressure of the atmosphere at various temperatures to its specific heat when its volume remains constant.' Laplace also points out that we can use the observed value of the speed of sound to experimentally determine γ. Of course, once we know γ and the equations for the isothermals, then we can regard equation (22.3) as a differential equation for the adiabatic curves. For example, for a large variety of gases under low pressure, γ is essentially a constant. (Hydrogen is an example of an exception to this rule.) Also, equation (22.7) holds at constant temperatures. If we assume that equation (22.7) are the equations for isothermals and that γ is constant then we can *conclude* from equation (22.3) that the adiabatic curves are given by

$$pV^{\gamma} = \text{const.} \tag{22.9}$$

Notice that in our discussion so far in this section we have not used either the first or the second law of thermodynamics. We have only used the hypothesis that the set of equilibrium states is a two-dimensional manifold and that the heat is given by a linear differential form – 'the doctrine of latent and specific heats'.

Let us now introduce both laws, the first law, $\mathrm{d}U = \alpha - p\mathrm{d}V$, and the second law, $\alpha = T\mathrm{d}S$, and combine them as

$$\mathrm{d}U = T\mathrm{d}S - p\mathrm{d}V. \tag{22.10}$$

If we take the exterior derivative of this equation we get

$$\mathrm{d}T \wedge \mathrm{d}S = \mathrm{d}p \wedge \mathrm{d}V. \tag{22.11}$$

How do we determine the functions T, S and U by observations, say in terms of the p, V coordinates? To illustrate the method, suppose that the observed isothermals and adiabatics are given by equations (22.7) and (22.9). So if we introduce the functions

$$t(p, V) = pV \tag{22.12}$$

and

$$a(p, V) = pV^{\gamma} \tag{22.13}$$

then

$$\text{the isothermals are the level curves of } t \tag{22.14}$$

and

$$\text{the adiabatics are the level curves of } a. \tag{22.15}$$

Since t and T have the same level curves, and since both have nowhere zero differentials, we conclude that T can be expressed as a function of t, i.e. that $T = T(t)$ with $T'(t)$ nowhere zero. Similarly, $S = S(a)$. Our problem is to determine the functions $T(t)$ and $S(a)$. Now

$$\mathrm{d}T \wedge \mathrm{d}S = T'(t)S'(a)\mathrm{d}t \wedge \mathrm{d}a = \mathrm{d}p \wedge \mathrm{d}V$$

by equation (22.11). So we must use the explicit expressions for t and a as functions of

p and V to express $\mathrm{d}t \wedge \mathrm{d}a$ in terms of $\mathrm{d}p \wedge \mathrm{d}V$. We will then find that

$$\frac{\mathrm{d}p \wedge \mathrm{d}V}{\mathrm{d}t \wedge \mathrm{d}a}$$

is a product of a function of t and a function of a, and this will determine T' and S'. Explicitly

$$\begin{aligned}\mathrm{d}t \wedge \mathrm{d}a &= (p\mathrm{d}V + V\mathrm{d}p) \wedge (\gamma p V^{\gamma-1}\mathrm{d}V + V^{\gamma}\mathrm{d}p)\\ &= (\gamma - 1)pV^{\gamma}\mathrm{d}p \wedge \mathrm{d}V = (\gamma - 1)a\mathrm{d}p \wedge \mathrm{d}V\end{aligned}$$

so

$$\frac{\mathrm{d}p \wedge \mathrm{d}V}{\mathrm{d}t \wedge \mathrm{d}a} = [(\gamma - 1)a]^{-1}.$$

This shows that

$$T'(t)S'(a) = [(\gamma - 1)a]^{-1}.$$

In this last equation we can multiply T' by any constant k provided that we multiply S' by k^{-1}. This of course reflects the fact that the absolute temperature is only determined up to a multiplicative constant. So with no loss of generality we may assume that $T' \equiv 1$ and then $S'(a) = [(\gamma - 1)a]^{-1}$ and so

$$T = t + T_0 \quad \text{and} \quad S = (\gamma - 1)^{-1}\log a + S_0,$$

where T_0 and S_0 are constants. Remember that S is only determined up to an additive constant. But we still must determine T_0 and the function U. For this purpose we substitute the above values into (22.10) to get, after some computation,

$$\mathrm{d}U = (\gamma - 1)^{-1}[T_0\mathrm{d}\log a + \mathrm{d}t].$$

We now call on an additional experiment – the Joule–Thompson experiment – which shows that if we allow a gas to expand adiabatically into a vacuum there is (essentially) no change in temperature. In such an expansion there is no heat added nor is any work done, so the internal energy, U, is unchanged. But the expansion is irreversible and so S definitely increases. The only way that this can happen is if the constant T_0 vanishes. Thus

$$T_0 = 0$$

and

$$U = (\gamma - 1)^{-1}t. \tag{22.16}$$

Of course the internal energy is proportional to the total amount of gas present, and this has to do with the constant relating t to a choice of the universal temperature scale. So let us write

$$t = nRT \tag{22.17}$$

where n denotes the number of moles present, and T the universal temperature scaled in kelvins. (This is a conventional choice of scale so that the difference between the freezing and boiling points of water is about 100°. The precise agreement going into this scale need not detain us.) Then R is a constant conversion factor from units of temperature to units of energy known as the *gas constant*. It follows from (22.10) that

$$dU \wedge dV = \alpha \wedge dV = C_V dT \wedge dV$$

and then from (22.16) and (22.17) that C_V is a constant given by

$$C_V = (\gamma - 1)^{-1} nR.$$

Of course then C_p must also be a constant and

$$C_p - C_V = nR.$$

All of the above properties were derived on the basis of three assumptions which define an *ideal gas*:

(i) the law of Boyle, Gay-Lussac (equation (22.7)) which says

$$pV = nRT,$$

(ii) That U is a function of T alone (the Joule experiment), and

(iii) that γ is a constant.

Let us summarize our conclusions, but state them per mole. That is, assume $n = 1$ and write lower case letters, v for V, c_p for C_p etc. to indicate that we are dealing in molar quantities. We then have

Ideal gas laws.

the isotherms are $pv = RT$,

the adiabatics are $pv^{\gamma} = \text{const.}$,

$u = (\gamma - 1)^{-1} RT$,

$c_v = (\gamma - 1)^{-1} R$,

$c_p = [\gamma/(\gamma - 1)]R$ so $c_p - c_v = R$,

$s = (\gamma - 1)^{-1} R \log pv^{\gamma} + s_0 = R \log v + c_v \log T + s_0'$.

The quantities that can be directly measured for real gases are p, v, T, c_v, c_p and γ. To measure c_v the gas is put in a thin walled steel flask with a heating coil wrapped around it. The heat delivered is measured by the known current flowing through the coil and the temperature change of the gas is measured. To measure c_p the gas flows at constant pressure through a similar heating coil arrangement and the difference between inlet and outlet temperatures is measured. The results of these experiments at low pressures, where the gas behaves approximately like an ideal gas, are as follows:

All gases

$$c_v \text{ and } c_p \text{ are functions of } T \text{ only}$$

and

$$c_p - c_v = R.$$

Monatomic gases such as He, Ne, A, and most metallic vapors

c_v is constant over a wide range of T and is very nearly equal to $\frac{3}{2}R$. So c_p is very nearly equal to $\frac{5}{2}R$ and γ is constant over a wide range of T and close to $\frac{5}{3}$.

Permanent diatomic gases such as O_2, N_2, NO, CO, and air

> c_v is constant at ordinary temperatures and approximately equal to $\frac{5}{2}R$. So c_p is approximately $\frac{7}{2}R$ and γ is constant at ordinary temperatures and approximately $\frac{7}{5}$, and decreases as the temperature is raised.

It is this value of $\frac{7}{5}$ for γ which gives Laplace's account of the speed of sound in air. The true meaning of the fractions $\frac{3}{2}, \frac{5}{2}$, etc. will only become apparent in the framework of statistical mechanics.

Now for *real* gases, we can carry out the entire discussion starting with equation (22.14). We no longer make the assumption that the isothermals and adiabatics are given by equations (22.12) and (22.13) or that γ is constant. But one can experimentally determine the isothermals and γ and hence solve equation (22.3) (perhaps only numerically) to get the adiabatics, and then find $T'(t)$ and $S'(a)$ to get S and then U by solving equation (22.10). Over the past 150 years this information has been accumulated for various substances and is available in chart form. Furthermore certain functions in addition to the ones we have been considering are useful for special purposes. For example, the *enthalpy* H is defined as

$$H = U + pV.$$

It is important for the following reason. Equation (22.10) says that the heat added to a system which is held at constant volume is equal to the change in the values of U. But in a chemical laboratory, it is much more convenient to keep the pressure constant (say everything at atmospheric pressure) and let the volume vary. Now differentiating the above equation gives

$$\mathrm{d}H = \alpha + V\mathrm{d}p$$

so differences in H give the amount of heat added under constant pressure. Thus heats of chemical reaction are recorded as differences in enthalpy.

A particularly useful type of emprical chart is a Mollier diagram in which p and H are used as coordinates and the level curves of S, V, and T are then drawn in. An example is given opposite

There are several other combinations of U, p, V and S which give rise to functions

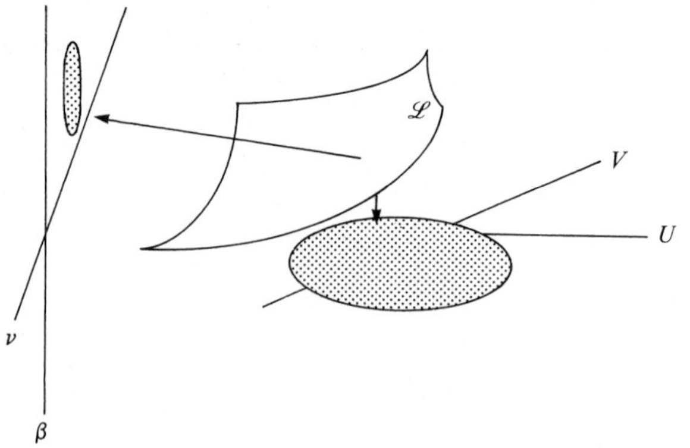

Figure 22.7 The submanifold $\mathscr{L}$ is Lagrangian. Projection to the β, v plane allows the introduction of β, v as local coordinates. Similarly, projection to the U, V plane allows us to use U and V as coordinates.

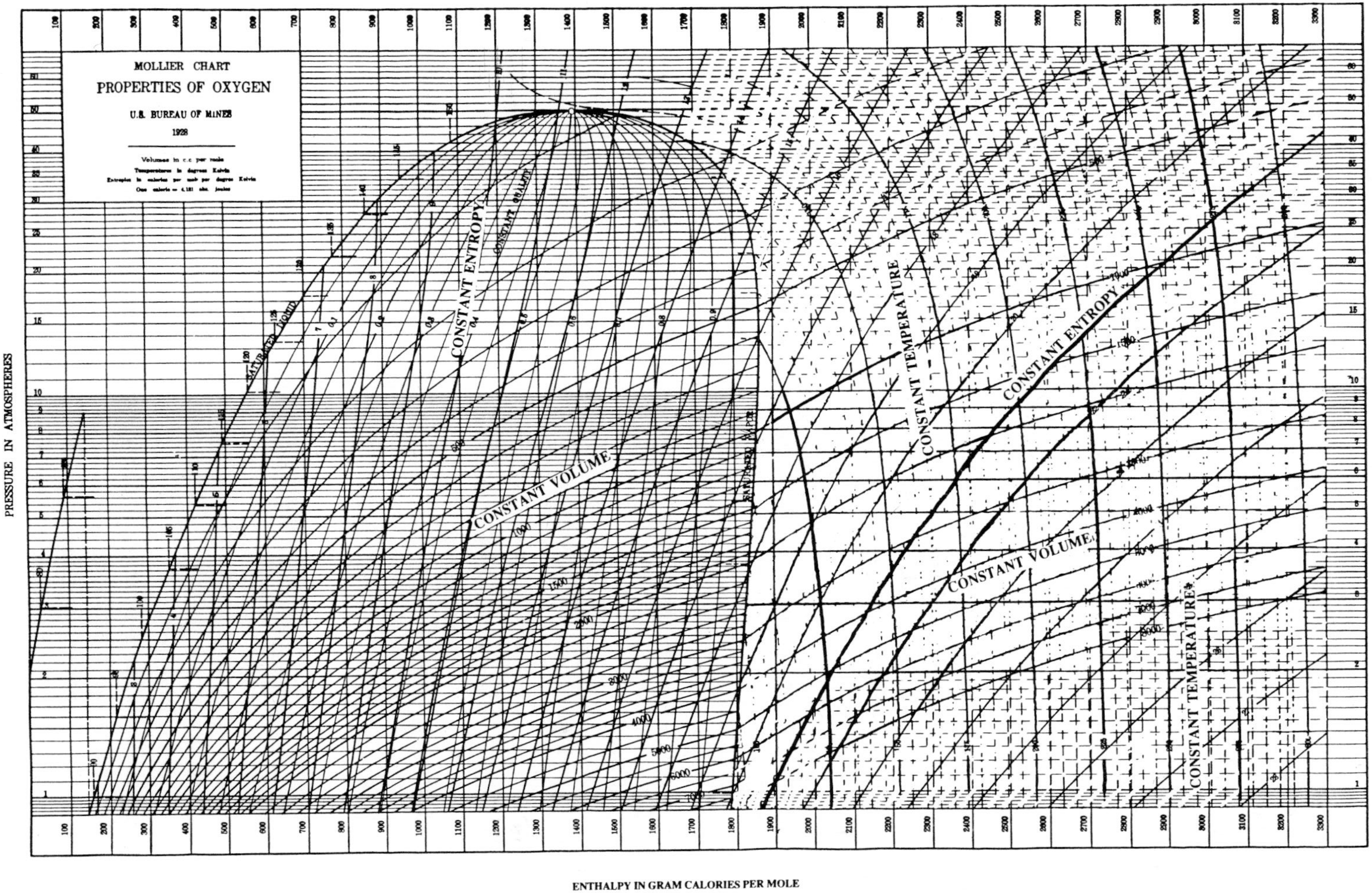

Mollier diagram for oxygen. Notice that if values of any two of the five functions p, V, H, T or S are given, we can locate the point taking on these values by finding the intersection of the appropriate level curves. We can then determine the values of the remaining three functions from the chart and U by $U = H - pV$.

which are each very useful for special purposes. The notation for them varies and we need not go into their description here. However, there is one function of some practical application but with immense theoretical significance. We will describe it several times in increasing generality over the next several sections. What we are about to do may look strange here, but will become convincing when we get to statistical mechanics. First consider the *Planck function* Y defined as follows:

$$Y = S - (RT)^{-1}U - (RT)^{-1}pV.$$

To write this function in more convenient form let us introduce

$$\beta = (RT)^{-1} \tag{22.18}$$

and

$$\nu = \beta p.$$

Notice (from equations (22.16) and (22.17) for example), that R is a conversion factor going from temperature to energy, so β has units of inverse energy, say $(\text{joule})^{-1}$, and ν has units of inverse volume, so that $\beta \mathrm{d}U$ and $\nu \mathrm{d}V$ are dimensionless. Let us assume that we had absorbed R into the definition of T from the very beginning, so that we choose to measure temperature in units of energy and the gas constant disappears. Since $\alpha = T\mathrm{d}S$ has units of energy, and we are measuring T in units of energy, we see that S also becomes dimensionless with this choice of temperature scale. Thus both S and the Planck function

$$Y = S - \beta U - \nu V \tag{22.19}$$

are dimensionless. We can write the combined first and second laws, equation (22.10), as

$$\mathrm{d}S = \beta \mathrm{d}U + \nu \mathrm{d}V. \tag{22.20}$$

It then follows from equation (22.19) that

$$\mathrm{d}Y = -U\mathrm{d}\beta - V\mathrm{d}\nu. \tag{22.21}$$

We wish to interpret equations such as (22.20) and (22.21) as follows: Consider a four-dimensional space $\mathbb{R}^4$ with coordinates (β, ν, U, V). On this space introduce the exterior two-form

$$\Omega = \mathrm{d}\beta \wedge \mathrm{d}U + \mathrm{d}\nu \wedge \mathrm{d}V. \tag{22.22}$$

Notice that on all of $\mathbb{R}^4$ we have

$$\Omega = \mathrm{d}(\beta \mathrm{d}U + \nu \mathrm{d}V)$$

and also

$$\Omega = \mathrm{d}(-U\mathrm{d}\beta - V\mathrm{d}\nu).$$

A two-dimensional submanifold, $\mathscr{L}$, of $\mathbb{R}^4$ is called *Lagrangian* if the two-form Ω vanishes identically when restricted to $\mathscr{L}$. For any two-dimensional submanifold of $\mathbb{R}^4$ we can consider β and ν as functions. If their differentials are linearly independent when restricted to the submanifold, we can use β and ν as local coordinates. Similarly we can consider the functions U and V as local coordinates if their differentials are linearly independent. See figure 22.8. If the submanifold $\mathscr{L}$ is

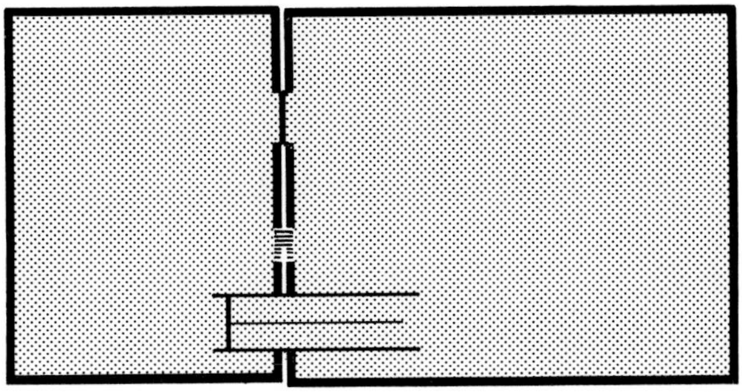

Figure 22.8 Two systems with three kinds of contact. They can exchange heat (top) material (middle) and volume (bottom).

Lagrangian then

$$d(\beta dU + v dV) = 0 \text{ when restricted to } \mathscr{L}$$

and hence (locally) there is a function S defined on such that equation (22.20) holds on $\mathscr{L}$. Similarly, since

$$d(-U d\beta - V dv) = 0 \text{ when restricted to } \mathscr{L}$$

we conclude that (locally) there is a function Y such that equation (22.21) holds on . Thus we can formulate the combined first and second laws as saying that

> the manifold of equilibrium states is a Lagrangian submanifold of (β, v, U, V) space.

But let us carry the mathematical discussion a little further. Suppose we introduce U and V as local coordinates as in figure 22.8, and suppose that we know the function S. This means that we are given S as an explicit function of U and V. If we compare the equation

$$dS = (\partial S/\partial U)dU + (\partial S/\partial V)dV$$

with equation (22.20) we see how to express β and v in terms of U and V, namely,

$$\beta = \partial S/\partial U \text{ and } v = \partial S/\partial V. \tag{22.23}$$

This of course, determines the submanifold $\mathscr{L}$. In other words, if we had a Mollier type diagram with U and V as coordinates and with the level curves and values of just the single function S, we could reconstruct $\mathscr{L}$ and hence, at least in principle, all the relevant thermodynamic functions. All the information about $\mathscr{L}$ is coded into S when S is given as a function of U and V. We say that S is a *generating function* of the Lagrangian submanifold, $\mathscr{L}$.

Similarly if we start with β and v as local coordinates and are given Y as a function of β and v, then it follows from equation (22.21) that can we recover U and V as functions of β and v by

$$U = -\partial Y/\partial \beta \text{ and } V = -\partial Y/\partial v.$$

Thus Y is a generating function of relative to the coordinates β and v. (Notice that this gives the inverse of the map expressing β and v as functions of U and V.)

We can play this game with still other choices of variables as local coordinates. For example, we can write (on all of $\mathbb{R}^4$)

$$\Omega = \mathrm{d}(-U\mathrm{d}\beta + v\mathrm{d}V)$$

and, correspondingly

$$\mathrm{d}Z = -U\mathrm{d}\beta + v\mathrm{d}V \quad \text{on } \mathscr{L}$$

where the *Massieu function* Z is defined by

$$Z = S - \beta U. \tag{22.24}$$

Thus the Massieu function is a generating function for in terms of the variables β and V. Coded into (22.24) are the first and second laws of thermodynamics together with a complete description of the equilibrium states of our system!

When we get to the subject of equilibrium statistical mechanics, we will find that the functions S and Z are the truly basic ones. The function S will be given an interpretation in terms of probability theory whose significance extends far beyond the domain of thermodynamics. The function Z will provide the link between the microscopic theory and the observed macroscopic phenomena. That is, a very general construction will show how a model of the energy at the atomic or molecular level leads to a definite expression for Z as a function of β and V. (Much of the technical aspects of the subject then becomes the purely mathematical question of how to evaluate or approximate this expression.) Of course we will develop the theory so as to apply to systems with more than one configurational variable.

Before getting to statistical mechanics we need to develop some further ideas about equilibrium, and will do so in the next section. But let us close this section with an amplification on a remark we made at the beginning. We mentioned that many physically measurable quantities are given as quotients of two-forms on the surface of equilibrium states, and illustrated this with the two 'specific heats'. Here are two more:

coefficient of thermal expansion at constant pressure	$\dfrac{\mathrm{d}V \wedge \mathrm{d}p}{V\mathrm{d}T \wedge \mathrm{d}p}$
coefficient of compressibility at constant temperature	$\dfrac{\mathrm{d}V \wedge \mathrm{d}T}{V\mathrm{d}T \wedge \mathrm{d}p}$

We will leave some others and manipulations with them to the exercises.

22.5. Conditions for equilibrium

We return to the study of systems with an arbitrary number of configurational variables. Up until now we have defined the function S only on the manifold of equilibrium states. It is important to observe that certain functions, like temperature or pressure, are only defined on the manifold of equilibrium states. It does not make sense to talk of the temperature of a gas (as a whole) unless it is in equilibrium. It was one of the great discoveries of the subject that the entropy is a function which is defined on the set of all states. That is, there is a function, Ent, defined on the set of all states of the system such that

$$\mathrm{Ent}(\rho) = S(\rho)$$

whenever ρ is an equilibrium state. We shall not write down a definition of Ent now, since this would require us to have a workable description of the general states of a system. Once we get to statistical mechanics, where we will give such a description, we will write down an explicit formula for Ent, due essentially to Boltzmann. However, many of the ideas of this section were developed by Clausius and others prior to the discovery of the statistical interpretation.

We shall make use of one property of the entropy function for two systems in contact. (We can regard this property as an axiom at present – it will be obviously verified in the statistical mechanical model.) By two systems in contact we mean that we have a larger system built out of the two smaller system so that a state ρ of the new system is a pair of states $\rho = (\rho_1, \rho_2)$ of the individual component systems and so that

$$U(\rho) = U_1(\rho_1) + U_2(\rho_2).$$

In other words the 'interaction' energy is negligible in comparison with the sum of the individual energies. We are *not* assuming that the contact is only thermal. For example, there might be a movable piston between the two systems so that they can 'exchange volume'. Or there might be a permeable wall between them so that they can exchange gas molecules, etc. The property that we shall use is that the entropy for systems in contact is also additive, i.e. that

$$\mathrm{Ent}(\rho) = \mathrm{Ent}(\rho_1) + \mathrm{Ent}(\rho_2) \quad \text{if} \quad \rho = (\rho_1, \rho_2).$$

Let us now go back to the second law of thermodynamics. Once we have an entropy function defined on all states, then it is reasonable to formulate a strengthened version of the second law which says

> if a system is placed in an adiabatic enclosure, the entropy of any state will not decrease as it evolves in time, and in fact, the entropy will increase unless the system is already in an equilibrium state.

Since energy is conserved, these lead to an important characterization of the equilibrium states, one that will be central in statistical mechanics:

Maximum entropy principle.

> among all states with a given internal energy the equilibrium state has maximum entropy.

Although this is the basic underlying principle, a variant of this principle due to Gibbs is also very useful in practice:

Minimum energy principle of Gibbs.

> If ρ is an equilibrium state, then if σ is any other state with the same entropy as ρ, i.e., $\mathrm{Ent}(\sigma) = \mathrm{Ent}(\rho)$, we must have $U(\rho) \leqslant U(\sigma)$.

In order to derive Gibbs' rule from the maximum entropy principle we make use of the physical observation that we can always add heat to a system and so continuously increase both the energy and the entropy, and we can do so without changing the configurational variables. Suppose that ρ is an equilibrium state, and suppose (contrary to fact) that there were some other state σ near ρ with $\mathrm{Ent}(\sigma) = \mathrm{Ent}(\rho)$ and $U(\sigma) < U(\rho)$. By adding heat we could gradually increase both the entropy and the internal energy until we reach a new state having the same internal energy as ρ but having a larger entropy, contradicting the maximal entropy principle. This establishes Gibbs' principle.

To apply Gibbs' principle we make a number of preliminary remarks. Suppose that $f_1,\ldots,f_k$ are functions of the configurational variables of a system. We can then obtain a *new* system by fixing the values of these functions. For example, imagine that the two systems in figure 22.9 were initially separated (and each had its own piston). Before the systems were brought together, we had a system which was simply the direct product of the two subsystems, and so there were four configurational variables, V_1 and V_2, the volumes of each, and M_1 and M_2, the total mass of the gas in each. (Let us assume for the sake of the discussion that we have one and the same type of gas in each subsystem.) By bringing the systems into contact as in figure 22.9 we have fixed the total volume and the total mass. In other words we have imposed constraints of the form $f = \text{const}$ and $g = \text{const}$ where

$$f = V_1 + V_2 \quad \text{and} \quad g = M_1 + M_2,$$

cutting the number of independent variables down from four to two.

Let us start with some system and obtain a new system by imposing a number of constraints on the configurational variables. The constrained system will have its own equilibrium states which can be characterized by maximum entropy or minimum internal energy, but of course, relative to the more restricted class of states – those satisfying the constraints. Suppose that we are given a subset, $\mathcal{M}$, of the states of the unconstrained, original system which contains all the equilibrium states of the original system and suppose that $\mathcal{M}$ is a manifold. Furthermore assume that the function Ent and the constraint functions $f_1,\ldots,f_k$ are differentiable functions on $\mathcal{M}$. Let ρ be an equilibrium state of the constrained system, where the fs are held constant. The Gibbs minimum energy principle asserts that U must take a local minimum at ρ subject to the constraints

$$\mathrm{Ent} = \text{const}, \quad f_1 = \text{const},\ldots, \quad f_k = \text{const}.$$

Then the method of Lagrange multipliers implies that there exist functions $\lambda_0, \lambda_1,\ldots,\lambda_k$ such that

$$\mathrm{d}U = \lambda_0 \mathrm{d}\,\mathrm{Ent} + \lambda_1 \mathrm{d}f_1 + \cdots + \lambda_k \mathrm{d}f_k. \tag{22.25}$$

For example, suppose we consider the case illustrated by figure 22.9, where we take $\mathcal{M}$ to be the set of all pairs of equilibrium states of the individual systems, if they were not in contact with one another. Each subsystem when considered separately has two configurational variables, V and M. Thus for the two subsystems we have

$$dU_1 = T_1 dS_1 - p_1 dV_1 + \mu_1 dM_1 \quad \text{and} \quad dU_2 = T_2 dS_2 - p_2 dV_2 + \mu_2 dM_2.$$

(The coefficient, μ, of dM, is called the 'chemical potential'.) The manifold $\mathcal{M}$ is six-dimensional since the set of equilibrium states of each system separately is three-dimensional. We are assuming that

$$U = U_1 + U_2,$$

and the additive property of the entropy function asserts that

$$\text{Ent} = S_1 + S_2.$$

Thus

$$dU = T_1 dS_1 + T_2 dS_2 - p_1 dV_1 - p_2 dV_2 + \mu_1 dM_1 + \mu_2 dM_2,$$

while condition (22.25) requires that

$$dU = \lambda_0(dS_1 + dS_2) + \lambda_1(dV_1 + dV_2) + \lambda_2(dM_1 + dM_2)$$

since the constraint functions are $V_1 + V_2$ and $M_1 + M_2$.

Since the differentials occurring on the right of these two equations are independent on $\mathcal{M}$, the only way that this can happen is that

$$T_1 = T_2, p_1 = p_2 \quad \text{and} \quad \mu_1 = \mu_2.$$

Thus the temperatures, pressures and chemical potentials must be equal. It is clear that this argument generalizes to the case where we have several (not just two) systems in contact and where each system can contain a mixture of a number of different species of substance. So if the expression for the internal energy on the equilibrium states of each separate subsystem is

$$dU_i = T_i dS_i - p_i dV_i + \mu_i^a dM_i^a + \mu_i^b dM_i^b + \cdots$$

the conditions for equilibrium become

$$\left.\begin{array}{l} T_1 = T_2 = \cdots, \\ p_1 = p_2 = \cdots, \\ \mu_1^a = \mu_2^a = \cdots, \\ \mu_1^b = \mu_2^b = \cdots, \end{array}\right\} \qquad (22.26)$$
$$\text{etc.}$$

These are then the general conditions for equilibrium and give us some further feeling about the meaning of the 'intensive variables' T, p, μ, etc.

For example, let us consider a gas consisting of a single type of substance. We can imagine each small region of the gas as being a subsystem. If we think that each small subsystem is in a state of equilibrium (considered separately) then we would get a value of T, p and μ for each such region. In other words, we could think of T, p and μ as being a function on three-dimensional space which assigns to each point the values of temperature, pressure and μ in a small region about that point. The condition that the gas as a whole be in equilibrium is that these functions be constant: that the same temperature, pressure and μ persist throughout. Now the gas as a whole, with a constant total mass, say, has only one configurational degree of freedom, V. Thus we can use p and T as coordinates on the two-dimensional space of equilibrium states of the gas as a whole. This means that the value of μ can be expressed as a function of p and T. Thus for a gas of a pure substance in equilibrium with itself, μ is some definite function of p and T, $\mu = \mu_{\text{gas}}(p, T)$. Suppose this same substance can exist in liquid form. For the liquid in equilibrium with itself we would get some other function, $\mu_{\text{liq}}(p, T)$. Now consider a combined system consisting of liquid and gas of the same substance. The gas and the liquid are separately in equilibrium and suppose that the system as a whole is in equilibrium. Then conditions (22.6) imply

$$p_{\text{liq}} = p_{\text{gas}},$$
$$T_{\text{liq}} = T_{\text{gas}}$$

and

$$\mu_{\text{liq}} = \mu_{\text{gas}}.$$

If we let p and T denote the common values in the first two equations, the third equation becomes

$$\mu_{\text{liq}}(p, T) = \mu_{\text{gas}}(p, T).$$

Since μ_{liq} and μ_{gas} are (usually) independent functions of p and T, this last equation defines a curve in the p, T plane. In other words a liquid and gas of the same substance can only coexist in equilibrium when there is a definite relation between temperature and pressure. If there were three phases, say gas, liquid and solid, then they could only exist at equilibrium if

$$\mu_{\text{sol}}(p, T) = \mu_{\text{liq}}(p, T) = \mu_{\text{gas}}(p, T)$$

and these equations then (in general) determine a point (the so called triple point) with a definite value of temperature and pressure. Finally one cannot have four distinct phases of the same substance coexisiting in equilibrium. This was Gibbs' famous derivation of his celebrated *phase rule*. It (and its obvious generalizations) are all contained in equation (22.26) which is an immediate consequence of equation (22.5) – Gibbs' principle of minimum energy.

22.6. Systems and states in statistical mechanics

We now want to examine microscopic theories that can serve as a model for thermodynamics. We begin with a discussion of classical statistical mechanics wherein a 'state' is defined as a 'probability measure' of a certain kind and the entropy of a state will be 'the amount of disorder in the state'. The mathematical foundations of the theory of probability were laid in the first third of this century, principally by Borel, Lebesgue and Kolmogorov. Again this was long after the basic ideas of statistical mechanics were put forth by Boltzmann and Gibbs, leading to another communications barrier. The basic language of the theory of probability is measure theory. We shall use this language without going into the theory. For any of the deeper results in probability theory which involve infinite numbers of alternatives the use of the full machinery of measure theory is essential. As we will only be using the most elementary facts, we will only have to use the most elementary notions which will be transparent in the examples that we develop.

A *measure space* is a set M together with a collection, $\mathscr{A}$ of subsets of M and *measure*, μ, which is a rule that assigns a non-negative number (or $+\infty$), $\mu(A)$, to each set A in the collection $\mathscr{A}$. The collection $\mathscr{A}$ and the measure μ are subject to certain axioms which need not detain us. (Essentially that $\mathscr{A}$ be closed under countable union, intersection and under complement and that $\mu(\bigcup A_i) = \sum \mu(A_i)$ for any countable union of disjoint sets, A_i.) The reason for not considering the collection of *all* subsets of M is to avoid certain contradictions which arise from the consideration of highly pathological sets.

In most of our applications the set M will be of a very simple kind. For example, if M is a finite set, we will take $\mathscr{A}$ to be the collection of all subsets of M. Then the sets consisting of single points of M are in our collection, and to each such set, $\{m\}$ we have assigned a non-negative number $\mu(\{m\})$. From the knowledge of these numbers we can reconstruct the measure μ because

$$\mu(A) = \sum_{m \in A} \mu(\{m\}).$$

If f is any (real or vector valued) function of M we defined its 'integral with respect to μ' as

$$\int f\mu = \sum_{m \in M} f(m)\mu(\{m\}).$$

(This is the same convention we made in our study of electrical network theory.)

We might take M to be $\mathbb{Z}$, the set of all integers, or $\mathbb{N}$, the set of all non-negative integers. Again we would take $\mathscr{A}$ to consist of all subsets, so assigning a measure means assigning a non-negative number to each integer (in the case of $\mathbb{Z}$) or to each non-negative integer (in the case of $\mathbb{N}$). We would get the same expression for $\mu(A)$ as above, but the sum on the right is now an infinite series which may or may not converge. If it converges, this is the number we assign to $\mu(A)$. If the series does not converge we assign the value $\mu(A) = +\infty$. Similarly, if f is a function we attempt to define its integral, $\int f\mu$ by the above expressions. It is also an infinite series. We will say that f is integrable if this series is absolutely convergent, and then $\int f\mu$ is defined as the sum of the series. Otherwise we do not assign a meaning to $\int f\mu$.

We might take M to be the real line, $\mathbb{R}$. Here the collection $\mathscr{A}$ should contain all intervals $[a, b] = \{x \mid a \leqslant x \leqslant b\}$, and the usual (Lebesgue) measure assigns to each interval its length:

$$\mu([a, b]) = b - a.$$

If f is any piecewise continuous function which vanishes sufficiently rapidly at infinity we know how to compute its integral

$$\int f\mu = \int_{-\infty}^{\infty} f(x)\mathrm{d}x.$$

The whole machinery of measure theory is devoted to constructing a suitably rich class of sets, $\mathscr{A}$, on which μ is defined, and a suitably broad class of functions whose integral is well defined. In all of our applications the sets and functions will be so simple-minded that we need not avail ourselves of the theory. As we have mentioned it is essential to use the full theory when one wants to probe a bit deeper.

A *probability measure* is a measure which assigns the value 1 to the set M, i.e. $\mu(M) = 1$. This corresponds to the convention in probability theory that 'probabilities' take on values between 0 to 1.

The fundamental concept of a 'system' in statistical mechanics is a measure space. The set M, the collection of subsets, $\mathscr{A}$, and the measure μ are given to us by the physics or the geometry underlying the theory. The measure μ will not, in general, be a probability measure. But it plays the role of providing the '*a priori* state of knowledge' of the system. We illustrate with a series of examples.

A. *Finite sample space, equal a priori probability*. This is the simplest example. M is a finite set, $\mathscr{A}$ is the collection of all subsets of M. Here $\mu(\{m\}) = 1$ for every $m \in M$. This describes the situation where there are a finite number of alternatives and we have no reason to prefer any one to any other.

B. $M =$ *the real line*, $\mathbb{R}$, *with its standard collection of (Lebesgue) measurable sets and its usual measure* μ. Thus $\mu([a, b]) = b - a$ for any interval $[a, b]$. The measure μ indicates that the '*a priori* probability' of any interval is proportional to its length. Of course μ is not a probability measure since $\mu(\mathbb{R}) = \infty$. Notice that if ρ is any integrable function with

$$\int_{-\infty}^{\infty} \rho(x)\mathrm{d}x = 1$$

then ρ determines a probability measure, $\rho \mathrm{d}x$, on $\mathbb{R}$.

$$P([a, b]) = \int_a^b \rho(x)\mathrm{d}x.$$

The measure $\mu(= \mathrm{d}x)$ thus determines a distinguished class of probability measures on $\mathbb{R}$. (In the technical language of measure theory, it is the class of those probability measures which are absolutely continuous with respect to $\mathrm{d}x$. Not all probability measures on $\mathbb{R}$ are of this form. For example, consider the probability measure, P, which assigns $P(\{0\}) = \frac{1}{2}$, $P(\{1\}) = \frac{1}{2}$, and $P(A) = 0$ if $\{0, 1\} \cap A = \varnothing$. Thus P corresponds to the probability measure in the case that either 0 or 1 is achieved, each one with equal probability. Clearly P is not of the form $\rho \mathrm{d}x$.)

For any measure μ on a measure space and any integrable non-negative function ρ we obtain a new measure, which we denote by $\rho\mu$ by setting

$$\rho\mu(A) = \int_A \rho\mu.$$

This suggests the following general definition:

Definition: Let $(M, \mathscr{A}, \mu)$ *be a measure space. A (statistical) state of* $(M, \mathscr{A}, \mu)$ *is a* *probability measure on M of the form $\rho\mu$ where $\rho \geqslant 0$ is an integrable function. In other words, a state corresponds to an integrable function ρ such that $\int \rho\mu = 1$.*

Examples, continued

C. *$M = \mathbb{R}$ and $\mu = H(\cdot)\mathrm{d}x$ where $H(x) = 0$ for $x \leqslant 0$ and $H(x) = 1$ for $x > 0$.* This corresponds to the situation where we are sure that the real number is positive but otherwise all positive numbers are equally probable. Of course, we could equally take $M = \mathbb{R}^+ = \{x | x > 0\}$ since the measure μ is concentrated there.

D. *Occupancy for classical particles. Here $M = \mathbb{N} = \{0, 1, 2, 3,...\}$* and $\mu(\{k\}) = 1/k!$. The measure μ assigns the weight $1/k!$ to the possibility of k particles occupying the box. Let us explain why this is an appropriate measure. Suppose we had many, say n, boxes and several, say N, particles. The number of ways of distributing the particles so that k_1 particles lie in the first box, k_2 in the second, etc., is

$$\frac{N!}{k_1!\dots k_n!}.$$

(Here, of course, the particles are considered as 'distinguishable': interchanging a pair of particles between different boxes given a different way of distributing the particles. There are $N!$ different permutations of the particles, but we must divide by $k_1!\dots k_n!$, since permuting particles within a box does not give a different way of distributing particles.) If we are unaware of N and n, the best we can say is that the number of ways of distributing the particles so that k_1 end up in box number 1 is proportional to $1/k_1!$. This is our *a priori* assignment of relative probabilities.

E. *Occupancy according to Fermi–Dirac.* We have a box which can contain at most one particle. Here $M = \{0, 1\}$ and $\mu(\{0\}) = \mu(\{1\}) = 1$. This, of course, is a subcase of A. We could also take $M = \mathbb{N} =$ the non-negative integers, with $\mu(\{0\}) = \mu(\{1\}) = 1$ and $\mu(\{k\}) = 0$ for all $k > 1$. The scheme represents occupancy with an 'exclusion principle'. No more than one particle can occupy the box. The existence of zero and one particles in the box are regarded as equally likely, *a priori.* Of course, in actual applications, the 'box' may be rather abstract. It may represent a single particle quantum state, for example, for particles obeying the Pauli exclusion principle.

F. *Occupancy for Bose–Einstein particles.* Here $M = \mathbb{N} = \{0, 1, 2, 3, \dots\}$ and $\mu(\{k\}) = 1$. Thus all numbers of particles in the box are equally likely. If, for example, a particle represents a disturbance of medium (a blip on a screen, for instance) and two particles represent a disturbance of twice the intensity, then in distributing 'N units of disturbance' among n boxes the 'particles' are 'indistinguishable' in the sense that all distributions are equally likely. Thus the *a priori* measure assigned to getting k_1 particles in box number 1 is 1.

In quantum mechanics all particles are of either Fermi–Dirac or Bose–Einstein type. The Fermi–Dirac type particles, for example electrons, protons, neutrons, are called fermions. Particles of Bose–Einstein type (for example photons, pions) are called bosons.

G. *Binomial measure.* Suppose that we have a large number, say N, of particles, each of which can be in one of two possible states 'up' and 'down'. Suppose we are affected only by the total number of particles in the 'up' position minus the total number in the 'down' position. Then $M = \{-N, -N+2, \ldots, N-2, N\}$. Suppose, for simplicity, that N is even. Then #(up) − #(down) = $2m$ means that $(N/2)+m$ particles are up and $(N/2)-m$ particles are down. This can happen in

$$\binom{N}{\frac{N}{2}+m} = \frac{N!}{\left(\frac{N}{2}+m\right)!\left(\frac{N}{2}-m\right)!}$$

different ways.

If we are interested in the relative frequency of occurrence of $2m$ we divide by the largest binomial coefficient which is $\binom{N}{N/2}$ and thus define

$$\mu(\{2m\}) = \binom{N}{\frac{N}{2}+m}\bigg/\binom{N}{\frac{N}{2}} = \frac{[(N/2)!]^2}{(N/2+m)!(N/2-m)!}.$$

An application of Stirling's formula $N! \equiv N(2\pi N) \cdot N^N e^{-N}$ shows that for large values of N this last expression is fairly closely approximated by $e^{-2m^2/N}$.

This approximation suggests

H. *Discrete Gaussian measure.* We let $M = \mathbb{Z}$ = the set of all integers and

$$\mu\{m\} = e^{-2m^2/N}.$$

Notice that when $m/N = (1/2N)^{1/2}$ the value of $\mu(\{m\})$ is reduced to e^{-1} times $\mu(\{0\})$. Suppose we are interested in the quantity m/N. (That is suppose each 'up' or 'down' state contributes an effect of order $\pm 1/N$ to the quantity that we measure.) Then m/N is extremely sharply peaked about the origin if N is large. Thus, for example, if $N \sim 10^{24}$ then $\mu(\{m\})$ will have decreased by a factor of e^{-1} by the time that m/N has moved from the origin by about 10^{-12}.

To actually 'see' the distribution we have to change the scale by a factor of $N^{1/2}$ rather than N. If we set $x_k = 2k/N^{1/2}$ then

$$\mu\{a \leqslant x_k \leqslant b\} = \sum_a^b e^{-x_k^2/2}.$$

The sum on the right grows as approximately $N^{1/2}$. Indeed, if we divide by $(2\pi N)^{-1/2}$ then the right hand side tends to

$$\frac{1}{\sqrt{(2\pi)}}\int_a^b e^{-x^2/2}dx.$$

This leads to example

I. $M = \mathbb{R}$ and μ is the (normalized) continuous Gaussian measure

$$\mu([a,b]) = \frac{1}{\sqrt{(2\pi)}}\int_a^b e^{-x^2/2}dx.$$

J. As our final example we consider the case of *classical mechanics*. Here the space M is the phase space of the mechanical system. Phase space has distinguished coordinates (canonical coordinates) q, p and a distinguished measure (Liouville measure)

$$\mu = \mathrm{d}p\mathrm{d}q.$$

For example, for a single classical particle moving in $\mathbb{R}^3$, the coordinates q_1, q_2, q_3 describe its position and the coordinates p_1, p_2, p_3 describe its momentum. Then $M = \mathbb{R}^6$ and

$$\mathrm{d}p\mathrm{d}q = \mathrm{d}p_1\mathrm{d}p_2\mathrm{d}p_3\mathrm{d}q_1\mathrm{d}q_2\mathrm{d}q_3,$$

that is $\mathbb{R}^6$ with its usual measure.

22.7. Products and images

In this section we are going to discuss various ways of constructing systems. The first notion we wish to discuss is that of product. Suppose that we are given two systems, $(M_1, \mathscr{A}_1, \mu_1)$ and $(M_2, \mathscr{A}_2, \mu_2)$. We can then form the product system $(M_1 \times M_2, \mathscr{A}_1 \times \mathscr{A}_2, \mu_1 \times \mu_2)$. For example, suppose $(M_1, \mathscr{A}_1, \mu_1)$ and $(M_2, \mathscr{A}_2, \mu_2)$ are as in E, that is, they represent the possible number of particles in a box (figure 22.10). Then

Figure 22.9.

$(M_1 \times M_2, \mathscr{A}_1 \times \mathscr{A}_2, \mu_1 \times \mu_2)$ represents the system consisting of the pair of boxes together (figure 22.11).

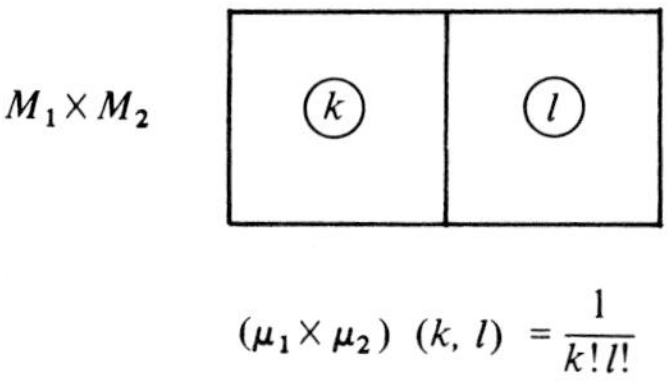

Figure 22.10

The measure $\mu_1 \times \mu_2$ is characterized by

$$(\mu_1 \times \mu_2)(A_1 \times A_2) = \mu_1(A_1)\mu_2(A_2)$$

for product sets $A = A_1 \times A_2, A_i \in \mathscr{A}_i, i = 1, 2$. A state ρ, on the product system need not, of course, be of the form $\rho_1 \times \rho_2$. If ρ *is* of the form $\rho = \rho_1 \times \rho_2$ then the corresponding probability measures $(\rho_1 \times \rho_2)(\mu_1 \times \mu_2) = \rho_1\mu_1 \times \rho_2\mu_2$ on $M_1 \times M_2$ is also a product measure. This corresponds to an assignment of *independent* probabilities to the events of M_1 and M_2.

Similarly, we can form the product of any collection of systems $(M_i, \mathscr{A}_i, \mu_i)$ where ι varies through an arbitrary index set, I.

Another way of constructing a system is via a map. Let $(M, \mathscr{A}, \mu)$ be a system and let $(N, \mathscr{B})$ be a measure space, that is N is a set with a distinguished class of subsets. Let $f\colon M \to N$ be a map. Suppose that for each $B \in \mathscr{B}$ the subset, $f^{-1}(B)$, of M belongs to the collection $\mathscr{A}$. We can then define the rule $f_*\mu$ which assigns to every $B \in \mathscr{B}$ the value

$$f_*\mu(B) = \mu(f^{-1}B).$$

If the function $f_*\mu$ on sets in $\mathscr{B}$ is a measure, then it is called the *pushforward* of the measure μ under the map f. Then we have a new system $(N, \mathscr{B}, f_*\mu)$ which we might call the pushed forward system. For example, suppose we begin with the product system $M = M_1 \times M_2$ of two systems of type E that we have just been considering. Let N consist of the large box with the partition removed (figure 22.12). where the map f is given by

$$f((k, l)) = k + l.$$

Figure 22.11

Thus the map f has the effect of ignoring the fact that there are k particles in the box on the left and l particles in the box on the right, and yields the less informative assertion that there are $k + l$ particles altogether in the larger region. Here, for any positive integer s,

$$(f_*\mu)(\{s\}) = \sum_{k+l=s} \frac{1}{k!l!} = \frac{1}{s!} \sum_{k+l=s} \frac{s!}{k!l!}.$$

Thus

$$(f_*\mu)(\{s\}) = \frac{2^s}{s!}.$$

Both the product construction and the map construction become more intuitive in this example if we modify example E slightly. Suppose we define M_V to be a system corresponding to a box of volume V by setting

$$M_V = N, \quad \mu_V(\{k\}) = \frac{V^k}{k!}.$$

Then if we have to systems corresponding to boxes of volume V_1 and V_2 (figure 22.13).

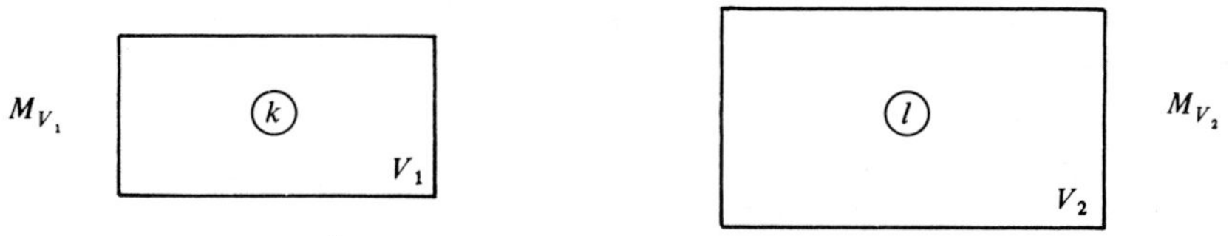

$$\mu_{V_1}(\{k\}) = \frac{V_1^k}{k!} \qquad \mu_{V_2}(\{l\}) = \frac{V_2^l}{l!}$$

Figure 22.12

then $M_{V_1} \times M_{V_2}$ corresponds to considering the two boxes together (figure 22.14)

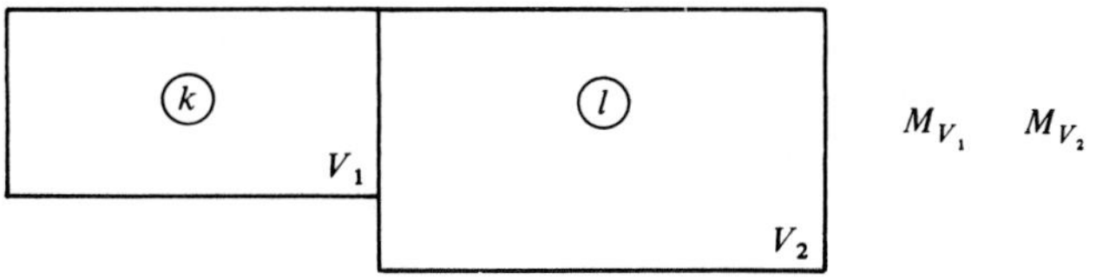

$$\mu_{V_1} \times \mu_{V_2}(\{k, l\}) = \frac{V_1^k V_2^l}{k!l!}$$

Figure 22.13

Then we can remove the partition and consider the box of volume $V_1 + V_2$ (figure 22.15)

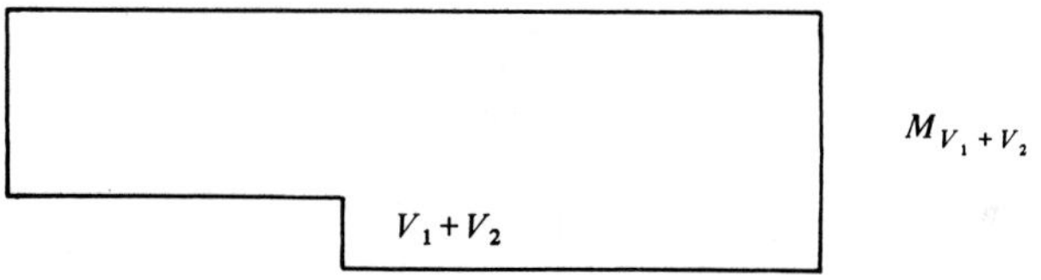

Figure 22.14

Again we set $N = M_{V_1+V_2}$ and $f: M_{V_1} \times M_{V_2} \to N$ by $f((k,l)) = k + l$. Then

$$\begin{aligned}(f_*\mu)(\{s\}) &= \sum_{k+l=s} \frac{V_1^k V_2^l}{k!l!} \\ &= \frac{1}{s!} \sum_{k+l=s} \frac{s!}{k!l!} V_1^k V_2^l \\ &= \frac{(V_1+V_2)^s}{s!}.\end{aligned}$$

Let us return to general considerations. Let $(M, \mathscr{A}, \mu)$ be a system and $F: M \to N$ give an image system $(N, \mathscr{B}, f_*\mu)$. If σ is any state on $(N, \mathscr{B}, F_*\mu)$ then the function $f^*\sigma = \sigma \circ f$ defines a state on $(M, \mathscr{A}, \mu)$. Indeed $f^*\sigma \geqslant 0$ and, it follows from the definition of integration that

$$\int f^*\sigma\mu = \int \rho(f_*\mu) = 1.$$

Thus states on N determines states on M.

A state ρ on M also determines a state $f_*\rho$ on N but the definition lies considerably deeper. Before stating the result in general let us illustrate by an example. Suppose $M_1 = \mathscr{R}^2$ with its standard choice of $\mathscr{A}$ and μ $(= \mathrm{d}x\,\mathrm{d}y)$. Let $f: \mathscr{R}^2 \to \mathscr{R}^+$ be given by $f(x, y) = (x^2 + y^2)^{1/2}$. Thus $f_*\mu = 2\pi r\,\mathrm{d}r$. Let $\rho(x, y)$ be any continuous function on the plane with $\int \rho \mathrm{d}x\mathrm{d}y = 1$. Thus ρ is a state. Define the state $f_*\rho$ by

$$(f_*\rho)(r) = \frac{1}{2\pi}\int_0^{2\pi} \rho(r\cos\theta, r\sin\theta)\,\mathrm{d}\theta.$$

Thus $f_*\rho(r)$ is the 'average' of ρ over the circle of radius r, i.e. the average of ρ over $f^{-1}(r)$. Then

$$\begin{aligned}\int_{\mathbb{R}^+} (f_*\rho)\,\mathrm{d}\mu &= \int_0^\infty (f_*\rho)(r) 2\pi r\,\mathrm{d}r \\ &= \int_0^\infty \int_0^{2\pi} \rho(r\cos\theta, r\sin\theta) r\,\mathrm{d}r\,\mathrm{d}\theta \\ &= \int_{-\infty}^\infty \int_{-\infty}^\infty \rho(x, y)\,\mathrm{d}x\,\mathrm{d}y = 1.\end{aligned}$$

Thus $f_*\rho$ is a state on N. Notice that $f_*f^*\sigma = \sigma$ but, in general, $f^*f_*\rho$ need not equal ρ. In the general situation, if ρ is a state on M then $\rho\mu$ is a probability measure on M. Therefore $f_*(\rho\mu)$ is a probability measure on N. Furthermore, if $B \in \mathscr{B}$ is any set such that

$$(f_*\mu)(B) = 0$$

then

$$\mu(f^{-1}(B)) = 0$$

and hence

$$(f_*(\rho\mu)(B) = (\rho\mu)(f^{-1}(B)) = \int_B \rho\mu = 0.$$

Thus $(f_*\mu)(B) = 0$ implies that $f_*(\rho\mu)(B) = 0$. A theorem in measure theory, called the Radon–Nikodym theorem, asserts that if v_1 and v_2 are measures such that $v_2(B) = 0$ implies $v_1(B) = 0$ for all B then there exists a function σ such that $v_1(B) = \int_B \sigma \mathrm{d}v_2$. Taking $v_1 = f_*(\rho\mu)$ and $v_2 = f_*\mu$ we conclude that there is a non-negative integrable function σ on N such that

$$(f_*(\rho\mu))(B) = \int_B \rho\mathrm{d}(f_*\mu)$$

for any B. Since $f_*(\rho\mu)$ is a probability measure we conclude that $\int_B \sigma \mathrm{d}(f_*\mu) = 1$. Thus σ is a state on $(N, \mathscr{B}, f_*\mu)$ and we define

$$f_*\rho = \sigma.$$

22.8. Observables, expectations and internal energy

A particular kind of map is an ordinary numerical valued function $f: M \to \mathbb{R}$. Such a map is called an *observable* of the system. Roughly speaking, an observable of a system is a numerical property of the system that we can measure. We shall, when discussing theoretical points, usually denote an observable by the letter J. Specific observables, as they arise, will be denoted by letters connected with their physical significance. Of course, as part of our definition of the concept of observable, we assume that $J^{-1}([a, b])$ belongs to the collection $\mathscr{A}$ for every interval $[a, b]$ on the real line. This is a regularity assumption about the map J.

Let J be an observable and let ρ be a state. For each interval $[a, b]$ we can consider the subset $J^{-1}([a,b])$ of M and then the integral

$$\int_{J^{-1}([a,b])} \rho\mu.$$

This is just the size of the set $J^{-1}([a,b])$ with respect to probability measure $\rho\mu$. In other words, it is the probability that a point of M lies in $J^{-1}([a,b])$ when we assign probabilities according to the probability measure $\rho\mu$. Or, which is just a different way of saying the same thing, it is the probability that the observable J take values in the interval $[a,b]$ when we assign probabilities on the set M according to the probability measure $\rho\mu$. We write this as

$$\text{Prob}\,(J\in[a,b],\ \text{when the system is in the state } \rho) = \int_{J^{-1}([a,b])} \rho\mu,$$

or more succinctly as

$$\text{Prob}\,(J\in[a,b];\rho) = \int_{J^{-1}([a,b])} \rho\mu. \tag{22.27}$$

Thus an observable and a state together give a probability assignment on the real line. The probability assigned to any interval being the probability of measuring the observable J in the interval when the system is in the state ρ; it is given by the preceding formula.

There is some useful language in probability theory that we shall adopt here. The integral

$$\int_M J\rho\mu$$

is called the *expectation*, or *expected value* of the function J with respect to the probability measure $\rho\mu$. It is an average of J where we weight subsets of M according to the probability assignment $\rho\mu$. Accordingly we shall define the *expected value of J in the state ρ* as

$$E(J;\rho) = \int_M J\rho\mu. \tag{22.28}$$

Here is an example. Let M be the phase space of a classical particle, so that $M = \mathbb{R}^6$ with its standard measure and with coordinates $q_1, q_2, q_3, p_1, p_2, p_3$. We will take J to be the energy function which we assume is the sum of a kinetic and a potential term, so that $J = H = K + \mathscr{V}$ where

$$H(q_1,q_2,q_3,p_1,p_2,p_3) = \tfrac{1}{2}(p_1^2 + p_2^2 + p_3^2) + \mathscr{V}(q_1,q_2,q_3)$$

where

$$K = \tfrac{1}{2}(p_1^2 + p_2^2 + p_3^2)$$

is the kinetic energy and $\mathscr{V}$ is the potential energy. A state is just a non-negative function, ρ, on $\mathbb{R}^6$ with total integral 1 and, by definition,

$$E(H;\rho) = \int_{\text{all } \mathbb{R}^6} H\rho\,\mathrm{d}q_1\mathrm{d}q_2\mathrm{d}q_3\mathrm{d}p_1\mathrm{d}p_2\mathrm{d}p_3. \tag{22.29}$$

Suppose we want to consider the energy to be that associated 'free particle constrained to lie in a box, B, of volume V'. We can do this by assuming an appropriate form for the potential $\mathscr{V}$. Let us take $\mathscr{V}$ to vanish when (q_1, q_2, q_3) lies inside the box B, and to have some extremely large value when (q_1, q_2, q_3) lies outside the box, B. The idea is that a particle has to overcome some huge potential barrier to get out of the box. We would then plug this choice of $\mathscr{V}$ into the expression for H in (22.29). Now suppose that the state ρ is such that it assigns negligible probability to extremely high values of the energy. That is, assume that $\mathrm{Prob}(H > E; \rho)$ is essentially equal to zero for E very large. To be precise, let us assume that this probability assignment is so small that in computing (22.29) we can ignore the contribution to the integral coming from values of (q_1, q_2, q_3) lying outside the box. Inside the box $\mathscr{V} = 0$ and the integral becomes

$$\int_{B \times \mathbb{R}^3} \tfrac{1}{2}(p_1^2 + p_2^2 + p_3^2)\rho(q_1, q_2, q_3, p_1, p_2, p_3)\mathrm{d}q_1\mathrm{d}q_2\mathrm{d}q_3\mathrm{d}p_1\mathrm{d}p_2\mathrm{d}p_3. \qquad (22.30)$$

To go further we need more information about the state ρ. For example, suppose that we expect that once we know that the particle is in the box, it is just as likely that it be at one location as at another, that is, ρ does not depend on (q_1, q_2, q_3) but is a function of (p_1, p_2, p_3) alone. Then the integral (22.30) becomes

$$V\int_{\mathbb{R}^3} \tfrac{1}{2}(p_1^2 + p_2^2 + p_3^2)\rho(p_1, p_2, p_3)\mathrm{d}p_1\mathrm{d}p_2\mathrm{d}p_3. \qquad (22.31)$$

Suppose that we are told that ρ is given by the so-called Maxwell velocity distribution law with parameter β; that is suppose that

$$\rho(p_1, p_2, p_3) = F_\beta^{-1}\exp[-(\beta\tfrac{1}{2})(p_1^2 + p_2^2 + p_3^2)]$$

where F_β is a constant of proportionality depending on β. We must choose F_β to make the total integral $\int_M \rho\mu = 1$. So we must choose $F_\beta = \int_M \rho\mu$, or

$$F_\beta = V(2\pi\beta)^{-3/2}$$

and (22.31) becomes

$$(\beta/2\pi)^{3/2}\int \tfrac{1}{2}(p_1^2 + p_2^2 + p_3^2)\exp[-(\beta\tfrac{1}{2})(p_1^2 + p_2^2 + p_3^2)]\mathrm{d}p_1\mathrm{d}p_2\mathrm{d}p_3.$$

This is a Gaussian integral which we can evaluate to get $(3/2)\beta^{-1}$. Thus with all these choices we would obtain

$$E(H; \rho) = (3/2)\beta^{-1}.$$

How are we supposed to think about all of this? From the point of view of mathematics, formulas (22.29)–(22.31) represent cut and dried integrations. They are standard procedures in the *mathematical* theory of probability. From the viewpoint of the *kinetic theory of gases* we might think of an enormous number of gas molecules confined to a box, but otherwise not interacting with one another. We then might think of $\rho(q_1, q_2, q_3, p_1, p_2, p_3)$ as measuring the relative frequency of gas molecules in a small region centered about the point $(q_1, q_2, q_3, p_1, p_2, p_3)$ in phase space. With this interpretation the averaging process going into the definition of $E(H;\rho)$ is an average of the gas molecules and we would then interpret

$E(H;\rho)$ as the average energy per gas molecule.

It took the genius of Gibbs to realize that we *don't have to commit ourselves to this interpretation.* We might be studying a system consisting of a single gas molecule in a box. The theory of probability applies just as well to this situation, and all the mathematics goes through unchanged. Having chosen some state, ρ, we can talk of the probability of the energy taking values in any interval, or of the particle having its position and momentum in some region of phase space once the state ρ is specified. Of course, if we want to probe any deeper we must investigate the general meaning of the word 'probability'. Unfortunately, although the mathematical foundations of the theory of probability were clarified over fifty years ago, the same cannot be said for the philosophical interpretation. Writing today, in 1987, there is still a raging controversy over what the 'true' meaning of probability is. At one extreme there are the frequentists who restrict the range of application of probability to situations in which the 'same' phenomenon recurs many times and probability is a form of frequency. At the other end are the extreme 'bayesians' who regard probability as providing rules of thought – the theory of probability expressing how every right minded 'rational' citizen should think. Extremists of each school regard the members of the other as being 'muddle-headed' etc. This philosophical argument has real consequences in actual statistical practice. The different views lead to different statistical procedures, especially when dealing with small samples. Fortunately, as far as statistical mechanics is concerned, because the samples involved are extremely large, these debates are irrelevant. All that matters is the mathematical aspect of the subject, that is, computations with the definitions we have given. However, Gibbs, writing a hundred years ago, before the mathematical theory had been developed, felt compelled to take a stand on the philosophical issue. He seems to have leaned toward the frequentist position. Therefore he would talk of an 'ensemble' in describing a probability measure such as $\rho\mu$, signifying the frequentist interpretation of the probability assignment. Unfortunately this terminology and the word 'ensemble' as a synonym for probability measure have gotten stuck in the standard texts.

Whatever the interpretation, we are now prepared to make the first link up between the notions of statistical mechanics and thermodynamics. If the observable we are considering is regarded as the energy, H, then

the internal energy of the state ρ *is defined to be* $E(H;\rho)$.

In other words, the internal energy is the expected value of the energy in the given state. Of course the energy function H is a fixed function on M. There are various different states, ρ, each giving a value to $E(H;\rho)$. In other words we write the above definition as

$$U(\rho) = E(H;\rho) \tag{22.32}$$

so the internal energy is a function on the space of states.

We have said that H is a fixed function on M. But of course we will, in general, want to allow H to depend on some additional parameters. For example, consider the situation we described above; a gas confined to a box. But now allow the 'box' to have various pistons and so its shape to vary according to the values $Q_1, Q_2, \ldots, Q_d$ which determine the position of the pistons. Then the potential energy function $\mathscr{V}$ (assumed to be 0 inside the box and essentially '$+\infty$' outside the box) depends on the shape of the box and hence on the parameters $Q_1, Q_2, \ldots, Q_d$.

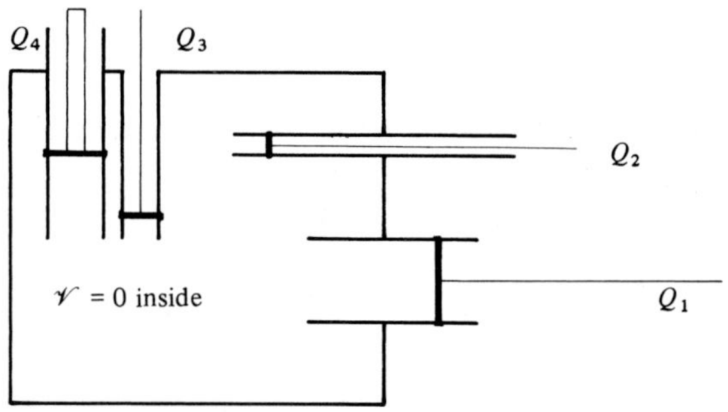

Figure 22.15 Dependence of $\mathscr{V}$ on parameters.

Thus U depends on these parameters as well,

$$U = U(\rho, Q_1, Q_2, \ldots, Q_d).$$

So we now have the notion of state and of internal energy. What are the equilibrium states? Once we have given the definition of entropy, we will use the maximum entropy principle to define and determine the equilibrium states. We will do this in general in section 22.11. But let us give a preview here for our special case of gas in a box.

For each positive value of β consider the function $e^{-\beta H}$ on $\mathbb{R}^6$. (Here β is a parameter which has units of inverse energy, so that βH is a numerical-valued function which can be exponentiated. Of course β will soon play the role we assigned to it in section 22.4). Since $\mathscr{V}$ and hence H take on the value $+\infty$ when (q_1, q_2, q_3) is outside the box, the function $e^{-\beta H}$ vanishes when (q_1, q_2, q_3) is outside the box, and is non-negative everywhere. So we could use $F^{-1}e^{-\beta H}$ as a state, where the constant F is chosen to make the total integral 1, that is

$$F = \int_{\mathbb{R}^6} e^{-\beta H} dq_1 dq_2 dq_3 dp_1 dp_2 dp_3. \tag{22.33}$$

We have essentially done this integration already. The q integration is only over the region where (q_1, q_2, q_3) lies in the box, and there H does not depend on (q_1, q_2, q_3). So the q integral gives a factor V, the volume of the box, while the p integral is a Gaussian integral and the final answer is

$$F = V(2\pi\beta)^{-3/2}. \tag{22.34}$$

Here F depends upon β and also upon $Q_1, Q_2, \dots, Q_d$, but only since V does. The dependence of F on the Qs is only through V. The basic definition (in this special case) is to assert that the states defined by

$$\rho_{\beta,Q} = F^{-1}\mathrm{e}^{-\beta H} \tag{22.35}$$

where $F = F(\beta, Q_1, Q_2, \dots, Q_d)$ is given by (22.33) are the equilibrium states.

Let us define

$$Z = \log F. \tag{22.36}$$

We may compute $\partial Z/\partial\beta$ by using (22.36) and differentiating under the integral sign in (22.33) to obtain

$$\begin{aligned}\partial Z/\partial\beta = F^{-1}\partial F/\partial\beta &= F^{-1}\int_{\mathbb{R}^6} -H\mathrm{e}^{-\beta H}\mathrm{d}q_1\mathrm{d}q_2\mathrm{d}q_3\mathrm{d}p_1\mathrm{d}p_2\mathrm{d}p_3 \\ &= -\int H\rho_{\beta,Q}\mathrm{d}q_1\mathrm{d}q_2\mathrm{d}q_3\mathrm{d}p_1\mathrm{d}p_2\mathrm{d}p_3 = -E(H;\rho_{\beta,Q})\end{aligned} \tag{22.37}$$

or

$$\partial Z/\partial\beta = -U. \tag{22.38}$$

Now let us return to the notion of reversible curve and work. Suppose that we slowly move one of the pistons, to change, say, Q_1. The corresponding change in H at some definite point of M, the function $\partial H/\partial Q_1$, would be the corresponding theoretical force term. At any given state ρ we must compute the average value to get the generalized force

$$E(\partial H/\partial Q_1;\rho).$$

Reversibility means that we are restricting the ρs to lie on the manifold of equilibrium states. Hence the work form is given by

$$\omega = E(\partial H/\partial Q_1;\rho)\mathrm{d}Q_1 + E(\partial H/\partial Q_2;\rho)\mathrm{d}Q_2 + \cdots + E(\partial H/\partial Q_k;\rho)\mathrm{d}Q_k.$$

Again, computing the derivative of Z with respect to the Q_i, we can write this as

$$-\beta\omega = (\partial Z/\partial Q_1)\mathrm{d}Q_1 + (\partial Z/\partial Q_2)\mathrm{d}Q_2 + \cdots + (\partial Z/\partial Q_k)\mathrm{d}Q_k \tag{22.39}$$

or, more symbolically as

$$\mathrm{d}_Q Z = -\beta\omega \tag{22.39a}$$

where the d_Q in (22.39a) means the right hand side of (22.39), i.e. that we compute the d operator in the Q directions (holding β fixed). Notice that our derivation of equations (22.37)–(22.39a) was quite general. It did not depend on the specific form of H. If we now assume that we are dealing with 'free gas in a box' where the function F and hence Z depend on the Qs only via the volume V, we can write $\partial Z/\partial Q_i = (\partial Z/\partial V)(\partial V/\partial Q_i)$ and (22.39) becomes

$$-\beta\omega = (\partial Z/\partial V)\mathrm{d}V.$$

Of course an increase in V corresponds to a decrease in H (since points that were outside the box where $\mathscr{V} = +\infty$ are now inside the box where $\mathscr{V} = 0$). With the usual convention to take pressure as positive this means that $\omega = -p\mathrm{d}V$. Thus the last equation becomes

$$\partial Z/\partial V = \beta p = v,$$

which shows that Z has all the properties we ascribed to the Massieu function.

In short we now have a rather general prescription for computing the Massieu function. Given a system $(M, \mathscr{A}, \mu)$ and a function H on M interpreted as the 'energy' define the *partition function* F by

$$F = \int e^{-\beta H}\mu. \tag{22.40}$$

The function H, and hence F, may depend on some auxiliary variables $Q_1, \ldots, Q_k$. Also the integral in (22.40) will converge only for some range in $(\beta, Q_1, \ldots, Q_k)$ space. Define the internal energy by (22.32), the equilibrium states by (22.35) (with F given by (22.40)) and the Massieu function by (22.36). Then (22.38) holds and the work form ω is given by (22.39a). This is the prescription for passing the 'microscopic model' coded into the function H to the 'macroscopic equilibrium phenomena' generated by the Massieu function. It is an explicit prescription, but we still must justify it. This will come once we undersand the probabilistic definition of entropy, a subject whose study we shall begin in the next section.

There is one slight generalization of what we have been doing which will prove convenient. Suppose that instead of considering a single observable, we want to consider several, $(J_1, \ldots, J_n)$, at once. We can regard $\mathbf{J} = (J_1, \ldots, J_n)$ as a single vector-valued observable with values in $\mathbb{R}^n$. Thus we want to allow W-valued observables where W is a vector space. A W-valued observable is just a function from M to W(satisfying the appropriate regularity conditions with respect to subsets as in the $\mathbb{R}$ valued case). The notion of an integral of a vector valued function makes good sense so the definition (22.28) of the expectation of a vector-valued observable with respect to a state carries through without change. Similarly we can talk of the probability (in a given state) for $\mathbf{J}$ to lie in some (nice) subset of W. We need not elaborate on these points.

22.9. Entropy

We now introduce the notion of the entropy of a state. Let $(M, \mathscr{A}, \mu)$ be a system. We wish to assign a number, $\mathrm{Ent}(\rho)$, to each state, ρ. The number $\mathrm{Ent}(\rho)$ is to be a measure of the 'disorder' of ρ relative to μ. Thus Ent is to be a function on the space of all states. In order to motivate our definition, we first examine the case of a finite sample space, case (A) of section 22.6. If $M = \{e_1, \ldots, e_k\}$ then a state ρ is specified by giving the k real numbers $p_i = \rho(e_i), i = 1, \ldots, k$. Thus p_i is the probability of the event $\{e_i\}$. In information theory, the entropy of the state $\rho = (p_1, \ldots, p_k)$ is defined as

$$\mathrm{Ent}_k(\rho) = -\sum_{i=1}^{k} p_i \log p_i \tag{22.41}$$

This function has various properties which make it very attractive as a measure of 'disorder' or lack of information. We list some of them:

(i) $\mathrm{Ent}_k(p_1,\ldots,p_k)$ is symmetric in the $(p_1,\ldots,p_k)$.

This corresponds to the requirement that our measure of disorder not depend on the way we label the outcomes.

(ii) $$\mathrm{Ent}_k(1,0,\ldots,0)=0$$

(Here we define $x\log x=0$ for $x=0$ by continuity.)

This corresponds to the assertion that a state in which we are sure of the outcome has zero disorder.

(iii) $$\mathrm{Ent}_{k+1}(p_1,\ldots,p_k,0)=\mathrm{Ent}_k(p_1,\ldots,p_k).$$

This corresponds to the idea that if we replace our system $M=\{e_1,\ldots,e_k\}$ with the system $M'=\{e_1,\ldots,e_k,e_{k+1}\}$ but only consider states in which e_{k+1} is impossible, then the entropy is unchanged. In other words, throwing in a fake alternative does not change the lack of information.

(iv) $$\mathrm{Ent}_k(p_1,\ldots,p_k)\leqslant \mathrm{Ent}\left(\frac{1}{k},\ldots,\frac{1}{k}\right)$$

with strict inequality if $p_i\neq 1/k$ for some i.

This says that the lack of information is a maximum when all alternatives are equally likely. For the proof of this assertion, observe that for all $x\geqslant 0$ we have

$$x\log x\geqslant x-1$$

with equality holding only at $x=1$, as can be verified by computing the derivatives of both sides. For $x\neq 0$ we can write this as

$$-\log x\leqslant x^{-1}-1.$$

Thus

$$-p_i\log p_i-\left(-p_i\log\frac{1}{k}\right)=-p_i\log\frac{p_i}{k^{-1}}$$

$$\leqslant -p_i\frac{k^{-1}}{p_i}-1=-\frac{1}{k}-p_i.$$

Summing over i the right side vanishes since $\sum(1/k)=1=\sum p_k$. Thus

$$-\sum p_i\log p_i\leqslant -\sum p_i\log\frac{1}{k}=-\log\frac{1}{k}=-\sum\frac{1}{k}\log\frac{1}{k}.$$

Notice that we can think of $\mathrm{Ent}(p_1,\ldots,p_k)$ as representing our ignorance *before* performing the experiment which tells us which of the k alternatives actually occurred. In this sense, $\mathrm{Ent}\,(p_1,\ldots,p_k)$ represents *the amount of information* obtained by performing the experiment.

(v) Suppose that $M=M_1\times M_2$ where M_1 consists of k elements $\{e_1,\ldots,e_k\}$ and M_2 consists of l elements, so that M consists of the kl elements $\{(e_i,f_j)\}$. Suppose that

$\rho = \rho_1 \times \rho_2$ so that $\dot{(}\{(e_i, f_j)\}) = p_i q_j$. Then

$$\begin{aligned}\mathrm{Ent}_{kl}(\rho) &= -\sum p_i q_j \log(p_i q_j)\\ &= -\sum p_i q_j (\log p_i + \log q_j)\\ &= -\sum p_i \log p_i - \sum q_j \log q_j.\end{aligned}$$

So since $\sum p_i = \sum q_j = 1$,

$$\mathrm{Ent}_{kl}(\rho_1 \times \rho_2) = \mathrm{Ent}_k(\rho_1) + \mathrm{Ent}_l(\rho_2).$$

In other words, if we conduct two independent experiments the total amount of information gained is the sum of the information gained in each.

It turns out that properties (i)–(iv), together with a slightly stronger version of (v), determine the function Ent (up to a multiplicative constant). Thus, for finite sample space, there is essentially only one way of measuring the 'disorder' of a state if this measure of disorder is to satisfy a few reasonable axioms. This measure of disorder is given by (22.41). We can rewrite (22.41) in such a way that it makes sense for any system.

For any system $(M, \mathscr{A}, \mu)$ and any state ρ we define

$$\mathrm{Ent}(\rho) = -\int \rho \log \rho \mathrm{d}\mu \tag{22.42}$$

provided that the integral converges.

22.10. Equilibrium in statistical systems

We now have a measure of 'disorder' for any state. Now suppose we are given a system and a sequence $J_1, \ldots, J_n$ of real valued observables. We can observe the expected values of these observables in any state. We collect the observables into a vector-valued observable $\mathbf{J} = (J_1, \ldots, J_n)$. Using the maximum entropy principle as motivation, we consider a state of 'statistical equilibrium' to be a state which maximizes 'disorder' subject to our knowledge of the expected value of $\mathbf{J}$. We can now pose a precise mathematical problem.

Let $\mathbf{J}: M \to V$ be a vector-valued observable of the system. If $\bar{\mathbf{J}}$ is any possible expected value of $\mathbf{J}$ we consider the following problem:

Among all states ρ for which

$$E(\mathbf{J}; \rho) = \int \rho \mathbf{J} \mu = \bar{\mathbf{J}}$$

find a state which maximizes the entropy.

We are thus looking for the solution of a maximum problem with constraints: *Maximize* Ent(ρ) *subject to the constraint that*

$$\int \rho \mathbf{J} \mu = \bar{\mathbf{J}}.$$

It turns out that if the system $(M, \mathscr{A}, \mu)$ and the observable $\mathbf{J}$ satisfy certain hypotheses which will be discussed below, this problem has a unique solution which we now proceed to describe.

Let V^* denote the dual space to V. Thus an element $\gamma \in V^*$ is a linear functional on V. We denote the value of γ at the vector $\mathbf{v}$ by $\gamma \cdot \mathbf{v}$. For any $\gamma \in V^*$ and $m \in M$ we can form $\gamma \cdot \mathbf{J}(m)$ which is a number depending on γ and m. Thus $\gamma \cdot \mathbf{J}$ is a numerical function on M. We can thus form the integral

$$F(\gamma) = \int_M e^{-\gamma \cdot \mathbf{J}} \mu.$$

Thus $0 < F(\gamma) \leqslant +\infty$ (the integral may diverge to $+\infty$). Suppose that γ is such that $F(\gamma)$ is finite. Then

$$\rho_\gamma = \frac{1}{F(\gamma)} e^{-\gamma \cdot \mathbf{J}} \tag{22.43}$$

defines a state of the system. Indeed ρ_γ is a positive function and by the definition of $F(\gamma)$ we have $\int \rho_\gamma \mu = 1$. For each value of γ where $F(\gamma) < +\infty$ we get a state ρ_γ. We thus have a collection of states, parametrized by a subset of V^*. Notice that if $F(\gamma) < +\infty$ then $\operatorname{Ent}(\rho_\gamma)$ is finite. Indeed

$$-\int \rho_\gamma \log \rho_\gamma \mu = -\int \rho_\gamma (-\log F(\gamma) - \gamma \cdot \mathbf{J}) \mu.$$

Now $F(\gamma)$ is a constant, when considered as a function of an M. Therefore

$$\int \rho_\gamma \log F(\gamma) \mu = \log F(\gamma).$$

Furthermore,

$$\int \rho_\gamma \mathbf{J} \mu = E(\mathbf{J}; \rho_\gamma)$$

is just the expectation of the vector-valued observable $\mathbf{J}$ in the state ρ_γ. Therefore we can write

$$\int \rho_\gamma \gamma \cdot \mathbf{J} \mu = \gamma \cdot E(\mathbf{J}; \rho_\gamma).$$

Thus

$$\operatorname{Ent}(\rho_\gamma) = \log F(\gamma) + \gamma \cdot E(\mathbf{J}; \rho_\gamma). \tag{22.44}$$

Let us take into account that the function $\mathbf{J}$ may also depend on some 'configurational' parameters, Q. If we write $S(\gamma, Q) = \operatorname{Ent}(\rho_\gamma, Q)$ and $Z(\gamma, Q) = \log F(\gamma, Q)$ then (22.44) becomes

$$S = Z + \gamma \cdot \mathbf{J}(\gamma, Q).$$

In the special case that $\mathbf{J} = H$ this is exactly (22.24). But (22.24) is the combined first and second law of thermodynamics and we derived.

Suppose that for a given value of $\bar{\mathbf{J}} \in V$ we can find a $\gamma \in V^*$ such that

$$\int \mathbf{J} \rho_\gamma \mu = E(\mathbf{J}; \rho_\gamma) = \bar{\mathbf{J}}.$$

(In particular we are assuming that $\int \mathbf{J} \rho_\gamma \mu$ converges absolutely.)

We claim that the state ρ is then the unique solution to the maximization problem. In other words, we have the following theorem.

Theorem: Let $\gamma \in V^*$ *be such that* $F(\gamma) < +\infty$, *and* (22.44) *holds, where* ρ_γ *is defined by* (22.43). *Let* ρ *be any other state with*

$$\int \rho \mathbf{J} \mu = \bar{\mathbf{J}}. \tag{22.45}$$

Then

$$\mathrm{Ent}(\rho) \leqslant \mathrm{Ent}(\rho_\gamma)$$

with strict inequality if ρ *and* ρ_γ *differ on a set of positive* μ*-measure.*

Proof The proof is a repetition of the argument we gave to show that $(1/k, \ldots, 1/k)$ maximizes, the entropy for a finite system. For any state ρ and any $m \in M$ we have

$$-\rho(m)\log\rho(m) - (-\rho(m)\log\rho_\gamma(m)) \leqslant \rho_\gamma(m) - \rho(m). \tag{22.46}$$

Indeed, if $\rho(m) = 0$ the left hand side vanishes and the right hand side is positive. If $\rho(m) > 0$ then

$$\begin{aligned} -\rho(m)\log\rho(m) - (-\rho(m)\log\rho_\gamma(m)) &= -\rho(m)\log[\rho(m)/\rho_\gamma(m)] \\ &\leqslant \rho(m)([\rho_\gamma(m)/\rho(m)] - 1) = \rho_\gamma(m) - \rho(m), \end{aligned}$$

the inequality holding since $-\log x \leqslant x^{-1} - 1$.

Integrating over M yields

$$-\int \rho \log \rho \mu - \int \rho \log \rho_\gamma \mu \leqslant 0$$

since

$$\int \rho \mu = \int \rho_\gamma \mu = 1.$$

But

$$\log \rho_\gamma = -\log F(\gamma) - \gamma \cdot \mathbf{J}$$

ans so

$$\begin{aligned} -\int \rho \log \rho_\gamma \mu &= \log F(\gamma) + \int \rho(\gamma \cdot \mathbf{J})\mu = \log F(\gamma) + \gamma \cdot \left(\int \rho \mathbf{J} \mu \right) \\ &= \log F(\gamma) + \gamma \cdot \mathbf{J} = \mathrm{Ent}(\rho_\gamma), \end{aligned}$$

since $E(\mathbf{J}; \rho_\gamma) = \bar{\mathbf{J}}$.

Thus

$$\mathrm{Ent}(\rho) \leqslant \mathrm{Ent}(\rho_\gamma).$$

(Notice also that inequality (20.46) shows that

$$-\rho \log \rho \leqslant -\rho[\log F(\gamma) + \gamma \mathbf{J}] + \rho_\gamma - \rho$$

and the right hand side of this inequality is an absolutely integrable function. Thus if $-\int \rho \log \rho \mu$ diverges, it must diverge to $-\infty$.)

Notice also that inequality (20.46) is strict if $\rho(m)/\rho_\gamma(m) \neq 1$. Thus if $\rho(m) \neq \rho_\gamma(m)$ for a set of positive measure we must have $\mathrm{Ent}(\rho) < \mathrm{Ent}(\rho_\gamma)$. This completes the proof of the theorem.

As a consequence of the above theorem we now give the (statistical) definition of an equilibrium state.

Definition: Let $\mathbf{J}$ *be an observable on* $(M, \mathscr{A}, \mu)$. *If* γ *is such that* $F(\gamma) = \int e^{-\gamma \cdot \mathbf{J}} \mu < \infty$ *then*

$$\rho_\gamma = \frac{1}{F(\gamma)} e^{-\gamma \cdot \mathbf{J}}$$

is called an equilibrium state of the system (relative to the observable $\mathbf{J}$).

We must show that (under suitable hypotheses) for any value of $\bar{\mathbf{J}}$ there exists a state ρ_γ such that $E(\mathbf{J}; \rho_\gamma) = \bar{\mathbf{J}}$. The proof of our theorem shows that if such a ρ_γ exists it must be unique. We defer this question to a later section. We first give various examples of the notion of equilibrium state, using the systems introduced in those examples of section 22.6.

A1. Let $M = \{e_1, \ldots, e_k\}$ and μ be as in example A of section 22.6. Let $V = \mathbb{R}$ be one-dimensional and let $\mathbf{J}: M \to \mathbb{R}$ be given by $\mathbf{J}(e_i) = \varepsilon_i$ where the ε_i are real numbers (which we may think of as 'energy levels'). With no loss of generality we may assume the labelling of the elements of M is such that

$$\varepsilon_1 \leqslant \varepsilon_2 \leqslant \cdots \varepsilon_k.$$

Then $V^* = \mathbb{R}$ and for any $\beta \in V^*$ we have

$$F(\beta) = \sum_{p=1}^{k} e^{-\beta \varepsilon_p}$$

so that the equilibrium state, ρ_β, is given by

$$\rho_\beta(e_i) = \frac{e^{-\beta \varepsilon_i}}{\sum_{p=1}^{k} e^{-\beta \varepsilon_p}}.$$

This is known as the 'Maxwell–Boltzmann distribution'.

We recall the computation of the expected value of the 'energy':

$$E(\mathbf{J}; \rho_\beta) = \frac{\sum_i \varepsilon_i e^{-\beta \varepsilon_i}}{\sum_i e^{-\beta \varepsilon_i}}$$

or more succintly,

$$E(\mathbf{J}; \rho_\beta) = -\frac{\partial \log F(\beta)}{\partial \beta}.$$

(This last equation represents a general rule, as we have seen.) Then

$$-\frac{\partial E(\mathbf{J};\rho_\beta)}{\partial\beta}=\frac{\sum\varepsilon_i^2 e^{-\beta\varepsilon_i}}{\sum e^{-\beta\varepsilon_i}}-\left(\frac{\sum\varepsilon_i e^{-\beta\varepsilon_i}}{\sum e^{-\beta\varepsilon_i}}\right)^2$$

$$=\int\|\mathbf{J}\|^2\rho_\beta\mu-\left(\int\mathbf{J}\rho_\beta\mu\right)^2$$

$$=\int(\mathbf{J}-\bar{\mathbf{J}})^2\rho_\beta\mu\leqslant 0$$

with strict inequality unless all the ε_i are equal. (If all the ε_i are equal there is only one possible value for $\mathbf{J}$, namely the common value of all the ε_i, and all the ρ_β are the same: $\rho_\beta(e_i)=1/k$. Thus, in this case, our theorem reduces to assertion (iii).) If $\varepsilon_1\neq\varepsilon_k$ then $E(\mathbf{J};\rho_\beta)$ is a strictly decreasing function of β. It is clear that

$$\lim_{\beta\to-\infty}E(\mathbf{J};\rho_\beta)=\varepsilon_k\quad\text{and}\quad\lim_{\beta\to\infty}E(\mathbf{J};\rho_\beta)=\varepsilon_1.$$

Thus, any value of $\bar{\mathbf{J}}$ such that $\varepsilon_1<\mathbf{J}<\varepsilon_k$ can be achieved by a unique choice of β. For any value of β we have

$$\mathrm{Ent}(\rho_\beta)=-\rho_\beta(e_i)\log\rho_\beta(e_i)$$

$$=-\sum_i\frac{e^{-\beta\varepsilon_i}}{\sum_p e^{-\beta\varepsilon_p}}\left[\log(e^{-\beta\varepsilon_i})-\log\left(\sum_j e^{-\beta\varepsilon_j}\right)\right]$$

$$=\frac{\sum\beta\varepsilon_i e^{-\beta\varepsilon_i}}{\sum e^{-\beta\varepsilon_p}}+\log F(\beta)$$

$$=\beta\cdot\bar{\mathbf{J}}+\log F(\beta).$$

This is of course a special case of (22.24).

A2. Let us again consider the case of a finite system M where, for purposes of convenience, we label the elements as $e_0,\ldots,e_k$ so that there are $k+1$ elements in M. Let $\mathbf{J}=(J_1,\ldots,J_k)$ map M into $\mathbb{R}^k$ where

$$J_i(e_j)=\delta_i^j.$$

Thus $\mathbf{J}$ maps M onto the vertices of the simplex in $\mathbb{R}^k$ spanned by the basis vectors.

In this case $\gamma=(\gamma_1,\ldots,\gamma_k)$ and

$$F(\gamma)=1+e^{-\gamma_1}+\cdots+e^{-\gamma_k}$$

clearly converges for all values of γ with

$$\rho_\gamma(e_0)=\frac{1}{F(\gamma)},\rho_\gamma(e_1)=\frac{e^{-\gamma_1}}{F(\gamma)},\ldots,\rho_\gamma(e_k)=\frac{e^{-\gamma_k}}{F(\gamma)}.$$

If we denote $\rho_\gamma(e_i)$ by p_i then it is clear that any point $\mathbf{p}=(p_1,\ldots,p_k)$ in the interior of the simplex can be achieved by a suitable choice of the γ_i. In this case all states corresponding to interior points $\mathbf{p}$ are equilibrium states.

The mapping $\gamma\to E(\mathbf{J}\rho_\gamma)$ maps all of $\mathbb{R}^n=V^*$ onto the interior of the unit simplex.

The reader can check that this map is one-to-one with a differentiable inverse.

B. Let $(\mathbb{R}, \mathscr{A}, \mu)$ be the real line with its standard measure. If we take $\mathbf{J}$ to be the identity map $\mathbf{J}(x)=x$ then the corresponding $Z(\gamma)=\int_{-\infty}^{\infty}\mathrm{e}^{-\gamma x}\mathrm{d}x$ diverges for all values of γ. However, let us consider the map $\mathbf{J}:\mathbb{R}\to\mathbb{R}^2$ given by $J(x)=(x,x^2)$. Then for all $\boldsymbol{\gamma}=(\gamma_1,\gamma_2)$ with $\gamma_2>0$ the function

$$F(\gamma)=\int_{-\infty}^{\infty}\mathrm{e}^{-\gamma_1 x-\gamma_2 x^2}$$

converges. The corresponding equilibrium state

$$\rho_\gamma=F(\gamma)^{-1}\mathrm{e}^{-\gamma_1 x-\gamma_2 x^2}=F(\gamma)^{-1}\mathrm{e}^{\gamma_1^2/4\gamma_2}\mathrm{e}^{-(x+\gamma_1/2\gamma_2)/2(1/2\gamma_2)}$$

is clearly a normal density with expectation $m=\gamma_1/2\gamma_2$ and variance $\sigma=(2\gamma_2)^{-1/2}$. It is clear that for any $\sigma>0$ and any m we can find $\gamma=(\gamma_1,\gamma_2)$ such that

$$\rho_\gamma(x)=\frac{1}{\sigma\sqrt{(2\pi)}}\mathrm{e}^{-(x-m)^2/2\sigma^2}.$$

Thus among all random variables with derivatives and having a given expectation and variance, the normal distributions maximize the entropy. To compute the entropy we observe that by translational invariance we may as well assume that $m=0$. Then

$$\begin{aligned}\mathrm{Ent}(\rho_\gamma)&=\frac{1}{\sigma\sqrt{(2\pi)}}\int\mathrm{e}^{-x^2/2\sigma^2}\left[\frac{1}{2}\log 2\pi+\log\sigma+\frac{x^2}{2\sigma^2}\right]\mathrm{d}x\\&=\log\sigma+\tfrac{1}{2}\log 2\pi+1.\end{aligned}$$

Notice that the entropy tends to $-\infty$ as $\sigma\to 0$. This corresponds to the fact as $\sigma\to 0$ the ρ is more and more concentrated about a point and an 'infinite amount of information is required to pick a point out of a continuum'.

C. Let us consider the observables $J_1=x$ and $J_2=\log x$ on $\mathbb{R}^+$. Then $\boldsymbol{\gamma}=(\gamma_1,\gamma_2)$, $\mathbf{J}=(J_1,J_2)$ and

$$\begin{aligned}F(\gamma)&=\int_0^{\infty}\mathrm{e}^{-\gamma_1 x}\mathrm{e}^{-\gamma_2\log x}\mathrm{d}x\\&=\int_0^{\infty}\mathrm{e}^{-\gamma_1 x}x^{-\gamma_2}\mathrm{d}x\end{aligned}$$

converges for $\gamma_1>0$ and $\gamma_2<-1$.

If we set $k=-\gamma_2+1$ and $y=\gamma_1 x$ the integral becomes

$$\begin{aligned}F(\gamma)&=\int_0^{\infty}\mathrm{e}^{-\gamma_1 x}x^{k-1}\mathrm{d}x\\&=\gamma_1^{-k}\int_0^{\infty}\mathrm{e}^{-y}y^{k-1}\mathrm{d}y\\&=\gamma_1^{-k}\Gamma(k).\end{aligned}$$

The corresponding distributions are called the *gamma distributions*. The density ρ_γ is now

$$\rho_\gamma = \frac{1}{\sqrt{Z(\gamma)}} e^{-\gamma_1 x - \gamma_2 \log x}.$$

We want to compute the vector valued expectation

$$E(\mathbf{J}) = (E(J_1), E(J_2)).$$

We have

$$E(J_1) = \int_0^\infty x e^{-\gamma_1 x - \gamma_2 \log x} dx$$

$$E(J_2) = \int_0^\infty \log x \cdot e^{-\gamma_1 x - \gamma_2 \log x} dx.$$

So,

$$E(J_1) = \int_0^\infty x^{1-\gamma_2} e^{-\gamma_1 x} dx.$$

Again, we set $y = \gamma_1 x$ and $k = 1 - \gamma_2$ which give

$$E(J_1) = \gamma_1^{k-1} \int_0^\infty e^{-y} y^k dy = \gamma_1^{k-1} \Gamma(k+1).$$

The evaluation of $E(J_2)$ is left as an exercise.

D. Let $\mathbf{J}$ be the identity observable given by $\mathbf{J}(k) = k$. Then for $\beta \in \mathbb{R}^1$ we have that

$$F(\beta) = \sum \frac{e^{-\beta \cdot k}}{k!} = e^{-\beta}$$

converges for all values of β. We set

$$e^{-\beta} = \lambda, \quad \text{then} \quad F(\beta) = e^{\lambda}.$$

(The parameter λ is frequently called the "activity".) Then

$$\rho_\beta(k) = \frac{1}{F(\beta)} e^{-\beta \cdot k}$$

and so

$$\text{Prob}\{J\rho_\beta = k\} = e^{-\lambda} \frac{\lambda^k}{k!}.$$

This assignment of probabilities is called the *Poisson* distribution. Notice that

$$E(\mathbf{J}; \rho_\beta) = -\partial \log F(\beta)/\partial\beta = -\partial e^{-\beta}/\partial\beta = e^{-\beta}$$

or

$$E(J, e_\beta) = \lambda.$$

In other words, the parameter λ measures the expected number of particles in the box. Let us pause in our list of examples to summarize the general theoretical results once more. In classical statistical mechanics

a *system* is a measure space $(M, \mathscr{A}, \mu)$,

a *state* is a non-negative function ρ on M such that $\int \rho\mu = 1$,
the *entropy* of a state is $\text{Ent}(\rho) = -\int \rho \log \rho\mu$,
an *observable* is a vector-valued function on M,
the *expectation* of an observable K in a state ρ is $E(K;\rho) = \int K\rho\mu$.

If J is a particular V-valued observable the *partition function* associated to J is the function defined for $\gamma \in V^*$ by

$$F(\gamma) = \int e^{-\gamma J}\mu$$

This is defined on the subset C of V^* where this integral converges.

The *Massieu* function Z is defined on C by $Z = \log F$. The *equilibrium states* relative to J are the states of the form

$$\rho_\gamma = F(\gamma)^{-1} e^{-\gamma \mathbf{J}}$$

for $\gamma \in C$. Whenever γ lies in the interior of C we can compute the expectation of $\mathbf{J}$ in the equilibrium state ρ_γ from the Massieu function by the formula

$$J_i = -\partial Z/\partial \gamma_i$$

which we may write as

$$\mathbf{J} = -\partial Z/\partial \gamma$$

for short. Then

$$S(\gamma) = \text{Ent}(\rho_\gamma) = Z + \gamma \cdot \mathbf{J}.$$

In particular, we can determine γ from $\mathbf{J}$ by the formula

$$\gamma = \partial S/\partial \mathbf{J}.$$

(Here we are thinking of S as a function of $\mathbf{J}$ by considering γ as a function of J by inverting the given function $\mathbf{J} = \mathbf{J}(\gamma)$.) The observable $\mathbf{J}$ may depend on some auxiliary variables, $Q_1, \ldots, Q_d$. That is $\mathbf{J}$ may be a function of $M \times B$ where B is some space with coordinates $Q_1, \ldots, Q_d$. Then the function F and the equilibrium states ρ_γ will also depend on these auxiliary parameters.

If one of the components of $\mathbf{J}$ is called H – the 'energy' – then the *internal energy* of a state ρ is defined to be $E(H;\rho)$. With these definitions, and the discussion of work that we gave in section 22.8 (cf. equation (22.40)) we now have all the ingredients to pass from the 'microscopic' model to the 'macroscopic' observed phenomena.

22.11 Quantum and classical gases

Let us begin by comparing the partition functions for the three systems, D, E, and F in section 22.8. For each of these systems we shall consider a two-dimensional observable $\mathbf{J} = (N, H)$ as before, where N is to be thought of as the 'number of occupancy' and H as the 'energy level', and where the functions N and H are related by

$$H = N\varepsilon$$

Table 22.2

System	$\mu(k)$	Partition function	$E(N)$
D	$1/k!$	$\exp[e^{-(\beta_1+\beta_2\varepsilon)}] = \exp[e^{(\mu-\varepsilon)/T}]$	$e^{(\mu-\varepsilon)/T}$
E	$\mu(0)=\mu(1)=1$ $\mu(k)=0$ if $k>1$	$1+e^{-(\beta_1+\beta_2\varepsilon)} = 1+e^{(\mu-\varepsilon)/T}$	$\dfrac{1}{1+e^{-(\mu-\varepsilon)/T}}$
F	1	$\dfrac{1}{1-e^{-(\beta_1+\beta_2\varepsilon)}} = \dfrac{1}{1-e^{(\mu-\varepsilon)/T}}$	$\dfrac{1}{e^{-(\mu-\varepsilon)/T}-1}$

where ε is thought as the 'energy of the single state occupancy'. We have already done the computation for case D in the preceding section, but let us now do them all at once in the form of a table: Table 22.2.

In all three cases we compute the last column by $E(N) = -\partial \log F/\partial\beta_1$. The sum involved in the computation in case F is just a geometric series. In case E it is just the sum of two terms. Notice that in all three cases, $E(N)$ is small if and only if $e^{-(\mu-\varepsilon)/T}$ is large, in which case the ± 1 occurring in the denominator of the expression for Z is negligible. When this happens the probability assignments to the observable N are quite close. In Table 22.3 we have set

$$\lambda = e^{(\mu-\varepsilon)/T}.$$

If we think of the expected occupancy number as a measure of 'concentration' then for a 'dilute system' the Boltzmann, Fermi–Dirac, and Bose–Einstein probability assignments will be quite close. That is, the terms in the right hand column of Table 22.3 will all be close to one another when λ is very small. But for high 'concentration' they can differ markedly. Figure 22.17 is a graph of the three functions e^{-t}, $1/(e^t+1)$ and $1/(e^t-1)$. Notice that for $t>2$ the curves are practically identical, but for small values of t (corresponding to high concentrations) they diverge from one another considerably.

Let us now consider the situation where we have d copies of one of these systems, each with a different value of the single occupancy energy, so that we have d values $\varepsilon_1, \ldots, \varepsilon_d$. Our system now consists of the direct product

$$M = M_1 \times \cdots \times M_d.$$

Table 22.3

System	Name	Prob $(N=k)$
D	Boltzmann–Poisson	$e^{-\lambda}\lambda^k/k!$
E	Fermi–Dirac	$\begin{cases} \text{Prob}(N=0) = 1/(1+\lambda) \\ \text{Prob}(N=1) = \lambda/(1+\lambda) \\ \text{Prob}(N>1) = 0 \end{cases}$
F	Bose–Einstein (geometric)	$(1-\lambda)\lambda^k$

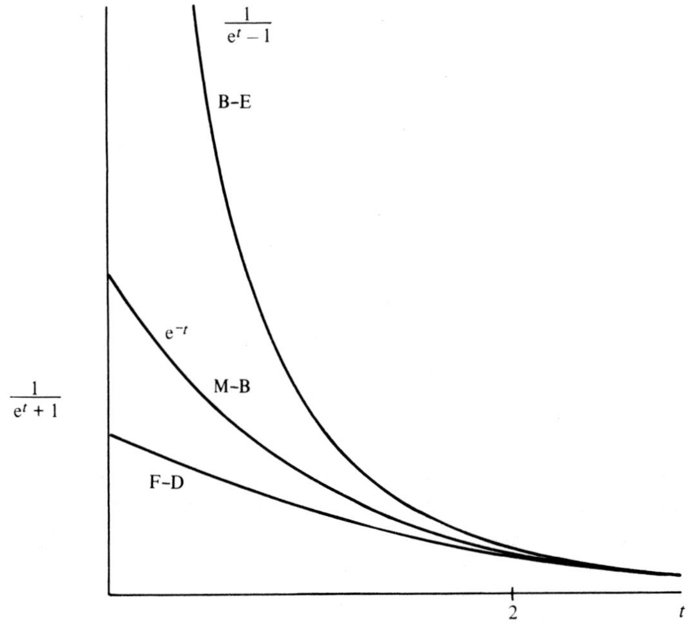

Figure 22.16 Plot of the Boltzmann, Fermi–Dirac and Bose–Einstein expectation values for the number of particles as a function of $(\mu - \varepsilon)/T = t$.

We assume that the 'interaction energy' is negligible – that is we assume that

$$H = H_1 + \cdots + H_d.$$

and define

$$N = N_1 + \cdots + N_d.$$

Then the partition function for the total system is just the product of the partition functions for each component system:

$$F = F_1 \times \cdots \times F_d$$

with

$$F_i = F_{\varepsilon_i}(\beta_1, \beta_2)$$

where we have to plug in ε_i for ε in our expression for the partition function from Table 22.2.

Let us say once more what this means. We have d systems. We have put them together to get one system. This corresponds to allowing them to 'interact'. The equilibrium states for the ith subsystem would be determined by its own two parameters, β_{1i} and β_{2i}. The fact that the systems are all in equilibrium means that we have a common value of β_1 and β_2 for all of them. We have recovered Gibbs' criterion (22.26).

So far we have been very abstract and not chosen any physical description of our component systems. Now consider the following model of 'gas in a box' (different from the approach we took before). Consider a subregion in phase space of the form $B \times \square$ where $B \subset \mathbb{R}^3$ is our box and $\square \subset \mathbb{R}^3$ is a little cube (say) centered about some point $\mathbf{p}$ in momentum space. To say that a gas molecule lies in our region means that it is in the box and that its momentum is close to $\mathbf{p}$. The occupation number for this subsystem tells us how many particles lie in the box and have momentum close to $\mathbf{p}$. The subsystems 'interact' by collisions, that is by interchanging momenta. Let us assume that the energy of a single particle in the subregion is its kinetic energy. Thus

$$\varepsilon(\mathbf{p}) = \tfrac{1}{2}\|\mathbf{p}\|^2 = \tfrac{1}{2}(p_x^2 + p_y^2 + p_z^2)$$

is the 'single occupancy' energy for the subsystem centered at $\mathbf{p}$. Thus the Boltzmann–Poisson prescription says that 'expected number of particles in our subsystem' is proportional to

$$\exp[(\mu - \tfrac{1}{2}\|\mathbf{p}\|^2)/T].$$

This is just the Maxwell-Boltzmann distribution, where the additional parameter μ adjusts for the density of the gas. Of course we have passed over two steps of a technical nature – allowing for an infinite number of subsystems to fill up all of phase space and letting the size of the phase space regions shrink to zero. Neither of these presents great difficulties.

But we could apply the same idea to the Fermi–Dirac or to the Bose–Einstein statistics. Thus for Fermi–Dirac the Maxwell–Boltzman distribution would be replaced by

$$f(\mathbf{p}) = \frac{1}{\exp[(\frac{1}{2}\|\mathbf{p}\|^2 - \mu)/T] + 1}.$$

with a similar replacement for the Bose–Einstein case.

22.12. Determinants and traces

We will now express the formula

$$F(\beta_1, \beta_2) = F_1(\beta_1, \beta_2) \times F_2(\beta_1, \beta_2) \times \cdots \times F_d(\beta_1, \beta_2)$$

(for each of our three cases) in a fashion that looks weird, but has profound consequences. Let $V = \mathbb{R}^d$ and let X be a diagonal matrix whose ith entry is ε_i.

Let us first consider the Boltzmann–Poisson case. Then since the product of exponentials is the exponential of a sum we have

$$\begin{aligned} F(\beta_1, \beta_2) &= \exp[\mathrm{e}^{-(\beta_1 + \beta_2\varepsilon_1)} + \cdots \mathrm{e}^{-(\beta_1 + \beta_2\varepsilon_d)}] \\ &= \exp(\mathrm{tr}\,\mathrm{e}^{-(\beta_1 I + \beta_2 X)}) \\ &= \mathrm{Det}(\exp \mathrm{e}^{-(\beta_1 I + \beta_2 X)}) \end{aligned}$$

since, for any matrix we have $\mathrm{Det}(\exp A) = \exp(\mathrm{tr}\, A)$.

Next let us look at the Fermi–Dirac case. Then since the determinant of any

Table 22.4

System	Partition function
Boltzmann	$\mathrm{Det}(\exp \mathrm{e}^{-(\beta_1 I+\beta_2 X)})$
Fermi–Dirac	$\mathrm{Det}(I+\mathrm{e}^{-(\beta_1 I+\beta_2 X)})$
Bose–Einstein	$\mathrm{Det}[I-\mathrm{e}^{-(\beta_1 I+\beta_2 X)}]^{-1}$

diagonal matrix is just the product of its diagonal entries,

$$F(\beta_1,\beta_2)=\det\begin{pmatrix} F_1(\beta_1,\beta_2) & 0 & \cdots & 0 \\ 0 & F_2(\beta_1,\beta_2) & \cdots & 0 \\ \cdots & \cdots & \cdots & \\ 0 & & & F_d(\beta_1,\beta_2)\end{pmatrix} \tag{22.48}$$

The matrix on the right of (22.48) is the same as the matrix

$$I+\mathrm{e}^{-(\beta_1 I+\beta_2 X)}$$

and so

$$F=\mathrm{Det}(I+\mathrm{e}^{-(\beta_1 I+\beta_2 X)}).$$

Similarly, in the Bose–Einstein case the corresponding equation is

$$F=\mathrm{Det}[I-\mathrm{e}^{-(\beta_1 I+\beta_2 X)}]^{-1}.$$

In other words, we get Table 22.4.

If we compare Table 22.2 with Table 22.4 we see that the change consists of replacing the scalar $\beta_1+\beta_2\varepsilon$ by the matrix $\beta_1 I+\beta_2 X$ and taking the determinant.

We will now express the process of taking the determinant in a slightly different fashion. Before doing so we make a number of preliminary remarks about linear algebra. Let V be a vector space and A a linear transformation of V. Then A induces a linear transformation $A\otimes A$ of $V\otimes V$ determined by

$$(A\otimes A)(u\otimes v)=Au\otimes Av.$$

If $A=\mathrm{e}^{tY}$ then differentiating the equation

$$\mathrm{e}^{tY}\otimes\mathrm{e}^{tY}(u\otimes v)=\mathrm{e}^{tY}u\otimes\mathrm{e}^{tY}v$$

with respect to t at $t=0$ gives

$$\begin{aligned}(\mathrm{d}/\mathrm{d}t)[\mathrm{e}^{tY}\otimes\mathrm{e}^{tY}(u\otimes v)]|_{t=0} &= Yu\otimes v+u\otimes Yv\\ &=(Y\otimes I+I\otimes Y)[u\otimes v]\end{aligned}$$

This shows that $\mathrm{e}^{tY}\otimes\mathrm{e}^{tY}=\exp tZ$ where Z is the operator

$$Y\otimes I+I\otimes Y.$$

In other words,

$$\mathrm{e}^{tY}\otimes\mathrm{e}^{tY}=\mathrm{e}^{t[Y\otimes I+I\otimes Y]}.$$

Similarly, if we consider the induced action on $V\otimes V\otimes V$ we obtain

$$\mathrm{e}^{tY}\otimes\mathrm{e}^{tY}\otimes\mathrm{e}^{tY}=\mathrm{e}^{t[Y\otimes I\otimes I+I\otimes Y\otimes I+I\otimes I\otimes Y]},$$

with a similar expression for $V\otimes V\otimes V\otimes V$ and so on. We shall use the general expression for the matrix that occurs in the exponential on the right hand side as $D_k(Y)$. Thus

$$D_2(Y)=Y\otimes I+I\otimes Y \qquad \text{on } V\otimes V,$$
$$D_3(Y)=Y\otimes I\otimes I+I\otimes Y\otimes I+I\otimes I\otimes Y \quad \text{on } V\otimes V\otimes V,$$

and so on. Thus, for example, setting $t=1$, on $V\otimes V\otimes V$ we would get

$$e^Y\otimes e^Y\otimes e^Y=e^{D_3(Y)}.$$

Taking the trace of the left-hand side of this equation gives $(\operatorname{tr} e^Y)^3$, since $\operatorname{tr}(A\otimes B\otimes C)=(\operatorname{tr} A)(\operatorname{tr} B)(\operatorname{tr} C)$ – traces multiply under tensor product. Thus

$$(\operatorname{tr} e^Y)^3=\operatorname{tr} e^{D_3(Y)}.$$

Let us also set

$$D_0(Y)=0$$

and

$$D_1(Y)=Y$$

so that

$$(\operatorname{tr} Y)^k=\operatorname{tr} e^{D_k(Y)}$$

for all k.

We can thus write the partition function for the Boltzmann composite system as

$$\begin{aligned}\operatorname{Det}(\exp e^{-(\beta_1 I+\beta_2 X)})&=\exp(\operatorname{tr} e^{-(\beta_1 I+\beta_2 X)})\\ &=\sum(\operatorname{tr}[e^{-(\beta_1 I+\beta_2 X)}])^k/k!\\ &=\sum(1/k!)\operatorname{tr}\{\exp(D_k[-(\beta_1 I+\beta_2 X)])\}\end{aligned}$$

by taking $Y=e^{-(\beta_1 I+\beta_2 X)}$. We shall write this formula in even more compact form. Let $T_k(V)$ denote the k-fold tensor product of V with itself, so

$$\begin{aligned}T_0(V)&=\mathbb{R},\\ T_1(V)&=V,\\ T_2(V)&=V\otimes V,\\ T_3(V)&=V\otimes V\otimes V,\ \text{etc.}\end{aligned}$$

Let us now consider the full tensor algebra

$$T(V)=T_0(V)\oplus T_1(V)\oplus T_2(V)\oplus\dots$$

(This is an infinite-dimensional space, but not to worry.) Any matrix on this space will have a block decomposition

$$A=\begin{pmatrix}A_{00} & A_{01} & A_{02} & A_{03} & \cdots\\ A_{10} & A_{11} & A_{12} & A_{13} & \cdots\\ A_{20} & A_{21} & A_{22} & A_{23} & \cdots\\ A_{30} & A_{31} & A_{32} & A_{33} & \cdots\end{pmatrix}$$
$$\cdots \quad \cdots \quad \cdots \quad \cdots \quad \cdots$$

where A_{00} is a transformation of $T_0(V)$ (i.e. a scalar), A_{11} is a linear transformation of

$T_1(V)$, A_{22} is a linear transformation of $T_2(V)$ and so on. Now define

$$\mathbf{tr}^{\otimes} A = \sum (1/k!) \operatorname{tr} A_{kk}.$$

Of course this expression need not converge. We will consider it defined only when the infinite series converges absolutely.

For any operator Y on V we can put all the $D_k(Y)$s together as

$$D(Y) = \begin{pmatrix} D_0(Y) & 0 & 0 & 0 & 0 & \cdots \\ 0 & D_1(Y) & 0 & 0 & 0 & \cdots \\ 0 & 0 & D_2(Y) & 0 & 0 & \cdots \\ 0 & 0 & 0 & D_3(Y) & 0 & \cdots \\ \cdots & \cdots & \cdots & \cdots & & \end{pmatrix}$$

Here we have set $D_0(Y) = 0$ and $D_1(Y) = Y$. We then have

$$\mathbf{tr}^{\otimes} \mathrm{e}^{D(Y)} = \sum (1/k!) \operatorname{tr} \exp\{D_k(Y)\}$$

so

$$F(\beta_1, \beta_2) = \mathbf{tr}^{\otimes} \mathrm{e}^{-D(\beta_1 I + \beta_2 X)}.$$

We now turn to an analogous construction for the Fermi–Dirac situation. Instead of $T(V)$, the tensor algebra, let us consider $\bigwedge(V)$, the exterior algebra. For example, let us look at $\bigwedge^2(V) = V \wedge V$, the space of exterior two-vectors over V. This time define $D_2(Y)$ as a linear transformation on $\bigwedge^2(V)$ by

$$D_2(Y)(u \wedge v) = Yu \wedge v + u \wedge Yu.$$

It is the same formula as before with $\otimes$ replaced by $\wedge$. As before we have

$$\mathrm{e}^Y \wedge \mathrm{e}^Y = \mathrm{e}^{D_2(Y)} \quad \text{on} \quad \textstyle\bigwedge^2(V).$$

In the same way we define

$$D_3(Y) = Y \wedge I \wedge I + I \wedge Y \wedge I + I \wedge I \wedge V \text{ on } \textstyle\bigwedge^3 V$$

and so on. Just as before we than define $D(Y)$ to be the direct sum of all the $D_i(Y)$ – the single matrix on $\bigwedge(V)$ whose diagonal components are the $D_i(Y)$. Of course, this is now a finite matrix (if V is a finite-dimensional vector space).

Suppose that Z is a (diagonalizable) matrix with eigenvalues $z_1, \ldots, z_d$ and corresponding eigenvectors $e_1, \ldots, e_d$. Then we know that the vectors $e_i \wedge e_j$ form a basis of $\bigwedge^2(V)$, and

$$(Z \wedge Z)(e_i \wedge e_j) = Ze_i \wedge Ze_j = z_i z_j e_i \wedge e_j,$$

so that

$$z_i z_j, \quad i < j,$$

are the eigenvalues of $Z \wedge Z$ on $\bigwedge^2(V)$. Similarly,

$$z_i z_j z_k, \quad i < j < k,$$

are the eigenvalues of $Z \wedge Z \wedge Z$ on $\bigwedge^3(V)$ and so on. So if we let $Z^\wedge$ denote the direct sum operator on $\bigwedge(V) = \mathbb{R} \oplus V \oplus (V \wedge V) \oplus \ldots,$

$$Z^\wedge = \begin{pmatrix} I & 0 & 0 & 0 & 0 & \cdots \\ 0 & Z & 0 & 0 & 0 & \cdots \\ 0 & 0 & Z\wedge Z & 0 & 0 & \cdots \\ 0 & 0 & 0 & Z\wedge Z\wedge Z & 0 & \cdots \\ \cdots & \cdots & \cdots & \cdots & \cdots & \cdots \\ \cdots & \cdots & \cdots & \cdots & \cdots & \cdots \end{pmatrix}$$

then

$$\begin{aligned} \operatorname{tr} Z^\wedge &= 1 + \sum z_i + \sum z_i z_j + \sum z_i z_j z_k + \cdots \\ &= \prod (1 + z_i) = \operatorname{Det}(I + Z). \end{aligned}$$

We apply this to $Z = \mathrm{e}^Y$. We use the formula

$$(\mathrm{e}^Y)^\wedge = \mathrm{e}^{D(Y)}$$

which is just the direct sum of all the formulas

$$\begin{aligned} 1 = \mathrm{e}^0 &= \mathrm{e}^{D_0(Y)} && \text{on } \mathbb{R}, \\ \mathrm{e}^Y &= \mathrm{e}^{D_1(Y)} && \text{on } V, \\ \mathrm{e}^Y{}^\wedge \mathrm{e}^Y &= \mathrm{e}^{D_2(Y)} && \text{on } \wedge^2(V), \text{ etc.} \end{aligned}$$

We obtain

$$\mathbf{tr}^\wedge \mathrm{e}^{D(Y)} = \det(1 + \mathrm{e}^Y)$$

where we have used a boldface $\mathbf{tr}^\wedge$ on the left of this equation to emphasize that we are computing the trace over $\wedge^2(V)$. If we take $Y = \mathrm{e}^{-(\beta_1 I + \beta_2 X)}$ we get

$$\det(I + \mathrm{e}^{-(\beta_1 I + \beta_2 X)}) = \mathbf{tr}^\wedge \mathrm{e}^{-D(\beta_1 I + \beta_2 X)}.$$

Let us summarize the strange mathematical manipulations that we have been performing so far. Suppose we start with a simple system that can exist in d states, with 'energies' $\varepsilon_1, \ldots, \varepsilon_d$. The corresponding partition function is

$$Z(\beta) = \mathrm{e}^{-\beta\varepsilon_1} + \cdots \mathrm{e}^{-\beta\varepsilon_d}.$$

If we let X be the diagonal matrix with eigenvalues $\varepsilon_1, \ldots, \varepsilon_d$ then we can write this formula as

$$Z(\beta) = \operatorname{tr} \mathrm{e}^{-\beta X}$$

We are letting V denote the vector space $\mathbb{R}^d$ and the trace is the usual trace of a matrix. We now want to consider a more complicated system where each of these 'energy levels' can be occupied by a varying number of 'occupants'. We now have two equilibrium observables – the total number of occupants and the total energy. The partition function now depends on two variables and the passage from the simple system to the occupancy system is:

replace

$$\begin{aligned} &V \text{ by } T(V) \text{ and then } Z(\beta_1, \beta_2) = \mathbf{tr}^\otimes \mathrm{e}^{-D[\beta_1 I + \beta_2 X]} && \text{(Boltzmann–Poisson)}, \\ &V \text{ by } \wedge(V), \text{ then } Z(\beta_1, \beta_2) = \mathbf{tr}^\wedge \mathrm{e}^{-D[\beta_1 I + \beta_2 X]} && \text{(Fermi–Dirac)}. \end{aligned}$$

Table 22.5

System	Vector space	Partition function
Simple	V	$\exp(-\beta X)$
Boltzmann–Poisson	$T(V)$	$\mathbf{tr}^{\otimes}\mathrm{e}^{-D[\beta_1 I+\beta_2 X]} = \mathrm{Det}\,(\exp \mathrm{e}^{-[\beta_1 I+\beta_2 X]})$
Fermi–Dirac	$\Lambda(V)$	$\mathbf{tr}^{\wedge}\mathrm{e}^{-D[\beta_1 I+\beta_2 X]} = \mathrm{Det}\,(I+\mathrm{e}^{-[\beta_1 I+\beta_2 X]})$
Bose–Einstein	$S(V)$	$\mathbf{tr}^{\mathrm{sym}}\mathrm{e}^{-D[\beta_1 I+\beta_2 X]} = \mathrm{Det}(I-\mathrm{e}^{-[\beta_1 I+\beta_2 X]})^{-1}.$

Thus the difference between Boltzmann–Poisson and Fermi–Dirac consists of using the exterior algebra, $\wedge(V)$ instead of the full tensor algebra, $T(V)$. For the Bose–Einstein case, we replace the exterior algebra by the symmetric algebra $S(V)$ (the algebra of all polynomials on V^*). The argument is the same as in the Fermi–Dirac case except that now we will have

$$\mathbf{tr}^{\mathrm{Sym}}Z = \mathrm{Det}(I-Z)^{-1}.$$

We summarize our results in Table 22.5.

22.13. Quantum states and quantum logic

Let us review the construction of the preceding section in the simplest case. We consider a system which is a finite set and for which μ assigns equal weight to all points. In other words system A of section 22.6. 'Integration' is just summation. An observable is a function on this finite set and a state is a non-negative function which sums up to 1. But we decide to write functions as diagonal matrices, so that sums become traces. Thus a state ρ is written as

$$\rho = \begin{pmatrix} \rho(1) & 0 & 0 & 0 & \cdots \\ 0 & \rho(2) & 0 & 0 & \cdots \\ 0 & 0 & \rho(3) & 0 & \cdots \\ \cdots & \cdots & \cdots & \cdots & \cdots \end{pmatrix}$$

and an observable $\mathbf{J}$ is written as

$$\mathbf{J} = \begin{pmatrix} \mathbf{J}(1) & 0 & 0 & 0 & \cdots \\ 0 & \mathbf{J}(2) & 0 & 0 & \cdots \\ 0 & 0 & \mathbf{J}(3) & 0 & \cdots \\ \cdots & \cdots & \cdots & \cdots & \cdots \end{pmatrix}$$

so that

$$E(\mathbf{J};\rho) = \mathrm{tr}\,\mathbf{J}\rho. \qquad (22.50)$$

The partition function was given as

$$F(\beta) = \mathrm{tr}\,\mathrm{e}^{-\beta X} \qquad (22.51)$$

where

$$X = \begin{pmatrix} \varepsilon_1 & 0 & 0 & 0 & \cdots \\ 0 & \varepsilon_2 & 0 & 0 & \cdots \\ 0 & 0 & \varepsilon_3 & 0 & \cdots \\ \cdots & \cdots & \cdots & \cdots & \cdots \end{pmatrix}$$

and then the equilibrium states are given by

$$\rho_\beta = F(\beta)^{-1} \begin{pmatrix} e^{-\beta\varepsilon_1} & 0 & 0 & 0 & \cdots \\ 0 & e^{-\beta\varepsilon_2} & 0 & 0 & \cdots \\ 0 & 0 & e^{-\beta\varepsilon_3} & 0 & \cdots \\ \cdots & \cdots & \cdots & \cdots & \cdots \end{pmatrix}$$

or

$$\rho_\beta = F(\beta)^{-1} e^{-\beta X}. \qquad (22.51)$$

The passage from the simple case to the occupancy case involved replacing the finite-dimensional vector space V by one of the larger spaces $T(V)$, $\wedge(V)$ or $S(V)$ depending on what kind of 'statistics' we wished to consider. In all of this, the only matrices we considered were diagonal matrices, so the use of matrices altogether was very artificial. But now let us take this matrix picture seriously.

Let V be a vector space with a scalar product. (For important physical reasons we will want V to be a complex vector space and $(\ ,\ )$ to be a positive definite scalar product, so that V is a Hilbert space. In actual applications one frequently will also want V to be infinite-dimensional. But for understanding the key ideas we do not need to get involved with the technical problems of Hilbert space theory and can illustrate the concepts in a finite-dimensional setting.) Recall that an operator A on V is called *self-adjoint* if

$$(A\mathbf{u}, \mathbf{v}) = (\mathbf{u}, A\mathbf{v}) \quad \text{for all} \quad \mathbf{u}, \mathbf{v} \in V$$

The arguments of section 4.3 (done in the complex vector space setting) show that any self-adjoint can be diagonalized – that is, that there is an orthonormal basis $\mathbf{e}_1, \ldots, \mathbf{e}_d$ of V such that

$$A\mathbf{e}_i = \lambda_i \mathbf{e}_i, \quad i = 1, \ldots, d,$$

where the λ_i are real numbers. (In the infinite-dimensional setting the corresponding theorem, suitably modified, is known as the spectral theorem.) Let us say that A is non-negative, or $A \geqslant 0$, if all the λ_i are $\geqslant 0$. What amounts to the same thing, A is non-negative if and only if

$$(A\mathbf{u}, \mathbf{u}) \geqslant 0 \quad \text{for all} \quad \mathbf{u} \in V$$

as can be seen by writing $\mathbf{u}$ as a linear combination of the $\mathbf{e}_i$. We can now make the following definitions.

A *quantum system* is a complex Hilbert space, V.

A *quantum (statistical) state* is a non-negative self-adjoint operator, ρ, such that

$$\operatorname{tr} \rho = 1.$$

The *entropy* of a state ρ is given by

$$\mathrm{Ent}(\rho) = -\operatorname{tr}\rho\log\rho.$$

A *quantum observable* is a self-adjoint operator.
The *expectation* of the observable A in the state ρ is given by

$$\mathrm{E}(A;\rho) = \operatorname{tr} A\rho.$$

Let $J_1, \ldots, J_k$ be k *commuting* observables. That is, assume that

$$J_iJ_j = J_jJ_i \quad \text{for all } i \text{ and } j.$$

Let $\beta = (\beta_1, \ldots, \beta_k)$ where the β_i are real numbers, and define $\beta\mathbf{J}$ to be the observable

$$\beta\mathbf{J} = \beta_1 J_1 + \cdots + \beta_k J_k.$$

Then the corresponding *partition function* is defined by

$$F(\beta) = \operatorname{tr} \mathrm{e}^{-\beta\mathbf{J}} \tag{22.52}$$

and the *equilibrium states* ρ_β are defined by

$$\rho_\beta = F(\beta)^{-1}\mathrm{e}^{-\beta\mathbf{J}}.$$

Thus the expectation of an observable A in the equilibrium state corresponding to β is given by

$$E(A;\rho_\beta) = F(\beta)^{-1}\operatorname{tr} A\mathrm{e}^{-\beta\mathbf{J}}. \tag{22.53}$$

Equations (22.52) and (22.53) represent the basic theoretical content of quantum statistical mechanics. To quote from page 1 of Feynman's book *Statistical mechanics*: 'This fundamental law is the summit of statistical mechanics, and the entire subject is either the slide-down from this summit, as the principle is applied to various cases, or the climb-up to where the fundamental law is derived...'.

In the case when V is infinite-dimensional, the trace of an operator becomes an infinite series, which may not converge. So not all operators have traces. Correspondingly, the function $F(\beta)$ will only converge for a given range of β. We will leave this question aside and concentrate on the case where V is finite-dimensional as a model.

The basic difference between the quantum system and the classical system is that in the classical system an observable was a function (on a finite set) which we chose to regard as a *diagonal* matrix while in the quantum system we allow *all* (self-adjoint) matrices. This might appear, at first glance, as a technical modification (one that has been amply verified by experiment over the past sixty years). In fact, it represents the most profound revolution in the history of science because it modifies the elementary rules of logic. Let us explain.

We begin with a classical system again. Let us define a 'yes or no observable' to be a function, f, that can take on only the values 0 and 1. This corresponds to any experiment in which the final answer is given by a certain indicator (say a light or a click) being on or off. To say that a function f takes only the values 0 and 1 is the same as to say that

$$f(m)^2 = f(m)$$

for all m, or, more succinctly, that,

$$f^2 = f. \tag{22.54}$$

Every such function corresponds to a subset, $B \subset M$, the set where $f = 1$. So given any subset b we get a function, f_B, where

$$f_B(m) = 1 \text{ if } m \in B \text{ and } f_B(m) = 0 \text{ if } m \notin B.$$

If f and g are two 'yes or no observables' then, in a classical system,

$$fg \text{ is again a 'yes or no observable'}. \tag{22.55}$$

Indeed, if C and D are subsets then

$$f_C f_D = f_{C \cap D} \tag{22.56}$$

since $f_C(m) f_D(m) = 1$ if and only if both $f_C(m) = 1$ and $f_D(m) = 1$. Thus multiplication of functions corresponds to intersection of subsets, i.e. to logical conjunction – a point belongs to $C \cap D$ if and only if it belongs to C *and* it belongs to D. For example, C and D are disjoint, i.e.,

$$C \cap D = \varnothing \quad \text{if and only if} \quad f_C f_D = 0.$$

If C and D are disjoint then

$$F_C + F_D = F_{C \cup D}. \tag{22.57}$$

since both sides take on the value 1 when the argument belongs either to C or to D (and no point belongs to both). The distributive law for *multiplication*

$$f_B(f_C + f_D) = f_B f_C + f_B f_D \tag{22.58}$$

is just a translation of the distribution law in *set theory*

$$B \cap (C \cup D) = (B \cap C) \cup (B \cap D). \tag{22.59}$$

This, in turn, is just a version of the distributive law in logic. If we let **B** denote the assertion that $m \in B$ etc., then $\cap$ denotes conjunction: $\mathbf{B} \cap \mathbf{C}$ is the assertion that *both* **B** *and* **C** are true. Similarly $\cup$ denotes the (inclusive) *or*: $\mathbf{C} \cup \mathbf{D}$ means that either **C** or **D** (or both) is true. (We will use the word 'or' inclusively, so we don't have to keep saying 'or both'.) Thus the distributive law above just reflects the elementary principle of logic:

> To say that both **B** and (either **C** and **D**) are true is the same as saying that either (**B** and **C** are both true) or (**B** and **D** are both true).

Let us now examine the corresponding situation in a quantum system. Our observables are no longer functions but self-adjoint operators. We can not talk about the 'values' of an observable, but we can talk about its eigenvalues. So we will call a self-adjoint operator, π, a 'yes or no observable' if it has only 0 or 1 as eigenvalues. This will clearly happen if and only if

$$\pi^2 = \pi \tag{22.60}$$

which is the analogue of (22.54). Now condition (22.60) says that π is a projection. In fact, if $\mathbf{e}_1, \ldots, \mathbf{e}_d$ are the eigenvectors of π, then π is *orthogonal projection* onto the subspace spanned by the eigenvectors corresponding to the eigenvalue 1. Thus each yes or no question corresponds to a *subspace* (instead of a subset) and we can write π_A for orthogonal projection onto the subspace A. Notice that now the zero operator corresponds to orthogonal projection onto the zero subspace, $\{0\}$, and

$$\pi_C\pi_D = 0 \text{ if and only if } C \cap D = \{0\}, \tag{22.61}$$

so the zero subspace, $\{0\}$, plays a role analogous to the empty set. Notice also that if $\pi_C\pi_D = 0$ then, taking adjoints, we have

$$0 = (\pi_C\pi_D)^* = \pi_D^*\pi_C^* = \pi_D\pi_C$$

so $\pi_D\pi_C = 0$ and

$$(\pi_C + \pi_D)^2 = \pi_C^2 + \pi_D^2 = \pi_C + \pi_D$$

since the cross terms vanish. Thus $\pi_C + \pi_D$ is again a yes or no observable. It is clearly orthogonal projection onto the direct sum, $C \oplus D$, of the orthogonal spaces C and D. Thus

$$\pi_C + \pi_D = \pi_{C\oplus D} \text{ if } \pi_C\pi_D = 0 \tag{22.62}$$

in complete analogy with (22.57) where $\oplus$ replaces $\cup$. What fails drastically is (22.55). Given two subspaces C and D, it is *not* true, in general, that $\pi_C\pi_D = \pi_D\pi_C$ and therefore the operator $\pi_C\pi_D$ is not self-adjoint and hence is *not an observable*. For example, if C and D are two non-orthogonal lines in the plane, then $\pi_C\pi_D \neq 0$. But the image of $\pi_C\pi_D$ is C, while the image of $\pi_D\pi_C$ is D. So $\pi_C\pi_D \neq \pi_D\pi_C$. The problem stems from the noncommutativity of matrix multiplication. Notice that the analogue of the distributive law in set theory – the would-be assertion that

$B \cap (C \oplus D) = (B \cap C) \oplus (B \cap D)$ – *is not true in general.*

Indeed, if we take C to be the x-axis and D to be the y-axis in the plane, then $C \oplus D$ is the whole plane and therefore $B \cap (C \oplus D) = B$ for any line B in the plane. But if B is any line other than the x or y axis, then $B \cap C = B \cap D = \{0\}$ so $(B \cap C) \oplus (B \cap D) = \{0\} \oplus \{0\} = \{0\}$, and the two sides are not equal.

Thus the distributive law does not hold in quantum logic. As we mentioned above, the validity of quantum mechanics has been experimentally demonstrated over and over again during the past sixty years. So experiment has shown that one must abandon one of the most cherished principles of logic when dealing with quantum observables.

Summary

A Carathéodory's formulation of the Second Law

You should be able to explain under what circumstances there exist points near a point P which cannot be joined to P by null curves of a one-form α.

You should be able to state and explain Carathéodory's formula of the Second Law.

You should be able to explain the concepts of absolute temperature and entropy in terms of differential forms and line integrals and to describe how these quantities may be computed from empirical data.

Exercises

22.1. Let $\alpha = (y^3 + y)\mathrm{d}x + (xy^2 + x)\mathrm{d}y$. Characterize the set of points in $\mathbb{R}^2$ that can be joined to $\begin{pmatrix}1\\1\end{pmatrix}$ by a null curve.

22.2. Let $\alpha = y\mathrm{d}x + x\mathrm{d}z$. Find a null curve of α that joins the origin to the point $\begin{pmatrix}1\\b\\c\end{pmatrix}$, where $b > 0$ and $c > 0$. With the exception of the origin, the curve should not pass through any points where x, y, or z equals zero.

22.3. Let $\alpha = x\mathrm{d}y$. Show that any two points in $\mathbb{R}^3$ can be joined by a null curve of α, even though $\alpha \wedge \mathrm{d}\alpha = 0$. Show that for the form $\beta = (1 + x^2)\mathrm{d}y$, two points with different y coordinates cannot be joined by a null curve.

22.4. Write to form $\alpha = 2y\,\mathrm{e}^x\mathrm{d}x$ as $\alpha = f\mathrm{d}g$ and as $\alpha = F\mathrm{d}\Omega$ in such a way that F is not simply a constant multiple of f.

22.5. Consider one mole of a monatomic ideal gas, for which $pV = RT$ and $U = \frac{3}{2}pV$. Suppose the gas in initially in the state $p = 32$, $V = 1$ and expands to the state $p = 1$, $V = 8$. Calculate the work done by the gas and the heat absorbed by the gas for each of the following processes:
(a) First p is reduced to 1 by cooling the gas at constant volume, then V is increased to 8 at constant pressure.
(b) First the gas expands isothermally to $V = 8$, then p is reduced to 1 at fixed volume.
(c) Throughout the process, $pV^{5/3} = 32$.

22.6. For the monatomic ideal gas of exercise 22.5, write the heat form as $\alpha = T\mathrm{d}S$. Thereby determine S (up to an additive constant) as a function of p and V. Confirm that $\mathrm{d}T \wedge \mathrm{d}S = \mathrm{d}p \wedge \mathrm{d}V$.

22.7. For a container of volume V filled only with electromagnetic radiation at temperature T, the pressure is $p = \frac{1}{3}CT^4$, while the energy is $U = 3pV$. (Here C is constant.)
(a) By expressing the heat form $\mathrm{d}U + p\mathrm{d}V$ as $\alpha = T\mathrm{d}S$, find an expression for S as a function of p and V.
(b) Suppose that initially the volume of the container is V_0, the temperature T_0. The volume is now increased adiabatically to $64\,V_0$. Determine the final temperature and the work performed during the expansion.

22.8. (a) In a Carnot engine, heat is absorbed at temperature $4T_0$ and exhausted at temperature T_0. What maximum fraction of the heat absorbed can be delivered as useful work by the engine?
(b) Suppose this same engine were operated backwards as a refrigerator. If heat Q is absorbed at the lower temperature T_0, how much work must be performed on the refrigerator?

22.9. For a system of N protons, each of magnetic dipole moment μ, placed in an external megnetic field H, it is reasonable to regard H as the only configuration variable. The magnetization M of the protons, and the associated energy U, are well approximated, in the limit of large T, by

$$M = \frac{N\mu^2 H}{kT}, \quad U = -\frac{N(\mu H)^2}{kT}.$$

The work done by the system is given by $W = \int M \mathrm{d}H$. The heat absorbed is of course $\Delta U + W$. Find an expression for entropy S as a function of T and H.

22.10. Consider a system for which the volume V is the only configurational variable. Prove the following relationships:

(a) $$\left(\frac{\partial T}{\partial V}\right)_{\text{adiabatic}} = \left(\frac{\partial p}{\partial S}\right)_{\text{fixed volume}};$$

(b) $$\left(\frac{\partial S}{\partial V}\right)_{\text{isothermal}} = \left(\frac{\partial p}{\partial T}\right)_{\text{fixed volume}}.$$

(Hint: make use of $\mathrm{d}T \wedge \mathrm{d}S = \mathrm{d}p \wedge \mathrm{d}V$.)

22.11. The 'Helmholtz free energy' F of a system is defined in terms of the internal energy U as $F = U - TS$.
(a) Show that if a system interacts isothermally with its surroundings, the increase in F equals the work done on the system.
(b) Suppose F is expressed as a function of V and T. Determine its partial derivatives with respect to these two variables.
(c) Determine the function $F(V, T)$ for the ideal monatomic gas of exercise 22.5.

22.12. Let $K = \dfrac{\mathrm{d}V \wedge \mathrm{d}T}{V \mathrm{d}T \wedge \mathrm{d}p}$ denote isothermal compressibility

and $\alpha = \dfrac{\mathrm{d}V \wedge \mathrm{d}p}{V \mathrm{d}T \wedge \mathrm{d}p}$ denote coefficient of thermal expansion at constant pressure.

Prove the relation $C_p = C_v + VT\dfrac{\alpha^2}{K}$.

22.13. For the monatomic gas of exercise 22.5, find an explicit expression for the function $Z(\beta, V)$. Evaluate its partial derivatives and show that they equal $-U$ and βp respectively.

22.14. Suppose that entropy S of a system is a function only of its energy U, in accordance with the formula $S = N(U/U_0)^{1/2}$. Find expressions for the heat capacities of this system as a function of temperature.

22.15. Consider a system that consists of three particles. Two of them are indistinguishable bosons, while the third is distinguishable from the other two. Available to each particle are three states, all of the same energy.
(a) What is the probability that all three particles occupy the same state?
(b) What is the probability that each particle occupies a different state?

22.16. For a classical magnetic dipole of moment $\boldsymbol{\mu}$ in a magnetic field $\mathbf{H}$, the energy is $U = -\boldsymbol{\mu}\mathbf{H}\cos\theta$, where θ is the angle between the vectors.

By integrating over all possible orientations, evaluate the partition function for this system.

22.17. Consider a system with three states whose energies are $-\varepsilon, 0$, and ε respectively. Suppose the expected value of the energy is $-\frac{3}{7}\varepsilon$. Determine the probabilities of the three states.

22.18. Suppose that a system has one state of energy $-\varepsilon$, two states of energy 0, and one state of energy ε. If the expected value of the energy is $-\frac{1}{5}\varepsilon$, what are the probabilities of the various states?

22.19. Consider a system with just two states, of energy $-\varepsilon$ and ε respectively. Write down its partition function and use it to determine the energy U and entropy S as functions of temperature T.

Appendix. Proof of the normal theorem

1. Rank of a one-form

We first define the rank of a two-form. Let Ω be a two-form. For each positive integer k we can consider the $2k$-form

$$\Omega^k = \Omega \wedge \Omega \wedge \cdots \wedge \Omega \quad (k \text{ times}).$$

For each point x let k be that non-negative integer such that $(\Omega^k)_x \neq 0$ but $(\Omega^{k+1})_x = 0$. This integer k will, in general, vary with x. We define the *rank* of Ω at x to be the integer $2k$. We will, in the main, be interested in forms of constant rank where $2k$ is independent of x. To say that $2k = 0$ means that $\Omega = 0$. Then form $\Omega = \mathrm{d}x \wedge \mathrm{d}y$ has rank 2 since $\Omega \neq 0$ while $\Omega^2 = 0$. The form

$$\Omega = \mathrm{d}x \wedge \mathrm{d}y + \mathrm{d}u \wedge \mathrm{d}v$$

has rank 4 since

$$\Omega^2 = \mathrm{d}x \wedge \mathrm{d}y \wedge \mathrm{d}u \wedge \mathrm{d}v \neq 0$$

while $\Omega^3 = 0$. (Here we are assuming that x, y, u, v, etc. are independent functions.)

We now define the rank of a one-form. This definition is a little trickier, so we begin inductively. If $\omega = 0$ at x we say that ω has rank zero at x. If $\omega_x \neq 0$ but $\mathrm{d}\omega_x = 0$ we say that ω has rank one at x. If $\omega_x \neq 0$ and $\mathrm{d}\omega_x \neq 0$ but $\omega_x \wedge \mathrm{d}\omega_x = 0$ we say that ω has rank two at x. In general, look for the positive integer k such that

$$\omega_x \wedge (\mathrm{d}\omega)_x^{k-1} \neq 0$$

but

$$\omega_x \wedge (\mathrm{d}\omega)_x^k = 0.$$

There are then two possibilities:

If $(\mathrm{d}\omega)_x^k \neq 0$ we say that ω has rank $2k$ at x.

If $(\mathrm{d}\omega)_x^k = 0$ we say that ω has rank $2k - 1$ at x.

Of course, these alternatives exist at each point x. We will only be interested in the case where the rank is constant. So what we really want to say is that ω has rank $2k + 1$ if $\omega \wedge (\mathrm{d}\omega)^k$ is nowhere zero and $(\mathrm{d}\omega)^{k+1}$ is identically zero. We want to say that ω has rank $2k$ if $(\mathrm{d}\omega)^k$ is nowhere zero but

$$\omega \wedge (\mathrm{d}\omega)^k = 0 \quad \text{identically.}$$

If we apply d to this equation we see that then $(\mathrm{d}\omega)^{k+1}$ is identically zero, in other words $\mathrm{d}\omega$ has rank $2k$.

Examples

The zero-form has rank zero.

$$\mathrm{d}x_0 \text{ has rank 1 since } \mathrm{d}(\mathrm{d}x_0) = 0 \quad \text{so} \quad k = 0.$$

$x_1\mathrm{d}x_2$ has rank 2 in the region $x_1 \neq 0$ since $\mathrm{d}(x_1\mathrm{d}x_2) = \mathrm{d}x_1 \wedge \mathrm{d}x_2$ has $k = 1$ and $x_1\mathrm{d}x_2 \wedge \mathrm{d}x_1 \wedge \mathrm{d}x_2 = 0$.

$\mathrm{d}x_0 + x_1\mathrm{d}x_2$ has rank 3 since $\mathrm{d}(\mathrm{d}x_0 + x_1\mathrm{d}x_2) = \mathrm{d}x_1 \wedge \mathrm{d}x_2$ has $k = 1$ while

$$(\mathrm{d}x_0 + x_1\mathrm{d}x_2) \wedge \mathrm{d}x_1 \wedge \mathrm{d}x_2 = \mathrm{d}x_0 \wedge \mathrm{d}x_1 \wedge \mathrm{d}x_2 \neq 0.$$

In general it is clear that

$$\mathrm{d}x_0 + x_1\mathrm{d}x_2 + x_3\mathrm{d}x_4 + \cdots + x_{2k-1}\mathrm{d}x_{2k} \text{ has rank } 2k+1$$

while

$$x_1\mathrm{d}x_2 + x_3\mathrm{d}x_4 + \cdots + x_{2k-1}\mathrm{d}x_{2k} \text{ has rank } 2k$$

in the region where not all the x_{2k-1} vanish.

We wish to prove, in this appendix, that these are the only examples: that if ω is a linear differential form of constant rank $2k$ then we can always find local coordinates $x_1, x_2, \ldots, x_{2k}$ so that

$$\omega = x_1\mathrm{d}x_2 + x_3\mathrm{d}x_4 + \cdots + x_{2k-1}\mathrm{d}x_{2k},$$

while if the rank of ω is $2k+1$ then we can always find coordinates $x_0, x_1, x_2, \ldots, x_{2k}$, so that

$$\omega = \mathrm{d}x_0 + x_1\mathrm{d}x_2 + x_3\mathrm{d}x_4 + \cdots + x_{2k-1}\mathrm{d}x_{2k}.$$

It was this fact that we used to prove Carathéodory's theorem.

2. Reduction to Darboux's theorem

If we believe the above theorem, then we can always introduce coordinates so that

$$\mathrm{d}\omega = \mathrm{d}x_1 \wedge \mathrm{d}x_2 + \mathrm{d}x_3 \wedge \mathrm{d}x_4 + \cdots + \mathrm{d}x_{2k-1} \wedge \mathrm{d}x_{2k}.$$

Darboux's theorem, which we shall prove later on in this appendix, asserts that if Ω is any closed two-form of constant rank $2k$, then we can always introduce coordinates so that

$$\Omega = \mathrm{d}x_1 \wedge \mathrm{d}x_2 + \mathrm{d}x_3 \wedge \mathrm{d}x_4 + \cdots + \mathrm{d}x_{2k-1} \wedge \mathrm{d}x_{2k}.$$

Let us assume Darboux's theorem for the moment and derive the normal form theorem of the preceding section. If the rank of ω is odd, there is no effort at all. Indeed, $\mathrm{d}\omega$ is a closed two-form of rank $2k$, so we can write $\mathrm{d}\omega$ as above. Then

$$\mathrm{d}\omega = \mathrm{d}(x_1\mathrm{d}x_2 + x_3\mathrm{d}x_4 + \cdots + x_{2k-1}\mathrm{d}x_{2k})$$

or

$$\mathrm{d}(\omega - x_1\mathrm{d}x_2 + x_3\mathrm{d}x_4 + \cdots + x_{2k-1}\mathrm{d}x_{2k}) = 0.$$

So

$$\omega - x_1 dx_2 + x_3 dx_4 + \cdots + x_{2k-1} dx_{2k} = dx_0$$

by the Poincaré lemma, where x_0 is some function. Thus

$$\omega = dx_0 + x_1 dx_2 + x_3 dx_4 + \cdots + x_{2k-1} dx_{2k}.$$

But

$$\omega \wedge (d\omega)^k = dx_0 \wedge dx_1 \wedge dx_2 \wedge \cdots \wedge dx_{2k-1} \wedge dx_{2k}$$

is not equal to zero by assumption. Hence dx_0 is independent of the remaining dx_i and so we can use $x_0, x_1, \ldots, x_{2k}$ as part of a coordinate system. This completes the proof of the normal form theorem for linear differential forms of odd rank, assuming Darboux's theorem about closed two-forms. For even rank we have to work a little harder. What we shall prove is that if ω has (constant) rank $2k$, then we can find a positive function, f, so that

$$\omega = f\sigma$$

where σ has rank $2k - 1$. If we then apply the normal form theorem for forms of odd rank, we can write

$$\sigma = dx_0 + w_1 dx_2 + \cdots + w_{2k-3} dx_{2k-2}$$

where x_0, w_1, x_2, w_3 etc. are coordinates. Now

$$\begin{aligned}\omega &= f dx_0 + f w_1 dx_2 + \cdots \\ &= x_1 dx_2 + x_3 dx_4 + \cdots + x_{2k-3} dx_{2k-2} + x_{2k-1} dx_{2k}\end{aligned}$$

if we set

$$x_1 = f w_1, \quad x_3 = f w_3, \ldots, x_{2k-3} = f w_{2k-3}, \quad x_{2k-1} = f \text{ and } x_{2k} = x_0.$$

The fact that these xs are part of a coordinate system follows from the fact that $(d\omega)^k \neq 0$, which says that the exterior product of all the dx's does not vanish, so that the dx's are linearly independent at all points. So we must prove that $\omega = f\sigma$ where σ has rank $2k - 1$ and f is positive. For this purpose, we make use of Darboux's theorem (still to be proved), which allows us to write

$$d\omega = dy_1 \wedge dy_2 + dy_3 \wedge dy_4 + \cdots + dy_{2k-1} \wedge dy_{2k}$$

where $y_1, \ldots, y_n$ are suitable local coordinates. If we use these coordinates and write

$$\omega = a_1 dy_1 + a_2 dy_2 + \cdots + a_n dy_n$$

then, since $\omega \wedge (d\omega)^k \equiv 0$, all the $a_i = 0$ for $i > 2k$. The form of $d\omega$ implies that the a_is for $i \leqslant 2k$ cannot depend on the variables $y_{2k+1}, \ldots, y_n$, since otherwise we would then get non-zero coefficients of $dy_i \wedge dy_j$, with $i \leqslant 2k$ and $j > 2k$ in the expression for $d\omega$. So in proving our normal form theorem in the case of a form of even degree $2k$, we may as well assume that we are dealing with a form on $\mathbb{R}^{2k}$. All other variables are irrelevant.

Now

$$\omega \wedge (d\omega)^{k-1}$$

is a nonvanishing form of degree $2k - 1$ on $\mathbb{R}^{2k}$. Hence, at every point in its domain of definition it defines a line, namely the one dimensional space of all solutions of

$$i(\zeta)[\omega \wedge (\mathrm{d}\omega)^{k-1}] = 0.$$

Thus $\omega \wedge (\mathrm{d}\omega)^{k-1}$ defines a 'line element field' – in other words a system of ordinary differential equations. Let us choose a $(2k-1)$-dimensional surface and use points on this surface as initial conditions. Then each point near the hypersurface lies on a unique solution curve, and so corresponds to a definite time, t (the parameter along the curve), and a definite point (initial condition) on the hypersurface.

If we introduce coordinates $z_1, \ldots, z_{2k-1}$ on the hypersurface then $t, z_1, \ldots, z_{2k-1}$ are local coordinates on $\mathbb{R}^{2k}$. In terms of these coordinates $\xi = \partial/\partial t$. In other words, the forms

$$\omega \wedge (\mathrm{d}\omega)^{k-1} \quad \text{and} \quad \mathrm{d}z_1 \wedge \mathrm{d}z_2 \wedge \cdots \wedge \mathrm{d}z_{2k-1}$$

define the same line element field. Thus

$$\omega \wedge (\mathrm{d}\omega)^{k-1} = g\mathrm{d}z_1 \wedge \cdots \wedge \mathrm{d}z_{2k-1}$$

where g is a nowhere vanishing function. Replacing z_1 by $-z_1$ if necessary allows us to assume that g is positive. Take f to satisfy

$$f^k = g.$$

If we then define σ by $\omega = f\sigma$ then

$$\mathrm{d}\omega = \mathrm{d}f \wedge \sigma + f\mathrm{d}\sigma$$

so

$$\omega \wedge \mathrm{d}\omega = f\sigma \wedge \mathrm{d}\sigma$$

and, more generally,

$$\omega \wedge (\mathrm{d}\omega)^{k-1} = f^k\sigma \wedge (\mathrm{d}\sigma)^{k-1} = g\mathrm{d}z_1 \wedge \mathrm{d}z_2 \wedge \cdots \wedge \mathrm{d}z_{2k-1}.$$

Thus

$$\sigma \wedge (\mathrm{d}\sigma)^{k-1} = \mathrm{d}z_1 \wedge \mathrm{d}z_2 \wedge \cdots \wedge \mathrm{d}z_{2k-1}.$$

Taking d of both sides of this equation shows that

$$(\mathrm{d}\sigma)^k = 0.$$

In other words, σ is a form of rank $2k-1$. This completes the normal form theorem, assuming Darboux's theorem.

3. Proof of Darboux's theorem

Let us make some preliminary reductions for the proof of Darboux's theorem. We are starting with a closed two-form Ω of rank $2k$ on an n-dimensional space. Suppose that $n > 2k$. Then we can choose a vector field ξ so that $i(\xi)\Omega$ vanishes identically. As before, we can then introduce coordinates (by solving a system of ordinary differential equations) $y_1, \ldots, y_{n-1}, t$ so that $\xi = \partial/\partial t$. The fact that $i(\xi)\Omega$ vanishes identically then means that Ω does not involve $\mathrm{d}t$ in its local expression in these coordinates, i.e. that

$$\Omega = \sum a_{ij}\mathrm{d}y_i \wedge \mathrm{d}y_j$$

But then

$$d\Omega = \sum (\partial a_{ij}/\partial t) dt \wedge dy_i \wedge dy_j$$

and so the equation $d\Omega = 0$ implies that all the $(\partial a_{ij}/\partial t) = 0$. Thus the a_{ij} depend only on the ys. In other words, Ω is really a two-form defined on the $(n-1)$-dimensional space of the ys. If $n - 1 > 2k$ we can go through the same procedure to cut down an additional dimension. In other words, in the proof of Darboux's theorem we may assume that $2k = n$.

We claim that this condition implies that the form Ω is non-singular at every point in the sense that if

$$i(\xi_p)\Omega_p = 0$$

for a tangent vector ξ_p at some point p, then $\xi_p = 0$. Indeed, the above equation clearly implies that

$$i(\xi_p)[\Omega \wedge \Omega \wedge \cdots \wedge \Omega]_p = 0.$$
$$(k \text{ times})$$

But $[\Omega \wedge \Omega \wedge \cdots \wedge \Omega]_p$ is an n-form on an n-dimensional space, and the space of n-forms is one-dimensional. Since, by assumption, the n-form $[\Omega \wedge \Omega \wedge \cdots \wedge \Omega]_p \neq 0$, we must have $\xi_p = 0$. (In fact, it will follow from the ensuing discussion that, on an n-dimensional space, the condition of non-singularity of Ω is equivalent to the condition that it be of rank n.) Another way of saying that Ω is non-singular is simply to say that the n by n matrix

$$(a_{ij}(y))$$

that occurs in the expression

$$\Omega = \sum a_{ij} dy_i \wedge dy_j$$

is a non-singular matrix at all y.

We will break the proof of Darboux's theorem into two parts:

(a) a differential geometric part which asserts that it is always possible to introduce a local change of coordinates so that in the new coordinates the matrix

$$(a_{ij}(y))$$

is constant, i.e. that all the $a_{ij}(y)$ are independent of y;

(b) an algebraic part which asserts that we can make a further linear change of coordinates so that the constant matrix (a_{ij}) takes the form

$$\begin{pmatrix} \begin{matrix} 0 & -1 \\ 1 & 0 \end{matrix} & 0 & \cdots & 0 \\ 0 & \ddots & \cdots & 0 \\ \vdots & \vdots & \ddots & \vdots \\ 0 & \cdots & \cdots & \begin{matrix} 0 & -1 \\ 1 & 0 \end{matrix} \end{pmatrix}$$

This is a theorem about antisymmetric bilinear forms on a vector space which we shall prove by induction on the dimension.

Suppose that V is a two-dimensional vector space and Ω is a nondegenerate bilinear form on V. Let u be any non-zero vector in V. Then the linear form

$$i(u)\Omega = \Omega(u,\cdot)$$

is a non-zero linear form, by the nondegeneracy of Ω. Hence we can find a vector v so that

$$\Omega(u,v) = 1.$$

If we take u, v to be a basis of V, and $\mathrm{d}y_1, \mathrm{d}y_2$ the dual basis then

$$\Omega = \mathrm{d}y_1 \wedge \mathrm{d}y_2$$

which is our desired result. If V has dimension n where n is some even integer > 2, choose u and v as above, and let V_1 be the two-dimensional space spanned by u and v. Then

$$V = V_1 \oplus V_1^{\perp}$$

where $V_1^{\perp}$ denotes the orthogonal complement with respect to Ω. The restriction of Ω to $V_1^{\perp}$ is non-degenerate, and so we can find the desired basis of $V_1^{\perp}$ by induction. The direct sum decomposition together with the two-dimensional result then establishes the theorem for V. We now turn to the geometrical result, part a). If we freeze the coefficients $a_{ij}(y)$ at any particular value, y_0, we get a nondegenerate two-form Ω_0 which agrees with W at the point y_0. So it is enough to prove the following result.

Let Ω_1 and Ω_0 be two closed nondegenerate two-forms which agree at a point, say 0. Then there exists a diffeomorphism f defined in some neighborhood of 0 so that

$$f^*\Omega_1 = \Omega_0.$$

Once again we shall prove this result by solving some differential equations; but first some preliminaries. Set

$$\Omega_t = (1-t)\Omega_0 + t\Omega_1 = \Omega_0 + t\sigma \quad \text{where} \quad \sigma = \Omega_1 - \Omega_0.$$

Then $\mathrm{d}\sigma = 0$ and σ vanishes at 0. Since Ω_0 is non-singular, it follows that $\Omega_t(0) = \Omega_0$ is a non-singular for all t. Hence we can find a neighborhood of the origin on which Ω_t is non-singular for all $0 \leqslant t \leqslant 1$. By passing to a smaller neighborhood if necessary, we may assume that our neighborhood is star shaped, and hence there is a one-form β, defined in this neighborhood such that

$$\sigma = \mathrm{d}\beta \quad \text{and} \quad \beta(0) = 0.$$

Since the form Ω_t is nondegenerate, there exists a unique (t-dependent) vector field ξ_t with

$$i(\xi_t)\Omega_t = -\beta.$$

We think of ξ_t as defining a time-dependent system of ordinary differential equations. Solving these equations gives us a one-parameter family of diffeomorphisms, f_t, such that $f_0 = \mathrm{jd}$ and such that the tangent vector to the curve $f_t(x)$

at time t is $\xi_t(f_t(x))$. Furthermore $f_t(0) = 0$ for all t. Now by the fundamental formula of the differential calculus of forms (Chapter 18) we have

$$\frac{\mathrm{d}}{\mathrm{d}t} f_t^* \Omega_t = f_t^* \left(\frac{\mathrm{d}}{\mathrm{d}t} \Omega_t + i(\xi_t)\mathrm{d}\Omega_t + \mathrm{d}i(\xi_t)\Omega_t \right) = 0$$

since

$$\mathrm{d}\Omega_t = 0,$$

$$\frac{\mathrm{d}}{\mathrm{d}t} \Omega_t = \sigma,$$

and

$$\mathrm{d}i(\xi_t)\Omega\mathrm{t} = -\mathrm{d}\beta = -\sigma.$$

Hence

$$f_1^* \Omega_1 - \Omega_0 = f_1^* \Omega_1 - f_0^* \Omega_0 = \int_0^1 \frac{\mathrm{d}}{\mathrm{d}t} (f_t^* \Omega_t)\, \mathrm{d}t = 0.$$

Thus f_1 is our desired diffeomorphism. This completes the proof of Darboux's theorem, and with it the normal form theorem used in the text to derive Carathéodory's theorem.

Further reading

Once again, the list of books that we give at the end of this section is not meant as a bibliography, but as a guide to further reading. The book by **Loomis and Sternberg** has been out of print for a while. It has just been republished by Jones and Bartlett and so is available once again. Even more than was the case for volume I, it should be considered as a companion text. We refer to it on occasion for precise (or more abstract) mathematical proofs. Similarly the **Feynman Lectures** once again provide a general reference for the physics.

The main algebraic content of volume II is the passage from linear algebra to multilinear algebra. The book by **Greub** is a clear and careful introduction to this subject. In particular such subjects as the tensor algebra, the symmetric algebra, and the Clifford algebras, which we touch briefly, are given a leisurely and detailed treatment.

Chapters 12 and 13 treat electrical network theory. The classical old school circuit theory text is **Guillemin**. The book by **Lorrain, Corson, and Lorrain** discusses the circuit theorems in Chapter 8. Once again we recommend **Doyle and Snell** for a delightful introduction to the probabilistic aspects of network theory and **Kelly** for a more advanced treatment. For the subject of graph theory in its own right and with extensive applications we recommend the book by **Bondy and Murty**.

In Chapter 14 we barely give the definitions of homology and cohomology, so as to introduce some language. We don't really treat the subject at all. For an introductory treatment at the level of this book see **Giblin**. For an introduction at a higher level of mathematical sophistication see the book by **Munkres**. After finishing this text, you should be able to read **Bott and Tu** which is a brilliant and elegant treatment of many of the deep topics in algebraic topology from the viewpoint of the exterior calculus. For a geometrically intuitive and mathematically precise treatment of differential topology we recommend **Guillemin and Pollack** once again. Application of the exterior calculus, which we study in Chapter 15, to differential geometry can be found in **Sternberg**.

Chapters 16–19 treat electricity and magnetism. There are many excellent books on this subject, and we have listed a selection. The one most accessible at the level of this book is **Purcell** which is elegant and extremely clearly written and which develops the subject from a point of view different from ours. The book by **Misner *et al.*** discusses electromagnetism using differential forms.

Two series of books which develop mathematical physics at level substantially above this text are those of **Reed and Simon** and of **Thirring**.

The first is mathematical and the second is a physics text series. Both are rough going but worth the effort. The text by **Thirring** uses the exterior calculus.

The classical text on complex analysis is that by **Ahlfors** which can be read with profit after finishing our Chapter 20. Another good book is **Pólya and Latta**. Of course there are many others. A thorough book on asymptotic methods in relation to complex analysis is **Olver** which comes in a big and a small version. The book by **Guillemin and Sternberg** takes the discussion in Chapter 20 of the asymptotics of radiation as a starting point to develop the modern theory of Fourier integral operators and the geometric theory of asymptotics. It makes rather heavy mathematical demands on the reader.

There are hundreds of books on thermodynamics and on statistical mechanics. For an iconoclastic treatment of the history of the subject see **Truesdell**. By the way, he violently opposes the point of view we adopt here towards thermodynamics. The book by **Born** is a classic semipopular account of related topics. Our treatment of statistical mechanics is rather close to the viewpoint of **Tribus** which is a rather thick book. **Kittel** is thinner and more elementary in its outlook than our treatment. A standard text is **Reif** and a good treatment of various topics from a more mathematical angle is **Thompson**.

We never got to quantum mechanics in this course except for a brief mention of quantum logic at the very end. The best introduction to the subject from the mathematical point of view is still **Mackey**. A detailed scholarly treatment of quantum logic at an advanced level is **Varadarajan**. A balanced well thought out treatment with many applications is **Bohm**. Another user-friendly introduction is **Sudbery**.

References

Ahlfors, L. *Complex Analysis*, McGraw-Hill, 1979

Bohm, A. *Quantum Mechanics*, Springer, 1986

Bondy, J.A. and Murty, U.S. *Graph Theory with Applications*, Macmillan, 1976

Born, M. *The Natural Philosophy of Cause and Chance*, Oxford Univ. Press, 1964

Bott, R. and Tu, L. *Differential Forms in Algebraic Topology*, Springer, 1982

Doyle, P.G. and Snell, J.L. *Random Walks and Electric Networks*, Math. Assoc. of America, 1985

Feynman, R.P., Leighton, R.B. and Sands, M. *The Feynman Lectures on Physics*, Addison-Wesley, 1963

Feynman, R. *Statistical Mechanics*, Benjamin, 1972

Giblin, P.J. *Graphs, Surfaces and Homology*, 2nd edn, Chapman & Hall, 1981

Greub, W. *Multilinear Algebra*, Springer, 1978

Guillemin, E.A. *Introductory Circuit Theory*, Wiley, 1953
Guillemin, V. and Sternberg, S. *Geometric Asymptotics*, Am. Math. Soc., 1978
Guillemin, V.W. and Pollack, A. *Differential Topology*, Prentice-Hall, 1974
Kelly, F.P. *Reversibility and Stochastic Networks*, Wiley, 1979
Khinchin, A. *Mathematical Foundations of Information Theory*, Dover, 1957
Kittel, C. *Thermal Physics*, Freeman, 1980
Loomis, L. and Sternberg, S. *Advanced Calculus*, Jones & Bartlett, 1989
Lorrain, Corson, and Lorrain *Electromagnetic Fields and Waves*, Freeman, 1988
Mackey, G.W. *The Mathematical Foundations of Quantum Mechanics*, Benjamin, 1963
Marion, J. *Classical Electromagnetic Radiation*, Academic Press 1980
Misner, C.W., Thorne, K.S. and Wheeler, J. *Gravitation*, Freeman, 1973
Munkres, J. *Elements of Algebraic Topology*, Addison-Wesley, 1984
Olver, F.W. *Asymptotics and Special Functions*, Academic Press, 1974
Pólya, G. and Latta, G. *Complex Variables*, Wiley, 1974
Purcell, E. *Electricity and Magnetism*, McGraw-Hill, 1965
Reed, M. and Simon, B. *Methods of Modern Mathematical Physics*, Vol. I-IV, Academic Press, 1972
Reif, F. *Fundamentals of Statistical and Thermal Physics*, McGraw-Hill, 1965
Ross, S.M. *Introduction to Probability and Statistics*, Wiley, 1987
Scott, W. *Physics of Electricity and Magnetism*, Wiley, 1977
Sommerfeld, A. *Electrodynamics*, Academic Press, 1952
Sternberg, S. *Lectures on Differential Geometry*, Chelsea, 1983
Sudbery, A. *Quantum Mechanics and the Particles of Nature*, Cambridge Univ. Press, 1986
Thirring, W. *A Course in Mathematical Physics*, 4 vols., Springer, 1978
Thompson, C.J. *Mathematical Statistical Mechanics*, Princeton Univ. Press, 1972
Tribus, M. *Thermostatics and Thermodynamics*, van Nostrand, 1961
Truesdell, C. *The Tragicomical History of Thermodynamics*, Springer, 1980
Varadarajan, V.S. *Geometry of Quantum Theory*, 2nd edn, Springer, 1985

INDEX